78,-
60,-

Numerical Schemes for
Conservation Laws

Numerical Schemes for Conservation Laws

Dietmar Kröner
Institut für Angewandte Mathematik, Albert-Ludwigs-Universität, Freiburg, Germany

WILEY TEUBNER

A Partnership between John Wiley & Sons and B. G. Teubner Publishers
Chichester · New York · Brisbane · Toronto · Singapore · Stuttgart · Leipzig

Copyright © 1997 jointly by John Wiley & Sons Ltd and B. G. Teubner

John Wiley & Sons Ltd.
Baffins Lane, Chichester
Chichester
West Sussex PO19 1UD
England

B.G. Teubner
Industriestraβe 15
70565 Stuttgart (Vaihingen)
Postfach 80 10 69
70510 Stuttgart
Germany

National Chichester (01243) 779777 *National* Stuttgart (0711) 789 010
International +44 1243 779777   *International* +49 711 789 010

All rights reserved

No part of this book may be reproduced by any means,
or transmitted, or translated into a machine language
without the written permission of the publisher.

Other Wiley Editorial Offices

John Wiley & Sons, Inc., 605 Third Avenue,
New York, NY 10158-0012, USA

Brisbane · Toronto · Singapore

Other Teubner Editorial Offices

B.G. Teubner, Verlagsgesellschaft mbH, Johannisgasse 16,
D-04103 Leipzig, Germany

British Library Cataloguing in Publication Data
A catalogue record for this book is available from the British Library

ISBN Wiley 0-471-96793-9
ISBN Teubner 3-519-02720-8

Produced from camera-ready copy supplied by the authors
Printed and bound in Great Britain by Bookcraft (Bath) Ltd
This book is printed on acid-free paper responsibly manufactured from sustainable forestation, for which at least two trees are planted for each one used for paper production.

Contents

Preface		vii
1	Introduction	1
2	**Initial value problems for scalar conservation laws in 1-D**	**15**
2.1	Theoretical background	15
2.2	Finite difference schemes of first order for scalar equations in one spatial dimension	39
2.3	Convergence of finite difference schemes of first order in 1-D	53
2.4	Stability and consistency for linear initial value problems	87
2.5	Finite difference schemes of higher order for scalar equations in one spatial dimension	98
2.6	Streamline diffusion method	117
3	**Initial value problems for scalar conservation laws in 2-D**	**141**
3.1	Dimensional splitting schemes for scalar conservation laws in 2-D	141
3.2	Finite volume schemes in 2-D	155
3.3	Convergence of finite volume schemes in 2-D on unstructured grids	164
3.4	Error estimates for first-order schemes	191
3.5	Higher-order finite volume schemes for scalar equations	212
3.6	Proof of convergence of higher-order finite volume schemes for scalar equations	225
3.7	Essentially non-oscillatory schemes	238

3.8	Fluctuation splitting schemes	254
3.9	Adaptivity and local grid alignment: numerical experiments	258

4 Initial value problems for systems in 1-D   276

4.1	Basic results for systems in 1-D	276
4.2	Chorin's method for solving the Riemann problem for systems	319
4.3	The basis of Glimm's existence proof for systems in 1-D	325
4.4	Numerical schemes for hyperbolic systems in 1-D	333
4.5	The Osher–Solomon scheme	355

5 Initial value problems for systems of conservation laws in 2-D   372

5.1	Finite volume methods for systems in several space dimensions	372
5.2	The discretization matrices for systems in two and three bla spatial dimesions	384
5.3	Numerical examples for the Euler equations of gas dynamics	394

6 Initial boundary value problems for conservation laws   422

6.1	Boundary conditions for linear systems	422
6.2	Boundary conditions for nonlinear scalar equations	437

7 Convection-dominated problems   456

7.1	Singular perturbation problems for elliptic equations	456
7.2	Discretization of the compressible Navier–Stokes equations	473

List of figures   479

References   482

Index   503

PREFACE

Systems of conservation laws are very important and powerful mathematical models for a variety of physical phenomena that appear in fluid mechanics, astrophysics, groundwater flow, meteorology, semiconductors, reactive flows and several other areas. Since conservation laws may have discontinuous solutions, they pose a special challenge for theoretical and numerical analysis. Similar numerical problems as for conservation laws also arise for convection dominated flows.

In this book we shall describe numerical methods for solving the initial value as well as the initial boundary value problem for scalar conservation laws and systems of conservation laws in many-dimensions. In particular, modern developments concerning higher order upwind finite volume schemes on unstructured grids and a priori error estimates are taken into account. The details of the mathematical proofs concerning the most important results are included. Theoretical results are presented as far as necessary for the numerical treatment. For instance we have included the theory for solving the Riemann problem for the Euler equations for gas dynamics [218]. This is important for getting the exact solution in order to compare it with the numerical solution. Furthermore, the Riemann problem is a basic tool for several numerical schemes.

The arrangement of the book is as follows. In the Introduction we shall point out the possibility of discontinuous solutions for a simple conservation law in one space dimension, and we shall present several examples for the occurrence of conservation laws. In Chapters 2 and 3 we shall study the convergence to the entropy solution for finite difference and finite volume schemes of first and higher order on structured and unstructured grids as well as for the streamline diffusion method. Although there is no convergence proof for the essentially non-oscillatory schemes, we shall describe the algorithm and discuss some numerical examples. At the end of Chapter 3 different concepts of local adaptivity and grid alignment are discussed and applied to several examples in two and three space dimensions on structured and unstructured grids. Systems of conservation laws are considered in Chapters 4 and 5. At the beginning of Chapter 4 we shall discuss some basic results for systems in one space dimension; we describe the basic ideas of the Glimm scheme and formulate the convergence result for it. Since there are no further theoretical results concerning convergence of numerical schemes for systems in one and several space dimensions, we can only describe the most important algorithms. Chapter 7 is devoted to initial boundary value problems for conservation laws, and finally convection-dominated diffusion problems are considered in Chapter 7.

This book is a result of a one-year course for students in their third year in applied mathematics and physics held at the universities of Saarbrücken, Heidelberg, Bonn and Freiburg. But it can be recommended also for students in engineering science, astrophysics and meteorology or other fields related to this subject. In this monograph the reader will find the definitions of the most important numerical schemes as well as the details of the mathematical proofs, in particular for the convergence of the algorithms. He or she should be familiar with the basic contents of courses for beginners like the Arzelá–Ascoli Theorem and the Selection Principle of Helly. Furthermore, some elementary results concerning Young measures are required. But all other details that are used will be given in this book. Therefore it can form the basis for a lecture course on this subject.

The numerical experiments described in this book have been carried out by Diploma and PhD students in our group within the scope of their theses. At the end of this book there is a list of those responsible for the figures. Most of the figures have been made with GRAPE [204]. In particular, I should like to thank J. Becker, R. Beinert, T. Geßner, M. Küther, G. Lorse, T. Mackeben M. Schmieder, A. Schneider and Dr. M. Wierse for many postscript files that I have used for the illustration and demonstration of the results of many numerical examples. J. Becker, A. Dedner, T. Geßner, Prof. Dr W. Hackbusch, R.T. Happe, R. Kleinrensing, J. Koch, A. Koop, G. Lorse, T. Mackeben, M. Ohlberger, Prof. Dr R. Rannacher, C. Rhode, M. Rokyta, M. Schmieder, B. Schupp, Dr M. Wierse, Dr M. Rumpf, M. Wesenberg and in particular M. Küther read the manuscript very carefully, and gave me a lot of important and fruitful hints for improving it.

For writing the Latex files I am very grateful to Mrs. Fischbach in Saarbrücken, who wrote a first version in German, Th. Widmer in Bonn and in particular R. Axthelm and M. Tenhaeff in Freiburg for the final version.

I should like to thank my academic teachers Prof. Dr W. Jäger and Prof. Dr H.W. Alt, who gave me the opportunity to work in this area.

Last but not least, I have only been able to write this book because my family in particular my wife Karola, have always supported my work with great patience and intuitive understanding.

1 Introduction

Many problems in mechanics, in particular in gas dynamics lead us to the study of nonlinear hyperbolic conservation laws. The most interesting point about a hyperbolic equation is that its solution may become discontinuous, although the initial values or boundary values are smooth. That is why existence theory for nonlinear hyperbolic systems in higher (space) dimensions is not very well developed.

The simplest conservation law (or nonlinear hyperbolic differential equation) is the Burgers' equation

$$\partial_t u(x,t) + \partial_x \frac{u^2(x,t)}{2} = 0 \quad \text{in } \mathbb{R} \times \mathbb{R}^+ . \tag{1.0.1}$$

A simple physical motivation for this type of equation is given in Example 1.0.1. In contrast to elliptic or parabolic equations, this simple conservation law can have discontinuous solutions even for smooth initial data. This can be seen very easily. Let us assume that (1.0.1) has a smooth solution $u \in C^1(\mathbb{R} \times [0,T[)$. Consider the solution $\gamma \in C^1(]0,T[) \cap C^0([0,T])$ of the initial value problem for the following ordinary differential equation:

$$\begin{aligned} \gamma'(t) &= u(\gamma(t),t) \quad \text{in }]0,T[, \\ \gamma(0) &= a \end{aligned} \tag{1.0.2}$$

for a given constant $a \in \mathbb{R}$. Then we have

$$\frac{d}{dt} u(\gamma(t),t) = \gamma' \partial_x u + \partial_t u = u(\gamma(t),t) \partial_x u(\gamma(t),t) + \partial_t u(\gamma(t),t)$$

$$= \partial_x \frac{u^2}{2} + \partial_t u = 0$$

for all $t \in]0,T[$. This means that $u(\gamma(t),t)$ is constant for all $t \in [0,T]$, or u is constant along the curves $\{(\gamma(t),t) | t \in [0,T]\}$. These curves are called characteristics. From (1.0.2) we also get that

$$\gamma'(t) = u(\gamma(t),t) = \text{const} \tag{1.0.3}$$

for all $t \in]0, T[$. This means that the characteristics are straight lines. Altogether we have shown that the solution u is constant along straight lines. The slopes of these lines is given by (1.0.3):

$$\gamma'(t) = u(\gamma(t), t) = u(\gamma(0), 0) ,$$

i.e. by the initial values for u. Consider now the initial value problem

$$\partial_t u + \partial_x \frac{u^2}{2} = 0 \quad \text{in } \mathbb{R} \times \mathbb{R}^+ ,$$
$$u(x, 0) = u_0(x) \quad \text{in } \mathbb{R} ,$$

where u_0 is given as in Figure 1.0.1.

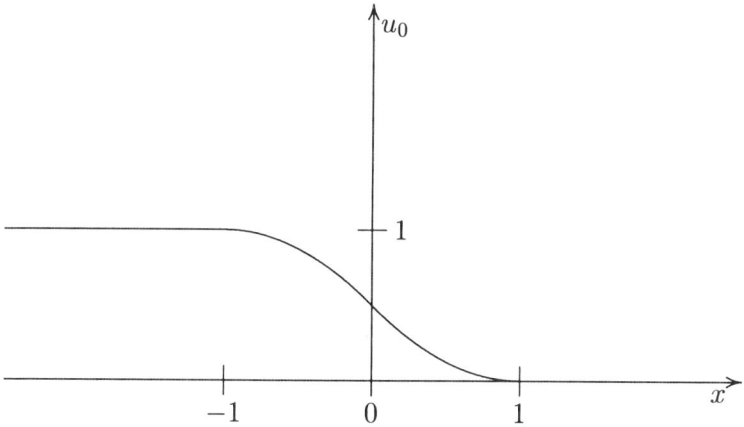

Figure 1.0.1

Then we obtain that u is constant along the characteristic curve $(\gamma(t), t)$, with

$$\gamma'(t) = u(\gamma(0), 0) = u_0(\gamma(0)) = \begin{cases} 1 & \text{if } \gamma(0) \leq -1 , \\ 0 & \text{if } \gamma(0) \geq 1 , \end{cases}$$

$$u(\gamma(t), t) = u(\gamma(0), 0) = u_0(\gamma(0)) = \begin{cases} 1 & \text{if } \gamma(0) \leq -1 , \\ 0 & \text{if } \gamma(0) \geq 1 . \end{cases}$$

This means, if T is sufficiently large, that the characteristics can meet each other (see Figure 1.0.2), and therefore u cannot be a classical solution up

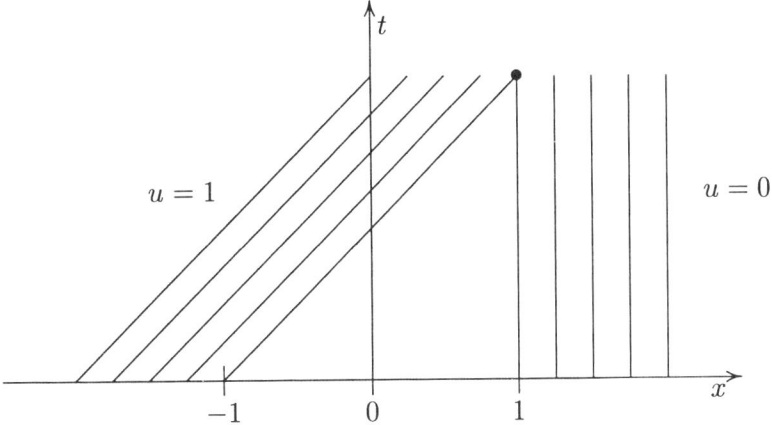

Figure 1.0.2

to this time. Then a new definition of solutions of conservation laws of type (1.0.1) will be introduced (weak solutions, see §2.1). This new type of solution can have discontinuities, as we shall see in §2.1.

For linear problems of the form

$$\partial_t u + a\partial_x u + b\partial_y u = 0 \quad \text{in } \mathbb{R}^2 \times \mathbb{R}^+,$$
$$u(x, y, 0) = u_0(x, y) \quad \text{in } \mathbb{R}^2 \tag{1.0.4}$$

the exact solution can be given explicitly. If $u_0 \in C^1(\mathbb{R}^2, \mathbb{R})$, we obtain

$$u(x, y, t) = u_0(x - at, y - bt). \tag{1.0.5}$$

For more general $u_0 \in L^1_{\text{loc}}(\mathbb{R}^2, \mathbb{R})$ the same formula as (1.0.5) still holds. This means the initial values are shifted with velocity $\binom{a}{b}$. Therefore the equation (1.0.4) is also refered to as the transport, advection or convection equation. Unfortunately, in the class of weak solutions uniqueness does not hold in general. Therefore an additional condition, the so–called entropy condition, has to be satisfied for the solution of conservation laws (see §2.1). This means that we have to deal with two basic problems in the numerical analysis. The algorithms should be able to treat discontinuous solutions and to select those that satisfy the entropy condition. We have just demonstrated these difficulties for a scalar equation in one space dimension, but they also appear for problems in many dimensions and for systems.

Let us discuss some basic problems in mechanics that can be modelled by hyperbolic conservation laws.

EXAMPLE 1.0.1 (Traffic flow) Let us consider traffic flow on a highway. Denote the car density (cars per mile) by $u = u(x,t)$ and the traffic flow (cars per hour) by $q = q(x,t)$, where x denotes the one-dimensional space coordinate on the highway and t the time. Now we fix an interval $[a,b]$ on the highway and two times t_1 and t_2, where $\Delta t := t_2 - t_1 \ll 1$. If the numbers of cars $\int_a^b u(x,t_1)\,dx$, $\int_a^b u(x,t_2)\,dx$ for different times in the interval $[a,b]$ are not the same then some cars have to enter this interval in the point a or they have to leave it in b. In the mathematical context this means

$$\int_a^b u(x,t_2)\,dx - \int_a^b u(x,t_1)\,dx = \int_{t_1}^{t_2} q(a,s)\,ds - \int_{t_1}^{t_2} q(b,s)\,ds, \quad (1.0.6)$$

where $\int_{t_1}^{t_2} q(a,s)\,ds$ and $\int_{t_1}^{t_2} q(b,s)\,ds$ denote the number of cars, entering the interval $[a,b]$ in a or leaving it in b respectively during the time $t_2 - t_1$. Now we divide both sides in (1.0.6) by $t_2 - t_1$ and $(b-a)$ and let t_2 tend to t_1 and $b, a \to x$. Then we obtain

$$\partial_t u(x,t_1) = -\partial_x q(x,t_1). \quad (1.0.7)$$

There exist some measurements [184] of the flow $f(u)$ versus the car concentration u. It turns out (see Figure 1.0.3) that the flow increases with increasing car concentration up to some value u_m and above that the flow is decreasing. f can be written approximately as

$$f(u) = -(u - u_m)^2 + f_0,$$

with a suitable constant f_0.

If we define $q(x,t) := f(u(x,t))$, we obtain from (1.0.7)

$$\partial_t u + \partial_x f(u) = 0 \quad \text{in} \quad \mathbb{R} \times \mathbb{R}^+,$$
$$u(x,0) = v(x) \quad \text{in} \quad \mathbb{R},$$

for given initial values v. Everybody knows that if one driver pulls up suddenly then a perturbation will run through the column of cars. Later on, we shall call this a shock.

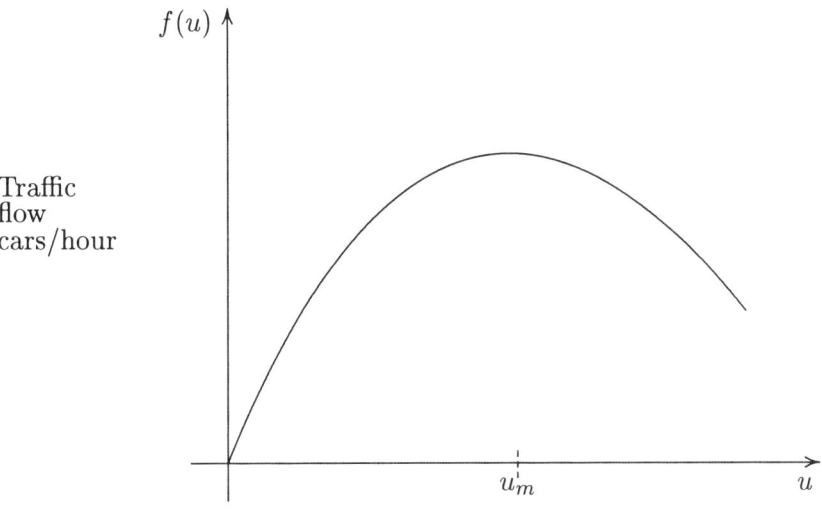

Figure 1.0.3

EXAMPLE 1.0.2 (Buckley–Leverett equation [73]) Another physical example for (1.0.1) is the mathematical model for a one-dimensional two-phase flow of immiscible fluids such as water and oil through porous media. It is given by

$$\partial_t s + \partial_x f(s) = 0$$

where s is the water saturation,

$$f(s) = \frac{k_1(s)/\mu_1}{k_1(s)/\mu_1 + k_2(s)/\mu_2},$$

μ_1 and μ_2 are the viscosities and

$$k_1(s) = s^2, \quad k_2(s) = (1-s)^2$$

are the relative permeability for water and oil respectively. In this model the porosity is taken to be 1 and capillarity and gravity effects are ignored. For a more realistic model see [118].

EXAMPLE 1.0.3 (Shock tube problem) Let us consider a tube as in Figure 1.0.4 that is filled by an inviscid, compressible fluid. We assume that initially ($t = 0$) the tube is separated into two parts by a membrane. On the left-hand side initially we prescribe the density $\rho = 1$, the pressure $p = 1$, the velocity u in the x-direction and the velocity v in the y-direction $u = 0$, $v = 0$. On the right-hand side we have $\rho = 0.125$, $p = 0.1$, $u = 0$, $v = 0$ initially. Now at time $t = 0$ we instantaneously destroy the membrane and the initial state will develop in time. We expect that there will be a "density shock" running to the right.

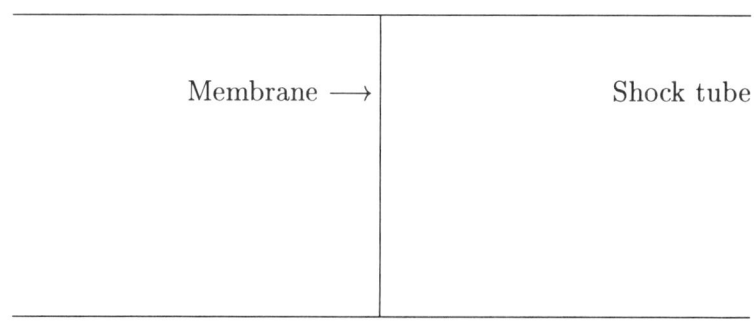

Figure 1.0.4

We shall derive a mathematical model for this physical experiment in order to simulate the time evolution of the initial values as described in Example 1.0.3. It is the aim of this book to present the mathematical methods and tools that are used to study this kind of model in particular from the numerical point of view. The situation as described in Example 1.0.3 was already considered by Riemann [192], and therefore the corresponding mathematical problem is called the "Riemann problem". Using some suitable numerical algorithms, we obtain the result shown in Figure 1.0.5. Here we have plotted the density ρ as a function of the length parameter x of the tube. The data has been obtained from a very good approximation of the exact solution (see Section 4.1). It consists (from left to right) of a rarefaction wave, a contact discontinuity and a shock. The last two are discontinuous parts of the solution. Now let us consider a real two-dimensional problem.

EXAMPLE 1.0.4 (Mach-three wind tunnel) We consider a tube with a forward facing step initially filled with a non-viscous compressible fluid. For $t = 0$ the functions ρ, p, u and v are assumed to be constant: $\rho = 1$,

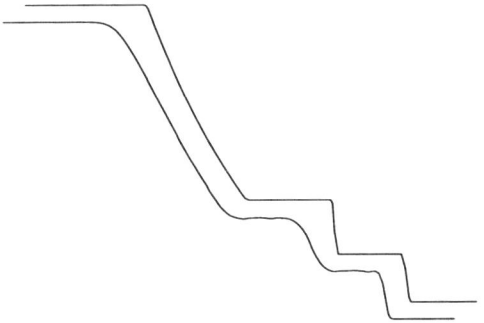

Figure 1.0.5

$p = 1$, $u = 3$, $v = 0$. For this data we have $m = u/a = 3$ where $a = \sqrt{\gamma p / \rho}$ is the sound velocity, m the Mach number and γ a physical constant (here $\gamma := 1$). After some time there will be a "bow shock" in front of the corner, which is reflected first at the upper wall of the tube and then on the lower part of the boundary (see Figure 1.0.6).

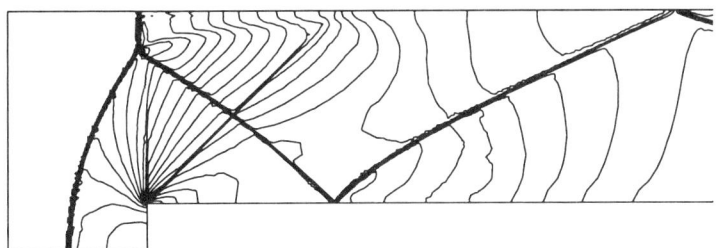

Figure 1.0.6

REMARK 1.0.5 (The mathematical model, Euler equations of gas dynamics) *Next we should like to derive the mathematical model for the flow of an inviscid compressible fluid in a general domain. Let $V \subset \Omega$ be a control volume in Ω. The basic physical laws governing the flow of a fluid are the conservation laws for mass, momentum and energy. We assume that u, v, ρ, p and e are smooth functions of $x \in \Omega$ and $t \in [0, \infty[$. First, let us consider the **conservation of mass**. The total mass in V at time t is $m(t) := \int_V \rho(\cdot, t)$. If we assume that there are no sources then the change in* mass $\Delta m = m(t_2) - m(t_1)$ during the time $\Delta t := t_2 - t_1 > 0$ is equal to the flow $-\int_{t_1}^{t_2} \int_{\partial V} (\rho \vec{u}) \nu$ where $\vec{u} = \binom{u}{v}$ and ν is the outer normal to V:

$$\Delta m = - \int_{t_1}^{t_2} \int_{\partial V} (\rho \vec{u}) \nu .$$

Divide by Δt and let $\Delta t \to 0$:

$$\partial_t \int_V \rho = - \int_{\partial V} (\rho \vec{u}) \nu .$$

Applying Gauss's theorem,

$$\partial_t \int_V \rho = - \int_V \mathrm{div}(\rho \vec{u}) . \tag{1.0.8}$$

Since V was arbitrary, (1.0.8) holds for all $V \subset \Omega$, and therefore

$$\partial_t \rho + \mathrm{div}(\rho \vec{u}) = 0 \quad \text{in } \mathbb{R}^+ \times \Omega. \tag{1.0.9}$$

Similarly, one obtains from the **conservation of momentum**

$$\left. \begin{aligned} \partial_t(\rho u) + \partial_x(\rho u^2 + p) + \partial_y(\rho u v) &= 0 \\ \partial_t(\rho v) + \partial_x(u v \rho) + \partial_y(\rho v^2 + p) &= 0 \end{aligned} \right\} \quad \text{in } \Omega \times \mathbb{R}^+ \tag{1.0.10}$$

and from the **conservation of energy**.

$$\partial_t e + \partial_x(u(e+p)) + \partial_y(v(e+p)) = 0 \quad \text{in } \Omega \times \mathbb{R}^+ , \tag{1.0.11}$$

where e is the total energy density. (For more details see Courant and Friedrichs [43].)

For the five unknown functions ρ, u, v, e and p we have derived four equations (1.0.9), (1.0.10) and (1.0.11). Therefore we need one additional relation, for instance in this case between the pressure p and the other functions ρ, u, v and e. This is given by the equation of state

$$p = f(\rho, u, v, e) \ . \tag{1.0.12}$$

For an ideal gas we obtain

$$p = (\gamma - 1)\left[e - \frac{\rho}{2}(u^2 + v^2)\right] \tag{1.0.13}$$

where γ is the adiabatic constant.

Equations (1.0.9), (1.0.10) and (1.0.11) are called the "Euler equations of gas dynamics". The functions ρ, u, v and p are called primitive variables and ρ, $u\rho$, $v\rho$ and e conservative variables.

The mathematical model for Examples 1.0.3 and 1.0.4 consists in the system (1.0.9), (1.0.10),(1.0.11), the equation of state (1.0.13), initial conditions as described in Example 1.0.3 and 1.0.4 respectively and suitable boundary conditions. (We shall carefully discuss the latter in Chapter 6.)

In primitive variables (ρ, u, p) (1.0.9), (1.0.10) and (1.0.11) in one space dimension look like $\left(a := \sqrt{\gamma p/\rho}\right)$

$$\partial_t \rho + \partial_x(u\rho) = 0 \ , \tag{1.0.14}$$

$$\partial_t u + u\partial_x u + \frac{\partial_x p}{\rho} = 0 \ , \tag{1.0.15}$$

$$\partial_t p + \rho a^2 \partial_x u + u\partial_x p = 0 \ . \tag{1.0.16}$$

EXAMPLE 1.0.6 (Lagrangian coordinates) Now we should like to write (1.0.14), (1.0.15) and (1.0.16) in Lagrangian coordinates. They are defined as follows (see [43, page 30]).

Let $x_0 \in \mathbb{R}$ be fixed and $y(t)$ be a solution of

$$y' = u(y, \cdot) \ ,$$
$$y(0) = u(x_0, 0) \ .$$

1 Introduction

Then define $x(0, t) := y(t)$, and for any $h \in \mathbb{R}$ let $x(h, t)$ be given by

$$h = \int_{x(0,t)}^{x(h,t)} \rho(y, t) \, dy \, . \tag{1.0.17}$$

Then the integral on the right-hand side is just the mass between $x(0, t)$, and $x(h, t)$, and therefore h has the meaning of mass. The conservation of mass ensures that h is independent of t (see [43, page 30]). Differentiating (1.0.17) with respect to h, we obtain

$$1 = \rho(x(h, t), t) \partial_h x(h, t) \, .$$

Applying the operator d/dt to (1.0.17) and using (1.0.14), we have

$$\partial_t x = u \, . \tag{1.0.18}$$

Let us define

$$\bar{u}(h, t) := u(x(h, t), t) \, , \quad \bar{\rho}(h, t) := \rho(x(h, t), t) \, .$$
$$\bar{v} = \frac{1}{\bar{\rho}}, \quad \bar{E}(h, t) := E(x(h, t), t) \tag{1.0.19}$$

where $E = \frac{p}{\rho(\gamma-1)} + \frac{u^2}{2}$. The Lagrangian coordinates are now given by t and $x(h, t)$ Now let us assume that we have a γ-law gas, i.e.

$$p(\rho) = k\rho^\gamma \, .$$

Then we obtain, using (1.0.14) and (1.0.15),

$$\partial_t \bar{v} - \partial_h \bar{u} = -\frac{\partial_t \bar{\rho}}{\bar{\rho}^2} - \partial_x u \partial_h x$$
$$= -\frac{1}{\bar{\rho}^2} (\partial_x \rho \partial_t x + \partial_t \rho + \rho \partial_x u) = 0 \, ,$$
$$\partial_t \bar{u} + \partial_h (k\bar{v}^{-\gamma}) = \partial_t \bar{u} + k \partial_h \bar{\rho}^\gamma = u \partial_x u + \partial_t u + k \gamma \bar{\rho}^{\gamma-1} \partial_h \bar{\rho}$$
$$= u \partial_x u + \partial_t u + k \gamma \bar{\rho}^{\gamma-2} \partial_x \rho \, . \tag{1.0.20}$$

Since $p(\rho) = k\rho^\gamma$, we can continue:
$$\partial_t \bar{u} + \partial_h(k\bar{v}^{-\gamma}) = u\partial_x u + \partial_t u + \frac{\partial_x p}{\rho} = 0.$$

Furthermore, using (1.0.18), (1.0.19) and (1.0.20),
$$\partial_t \bar{E} + \partial_h(\bar{u}\bar{p}) = \partial_x E \partial_t x + \partial_t E + \frac{1}{\rho}\partial_x(up)$$
$$= \frac{1}{\rho^2(\gamma-1)}(u\rho\partial_x p - up\partial_x \rho + \rho\partial_t p - p\partial_t \rho)$$
$$+ u^2 \partial_x u + u\partial_t u + \frac{p}{\rho}\partial_x u + \frac{u}{\rho}\partial_x p$$
$$= k\rho^{\gamma-2}(u\partial_x \rho + \partial_t \rho) + k\rho^{\gamma-1}\partial_x u = 0.$$

Thus we have shown that
$$\begin{aligned}\partial_t \bar{v} - \partial_h \bar{u} &= 0 \\ \partial_t \bar{u} + \partial_h(k/\bar{v}^\gamma) &= 0 \\ \partial_t \bar{E} + \partial_h(\bar{u}\bar{p}) &= 0.\end{aligned} \quad (1.0.21)$$

These are the Lagrangian equations for a γ-law gas. Equations (1.0.18) and (1.0.19) mean that we describe the motion in a coordinate system that is moving with the flow.

EXAMPLE 1.0.7 (Isentropic gas) A model for an isentropic gas (i.e. one of constant entropy) in one space dimension can be obtained from (1.0.9), (1.0.10) and (1.0.11). For the entropy we have (see [43, page 7])
$$S = S_0 + c_v \log \frac{p}{\rho^\gamma(\gamma-1)}$$

and, since S is assumed to be constant we obtain $p = \text{const} \cdot \rho^\gamma$. It can be seen very easily that $\partial_t S = \frac{\partial_t p}{p} - \frac{\partial_t \rho}{\rho}\gamma$, $\partial_x S = \frac{\partial_x p}{p} - \frac{\partial_x \rho}{\rho}\gamma$. Then (1.0.16) is equivalent to $p(\partial_t S + u\partial_x S) = 0$, since
$$\partial_t p + \rho a^2 \partial_x u + u\partial_x p = \partial_t p + \gamma p \partial_x u + u\partial_x p$$
$$= \partial_t p + u\partial_x p + \frac{\gamma p}{\rho}(-\partial_t \rho - u\partial_x \rho)$$
$$= \left(\frac{\partial_t p}{p} + \frac{u\partial_x p}{p} - \gamma\frac{\partial_t \rho}{\rho} - \gamma\frac{u\partial_x \rho}{\rho}\right)p$$
$$= p(\partial_t S + u\partial_x S).$$

12   1 Introduction

Since the entropy S is assumed to be constant, we can neglect (1.0.16).

Then in Lagrangian coordinates the model for an isentropic gas is

$$\partial_t \bar{v} - \partial_h \bar{u} = 0$$
$$\partial_t \bar{u} + \partial_h (k/\bar{v}^\gamma) = 0 \ . \tag{1.0.22}$$

This system is also called the p-system (see [218]).

EXAMPLE 1.0.8 (Nonlinear wave equation) The system (1.0.22) can be written as a single equation of second order. Assume that $\bar{u}, \bar{v} \in C^1$. In a simply connected region the first equation in (1.0.22) implies the existence of a function φ such that

$$\partial_h \varphi = \bar{v}, \quad \partial_t \varphi = \bar{u} \ .$$

Therefore, using $K(\bar{v}) := -k/\bar{v}^\gamma$ we obtain from (1.0.22)

$$\partial_t^2 \varphi = \partial_t \bar{u} = \partial_h K(\bar{v}) = \partial_h K(\partial_h \varphi) \ .$$

This means that φ satisfies the nonlinear wave equation

$$\partial_t^2 \varphi - \partial_h K(\partial_h \varphi) = 0 \ .$$

EXAMPLE 1.0.9 (Shallow water waves) Let $\varphi : \mathbb{R}^2 \times [0, \tau] \to \mathbb{R}$ be a function and $\Omega_t := \{(x, y, z) \in \mathbb{R}^3 | (x, y) \in \mathbb{R}^2,\ 0 < z < \varphi(x, y, t)\}$. We assume that Ω_t is "filled" with water. Then $S_t := \{(x, y, z) | (x, y) \in \mathbb{R}^2, z = \varphi(x, y, t)\}$ can be considered as the free surface of the water at time t. If we denote the velocity components with respect to x and y by u and v, the acceleration of gravity by g and the density of the fluid by ρ, the governing system for the flow of the water in Ω_t is given by

$$\partial_t \varphi + \partial_x (\varphi u) + \partial_y (\varphi v) = 0 \ ,$$
$$\rho(\partial_t u + u \partial_x u + v \partial_y u) = -g\rho \partial_x \varphi \ ,$$
$$\rho(\partial_t v + u \partial_x v + v \partial_y v) = -g\rho \partial_y \varphi \ ,$$
$$\partial_x u + \partial_y v = 0 \ .$$

The given data in this model are the constants g and ρ and the initial values for φ, u and v. The unknowns are φ, u and v as functions of $(x, y, t) \in \mathbb{R}^2 \times [0, \tau]$ (see [190], [228] and [43, page 32]).

EXAMPLE 1.0.10 (Reactive flows) We consider the flow of two different fluids, for instance burnt and unburnt gas. We assume that the reaction of the unburnt gas to burnt gas is a one-step, irreversible exothermic one with Arrhenius kinetics. In addition to the notation of Example 1.0.3 or 1.0.4 we denote the mass fraction of unburnt gas by Z, the density of the unburnt gas ρ and the temperature T. Then the mathematical model consists of the system of Euler equations (1.0.9), (1.0.10),(1.0.11) and the following first-order equation

$$\partial_t(\rho Z) + \mathrm{div}(\rho u Z) = -\rho W(Z,T), \tag{1.0.23}$$

where $W(Z,T) := K e^{-(A_0/T)} Z$ for some constants K and A_0, and two further equations of state:

$$e = F_1(T,Z), \quad T = F_2(p,\rho). \tag{1.0.24}$$

Equation (1.0.23) follows from the conservation of mass, and in (1.0.24) we can have for instance

$$F_1(T,Z) = \frac{T}{\gamma - 1} + q_0 Z, \quad F_2(p,\rho) = \frac{p}{\rho}.$$

For determining the seven unknown functions ρ, u, v, e, p, Z and T we have five partial differential equations and two equations of state (see [153]).

Now let us briefly discuss how this book is organized.

The system of equations (1.0.9) (1.0.10),and (1.0.11) can be written in the form

$$\partial_t u + \sum_{j=1}^{n} \partial_j F_j(u) = S(u,x,t), \quad u \in \mathbb{R}^m, \ x \in \mathbb{R}^n, \ t \in [0,T] \tag{1.0.25}$$

for some nonlinearities $F_j : \mathbb{R}^m \to \mathbb{R}^m$, $j = 1, \ldots, n$, and some given source term $S : \mathbb{R}^m \times \mathbb{R} \times \mathbb{R}^+ \to \mathbb{R}^m$. In this book we shall study some special cases of (1.0.25) from the theoretical and numerical points of view. In Chapter 2 we shall investigate scalar equations ($m = 1$) in one space dimension ($n = 1$). Scalar equations ($m = 1$) in two space dimensions ($n = 2$) will be considered in Chapter 3. In both chapters we discuss the details of proving the convergence and for getting error estimates. We treat finite difference, finite volume and finite element (streamline diffusion) methods. Systems in one space dimension are considered in Chapter 4. In Chapter 5 we shall discuss numerical schemes for systems ($m > 1$) in two or more space dimensions ($n > 1$). For this kind of problem, there are several numerical schemes that give results in good agreement with physical measurements, but up to now there is no general existence or convergence proof (globally in time) for systems in higher dimensions. The general problem of boundary conditions will be treated in Chapter 6. Special discretizations for convection dominated problems and for the compressible Navier–Stokes equations are discussed in Chapter 7.

At the beginning of each chapter we give an overview of the topics that will be treated there.

2 Initial value problems for scalar conservation laws in 1-D

In this chapter we consider the initial value problem for scalar conservation laws in 1-D [125]. We start with the theoretical background (§2.1) that is necessary to go through the corresponding numerical analysis. The definition of stable finite difference schemes and the proof of their convergence to the uniquely defined entropy condition will be given in §2.2. The main tools are the TVD boundedness and the compactness condition in L^1. As a sufficient condition for stability of the explicit scheme, the well-known CFL condition will be derived. The same condition can be derived for the stability of linear equations using the Fourier stability method (see §2.4). In §2.5 the results will be extended to higher-order schemes. Finally in this chapter we prove the convergence for the streamline diffusion method applied to the Burgers equation in 1-D.

2.1 Theoretical background

The purpose of this section is to present some basic theoretical results for the scalar conservation law

$$\partial_t u + \partial_x f(u) = 0 \quad \text{for } t > 0 \text{ and } x \in \mathbb{R} \tag{2.1.1}$$

with the initial data

$$u(x, 0) = u_0(x) \quad \text{for } x \in \mathbb{R}. \tag{2.1.2}$$

We show that in general classical solutions cannot exist, and we define weak solutions and the entropy condition, which guarantees uniqueness. The connection between the viscosity limit and different entropy conditions will be explained. Special entropy solutions like shocks and rarefaction waves will be shown. Furthermore, we consider the Lax representation formula. For more details concerning theoretical results we refer to [80] and [218].

First let us show that in general there are no classical solutions of (2.1.1) and (2.1.2), even for smooth data. Let us assume that $f \in C^2(\mathbb{R})$ with $f'' > 0$. The first interesting property of C^1 solutions of (2.1.1) is the fact that they are constant along Γ_a. A characteristic Γ_a is a curve $(\gamma(t), t)$ such that

$$\gamma'(t) = f'(u(\gamma(t), t)) \quad \text{for } t > 0 ,$$
$$\gamma(0) = a . \tag{2.1.3}$$

This property can be seen very easily. We assume that $u \in C^1(\mathbb{R} \times [0, T])$ is a solution of (2.1.1) with initial data (2.1.2). Then we have

$$\frac{d}{dt} u(\gamma(t), t) = \partial_t u + \gamma' \partial_x u = \partial_t u + \partial_x f(u) = 0 .$$

Therefore

$$\gamma'(t) = f'(u(\gamma(t), t)) = f'(u(\gamma(0), 0)) = f'(u_0(a)) = \text{const} .$$

This means that u is constant along the characteristic Γ_a for all $a \in \mathbb{R}$, and Γ_a is a straight line.

Let us consider the problem of existence of classical solutions. The following example shows that even in the case of smooth data f and u_0 we cannot expect that there are continuous solutions that exist globally in time.

EXAMPLE 2.1.1 (Local existence of classical solutions) Let $f(u) := \frac{1}{2}u^2$ and let the initial value $u_0 \in C^\infty(\mathbb{R})$ be such that $u_0(x) = 1$ if $x \in \,]-\infty, -1]$, $u_0(x) = 0$ if $x \in [1, \infty[$ and $u_0'(x) \leq 0$ (see Figure 2.1.1). Then we obtain that u is constant along the characteristic curve $(\gamma(t), t)$, with

$$\gamma'(t) = f'(u_0(\gamma(0))) = u_0(\gamma(0)) ,$$
$$u(\gamma(t), t) = u(\gamma(0), 0) = u_0(\gamma(0)) .$$

It turns out that u is no longer uniquely defined for all $(x, t) \in \mathbb{R} \times \mathbb{R}^+$. The characteristics starting in $]-\infty, -1]$ have slope 1 and $u = 1$ along these lines. On the other side, the characteristics starting in $[1, \infty[$ have slope 0 and $u = 0$ along them. Both types of characteristics meet each other (see Figure 2.1.2), and therefore u cannot be in $C^0(\mathbb{R} \times \mathbb{R}^+)$.

2.1 Theoretical background   17

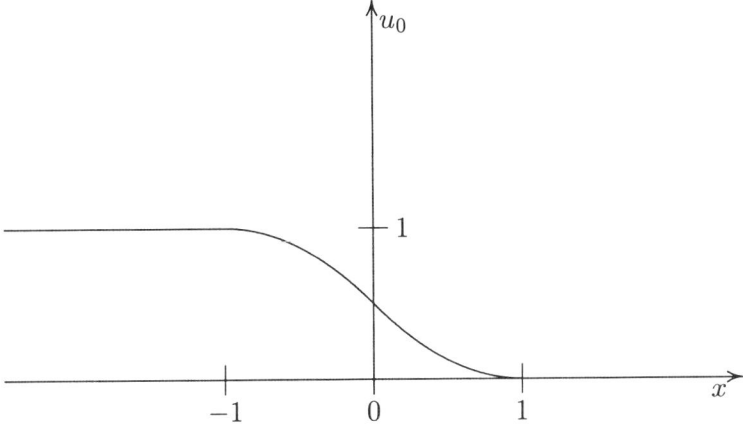

Figure 2.1.1

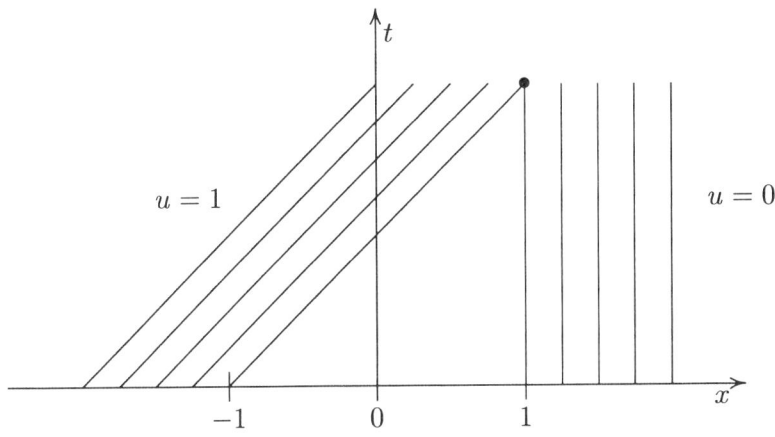

Figure 2.1.2

This is a problem of global existence (globally in time). Locally in time it is possible to show the existence of classical solutions.

LEMMA 2.1.2 *Assume $f \in C^2(\mathbb{R})$ and $u_0 \in C^1(\mathbb{R})$, $|f''|, |u_0'| \leq M$ on $\mathbb{R}$. Then there exists a time $T_0 > 0$ such that (2.1.1), (2.1.2) have a classical solution $u \in C^1(\mathbb{R} \times [0, T_0[)$.*

Proof Let $(x_0, t_0) \in \mathbb{R} \times \mathbb{R}^+$, let Γ be the characteristic through (x_0, t_0) and let $(x_1, 0)$ be the point where Γ, crosses the x-axis. Then

$$\Gamma = \left\{ (x, t) \in \mathbb{R} \times \mathbb{R}^+ \left| \frac{x - x_1}{t} = f'(u(x_0, t_0)) \right. \right\}.$$

In particular $(x_0, t_0) \in \Gamma$ and therefore $x_1 = x_0 - t_0 f'(u(x_0, t_0))$. Since u is constant along Γ we get

$$u(x_0, t_0) = u(x_1, 0) = u_0(x_0 - t_0 f'(u(x_0, t_0))). \tag{2.1.4}$$

This holds for all $(x_0, t_0) \in \mathbb{R} \times \mathbb{R}^+$ if t_0 is sufficiently small. Equation (2.1.4) is a necessary condition for u. Now we shall show that there exists a u satisfying (2.1.4). Let $F(s, x, t) := s - u_0(x - tf'(s))$. Since

$$F(u_0(x_0), x_0, 0) = 0 \quad \text{and} \quad \partial_s F = 1 + tu_0' f''(s) \neq 0,$$

if t is sufficiently small, the implicit function theorem implies that there is a solution u of (2.1.4). Differentiating (2.1.4) with respect to x and t respectively, we obtain

$$\partial_t u = -\frac{u_0' f'(u)}{1 + tu_0' f''(u)}, \quad \partial_x u = \frac{u_0'}{1 + tu_0' f''(u)}$$

for $(x, t) \in \mathbb{R} \times [0, t_0[$ and t_0 sufficiently small. But this means that u is a classical solution of (2.1.1), (2.1.2). $\square$

Since in general we can only prove local existence (see Lemma 2.1.2 and Example 2.1.1), we have to generalize the definition of solutions of conservation laws.

DEFINITION 2.1.3 *Let* $u_0 \in L^\infty(\mathbb{R})$. *Then* u *is called a weak solution of (2.1.1), (2.1.2) or a solution in the distributional sense if and only if* $u \in L^\infty(\mathbb{R} \times \mathbb{R}^+)$ *and*

$$\int\int_{\mathbb{R}\,\mathbb{R}^+} [u\partial_t \varphi + f(u)\partial_x \varphi] \, dt \, dx + \int_{\mathbb{R}} \varphi(x, 0) u_0(x) \, dx = 0 \tag{2.1.5}$$

for all $\varphi \in C_0^\infty(\mathbb{R} \times [0, \infty[)$.

2.1 Theoretical background

It can be seen very easily that a weak solution that lies in $C^1(\mathbb{R} \times [0, \infty[)$ satisfies (2.1.1), (2.1.2) in the classical sense.

Now we should like to study piecewise-smooth solutions near discontinuities. If in addition they are also weak solutions, they have to satisfy special jump conditions.

LEMMA 2.1.4 (Rankine–Hugoniot) *Let us assume that the half-space $\mathbb{R} \times \mathbb{R}^+$ is separated by a smooth curve $S: t \to (\sigma(t), t)$ into two parts M_l and M_r. Furthermore, let $u \in L^1_{loc}(\mathbb{R} \times \mathbb{R}^+)$ such that $u_l := u|_{M_l} \in C^1(\overline{M_l})$ and $u_r := u|_{M_r} \in C^1(\overline{M_r})$ and u_l, u_r satisfy (2.1.1) locally in M_l and M_r respectively in the classical sense. Then (2.1.5) holds for u and for all $\varphi \in C_0^\infty(\mathbb{R} \times \mathbb{R}^+)$ if and only if*

$$[u_l(\sigma(t), t) - u_r(\sigma(t), t)]\sigma'(t) = f(u_l(\sigma(t), t)) - f(u_r(\sigma(t), t)) \quad (2.1.6)$$

for all $t > 0$. Instead of (2.1.6), we shall often write

$$(u_l - u_r)s = f_l - f_r, \quad \text{where } s := \sigma'(t). \quad (2.1.7)$$

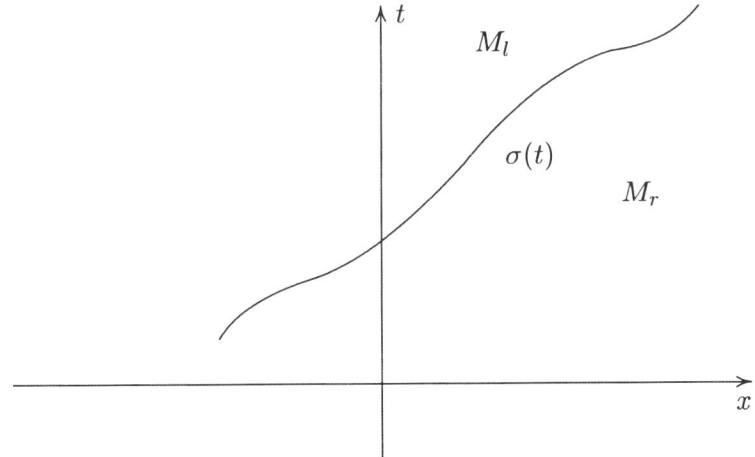

Figure 2.1.3

The condition (2.1.6) (or (2.1.7)) is called a jump condition or Rankine–Hugoniot condition, and $s(t) := \sigma'(t)$ is called the propagation speed of the discontinuity.

Proof Let ν denote the outer normal of M_l. Then we have for all $\varphi \in C_0^\infty(\mathbb{R} \times \mathbb{R}^+)$

$$0 = \int\int_{\mathbb{R} \ \mathbb{R}^+} [u\partial_t \varphi + f(u)\partial_x \varphi] \tag{2.1.8}$$

$$= \int_{M_l} [u\partial_t \varphi + f(u)\partial_x \varphi] + \int_{M_r} [u\partial_t \varphi + f(u)\partial_x \varphi]$$

$$= \int_{\partial M_l} \nu\varphi \begin{pmatrix} f(u_l) \\ u_l \end{pmatrix} - \int_{\partial M_r} \nu\varphi \begin{pmatrix} f(u_r) \\ u_r \end{pmatrix}$$

$$= \int_S \nu\varphi \left(\begin{pmatrix} f(u_l) \\ u_l \end{pmatrix} - \begin{pmatrix} f(u_r) \\ u_r \end{pmatrix} \right) \tag{2.1.9}$$

$$= \int_S c\varphi(-su_l + f(u_l) + su_r - f(u_r)) ,$$

where $\nu = c\begin{pmatrix} 1 \\ -s \end{pmatrix}$ and $c > 0$. This proves Lemma 2.1.4. □

The following example shows that in general there is no uniqueness for the weak solution.

EXAMPLE 2.1.5 Consider the following initial value problem:

$$\partial_t u + u\partial_x u = 0 \quad \text{for } t > 0,\ x \in \mathbb{R},$$
$$u_0(x) = \begin{cases} 0 & \text{for } x \leq 0, \\ 1 & \text{for } x > 0. \end{cases}$$

Then define

$$u_1(x,t) := \begin{cases} 0 & \text{for } x < t/2, \\ 1 & \text{for } x > t/2, \end{cases}$$

$$u_2(x,t) := \begin{cases} 0 & \text{for } x < 0, \\ x/t & \text{for } 0 \leq x \leq t, \\ 1 & \text{for } t < x. \end{cases}$$

Then u_1 is piecewise-constant and satisfies the Rankine–Hugoniot condition (2.1.7), because

$$s = \frac{1}{2} \quad \text{and} \quad \frac{f(u_l) - f(u_r)}{u_l - u_r} = \frac{1}{2}.$$

Furthermore, it satisfies the initial conditions, and therefore it is a weak solution of (2.1.1), (2.1.2). The function u_2 is a piecewise-classical solution, satisfying the Rankine–Hugoniot condition along the lines $\{(x,t)|x = t\}$, $\{(x,t)|x = 0\}$ and the initial conditions. Therefore the initial value problem (2.1.1), (2.1.2) has two different weak solutions. Now the question arises whether it makes sense to look for an additional condition to select a special weak solution among the whole set of weak solutions. Since (2.1.1) is not a realistic mathematical model for a physical flow problem, we consider the equations of gas dynamics for a viscous isentropic flow.

EXAMPLE 2.1.6 Consider the equations of gas dynamics for a viscous isentropic flow:

$$\partial_t \rho + \partial_x(\rho u) = 0 \quad \text{in } \mathbb{R} \times]0, T[$$
(conservation of mass),
$$\partial_t(\rho u) + \partial_x(\rho u^2 + p) = \varepsilon \partial_x^2 u \quad \text{in } \mathbb{R} \times]0, T[$$
(conservation of momentum),
$$(\rho, u, p) = (\bar{\rho}, \bar{u}, \bar{p}) \quad \text{in } \mathbb{R} \times \{0\}$$
(initial values), \hfill (2.1.10)

where ε is the viscosity coefficient and $(\bar{\rho}, \bar{u}, \bar{p})$ are prescribed initial values. For $\varepsilon = 0$ we obtain the system (2.1.11) of two conservation laws. This is a mathematical model, for inviscid isentropic, compressible flow:

$$\begin{aligned}
\partial_t \rho + \partial_x(\rho u) &= 0 & \text{in } \mathbb{R} \times \mathbb{R}^+, \\
\partial_t(\rho u) + \partial_x(\rho u^2 + p) &= 0 & \text{in } \mathbb{R} \times \mathbb{R}^+, \\
(\rho, u, p) &= (\bar{\rho}, \bar{u}, \bar{p}) & \text{in } \mathbb{R} \times \{0\}.
\end{aligned} \quad (2.1.11)$$

Let us assume, that there is a unique solution $(\rho, u, p)^\varepsilon$ of (2.1.10) for all $\varepsilon > 0$. Now, in order to have a convenient model one should expect that $(\rho, u, p)^\varepsilon$ converges for $\varepsilon \to 0$ (in some sense) to a solution (ρ, u, p) of the inviscid problem (2.1.11). Then for physical reasons we should like to define this limit

22 Initial value problems for scalar conservation laws in 1-D

as the uniquely defined weak solution of the initial value problem (2.1.11). This method to approximate (ρ, u, p) by $(\rho, u, p)^\varepsilon$ is called the viscosity method. Unfortunately the convergence of $(\rho, u, p)^\varepsilon \to (\rho, u, p)$ can only be proved for special cases (see Theorem 2.1.7). For general systems this is still an open problem.

Now we shall study the mathematical background of the viscosity limit.

THEOREM 2.1.7 *Let $u_0 \in L^\infty(\mathbb{R}^n) \cap L^1(\mathbb{R}^n)$ and $\varphi_i, \psi : \mathbb{R}^n \times \mathbb{R}^+ \times \mathbb{R} \to \mathbb{R}, i = 1, \ldots, n$, be such that all derivatives up to second order of φ_i and ψ exist and are bounded. Then for any $\varepsilon > 0$ there exists a uniquely defined classical solution u_ε of*

$$\begin{aligned} \partial_t u + \sum_{i=1}^n \partial_i \varphi_i(x,t,u) + \psi(x,t,u) &= \varepsilon \Delta u & \text{in } \mathbb{R}^n \times \mathbb{R}^+ \\ u(x,0) &= u_0(x) & \text{in } \mathbb{R}^n \end{aligned}, \quad (2.1.12)$$

such that u_ε converges almost everywhere in $\mathbb{R}^n \times \mathbb{R}^+$ as $\varepsilon \to 0$ to a function u that is a weak solution of

$$\begin{aligned} \partial_t u + \sum_{i=1}^n \partial_i \varphi_i(x,t,u) + \psi(x,t,u) &= 0 & \text{in } \mathbb{R}^n \times \mathbb{R}^+ \\ u(x,0) &= u_0(x) & \text{in } \mathbb{R}^n \end{aligned}. \quad (2.1.13)$$

REMARK 2.1.8 *This theorem is important for two reasons. First it is an existence result for (2.1.13) and secondly it controls the convergence of u_ε for $\varepsilon \to 0$. Since we shall prove the existence for the conservation law by proving the convergence of discrete solutions we omit the proof of this theorem and refer to [123] and [171].*

REMARK 2.1.9 *The following example is due to M. Rokyta and illustrates the problems that arise for single equations and systems of equations depending on a parameter ε. The situation is similar to the viscosity limit $\varepsilon \to 0$. Consider the linear system of equations*

$$\begin{aligned} x + y &= \varepsilon, \\ x + y(1 - \varepsilon) &= \varepsilon a \end{aligned} \quad (2.1.14)$$

for given numbers $\varepsilon, a \in \mathbb{R}$. The exact solution $y = 1 - a$, $x = \varepsilon - 1 + a$ is unique for any $\varepsilon > 0$. But in the limit $\varepsilon = 0$ we have instead of (2.1.14)

$$\begin{aligned} x + y &= 0, \\ x + y &= 0, \end{aligned}$$

and this system has an infinite number of solutions.

2.1 Theoretical background

Now we should like to derive a necessary condition for the limit u to be the viscosity limit of a higher-order problem like (2.1.10) or (2.1.12).

In order to get an idea of this, let us assume that $U, F \in C^2(\mathbb{R})$ such that U is convex and $F' = U'f'$. Let $u_\varepsilon \in C^2(\mathbb{R} \times \mathbb{R}^+)$ be the solution of

$$\partial_t u_\varepsilon + \partial_x f(u_\varepsilon) = \varepsilon \partial_x^2 u_\varepsilon \quad \text{in } \mathbb{R} \times \mathbb{R}^+ .$$

Then we obtain

$$\partial_t U(u_\varepsilon) + \partial_x F(u_\varepsilon) - \varepsilon \partial_x^2 U(u_\varepsilon) = U'(u_\varepsilon)\partial_t u_\varepsilon + F'(u_\varepsilon)\partial_x u_\varepsilon$$
$$- \varepsilon \partial_x(U'(u_\varepsilon)\partial_x u_\varepsilon)$$
$$= -\varepsilon U''(u_\varepsilon)(\partial_x u_\varepsilon)^2 \leq 0 ,$$

since U is convex.

For all test functions φ as in (2.1.5) and $\varphi \geq 0$ we get

$$\int_\mathbb{R} \int_{\mathbb{R}^+} [U(u_\varepsilon)\partial_t \varphi + F(u_\varepsilon)\partial_x \varphi - \varepsilon U(u_\varepsilon)\partial_x^2 \varphi] \geq 0 . \tag{2.1.15}$$

Now let us assume (and this can be proved, see [123]) that

$$u_\varepsilon \to u , \quad U(u_\varepsilon) \to U(u) , \quad F(u_\varepsilon) \to F(u) ,$$

in L^1, $|U(u_\varepsilon)| \leq$ const as $\varepsilon \to 0$. Then taking the limit $\varepsilon \to 0$ in (2.1.15), we obtain

$$\partial_t U(u) + \partial_x F(u) \leq 0 \tag{2.1.16}$$

in the distributional sense. This means that the limit u has to satisfy (2.1.16). The pair of functions (U, F) is called an entropy pair or pair of entropy functions.

DEFINITION 2.1.10 (Pair of entropy functions) *Let $U, F \in C^2(\mathbb{R})$ be such that*

$$F' = U'f' \text{ and } U \text{ is strictly convex.}$$

Then (U, F) is called a pair of entropy functions *for the equation*

$$\partial_t u + \partial_x f(u) = 0 .$$

EXAMPLE 2.1.11 *Let f be strictly convex. Then for the equation*

$$\partial_t u + \partial_x f(u) = 0$$

the functions $U(s) := f(s)$ and $F(s) := \int_0^s [f'(r)]^2 \, dr$ are an entropy pair, since

$$U'f' = (f')^2, \quad F' = (f')^2 \ .$$

Furthermore, for any convex function U an entropy pair is given by (U, F) where F is defined as

$$F(s) = \int_0^s U'(\tau) f'(\tau) \, d\tau \ .$$

Now let us show that (2.1.16) implies another interesting inequality, which is related to the characteristics.

THEOREM 2.1.12 *Let $\Gamma := \{(\sigma(t), t) \mid t > 0\}$ be a smooth curve with $s(t) := \sigma'(t)$ and let u be a weak solution of (2.1.1) that is piecewise-smooth outside Γ. Let $f, F, U \in C^2(\mathbb{R})$ be such that f and U are strictly convex and F satisfies $F' = U'f'$. Furthermore, we assume that u satisfies (2.1.16), i.e.*

$$\partial_t U(u) + \partial_x F(u) \leq 0 \tag{2.1.17}$$

in the distributional sense. Then across the curve Γ we have

$$f'(u_l) > s > f'(u_r) \ . \tag{2.1.18}$$

Proof Let (x_0, t_0) be a point on $\Gamma := \{(\sigma(t), t) \mid t > 0\}$, let V be a neighbourhood of (x_0, t_0), and let $V_l = V \cap \{(x, t) \mid x < \sigma(t)\}$ and $V_r = V \cap \{(x, t) \mid x > \sigma(t)\}$. Then it follows from (2.1.17) that we have for all test functions $\varphi \in C_0^\infty(V)$ and $\varphi \geq 0$ (see (2.1.9))

$$0 \geq \int_V [-U(u) \partial_t \varphi - F(u) \partial_x \varphi] = \int_{V_l} \ldots + \int_{V_r} \ldots \tag{2.1.19}$$

$$= \int_\Gamma c \Big\{ [U(u_l) - U(u_r)] s - [F(u_l) - F(u_r)] \Big\} \varphi \ . \tag{2.1.20}$$

Define
$$s(u) := \frac{f(u)-f(u_r)}{u-u_r},$$
$$E(u) := s(u)[U(u) - U(u_r)] - [F(u) - F(u_r)].$$
(2.1.21)

We should like to show that $u_l > u_r$. Then this implies the statement of the theorem, since $f'' > 0$.

From (2.1.19) we know that $E(u_l) \leq 0$ and $E(u_r) = 0$. Now it is easy to see that $E'(u_r) = 0$ and $s'(u) = \alpha f''(\xi)$ for some $\alpha > 0$ and ξ between u_r and u_l. Since $f'' > 0$, we have $s'(u) > 0$. It turns out that $E'(u) < 0$ if $u \neq u_r$. This can be derived as follows. Differentiating (2.1.21), we obtain

$$\begin{aligned} E'(u) &= s'(U - U_r) - s'U'(u - u_r) \\ &= s'U'(\xi)(u - u_r) - s'U'(u - u_r) \quad \text{for } \xi \text{ between } u \text{ and } u_r \\ &= s'(u - u_r)U''(\eta)(\xi - u) \quad \text{for } \eta \text{ between } u \text{ and } \xi. \end{aligned}$$

If $u < u_r$, we have $u < \xi < u_r$, $u < \eta < \xi < u_r$ and therefore $E'(u) < 0$. If $u > u_r$, we obtain $u_r < \xi < u$, $u_r < \xi < \eta < u$ and therefore $E'(u) < 0$. We have already mentioned that $E(u_l) \leq 0$, and since $E'(u) < 0$ ($u \neq u_r$), this implies $u_l > u_r$. From this fact and $s' > 0$ we get

$$f'(u_r) = s(u_r) < s(u_l) = \frac{f(u_l) - f(u_r)}{u_l - u_r} = f'(\xi) < f'(u_l)$$

for some $\xi \in]u_r, u_l[$. □

DEFINITION 2.1.13 (Entropy condition) *Let u be a weak solution of (2.1.1) and let S be a smooth curve in $\mathbb{R} \times \mathbb{R}^+$, where u has a discontinuity on S. Let $(x_0, t_0) \in S$, $u_l := \lim_{\varepsilon \to 0} u(x_0 - \varepsilon, t_0)$, $u_r := \lim_{\varepsilon \to 0} u(x_0 + \varepsilon, t_0)$ and*

$$s = \frac{f(u_l) - f(u_r)}{u_l - u_r}.$$

Then u satisfies the entropy condition *(or Lax entropy condition) in (x_0, t_0), if and only if (see Figure 2.1.4)*

$$f'(u_l) > s > f'(u_r).$$
(2.1.22)

A discontinuity that satisfies the jump condition (2.1.7) and the entropy condition (2.1.22) is called a **shock**. Then s is called the **shock velocity**.

26   Initial value problems for scalar conservation laws in 1-D

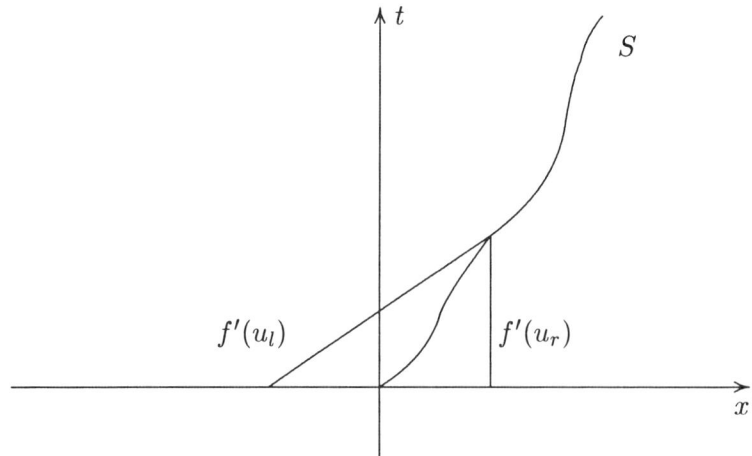

Figure 2.1.4

REMARK 2.1.14 *(a) If one generalizes this condition to the case of compressible flow, one can show that the physical entropy increases across the discontinuity. Therefore it is called the entropy condition.*

(b) Let S, s and u be as in Definition 2.1.13 and assume that f is convex. Suppose that u satisfies the entropy condition on S. Then the solution u can only jump down across S, that is $u_l > u_r$. This is obvious from Figure 2.1.5.

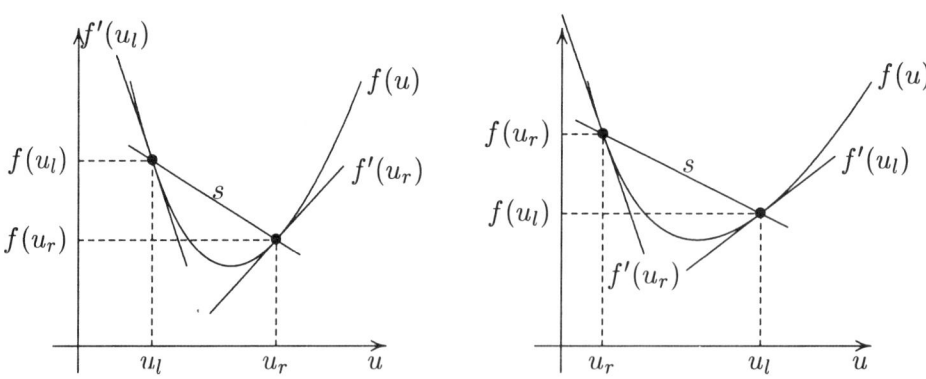

Figure 2.1.5

On the left we have $f'(u_l) < s < f'(u_r)$, which is not allowed, while on the right $f'(u_r) < s < f'(u_l)$.

Now let us mention that weak solutions satisfying the entropy condition are unique.

THEOREM 2.1.15 (Uniqueness of the entropy solution) *Let $f \in C^2(\mathbb{R})$ and $f'' > 0$ on $\mathbb{R}$, and let u and v be two weak solutions satisfying the entropy condition 2.1.13. Then $u = v$ almost everywhere in $t > 0$.*

Proof See [218, page 283].

THEOREM 2.1.16 (Kruzkov entropy condition) *Let u be a solution satisfying (2.1.1), (2.1.2) and (2.1.17) for all strict convex functions $U \in C^2(\mathbb{R})$. Then we have for all $k \in \mathbb{R}$ and any function $\varphi \in C_0^\infty(\mathbb{R} \times \mathbb{R}^+)$, $\varphi \geq 0$,*

$$\int_\mathbb{R} \int_{\mathbb{R}^+} \left\{ |u - k| \partial_t \varphi + \text{sign}(u - k) \left[f(u) - f(k) \right] \partial_x \varphi \right\} \geq 0 . \quad (2.1.23)$$

Proof Let $\eta \in C^\infty(\mathbb{R})$ be such that $\eta(x) = |x|$ for $|x| \geq 1$ and $\eta'' \geq 0$ otherwise. Now let

$$\eta_\varepsilon(x) := \varepsilon \eta \left(\frac{x - k}{\varepsilon} \right)$$

$$F_\varepsilon(x) := \int_k^x \eta'_\varepsilon(s) f'(s) \, ds .$$

Then η_ε is convex and $(\eta_\varepsilon, F_\varepsilon)$ is an entropy pair, since

$$F_\varepsilon'(x) = \eta'_\varepsilon(x) f'(x) .$$

By assumption, we have

$$\partial_t \eta_\varepsilon(u) + \partial_x F_\varepsilon(u) \leq 0 , \quad (2.1.24)$$

where u is the exact solution of (2.1.1). This implies that

$$\int_\mathbb{R} \int_{\mathbb{R}^+} [\eta_\varepsilon(u) \partial_t \varphi + F_\varepsilon(u) \partial_x \varphi] \geq 0 \quad (2.1.25)$$

for all $\varphi \in C_0^\infty(\mathbb{R} \times \mathbb{R}^+)$, $\varphi \geq 0$. Now we shall pass to the limit $\varepsilon \to 0$ in (2.1.25). Since $\eta'_\varepsilon(x) \to \text{sign}(x - k)$, we have

$$F_\varepsilon(x) = \int_k^x \eta'_\varepsilon(x) f'(s)\, ds \to \begin{cases} f(x) - f(k) & \text{if } x \geq k, \\ f(k) - f(x) & \text{if } x < k. \end{cases}$$

Therefore (2.1.25) implies (2.1.23). □

Conversely we can also show that (2.1.23) implies (2.1.17). This can be seen from the following.

LEMMA 2.1.17 *Let u be as in Theorem 2.1.16 and assume that (2.1.23) holds for any $k \in \mathbb{R}$ and all $\varphi \geq 0$, $\varphi \in C_0^\infty(\mathbb{R} \times \mathbb{R}^+)$. Then (2.1.17) holds for any entropy pair (U, F).*

Proof Let $a, b \in \mathbb{R}^+$ and define $h_m := (b - a)/(m + 1)$ for given $m \in \mathbb{N}$ and let $k_l := a + h_m l$ for $l = 0, \ldots, m + 1$. Let (U, F) be an entropy pair. Then there exist $\beta_1, \ldots, \beta_m \geq 0$, β_0, $\bar{k} \in \mathbb{R}$ such that

$$U_m \longrightarrow U,$$
$$U'_m \longrightarrow U'$$

uniformly on $[a, b]$ a. e., where

$$U_m(s) := \beta_0(s - \bar{k}) + \sum_{l=1}^m \beta_l |s - k_l|,$$
$$U'(k) = \lim_{\varepsilon \to 0} U'(k + \varepsilon) \quad \text{for } k = k_l.$$

This follows from Lemmas 2.1.18 and 2.1.19. Define

$$F_m(s) = \beta_0[f(s) - f(\bar{k})] + \sum_{i=1}^m \beta_i \,\text{sign}(s - k_i)[f(s) - f(k_i)] + c_m$$

and choose the constant c_m such that $F_m(0) = F(0)$. We obtain for $s \neq k_l$,

$$F'_m(s) = U'_m(s) f'(s) \longrightarrow U'(s) f'(s) = F'(s)$$

for almost all s (see Lemmas 2.1.18 and 2.1.19) and

$$F_m(s) = \int_0^s F'_m(t)\, dt + F_m(0) = \int_0^s F'_m(t)\, dt + F(0)$$

$$\longrightarrow \int_0^s F'(t)\, dt + F(0) = F(s).$$

Then we can conclude for $\varphi \in C_0^\infty(\mathbb{R} \times \mathbb{R}^+)$, $\varphi \geq 0$, that

$$\int_{\mathbb{R}} \int_{\mathbb{R}^+} [U(u)\partial_t\varphi + F(u)\partial_x\varphi]$$

$$= \lim_{m\to\infty} \int_{\mathbb{R}} \int_{\mathbb{R}^+} [U_m(u)\partial_t\varphi + F_m(u)\partial_x\varphi]$$

$$= \lim_{m\to\infty} \int_{\mathbb{R}} \int_{\mathbb{R}^+} \Big(\beta_0(u - \bar{k})\partial_t\varphi + \{\beta_0[f(u) - f(\bar{k})] + c_m\}\partial_x\varphi$$

$$+ \sum_{l=1}^m \beta_l\{|u - k_l|\partial_t\varphi + \mathrm{sign}(u - k_l)[f(u) - f(k_l)]\partial_x\varphi\}\Big).$$

Because of (2.1.23), the sum is greater than or equal to 0, and since u is a weak solution the remaining part is equal to zero. □

LEMMA 2.1.18 *Let $U \in C^2([a,b], \mathbb{R})$ be convex and $a \leq k_1 < \ldots < k_m \leq b$. Let U_m be the piecewise-linear interpolation of U such that*

$$U_m(k_i) = U(k_i), \quad i = 1, \ldots, m.$$

Then there exist $\beta_0, \bar{k} \in \mathbb{R}$ and $\beta_1, \ldots, \beta_m \in \mathbb{R}^+$ such that

$$U_m(x) = \beta_0(x - \bar{k}) + \sum_{i=1}^m \beta_i |x - k_i|.$$

Proof For given $\kappa \in \mathbb{R}$ and $\beta := (\beta_0, \beta_1, \ldots, \beta_m) \in \mathbb{R}^{m+1}$ let

$$S_\beta(x) := \beta_0(x - \kappa) + \sum_{i=1}^m \beta_i |x - k_i|.$$

For $k_i \leq x < k_{i+1}$, $i = 0, \ldots, m$, we have

$$S_\beta(x) = (\beta_0 + \beta_1 + \ldots + \beta_i - \beta_{i+1} - \ldots - \beta_m)x$$
$$- \beta_0 \kappa - \beta_1 k_1 - \ldots - \beta_i k_i + \beta_{i+1} k_{i+1} + \ldots + \beta_m k_m .$$

Choose $\beta_0, \beta_1, \ldots, \beta_m$ such that

$$S'_\beta(x) = U'_m(x) =: \alpha_i \quad \text{for } k_i \leq x \leq k_{i+1}, i = 0, \ldots, m . \qquad (2.1.26)$$

This is possible, since from (2.1.26) for $i = 0, \ldots, m$ we obtain the following linear system:

$$\begin{pmatrix} 1 & -1 & -1 & \ldots & -1 \\ 1 & 1 & -1 & \ldots & -1 \\ 1 & 1 & 1 & -1 & \ldots & -1 \\ \vdots & & & \ddots \\ 1 & 1 & 1 & 1 & \ldots & 1 \end{pmatrix} \begin{pmatrix} \beta_0 \\ \beta_1 \\ \vdots \\ \beta_m \end{pmatrix} = \begin{pmatrix} \alpha_0 \\ \alpha_1 \\ \vdots \\ \alpha_m \end{pmatrix} ,$$

and using Gaussian elimination

$$\begin{pmatrix} 1 & -1 & -1 & \ldots & -1 \\ 0 & 1 & 0 & \ldots & 0 \\ 0 & 0 & 1 & 0 & \ldots & 0 \\ \vdots & & & \ddots \\ 0 & 0 & 0 & 0 & \ldots & 1 \end{pmatrix} \begin{pmatrix} \beta_0 \\ \beta_1 \\ \vdots \\ \beta_m \end{pmatrix} = \begin{pmatrix} \alpha_0 \\ \frac{\alpha_1 - \alpha_0}{2} \\ \vdots \\ \frac{\alpha_m - \alpha_{m-1}}{2} \end{pmatrix} .$$

Since U is assumed to be convex, we have

$$\beta_i = \frac{\alpha_i - \alpha_{i-1}}{2} > 0 \quad \text{and } \beta_0 \in \mathbb{R}.$$

Choose κ such that

$$S_\beta(0) = U_m(0) .$$

Then we have for $i = 0, \ldots, m$

$$S'_\beta(x) = U'_m(x) \quad \text{on } k_i \leq x < k_{i+1}$$

and $S_\beta(0) = U_m(0)$. This implies

$$S_\beta(x) = U_m(x) .$$

Now we obtain the statement of Lemma 2.1.18 with $\bar{k} := \kappa$. □

LEMMA 2.1.19 *Let U and U_m be as in Lemma 2.1.18. Then we have*

$$U_m \longrightarrow U,$$
$$U'_m \longrightarrow U'$$

uniformly on $[a,b]$ a. e..

Proof This is left as an exercise.

When developing numerical algorithms for the conservation law (2.1.1), it will be the main problem to ensure that the corresponding scheme will select the entropy solution and nothing else. In the following example we shall see that we have to be very careful when defining a discretization for (2.1.1).

EXAMPLE 2.1.20 (Finite differences) (a) *Central differences.* In a numerical scheme for approximating the weak entropy solution of (2.1.1) we should not discretize $\partial_x f(u)$ using central differences. To see this let $f \in C^2(\mathbb{R})$ such that $f'' > 0$, $f(1) = f(-1)$, and $f'(1) > 0 > f'(-1)$, e.g. $f(u) = \frac{1}{2}u^2$.

As initial values we choose

$$u_0(x) = \begin{cases} 1 & \text{for } x > 0, \\ -1 & \text{for } x < 0. \end{cases} \qquad (2.1.27)$$

We shall use the following notation. We denote an approximation of $u(k\,\Delta x, n\,\Delta t)$ by u_k^n.

Then the numerical scheme

$$u_k^{n+1} - u_k^n = -\frac{\Delta t}{2\Delta x}\left[f(u_{k+1}^n) - f(u_{k-1}^n)\right]$$

reproduces the initial values, which do not satisfy the entropy condition (2.1.22). This is also true for the implicit formulation.

(b) *One-sided differences.* Consider the more simple equation

$$\partial_t u + a\partial_x u = 0 \quad \text{in } \mathbb{R} \times \mathbb{R}^+ \qquad (2.1.28)$$

for some $a > 0$ and let us apply the following discretization with $\lambda = \Delta t/\Delta x$:

$$u_k^{n+1} - u_k^n + a\lambda(u_{k+1}^n - u_k^n) = 0 \ .$$

This can be written in the form

$$u_k^{n+1} = (1 + a\lambda - a\lambda E)u_k^n \ ,$$

where we have used

$$Eu_k^n := u_{k+1}^n \ .$$

It follows that

$$u_k^n = (1 + a\lambda - a\lambda E)^n u_k^0$$

$$= \sum_{m=0}^{n} \binom{n}{m}(1 + a\lambda)^m(-a\lambda E)^{n-m} u_k^0$$

$$= \sum_{m=0}^{n} \binom{n}{m}(1 + a\lambda)^m(-a\lambda)^{n-m} u_{k+(n-m)}^0 \ .$$

Let us fix one point $(x_0, t_0) = (n\,\Delta x, k\,\Delta t) \in \mathbb{R} \times \mathbb{R}^+$. The scheme described above uses only values of u_0 on the right side of x_0 to define u_k^n. On the other hand, the exact solution u is equal to $u_0(x_0 - at_0)$ along the characteristic $x - x_0 = a(t - t_0)$. This line crosses the x-axis on the left side of x_0. Therefore we cannot expect that this scheme is stable. The reason for this is that we have used the forward difference quotient for discretizing $\partial_x u$. The same result can be obtained from Figure 2.1.6. The direction of the characteristics suggest that it should be more convenient to use backward differences if $a > 0$. In order to get a stable scheme in the case $a > 0$, we have to take backward difference quotients. Later on, we shall see that the corresponding scheme will converge.

REMARK 2.1.21 (Special entropy solutions) *We note that if u is a weak solution of (2.1.1) then $u_\lambda(x, t) := u(\lambda x, \lambda t)$ is also a solution for all $\lambda > 0$ (similarity solution). Therefore we look for weak solutions of (2.1.1) of*

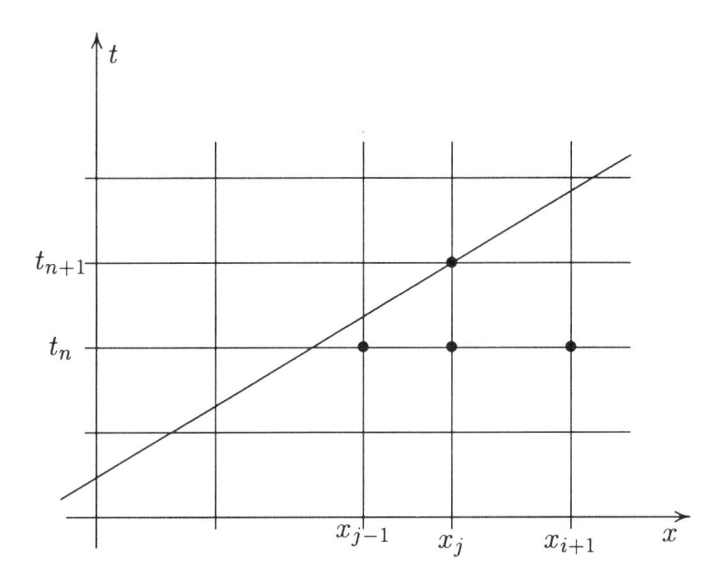

Figure 2.1.6

the form $v(x/t)$ satisfying the entropy condition. We suppose that u satisfies the Riemann initial data

$$u(x,0) = \begin{cases} u_l & \text{for } x < 0 \\ u_r & \text{for } x > 0 \end{cases}$$

where $u_l, u_r \in \mathbb{R}$ are given. In the region where u is smooth we have

$$-\frac{x}{t^2}v' + \frac{1}{t}f'(v)v' = 0 \ .$$

This implies

$$f'(v(\xi)) - \xi = 0 \quad \text{or} \quad v' = 0 \quad \text{for all } \xi = x/t \in \mathbb{R} \ . \tag{2.1.29}$$

Let us assume that $f'' > 0$. If $u_l = u_r$ then it is obvious that $u \equiv u_l$ is a solution of (2.1.1), (2.1.2). Now let us assume that $u_l > u_r$. Define

$$u(x,t) := \begin{cases} u_l & \text{if } x < st \\ u_r & \text{if } x > st \end{cases} \tag{2.1.30}$$

where $s := \frac{f(u_l)-f(u_r)}{u_l-u_r}$. This means the Rankine–Hugoniot condition is satisfied by definition and the entropy condition, since $u_l > u_r$. The solution

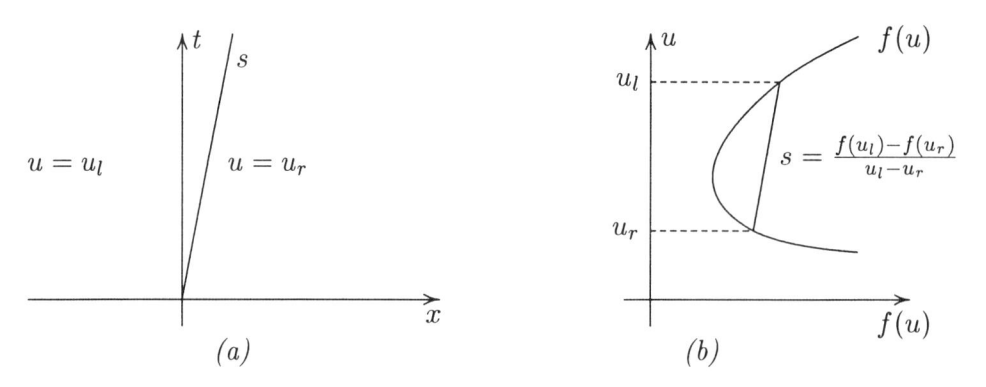

34 Initial value problems for scalar conservation laws in 1-D

Figure 2.1.7

u as defined in (2.1.30) is a shock. In order to construct the solution we should mention the following geometrical relationship (see Figure 2.1.7).

The shockline in (a) is parallel to the secant in (b). Finally we have to consider $u_l < u_r$. If we tried to treat this case similarly the case $u_l > u_r$, we would get a solution that does not satisfy the entropy condition. Then we apply $(f')^{-1}$ to the first part of (2.1.29) and use it as a motivation for the following definition:

$$u(\xi) := \begin{cases} u_r & \text{for } \xi \geq f'(u_r), \\ v(\xi) & \text{for } f'(u_l) < \xi < f'(u_r), \\ u_l & \text{for } \xi \leq f'(u_l), \end{cases} \qquad (2.1.31)$$

where $v(\xi) = f'^{-1}(\xi)$. The expression $f'^{-1}(\xi)$ exists since we have assumed $f'' > 0$. The function u is continuous. Since $f'' > 0$ implies $f'(u_l) < f'(u_r)$, the characteristics cannot intersect each other and u is the uniquely defined solution. In this case we can also give a geometrical interpretation of the definition of u (see Figure 2.1.8).

The tangents in $(f(u_l), u_l)$ and $(f(u_r), u_r)$ to the graph of f in (b) are determining the lines in (a) separating the regions where $u = u_l$, $u = f'^{-1}(\xi)$ and $u = u_r$. The solution in this case is called a rarefaction wave.

2.1 Theoretical background 35

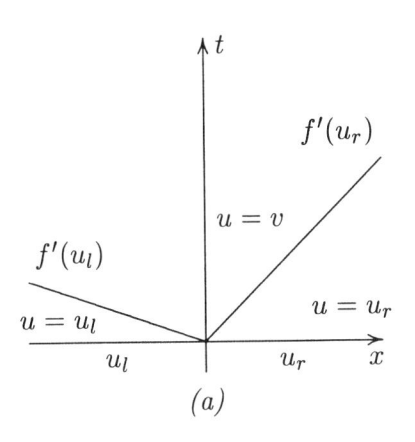

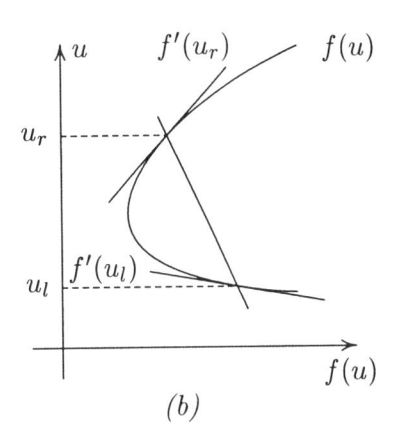

Figure 2.1.8

THEOREM 2.1.22 (Representation formula of Lax [126]) *Let $f \in C^2(\mathbb{R})$, $f(0) = 0$, $f'' > 0$, $f'(\mathbb{R}) = \mathbb{R}$, $u_0 \in C^0(\mathbb{R}) \cap L^\infty(\mathbb{R})$, $b(\xi) := f'^{-1}(\xi), g(\xi) := \xi b(\xi) - f(b(\xi))$ and*

$$v(x) := \int_0^x u_0(s)\, ds$$

for all $\xi, x \in \mathbb{R}$. Then for $(\xi, x, t) \in \mathbb{R} \times \mathbb{R} \times \mathbb{R}^+$ let

$$G(\xi, x, 0) := v(\xi) \quad \text{and} \quad G(\xi, x, t) := v(\xi) + t\, g\left(\frac{x-\xi}{t}\right) \quad \text{for } t > 0.$$

Then $G(\xi, x, t)$ as a function of ξ attains its minimum $y(x, t)$ in $\mathbb{R}$, i.e.

$$G(y(x,t), x, t) = \min_{\xi \in \mathbb{R}} G(\xi, x, t) \tag{2.1.32}$$

and

$$u(x,t) := b\left(\frac{x - y(x,t)}{t}\right) \tag{2.1.33}$$

is the uniquely defined entropy solution of (2.1.1),(2.1.2). Furthermore we have

$$u(x,t) = u_0(y(x,t)), \quad y(x,t) \to x \quad \text{for } t \to 0$$

and $y(x, t)$ is monotonically increasing in x.

Proof Since f is convex, we have

$$g' = b \quad \text{and} \quad g''(\xi) = \frac{1}{f''(b(\xi))} > 0 \ .$$

i.e. g is convex. Furthermore $f'(\mathbb{R}) = \mathbb{R}$ implies $\mathbb{R} = f'^{-1}(\mathbb{R}) = b(\mathbb{R}) = g'(\mathbb{R})$. From

$$g(z) = g(0) + g'(\xi)z$$

and the fact that g is convex we obtain

$$g(z) \longrightarrow \infty \quad \text{if} \quad z \longrightarrow \pm\infty \ .$$

In particular, g grows faster than any linear function since $|g'(z)| \longrightarrow \infty$ if $z \longrightarrow \infty$. Therefore

$$v(y) + tg\left(\frac{x-y}{t}\right) \longrightarrow \infty$$

if $y \longrightarrow \pm\infty$. Therefore $G(\cdot, x, t)$ attains its minimum $y(x, t)$ for any $(x, t) \in \mathbb{R} \times \mathbb{R}^+$:

$$G(y(x,t), x, t) = \min_{\xi} G(\xi, x, t) \ . \tag{2.1.34}$$

Now we shall show that $y(\cdot, t)$ is monotonically increasing in the first argument. To see this let

$$H(s) := g(\sigma + s) - g(\sigma) + g(\tau - s) - g(\tau)$$

for fixed σ, τ with $\sigma < \tau$ and for $0 < s < \tau - \sigma$. Since g is convex, we know that $H(s) < 0$ and furthermore we have

$$H(0) = H(\tau - \sigma) = 0 \ .$$

Now let $x_1 < x_2$ and t arbitrary and define

$$y_1 = y(x_1, t), \quad y_2 = y(x_2, t)$$

$$\tau = \frac{x_2 - y_2}{t}, \quad \sigma = \frac{x_1 - y_1}{t}, \quad s = \frac{x_2 - x_1}{t} \ .$$

2.1 Theoretical background

We assume $y_2 \leq y_1$, and obtain

$$g\left(\frac{x_2 - y_1}{t}\right) - g\left(\frac{x_1 - y_1}{t}\right) + g\left(\frac{x_1 - y_2}{t}\right) - g\left(\frac{x_2 - y_2}{t}\right) < 0 .$$

By definition of $y(x,t)$ this implies

$$G(y_1, x_2, t) = G(y_1, x_1, t) + tg\left(\frac{x_2-y_1}{t}\right) - tg\left(\frac{x_1-y_1}{t}\right)$$

$$\leq G(y_2, x_1, t) + tg\left(\frac{x_2-y_1}{t}\right) - tg\left(\frac{x_1-y_1}{t}\right)$$

$$< G(y_2, x_1, t) - tg\left(\frac{x_1-y_2}{t}\right) + tg\left(\frac{x_2-y_2}{t}\right)$$

$$= G(y_2, x_2, t) .$$

This contradicts the definition of $y_2(x_2, t)$. Therefore we have proved that

$$y_1(x_1, t) < y_2(x_2, t) \quad \text{if} \quad x_1 < x_2 .$$

Since $\partial_\xi G(\cdot, x, t)$ exists, we get

$$\begin{aligned} 0 = \partial_\xi G(y(x,t), x, t) &= u_0(y) - g'\left(\frac{x-y}{t}\right) \\ &= u_0(y) - b\left(\frac{x-y}{t}\right) = u_0(y) - u(x,t) . \end{aligned} \quad (2.1.35)$$

Furthermore we have to show that u is a weak solution of the conservation law. Let us use the following notation (where $\int := \int_{\mathbb{R}}$)

$$u_n(x,t) := \frac{\int b\left(\frac{x-y}{t}\right) e^{-nG(y,x,t)} dy}{\int e^{-nG(y,x,t)} dy} ,$$

$$f_n(x,t) := \frac{\int f\left(b\left(\frac{x-y}{t}\right)\right) e^{-nG(y,x,t)} dy}{\int e^{-nG(y,x,t)} dy} ,$$

$$V_n(x,t) := \log \int e^{-nG(y,x,t)} dy .$$

Using the definition of $G(y, x, t)$ and $g\left(\frac{x-y}{t}\right)$, it is easy to verify that

$$u_n = -\frac{1}{n} \partial_x V_n \quad f_n = \frac{1}{n} \partial_t V_n .$$

Consequently, we have for all $n \in \mathbb{N}$

$$\partial_t u_n + \partial_x f_n = 0 . \tag{2.1.36}$$

Because of the properties of the convolution with Dirac sequences, we have that

$$u_n \longrightarrow u, \quad f_n \longrightarrow f(u) \quad \text{in} \quad L^p_{\text{loc}}(\mathbb{R} \times \mathbb{R}^+) .$$

If we multiply (2.1.36) by a test function $\varphi \in C_0^\infty(\mathbb{R} \times \mathbb{R}^+)$, we find for $n \longrightarrow \infty$ that u is a weak solution of the conservation laws (2.1.1).

In the next step we shall prove that for $t \to 0$

$$y(x,t) \longrightarrow x \quad \text{a. e. in} \quad \mathbb{R} . \tag{2.1.37}$$

Then this implies $u(x,0) = u_0(x)$ by (2.1.35) of u. By definition, and (2.1.35) we have

$$u(x,t) = f'^{-1}\left(\frac{x - y(x,t)}{t}\right) = u_0(y(x,t))$$

and therefore

$$x - y(x,t) = tf'(u_0(y(x,t))).$$

Since u_0 is bounded by assumption, we obtain (2.1.37).

Finally we have to show that u satisfies the entropy condition. To prove this, assume that $x_1 < x_2$ and let $y_1 = y(x_1,t)$, $y_2 = y(x_2,t)$. Since b and $y(\cdot,t)$ are monotonically increasing, we obtain

$$u(x_1,t) = b\left(\frac{x_1 - y_1}{t}\right) \geq b\left(\frac{x_1 - y_2}{t}\right)$$

$$= b\left(\frac{x_2 - y_2}{t} - \frac{x_2 - x_1}{t}\right)$$

$$\geq b\left(\frac{x_2 - y_2}{t}\right) - \|b'\|_{L^\infty(\mathbb{R})} \frac{x_2 - x_1}{t}$$

$$= u(x_2,t) - \|b'\|_{L^\infty(\mathbb{R})} \frac{x_2 - x_1}{t} .$$

In particular,

$$\frac{u(x_2,t) - u(x_1,t)}{x_2 - x_1} \leq \frac{1}{t}\|b'\|_{L^\infty(\mathbb{R})}, \qquad (2.1.38)$$

where

$$\|b'\|_{L^\infty(\mathbb{R})} = \frac{1}{\inf f''(x)} < \infty.$$

This implies that u can only jump down at a discontinuity, and this is equivalent to the entropy condition since f is strictly convex. □

DEFINITION 2.1.23 (Oleinik entropy condition) *The condition (2.1.38) is called the Oleinik entropy condition.*

REMARK 2.1.24 *In Theorem 2.1.22 we have shown that the entropy solution of (2.1.1),(2.1.2) satisfies (2.1.38). It can also be shown that (2.1.38) implies the Lax entropy condition (see [218]).*

2.2 Finite difference schemes of first order for scalar equations in one space dimension

In this section we shall define and motivate the numerical schemes in conservation form that are adapted to the conservation law itself in an optimal way (see [125], [80], [161], [141] and [221]). In order to select the entropy solution and to remain stable in the case of shocks, it is necessary that these schemes contain enough numerical damping or numerical viscosity. A first justification of the form of these algorithms will be given by the theorem of Lax–Wendroff. This says if the numerical scheme in conservation form defines a convergent sequence of functions then it has to converge to the uniquely defined entropy solution of the corresponding conservation law. The main part of this section is devoted to the proof of the convergence of discrete solutions given by a numerical scheme in conservation form. The basic idea for the proof is to show that the discrete solutions are total variations diminishing in time. Then the compactness criterion in L^1 (Helly's selection principle) gives a convergent subsequence in L^1 for fixed t. By definition of the numerical scheme, a Lipschitz property with respect to t can be obtained. Then an argument similar as in the proof of the Theorem of Arzelá–Ascoli implies the existence of a convergent subsequence in $L^1(\mathbb{R} \times \mathbb{R}^+)$. Throughout this section, we assume that the scheme is monotone in order to ensure that the entropy solution will be approximated. Finally in this section we discuss several sufficient conditions for a scheme to be total variation diminishing (TVD).

LEMMA 2.2.1 *Let u be a weak solution of*

$$\partial_t u + \partial_x f(u) = 0 \quad \text{in } \mathbb{R} \times \mathbb{R}^+ \tag{2.2.1}$$
$$u(x,0) = u_0(x) \quad \text{in } \mathbb{R} \tag{2.2.2}$$

such that $u \in C^1((\mathbb{R}\times\mathbb{R}^+)\setminus\Gamma)$ where $\Gamma = \{(\sigma(t),t)|\, t > 0\}$ is a smooth curve and u satisfies the Rankine–Hugoniot condition along Γ. Let $(x_0,t_0) \in \Gamma$ and let V be a neighbourhood of (x_0,t_0) with a Lipschitz boundary. Then we have

$$\int_{\partial V} \nu \begin{pmatrix} f(u) \\ u \end{pmatrix} = 0, \quad \text{where } \nu \text{ is the outer normal to } V.$$

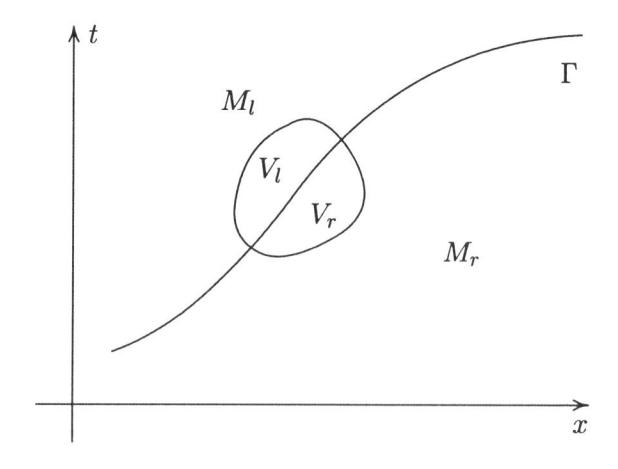

Figure 2.2.1

Proof Define $M_l := \{(x,t) |\, x < \sigma(t)\}$, $M_r := \{(x,t) |\, x > \sigma(t)\}$, $V_l := M_l \cap V$ and $V_r := M_r \cap V$. Since u is C^1 in V_l and V_r, and since u satisfies the differential equation in V_l and V_r in the classical sense, we obtain

$$0 = \int_{V_l} \partial_t u + \partial_x f(u) = \int_{V_l} \mathrm{div}\begin{pmatrix} f(u) \\ u \end{pmatrix} = \int_{\partial V_l} \nu \begin{pmatrix} f(u) \\ u \end{pmatrix},$$

and similarly

$$\int_{\partial V_r} \nu \begin{pmatrix} f(u) \\ u \end{pmatrix} = 0.$$

The Rankine–Hugoniot condition yields (see (2.1.7))

$$\int_{\Gamma \cap V} \nu_l \begin{pmatrix} f(u_l) \\ u_l \end{pmatrix} + \int_{\Gamma \cap V} \nu_r \begin{pmatrix} f(u_r) \\ u_r \end{pmatrix} = 0, \qquad (2.2.3)$$

and therefore we obtain

$$\int_{\Gamma \cap V} \ldots = \int_{\partial V_l} \ldots - \int_{\partial V \cap M_l} \ldots = -\int_{\partial V \cap M_l} \ldots$$

and

$$\int_{\Gamma \cap V} \ldots = \int_{\partial V_r} \ldots - \int_{\partial V \cap M_r} \ldots = -\int_{\partial V \cap M_r} \ldots$$

where the dots refer to the corresponding integrands in (2.2.3).

Then (2.2.3) implies

$$\int_{\partial V \cap M_r} \ldots + \int_{\partial V \cap M_l} \ldots = 0$$

but this is just the statement of the lemma. □

Now let us express the result of Lemma 2.2.1 in a slightly different form. For $V :=]a,b[\times]\sigma,\tau[$ we obtain

$$-\int_\sigma^\tau f(u(a,\cdot)) - \int_a^b u(\cdot,\sigma) + \int_\sigma^\tau f(u(b,\cdot)) + \int_a^b u(\cdot,\tau) = 0, \qquad (2.2.4)$$

or

$$\int_\sigma^\tau [f(u(b,s)) - f(u(a,s))]\,ds + \int_a^b [u(x,\tau) - u(x,\sigma)]\,dx = 0. \qquad (2.2.5)$$

This identity is also called the integral form of the conservation law. Assume that we have a uniform grid on $\mathbb{R} \times \mathbb{R}^+$ with $x_i := i\Delta x$ and $t^n := n\Delta t$. If we choose $\sigma := t^n$, $\tau := t^{n+1}$, $a = x_{i-1/2} := (i-1/2)\Delta x$, and $b := x_{i+1/2} := (i+1/2)\Delta x$, it follows that

$$\int_{t^n}^{t^{n+1}} [f(u(x_{i+\frac{1}{2}},s)) - f(u(x_{i-\frac{1}{2}},s))]\,ds \qquad (2.2.6)$$

$$+ \int_{x_{i-1/2}}^{x_{i+1/2}} [u(x,t^{n+1}) - u(x,t^n)]\,dx = 0.$$

This identity has an important physical meaning. Let us denote the integrals in (2.2.6) by I_1 and I_2 respectively. Then $-I_1$ is the flow difference through $x_{i+1/2}$ and $x_{i-1/2}$ during the time interval $[t^n, t^{n+1}]$ and I_2 is the change of mass in the volume $[x_{i-1/2}, x_{i+1/2}]$ during the time $[t^n, t^{n+1}]$. Then (2.2.6) means that the mass is conserved.

Let us try to find a function $g: \mathbb{R}^2 \to \mathbb{R}$ such that $u_i^{n+1}, u_i^n, g_{i+1/2}^n$ and $g_{i-1/2}^n$ are approximations of the following terms in (2.2.6):

$$u_i^{n+1} \sim \frac{1}{\Delta x} \int_{x_{i-1/2}}^{x_{i+1/2}} u(x,t^{n+1})\,dx,$$

$$u_i^n \sim \frac{1}{\Delta x} \int_{x_{i-1/2}}^{x_{i+1/2}} u(x,t^n)\,dx,$$

$$g_{i+\frac{1}{2}}^n := g(u_i^n, u_{i+1}^n) \sim \frac{1}{\Delta t} \int_{t^n}^{t^{n+1}} f(u(x_{i+\frac{1}{2}}, s)) \, ds \, ,$$

$$g_{i-\frac{1}{2}}^n = g(u_{i-1}^n, u_i^n) \sim \frac{1}{\Delta t} \int_{t^n}^{t^{n+1}} f(u(x_{i-\frac{1}{2}}, s)) \, ds \, .$$

Then (2.2.6) can be written in the following form:

$$u_i^{n+1} - u_i^n = -\frac{\Delta t}{\Delta x} \left(g_{i+\frac{1}{2}}^n - g_{i-\frac{1}{2}}^n \right) . \qquad (2.2.7)$$

The function g is called the numerical flux and is assumed to be consistent, i.e.

$$g(u, u) = f(u) \quad \text{for all } u \in \mathbb{R}. \qquad (2.2.8)$$

We need this property to estimate the local truncation error (see Lemma 2.2.4). Formally (2.2.8) can be seen as follows. Consider (2.2.1) and assume that the initial values u_0 are constant. Then $u = u_0$ is a solution of (2.2.1), and

$$g_{i+\frac{1}{2}}^0 = g(u_i^0, u_{i+1}^0) = g(u_0, u_0) \, ,$$

$$\frac{1}{\Delta t} \int_0^{\Delta t} f(u(x_{i+\frac{1}{2}}, s)) \, ds = \frac{1}{\Delta t} \int_0^{\Delta t} f(u_0) = f(u_0) \, .$$

Now we use (2.2.7) to define in general numerical schemes in conservation form.

DEFINITION 2.2.2 *Let $f \in C^1(\mathbb{R})$, $g \in C^{0,1}(\mathbb{R}^2)$ and suppose that g is consistent with the conservation law (2.2.1), i.e. $g(u,u) = f(u)$ for all $u \in \mathbb{R}$. Assume that we have a sequence $u_i^0 \in \mathbb{R}$, $i \in \mathbb{Z}$ of initial values and $\Delta t, \Delta x \in \mathbb{R}^+$. Then define successively for $n \geq 1$ and $i \in \mathbb{Z}$*

$$u_i^{n+1} = u_i^n - \frac{\Delta t}{\Delta x}(g_{i+\frac{1}{2}}^n - g_{i-\frac{1}{2}}^n) \, , \qquad (2.2.9)$$

where

$$g^n_{i+\frac{1}{2}} := g(u^n_i, u^n_{i+1}) , \quad g^n_{i-\frac{1}{2}} := g(u^n_{i-1}, u^n_i) .$$

Then we call g the numerical flux and (2.2.9) a numerical scheme in conservation form.

The definition (2.2.9) can be generalized to

$$u^{n+1}_i - u^n_i = -\frac{\Delta t}{\Delta x}\left[\theta(g^{n+1}_{i+\frac{1}{2}} - g^{n+1}_{i-\frac{1}{2}}) + (1-\theta)(g^n_{i+\frac{1}{2}} - g^n_{i-\frac{1}{2}})\right] \quad (2.2.10)$$

for $0 \le \theta \le 1$. For $\theta = 0$ this is the same as (2.2.9) and is an explicit scheme. If $\theta = 1$, the scheme (2.2.10) is implicit.

REMARK 2.2.3 The main reason for writing the numerical scheme in the form (2.2.10) is that this form also guarantees the conservation property for the discrete solution. We have (if the sums are finite)

$$\sum_i (g^n_{i+\frac{1}{2}} - g^n_{i-\frac{1}{2}}) = \sum_i [g(u^n_i, u^n_{i+1}) - g(u^n_{i-1}, u^n_i)] = 0 ,$$

and therefore

$$\sum_i u^{n+1}_i = \sum_i u^n_i .$$

The consistency (2.2.8) of g is used to estimate the local truncation error. For $u \in C^2(\mathbb{R} \times \mathbb{R}^+)$ and $u^n_i := u(x_i, t^n)$ let us define

$$Qu^n_i := u^n_i - \frac{\Delta t}{\Delta x}(g^n_{i+\frac{1}{2}} - g^n_{i-\frac{1}{2}}) .$$

LEMMA 2.2.4 *Assume that $u \in C^2(\mathbb{R} \times \mathbb{R}^+)$ is a classical solution of (2.2.1) and that $g \in C^2(\mathbb{R}^2)$, such that $|\partial_t^\beta u|$, $|\partial_x^\beta u|$ and $|\partial_1^\beta g|$, $|\partial_2^\beta g|$ are bounded for $0 \leq \beta \leq 2$. Furthermore, let*

$$\frac{\Delta t}{\Delta x} < \alpha \quad \text{for } \alpha > 0. \tag{2.2.11}$$

Let $u_i^n := u(i\,\Delta x, n\,\Delta t)$. Then

$$u_i^{n+1} - Qu_i^n = \mathcal{O}(\Delta x^2) + \mathcal{O}(\Delta t^2).$$

This means that the local truncation error is of order one.

Proof Let us fix i and $n \in \mathbb{N}$ and in the following let $u_i := u_i^n$ for $i \in \mathbb{Z}$. Then applying the Taylor expansion theorem several times, we obtain

$$u_i^{n+1} - u_i^n = \partial_t u_i \Delta t + \mathcal{O}(\Delta t^2),$$

$$\begin{aligned}
g(u_i, u_{i+1}) - g(u_{i-1}, u_i) &= \partial_2 g(u_i, u_i)(u_{i+1} - u_i) \\
&\quad - \partial_1 g(u_i, u_i)(u_{i-1} - u_i) + \mathcal{O}(\Delta x^2),
\end{aligned}$$

$$\begin{aligned}
f(u_{i+1}) - f(u_{i-1}) &= g(u_{i+1}, u_{i+1}) - g(u_{i-1}, u_{i-1}) \\
&= \partial_1 g(u_i, u_i)(u_{i+1} - u_{i-1}) \\
&\quad + \partial_2 g(u_i, u_i)(u_{i+1} - u_{i-1}) + \mathcal{O}(\Delta x^2).
\end{aligned}$$

The last two equations imply

$$\begin{aligned}
g(u_i, u_{i+1}) - g(u_{i-1}, u_i) &= \tfrac{1}{2}\partial_1 g(u_i, u_i)(u_{i+1} - u_{i-1}) \\
&\quad + \tfrac{1}{2}\partial_2 g(u_i, u_i)(u_{i+1} - u_{i-1}) \\
&\quad - \tfrac{1}{2}(\partial_1 g(u_i, u_i) - \partial_2 g(u_i, u_i)) \\
&\quad (u_{i+1} - 2u_i + u_{i-1}) + \mathcal{O}(\Delta x^2) \\
&= \tfrac{1}{2}(f(u_{i+1}) - f(u_{i-1})) + \mathcal{O}(\Delta x^2) \\
&= \partial_x f(u_i)\Delta x + \mathcal{O}(\Delta x^2).
\end{aligned}$$

Now we can estimate the truncation error:

$$\begin{aligned}
u_i^{n+1} - Qu_i^n &= u_i^{n+1} - u_i^n + \frac{\Delta t}{\Delta x}(g(u_i, u_{i+1}) - g(u_{i-1}, u_i)) \\
&= \partial_t u_i \Delta t + \partial_x f(u_i)\Delta t + \mathcal{O}(\Delta t^2) + \mathcal{O}(\Delta x^2) \\
&= \mathcal{O}(\Delta t^2) + \mathcal{O}(\Delta x^2).
\end{aligned}$$

□

EXAMPLE 2.2.5 Let us consider some special examples for the numerical flux g. Remember that

$$\delta f_i^n := \frac{1}{\Delta x}(g_{i+\frac{1}{2}}^n - g_{i-\frac{1}{2}}^n) \tag{2.2.12}$$

can be considered as an approximation for $\partial_x f(u(x_i, t_i))$. If

$$\begin{aligned} g(u,v) &= u & \text{then} && \delta f_i^n &= \frac{1}{\Delta x}(u_i^n - u_{i-1}^n), \\ g(u,v) &= v & \text{then} && \delta f_i^n &= \frac{1}{\Delta x}(u_{i+1}^n - u_i^n), \\ g(u,v) &= \frac{1}{2}(u+v) & \text{then} && \delta f_i^n &= \frac{1}{2\Delta x}(u_{i+1}^n - u_{i-1}^n). \end{aligned}$$

In this way we can define backward, forward and central differences respectively.

EXAMPLE 2.2.6 (Lax–Friedrichs scheme) In Example 2.1.20 we have seen that in general we cannot expect the convergence of the numerical scheme (2.2.9) to the entropy solution if we use central or (in the case where $f'(u) > 0$) forward difference quotients. Now it turns out that backward differences can be written as a central difference plus a diffusion term:

$$u_i^{n+1} - u_{i-1}^n = \frac{1}{2}(u_{i+1}^n - u_{i-1}^n) - \frac{1}{2}(u_{i+1}^n - 2u_i^n + u_{i-1}^n). \tag{2.2.13}$$

The last term $(u_{i+1}^n - 2u_i^n + u_{i-1}^n)$ is called the numerical viscosity or numerical damping term. It should be mentioned that the central difference quotient is an approximation of second order of $\partial_x u$. This property gives rise to the following definition of a numerical flux ($\lambda := \Delta t / \Delta x$)

$$g(u,v) := \frac{1}{2}[f(u) + f(v)] + \frac{1}{2\lambda}(u - v). \tag{2.2.14}$$

In detail for δf_i^n (see (2.2.12)) we obtain

$$\delta f_i^n = \frac{1}{2\Delta x}[f(u_{i+1}) - f(u_{i-1})] - \frac{1}{2\Delta t}(u_{i+1} - 2u_i + u_{i-1}), \tag{2.2.15}$$

which is similar to (2.2.13). The corresponding numerical scheme with the numerical flux (2.2.14) is called the Lax–Friedrichs scheme. It can be written in the form

$$u_i^{n+1} = u_i^n - \frac{\Delta t}{2\Delta x}[f(u_{i+1}^n) - f(u_{i-1}^n)] + \frac{1}{2}(u_{i+1}^n - 2u_i^n + u_{i-1}^n). \quad (2.2.16)$$

The numerical flux as defined in (2.2.14) is consistent and Lipschitz-continuous if f is Lipschitz-continuous. Later on we shall show that the scheme will converge. This is due to the damping or diffusion term in (2.2.15). Furthermore this term ensures that the numerical scheme will approximate the entropy solution and it plays the same role as the diffusion term $\partial_x^2 u_\varepsilon$ in (2.1.12). On the other hand, the damping term is responsible for smearing out the discontinuity of the solution over several gridpoints, as can be seen in Figures 2.2.2 and 2.2.3. Here we have solved (2.2.1), (2.2.2) with $f(u) = \frac{1}{2}u^2$ and

$$u_0 = \begin{cases} 1 & \text{if } x < 0, \\ 0 & \text{if } x \geq 0 \end{cases}$$

using the Lax–Friedrichs scheme as defined in (2.2.16). In the left part of Figure 2.2.2 you see the initial values and in the right part the solution for a later time.

EXAMPLE 2.2.7 (Engquist–Osher scheme) In Examples 2.1.20 and 2.2.5 we have seen that we should use one-sided differences and in particular backward differences if $f'(u) > 0$ and forward differences if $f'(u) < 0$. In more detail,

$$\partial_x f(u) \sim \frac{1}{\Delta x}[f(u_i^n) - f(u_{i-1}^n)] \quad \text{if } f' > 0$$

and

$$\partial_x f(u) \sim \frac{1}{\Delta x}[f(u_{i+1}^n) - f(u_i^n)] \quad \text{if } f' < 0.$$

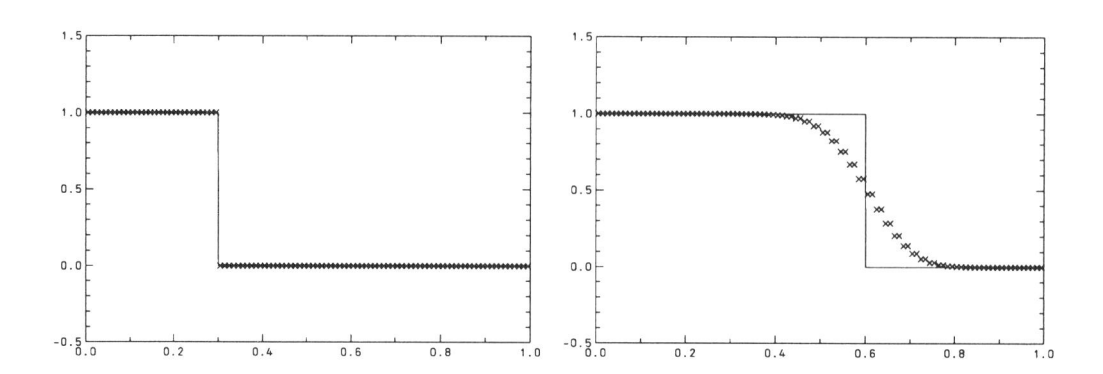

48 Initial value problems for scalar conservation laws in 1-D

Figure 2.2.2 Initial data, Lax–Friedrichs scheme.

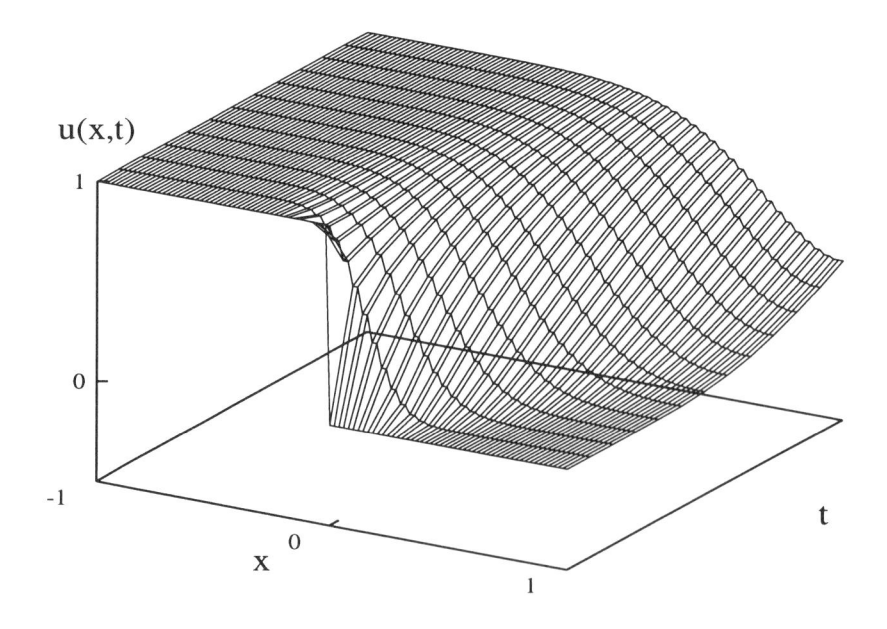

Figure 2.2.3

The other cases have to be defined such that the corresponding numerical flux is Lipschitz-continuous and consistent. This can be done in the following way. Let

$$f^+(u) := f(0) + \int_0^u \max(f'(s), 0)\,ds\ , \quad f^-(u) := \int_0^u \min(f'(s), 0)\,ds\ .$$

Then we have $f(u) = f^+(u) + f^-(u)$. We define the numerical flux

$$g(v, w) = f^+(v) + f^-(w)\ ,$$

and it turns out that (see (2.2.12))

$$\delta f_i^n = \frac{1}{\Delta x}[f^+(u_i^n) - f^+(u_{i-1}^n) + f^-(u_{i+1}^n) - f^-(u_i^n)]\ .$$

If f is Lipschitz-continuous then g is Lipschitz-continuous and consistent. The corresponding numerical scheme is called the Engquist–Osher scheme [62]. We have tested this scheme for the same problem as in Example 2.2.6. As one can see from Figures 2.2.4 and 2.2.5, this scheme smears out the shock over fewer grid points than the Lax–Friedrichs scheme.

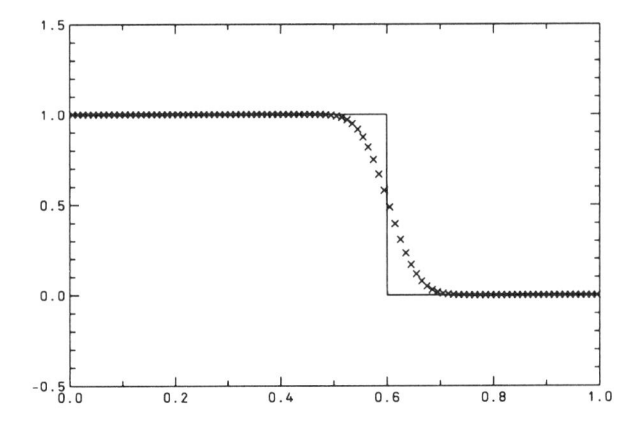

Figure 2.2.4 Engquist–Osher.

REMARK 2.2.8 (Interpretation of numerical damping) *The effect of the damping term (or the numerical viscosity) in (2.2.16) can be interpreted in the following way. Consider*

$$u_j^{n+1} - u_j^n = -\lambda a(u_j^n - u_{j-1}^n), \quad \lambda := \frac{\Delta t}{\Delta x}, \quad a \in \mathbb{R}, \ a > 0 \quad (2.2.17)$$

for given initial values $(u_j^0)_{j \in \mathbb{Z}}$. *The corresponding numerical flux* $g(u,v) = au$ *is consistent and Lipschitz-continuous. We can show that (2.2.17) is (formally, since Δx is fixed) a numerical scheme of second order with respect to*

$$\partial_t w + a \partial_x w = \Delta x D \, \partial_x^2 w, \quad (2.2.18)$$

where $D := \frac{1}{2} a(1 - \lambda a)$. *This can be seen as follows. Let w be a smooth solution of (2.2.18). Then*

$$w_i^{n+1} - w_i^n + \lambda a(w_i^n - w_{i-1}^n)$$
$$= w_i^{n+1} - w_i^n + \frac{\lambda a}{2}(w_{i+1}^n - w_{i-1}^n) - \frac{\lambda a}{2}(w_{i+1}^n - 2w_i^n + w_{i-1}^n)$$
$$= \partial_t w \Delta t + \partial_t^2 w \frac{\Delta t^2}{2} + \mathcal{O}(\Delta t^3) + \lambda a \, \Delta x \partial_x w + \mathcal{O}(\Delta x^3)$$
$$- \frac{\lambda a}{2} \Delta x^2 [\partial_x^2 w + \mathcal{O}(\Delta x^2)].$$

Using (2.2.18), we can continue:

$$= \Delta t (\Delta x D \partial_x^2 w - a \partial_x w) + \frac{\Delta t^2}{2}(\Delta x D \partial_t \partial_x^2 w - a \partial_t \partial_x w)$$
$$+ \lambda a \, \Delta x \partial_x w - \frac{\lambda a}{2} \Delta x^2 \partial_x^2 w + \mathcal{O}(\Delta x^3) + \mathcal{O}(\Delta t^3)$$
$$= \partial_x^2 w \left(\Delta t \Delta x D - \frac{\lambda a}{2} \Delta x^2 \right) - \frac{a \Delta t^2}{2} \partial_t \partial_x w + \mathcal{O}(\Delta x^3) + \mathcal{O}(\Delta t^3).$$

Since $\partial_t \partial_x w = -a \partial_x^2 w + \mathcal{O}(\Delta x)$, we continue:

$$= \partial_x^2 w \Delta x^2 \left(\lambda D - \frac{\lambda a}{2} + \frac{\lambda^2 a^2}{2} \right) + \mathcal{O}(\Delta t^3) + \mathcal{O}(\Delta x^3).$$

If $D > 0$, (2.2.18) is the well-known diffusion equation and $\Delta x D$ denotes the diffusion coefficient. Notice that $D > 0$ if and only if $1 - |a|\lambda > 0$, which is equivalent to

$$|a|\frac{\Delta t}{\Delta x} < 1 .$$

In the nonlinear case this has to be replaced by

$$\sup_{u \in \mathbb{R}} \frac{\Delta t |f'(u)|}{\Delta x} < 1 .$$

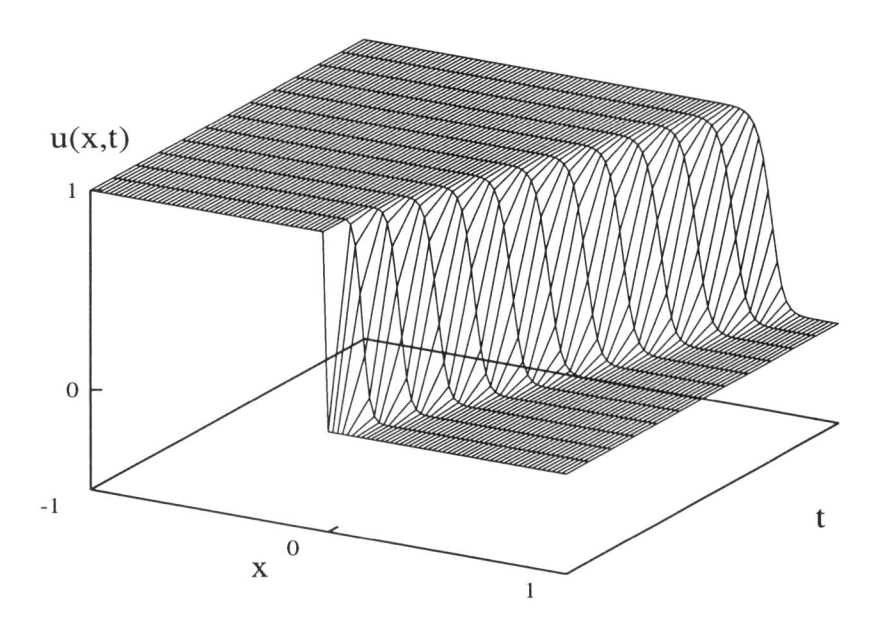

Figure 2.2.5 Engquist–Osher.

DEFINITION 2.2.9 *The condition*

$$\sup_{u \in \mathbb{R}} \frac{\Delta t |f'(u)|}{\Delta x} < 1$$

is called the CFL (Courant, Friedrichs, Lewy) condition *and* $\sup_{u \in \mathbb{R}} [\Delta t |f'(u)|/\Delta x]$ *is called the* CFL number *[44]*.

52   Initial value problems for scalar conservation laws in 1-D

In the next lemma we generalize Remark 2.2.8.

LEMMA 2.2.10 *Let $g \in C^2(\mathbb{R}^2)$ be the consistent numerical flux of a scheme in conservation form (2.2.9) and let (u_j^n) be the corresponding numerical solution. Then the truncation error for smooth solutions of*

$$\partial_t u + \partial_x f(u) = \Delta x \partial_x (b(u) \partial_x u) , \qquad (2.2.19)$$

where

$$b(u) := \tfrac{1}{2}[\partial_1 g(u,u) - \partial_2 g(u,u) - \lambda f'(u)^2] ,$$

is of second order and the truncation error of smooth solutions of

$$\partial_t u + \partial_x f(u) = 0 \qquad (2.2.20)$$

is of first order.

REMARK 2.2.11 *If one chooses central differences for g then $b(u)$ in (2.2.19) becomes nonpositive, and the term of second order in (2.2.19) no longer has the meaning of a diffusion or viscosity.*

Proof Without restriction, let us assume that $\Delta t = \lambda \Delta x$. Let u be an exact solution of (2.2.19) in $C^2(\mathbb{R} \times \mathbb{R}^+)$. Using the notation $u := u(i\Delta x, n\Delta t) =: u_i^n =: u_i$, $g := g(u,u)$, $\Delta_+ u = u_{i+1}^n - u_i^n$ and $\Delta_- u = u_i^n - u_{i-1}^n$ we get

$$\begin{aligned}
u_i^{n+1} - u_i^n &= \partial_t u \Delta t + \tfrac{1}{2}\partial_t^2 u \Delta t^2 + \mathcal{O}(\Delta t^3) \\
&= \Delta t \Delta x \partial_x(b \partial_x u) - \Delta t \partial_x f(u) \\
&\quad - \tfrac{1}{2}\Delta t^2 \partial_x(f'(u)\partial_t u) + \mathcal{O}(\Delta x^3) \\
&= \tfrac{1}{2}\lambda \Delta x^2 (\partial_x u)^2 (\partial_1^2 g - \partial_2^2 g) + \tfrac{1}{2}\lambda \partial_x^2 u (\partial_1 g - \partial_2 g) \Delta x^2 \\
&\quad - \tfrac{1}{2}\lambda[f(u_{i+1}) - f(u_{i-1})] + \mathcal{O}(\Delta x^3)
\end{aligned}$$

and

$$\begin{aligned}
&g(u_i, u_{i+1}) - g(u_{i-1}, u_i) - \tfrac{1}{2}(f(u_{i+1}) - f(u_{i-1})) \\
&= \partial_2 g \Delta_+ u + \tfrac{1}{2}\partial_2^2 g (\Delta_+ u)^2 + \partial_1 g (\Delta_- u) - \tfrac{1}{2}\partial_1^2 g (\Delta_- u)^2 \\
&\quad - \tfrac{1}{2}[\partial_1 g \Delta_+ u + \partial_2 g \Delta_+ u + \tfrac{1}{2}(\partial_1^2 g + \partial_{12} g + \partial_2^2 g)(\Delta_+ u)^2] \\
&\quad + \tfrac{1}{2}[-\partial_1 g \Delta_- u - \partial_2 g \Delta_- u \\
&\quad + \tfrac{1}{2}(\partial_1^2 g + \partial_{12} g + \partial_2^2 g)(\Delta_- u)^2] + \mathcal{O}(\Delta x^3) \\
&= \tfrac{1}{2}(\partial_2 g - \partial_1 g)(u_{i+1} - 2u_i + u_{i-1}) \\
&\quad + \tfrac{1}{4}(\partial_2^2 g - \partial_1^2 g)[(\Delta_+ u)^2 + (\Delta_- u)^2] \\
&\quad + \tfrac{1}{4}\partial_{12} g[(\Delta_- u)^2 - (\Delta_+ u)^2] + \mathcal{O}(\Delta x^3) \\
&= \tfrac{1}{2}(\partial_2 g - \partial_1 g)\partial_x^2 u \Delta x^2 + \tfrac{1}{2}(\partial_2^2 g - \partial_1^2 g)(\partial_x u)^2 \Delta x^2 + \mathcal{O}(\Delta x^3) ,
\end{aligned}$$

since

$$\begin{aligned}
(\Delta_+ u)^2 + (\Delta_- u)^2 &= \tfrac{1}{2}(u_{i+1} - u_{i-1})^2 + \tfrac{1}{2}(u_{i+1} - 2u_i + u_{i-1})^2 \\
&= 2(\partial_x u)^2 \Delta x^2 + \mathcal{O}(\Delta x^3) , \\
(\Delta_- u)^2 - (\Delta_+ u)^2 &= (u_{i+1} - 2u_i + u_{i-1})(u_{i-1} - u_{i+1}) = \mathcal{O}(\Delta x^3) .
\end{aligned}$$

Therefore

$$u_i^{n+1} - u_i^n + \frac{\Delta t}{\Delta x}(g(u_i, u_{i+1}) - g(u_{i-1}, u_i)) = \mathcal{O}(\Delta x^3) .$$

The statement (2.2.20) has been proved in Lemma 2.2.4. □

2.3 Convergence of finite difference schemes of first order in 1-D

Now we should like to study the convergence properties of numerical schemes in conservation form. First of all, we shall show that a convergent sequence that is uniformly bounded and that is defined by a numerical scheme in conservation form, will converge to a solution in the distributional sense of (2.2.1), (2.2.2). In order to state the corresponding theorem, we need some notation. Let $(k_m)_m$ and $(h_m)_m$ be sequences converging to zero and $\Delta t = k_m$, $\Delta x = h_m$. For given initial values $u_0 \in L^1(\mathbb{R})$ we define

$$u_i^0 := \frac{1}{\Delta x} \int_{x_{i-1/2}}^{x_{i+1/2}} u_0(x)\, dx \ . \tag{2.3.1}$$

We assume that u_i^n is the discrete solution of

$$u_i^{n+1} := u_i^n - \frac{\Delta t}{\Delta x}[g(u_i^n, u_{i+1}^n) - g(u_{i-1}^n, u_i^n)] \tag{2.3.2}$$

with respect to $\Delta t = k_m$, $\Delta x = h_m$ and a Lipschitz-continuous numerical flux g such that $g(u,u) = f(u)$. In order to extend the grid function u_i^n to all of $\mathbb{R} \times \mathbb{R}^+$, we define for $n \geq 0$

$$u_m(x,t) := u_i^n \quad \begin{array}{l} \text{for } n\Delta t < t \leq (n+1)\Delta t \\ \text{and } (i-\tfrac{1}{2})\Delta x < x \leq (i+\tfrac{1}{2})\Delta x\ , \end{array} \tag{2.3.3}$$

where $\Delta t = \lambda \Delta x$ for a fixed λ. Now we can formulate the important result of Lax and Wendroff [128].

THEOREM 2.3.1 (Lax–Wendroff theorem [128]) *Let $(u_m)_m$ be a sequence of discrete solutions as defined in (2.2.10) and (2.3.3) with respect to $\Delta x = h_m$, $\Delta t = \lambda k_m$ and the initial values u_i^0. Assume that $g \in C^{0,1}(\mathbb{R}^2)$ and that there exists a constant K such that*

$$\sup_m \sup_{\mathbb{R} \times \mathbb{R}^+} |u_m(x,t)| \leq K$$

and $u_m \longrightarrow u$ almost everywhere in $\mathbb{R} \times \mathbb{R}^+$ for $m \longrightarrow \infty$. Then u is a solution of

$$\partial_t u + \partial_x f(u) = 0 \quad \text{in } \mathbb{R} \times \mathbb{R}^+, \quad u(\cdot, 0) = u_0 \quad \text{in } \mathbb{R} \tag{2.3.4}$$

in the distributional sense.

2.3 Convergence of finite difference schemes of first order in 1-D

Proof For simplicity we consider only the explicit case. The more general case (2.2.10) can be treated similarly. Choose $\varphi \in C_0^\infty(\mathbb{R} \times [0,\infty[)$ and multiply (2.3.2) by φ:

$$(u_i^{n+1} - u_i^n)\varphi(x_i, t^n) = -\frac{\Delta t}{\Delta x}(g_{i+\frac{1}{2}}^n - g_{i-\frac{1}{2}}^n)\varphi(x_i, t^n).$$

Summing up for $i = 0, \ldots, \infty$ and $n = 0, \ldots, \infty$, we obtain

$$\Delta x \sum_i \sum_n (u_i^{n+1} - u_i^n)\varphi(x_i, t^n) \\ = -\Delta t \sum_i \sum_n (g_{i+\frac{1}{2}}^n - g_{i-\frac{1}{2}}^n)\varphi(x_i, t^n). \quad (2.3.5)$$

Since φ has compact support, we only have finite sums in (2.3.5). First let us consider the left-hand side of (2.3.5). Using partial summation, we can shift the differences to the test function φ:

$$\sum_{n=0}^{\infty}(u_i^{n+1} - u_i^n)\varphi(x_i, t^n)$$

$$= \sum_{n=1}^{\infty} u_i^n[\varphi(x_i, t^{n-1}) - \varphi(x_i, t^n)] - u_i^0 \varphi(x_i, 0)$$

$$= -\int_0^\infty u_m(x_i, t)\partial_t \varphi(x_i, t)\, dt - u_i^0 \varphi(x_i, 0)$$

This is true for fixed φ, since φ has compact support in $\mathbb{R} \times [0, \infty[$. Therefore we have

$$\Delta x \sum_i \sum_n (u_i^{n+1} - u_i^n)\varphi(x_i, t^n) \quad (2.3.6)$$

$$= \sum_i \Delta x \left(-\int_0^\infty u_m(x_i, t)\partial_t \varphi(x_i, t)\, dt - u_i^0 \varphi(x_i, 0)\right)$$

$$= -\int_{-\infty}^{\infty}\int_0^\infty u_m(x, t)\partial_t \varphi(x, t)\, dt\, dx - \int_{-\infty}^{\infty} u_0(x)\varphi(x, 0)\, dx + \mathcal{O}(\Delta t).$$

Here we have used the fact that the summation with respect to i is only finite. Next let us consider the right-hand side of (2.3.5).

56 Initial value problems for scalar conservation laws in 1-D

Using partial summation, we obtain

$$-\Delta t \sum_i \sum_n [g^n_{i+\frac{1}{2}} - g^n_{i-\frac{1}{2}}]\varphi(x_i, t^n) \qquad (2.3.7)$$
$$= -\Delta t \sum_i \sum_n g^n_{i+\frac{1}{2}}\left[\varphi(x_i, t^n) - \varphi(x_{i+1}, t^n)\right]$$
$$= \Delta x \Delta t \sum_i \sum_n g^n_{i+\frac{1}{2}} \partial_x \varphi(x_i, t^n) + \mathcal{O}(\Delta x) \ .$$

The term $\mathcal{O}(\Delta x)$ depends on the support of φ. If we define

$$g_m(x, t) := g^n_{i+\frac{1}{2}} \quad \text{if } x_i \leq x < x_{i+1},\ t^n < t \leq t^{n+1} \qquad (2.3.8)$$

then we can continue:

$$= \sum_i \sum_n \int_{t^n}^{t^{n+1}} \int_{x_i}^{x_{i+1}} g_m(x, t) \partial_x \varphi(x, t)\, dx\, dt + \mathcal{O}(\Delta x)$$
$$= \int_{\mathbb{R}} \int_{\mathbb{R}^+} g_m(x, t) \partial_x \varphi(x, t)\, dt\, dx + \mathcal{O}(\Delta x) \ . \qquad (2.3.9)$$

Because of the definition of u_m in (2.3.3), we have

$$g_m(x, t) = g(u_m(x - \tfrac{1}{2}\Delta x, t), u_m(x + \tfrac{1}{2}\Delta x, t)) \ .$$

Now we have to show that

$$g_m(x, t) \longrightarrow g(u(x, t), u(x, t)) \ .$$

To prove this, we neglect t and use $\sigma_m := \tfrac{1}{2}\Delta x$. Then we obtain

$$|g_m(x, t) - g(u(x), u(x))|$$
$$= |g(u_m(x - \sigma_m), u_m(x + \sigma_m)) - g(u(x), u(x))|$$
$$\leq L_1 |u_m(x - \sigma_m) - u(x)| + L_2 |u_m(x + \sigma_m) - u(x)| \ .$$

2.3 Convergence of finite difference schemes of first order in 1-D

Now we shall show that $u_m(\cdot \pm \sigma_m)$ converges to u almost everywhere. Let us consider the second term. The first can be treated similarly. Let $v_m := u_m(\cdot + \sigma_m)$. Now it turns out that

$$\int_{\mathbb{R}^+}\int_{\mathbb{R}} v_m(x)\varphi(x)\,dx\,dt = \int_{\mathbb{R}^+}\int_{\mathbb{R}} u_m(x)\varphi(x-\sigma_m)\,dx\,dt$$

$$\longrightarrow \int_{\mathbb{R}^+}\int_{\mathbb{R}} u\varphi\,dt\,dx\ .$$

The same arguments imply that

$$v_m^2 \longrightarrow u^2 \quad \text{weak-}*.$$

Now the L^2_{loc} convergence follows using

$$\int_{\mathbb{R}^+}\int_{\mathbb{R}} |v_m - u|^2 \varphi = \int_{\mathbb{R}^+}\int_{\mathbb{R}} v_m^2 \varphi - 2\int_{\mathbb{R}^+}\int_{\mathbb{R}} v_m u\varphi + \int_{\mathbb{R}^+}\int_{\mathbb{R}} u^2\varphi \longrightarrow 0\ .$$

Therefore (2.3.9) converges to

$$\int_{\mathbb{R}}\int_{\mathbb{R}^+} f(u(x,t))\partial_x\varphi(x,t)\,dt\,dx\ .$$

Replacing the sums in (2.3.5) by the integrals as shown in (2.3.6) and (2.3.9) we obtain the statement of the theorem for the limit $m \to \infty$. □

REMARK 2.3.2 *In Example 2.1.20 (a) we have seen that the numerical scheme (2.3.9) with*

$$g(u,v) := \tfrac{1}{2}(u+v)$$

(i.e. central differences) is in conservation form and the corresponding discrete solutions converge to some function u, but the limit u does not satisfy the entropy condition. This means that in general we cannot expect that a convergent sequence of approximating solutions, defined by a numerical scheme in conservation form (see Definition 2.2.2), converges to an entropy solution of (2.2.1), (2.2.2). The Lax–Wendroff theorem only tells us that the limit function is a weak solution of (2.2.1), (2.2.2). Therefore in order to get the entropy solution we have to define a discrete entropy condition.

DEFINITION 2.3.3 (Discrete entropy condition) *Assume that U, F and f satisfy the conditions of Theorem 2.1.12 and let $G \in C^{0,1}(\mathbb{R} \times \mathbb{R})$ be such that $G(u,u) = F(u)$, and let (u_j^n) be a solution of a numerical scheme in conservation form (2.2.10). Then we shall say that (u_j^n) satisfies a discrete entropy condition if*

$$U(u_j^{n+1}) - U(u_j^n) \leq -\lambda\theta(G_{j+\frac{1}{2}}^{n+1} - G_{j-\frac{1}{2}}^{n+1}) - \lambda(1-\theta)(G_{j+\frac{1}{2}}^n - G_{j-\frac{1}{2}}^n),$$

(2.3.10)

where $G_{j+\frac{1}{2}}^n := G(u_j^n, u_{j+1}^n)$. G is called the numerical entropy flux. *We shall say that the numerical scheme is* consistent with the entropy condition *if (2.3.10) holds uniformly for $\Delta t, \Delta x \longrightarrow 0$ for any U and G.*

In this case the limit will satisfy the entropy condition. This is stated in the following theorem.

THEOREM 2.3.4 (Convergence to the entropy solution) *Let us assume that the conditions of the Lax–Wendroff theorem (Theorem 2.3.1) are satisfied and that the corresponding scheme is consistent with the entropy condition, i.e. it satisfies (2.3.10). Then the scheme will converge to a weak solution of (2.2.1), (2.2.2) that satisfies the entropy condition (2.1.17).*

Proof In Theorem 2.3.1 we have already shown that the limit u is a solution of (2.2.1) in the distributional sense. In an analogous way, we can show that u satisfies the entropy condition (2.1.17). □

In Theorems 2.3.1 and 2.3.4 we have assumed that we already have a convergent sequence. Now we shall show how we can get it. Since we have to approximate discontinuous solutions, the choice of the function space in which we shall construct the converging sequence is the most important problem. It turns out that in spaces of functions with bounded variation the convergence of an approximating sequence can be proved. Before we can state the results, we have to fix some notation.

2.3 Convergence of finite difference schemes of first order in 1-D

DEFINITION 2.3.5 (Total variation) *For $f : [a, b] \to \mathbb{R}$ we define the total variation of f by*

$$\mathrm{TV}_{[a,b]}(f) := \sup_{\substack{a=x_0<x_1<\cdots<x_n=b \\ n\in\mathbb{N}}} \sum_{k=0}^{n} |f(x_{k+1}) - f(x_k)| \ .$$

In the discrete case we define for a sequence $v = (u_j)_{j\in\mathbb{N}}$ of discrete values u_j

$$\mathrm{TV}(v) := \sum_{j=0}^{\infty} |u_{j+1} - u_j| \ .$$

REMARK 2.3.6 *For $g \in L^1(\mathbb{R})$ define*

$$TV(g) := \sup_{h\in\mathbb{R}\setminus\{0\}} \frac{1}{h} \int_{\mathbb{R}} |g(x+h) - g(x)| \, dx \ .$$

Then this definition is equivalent to Definition 2.3.5 with f extended to $\mathbb{R}$ by setting

$$f(x) := \begin{cases} f(a) & \text{for } x < a \ , \\ f(b) & \text{for } x > b \ . \end{cases}$$

EXAMPLE 2.3.7 *Let $f \in C^{0,1}([a,b])$ be a monotonically increasing function. Then $\mathrm{TV}(f) = f(b) - f(a)$.*

More details about functions with bounded variations can be found in [64], [75] and [163].

In spaces of bounded variation there is a powerful compactness property, which is well-known as the "Selection principle of Helly".

60 Initial value problems for scalar conservation laws in 1-D

THEOREM 2.3.8 (Selection principle of Helly) *Let $(u_n)_{n \in \mathbb{N}}$ be a sequence of functions in $L^1(]a, b[)$ such that*

$$\|u_n\|_{L^\infty[a,b]} \leq M , \quad \mathrm{TV}_{[a,b]}(u_n) \leq M$$

uniformly for all $n \in \mathbb{N}$. Then there exists a subsequence $(u_{n'})_{n' \in \mathbb{N}}$ and an $u \in L^1(]a, b[)$ such that

$$u_{n'} \longrightarrow u \quad \text{in } L^1(]a, b[).$$

For the proof we refer to Natanson [163, page 250]. Now we can establish the main theorem of this section.

THEOREM 2.3.9 (Sufficient conditions for convergence) *Let $u_0 \in L^1(\mathbb{R}) \cap L^\infty(\mathbb{R})$ with compact support in $]a, b[$. Let $(u_m)_{m \in \mathbb{N}}$ be a sequence of discrete solutions as defined in Definition 2.2.2 and (2.3.3) with respect to h_m, k_m and the initial values u_i^0 such that the following hold.*

1. *For the numerical flux g we have $g \in C^{0,1}(\mathbb{R}^2)$ and $g(u, u) = f(u)$.*

2. *There exists a constant M_0 such that $|u_m(\cdot, t^n)| \leq M_0$ uniformly with respect to m and n.*

3. *There exists a constant M_1 such that $\mathrm{TV}_{[a,b]}(u_m(\cdot, t^n)) \leq M_1$ uniformly with respect to m and $n \in \mathbb{N}$.*

Then there exists an $u \in L^1_{\text{loc}}(\mathbb{R} \times \mathbb{R}^+)$ and a subsequence $(u_{m'})_{m'}$ such that

$$u_{m'} \longrightarrow u \quad \text{in } L^1_{\text{loc}}(\mathbb{R} \times \mathbb{R}^+)$$

and u is a solution in the distributional sense of (2.2.1),(2.2.2) .

Additionally, if the scheme is consistent with the entropy condition, i.e. it satisfies (2.3.10), then u satisfies the entropy condition. This solution is uniquely defined.

2.3 Convergence of finite difference schemes of first order in 1-D

Proof Let $[a,b] \subset \mathbb{R}$ and let $T, t_0 \in \mathbb{R}^+$ be fixed. Then Theorem 2.3.8 implies that there exists a subsequence of $(u_m)_m$ (we shall always denote subsequences by the same symbol as the original sequence) and a function $u(\cdot, t_0) \in L^1(]a, b[)$ such that

$$u_m(\cdot, t_0) \longrightarrow u(\cdot, t_0) \quad \text{in } L^1(]a, b[).$$

Using a diagonal procedure, we obtain another subsequence of $(u_m)_m$ such that

$$u_m(\cdot, t) \longrightarrow u(\cdot, t) \quad \text{in } L^1(]a, b[) \qquad (2.3.11)$$

for all $t \in M$, where $M = \{t \in \mathbb{R}^+ | t = ik_m,\, m \in \mathbb{N},\, 0 \leq i \leq T/k_m,\, i \in \mathbb{N}\}$. Since the numerical flux g is Lipschitz-continuous, we can establish the Lipschitz continuity of $t \longrightarrow u_m(\cdot, t) \in L^1(]a, b[)$. This can be seen as follows. First of all, we have

$$|u_i^{n+1} - u_i^n| \leq \lambda\theta|g_{i+\frac{1}{2}}^{n+1} - g_{i-\frac{1}{2}}^{n+1}| + \lambda(1-\theta)|g_{i+\frac{1}{2}}^n - g_{i-\frac{1}{2}}^n|$$
$$\leq L(\theta)\lambda(|u_{i+1}^{n+1} - u_i^{n+1}| + |u_i^{n+1} - u_{i-1}^{n+1}|$$
$$+ |u_{i+1}^n - u_i^n| + |u_i^n - u_{i-1}^n|), \qquad (2.3.12)$$

where $L(\theta)$ estimates the Lipschitz constant of the numerical flux g. If we sum up the inequality (2.3.12) with respect to i, assumption 3 of the theorem implies that

$$\sum_i \Delta x |u_i^{n+1} - u_i^n| \leq 4 M_1 L(\theta) \Delta t.$$

Of course M_1 will depend on $[a, b]$. Here the summation is taken over all i such that $x_i \in [a, b]$. Replacing $n+1$ by k, we can derive for $t^k = k\Delta t$, $t^n = n\Delta t$ ($\|\cdot\| := \|\cdot\|_{L^1(]a,b[)}$)

$$\|u_m(\cdot, t^k) - u_m(\cdot, t^n)\| = \int_a^b |u_m(x, t^k) - u_m(x, t^n)|\, dx$$
$$= \sum_i \Delta x (|u_m(x_i, t^k) - u_m(x_i, t^n)|)$$
$$= \sum_i \Delta x |u_i^k - u_i^n|$$
$$\leq 4 M_1 L(\theta) |t^k - t^n|. \qquad (2.3.13)$$

Since u_m is piecewise-constant, we have a similar estimate for $s, r \notin M$:

$$\|u_m(\cdot, s) - u_m(\cdot, r)\| \leq 4M_1 L(\theta)|s - r| + \mathcal{O}(k_m) \ . \tag{2.3.14}$$

To prove this, we choose $\bar{s} = n\Delta t$, $\bar{r} = k\Delta t$ such that $|\bar{s} - s| \leq \Delta t$ and $|r - \bar{r}| \leq \Delta t$. u_m is piecewise-constant and therefore the left-hand side of (2.3.14) can be estimated by

$$\|u_m(\cdot, \bar{s}) - u_m(\cdot, \bar{r})\| \leq 4\, M_1\, L(\Theta)\, |\bar{s} - \bar{r}|$$
$$\leq 4\, M_1\, L(\Theta)\, |s - r| + \mathcal{O}(\Delta t) \ .$$

Property (2.3.14) nearly means that $s \to u_m(\cdot, s) \in L^1(]a, b[)$ is equally continuous. Therefore we use similar arguments as in the proof of the theorem of Arzelá–Ascoli to show the existence of a convergent subsequence. We have for all $l, m \in \mathbb{N}$ and a suitable $t_1 \in M$

$$\|u_l(\cdot, t) - u_m(\cdot, t)\| \leq \|u_l(\cdot, t) - u_l(\cdot, t_1)\| + \|u_l(\cdot, t_1) - u_m(\cdot, t_1)\|$$
$$+ \|u_m(\cdot, t_1) - u_m(\cdot, t)\|$$
$$\leq C_1|t - t_1| + \|u_l(\cdot, t_1) - u_m(\cdot, t_1)\| + \mathcal{O}(\Delta t) \ .$$

Now we shall show that the convergence is uniform in t. Let $\varepsilon > 0$ be given and choose a finite subset $E \subset M$ such that for every $t \in [0, T]$ there is an $t_1 \in E$ such that

$$C_1|t - t_1| \leq \frac{\varepsilon}{3} \ .$$

Furthermore let Δt be sufficiently small and l and m be so large such that

$$\|u_l(\cdot, t_1) - u_m(\cdot, t_1)\| \leq \frac{\varepsilon}{3}$$

uniformly for all $t_1 \in E$. Then we obtain

$$\sup_{t \in [0,T]} \|u_l(\cdot, t) - u_m(\cdot, t)\| \leq \varepsilon$$

uniformly in t. We also have

$$\int_0^T \int_a^b |u_l - u_m| \leq \mathcal{O}(\varepsilon) \ . \tag{2.3.15}$$

2.3 Convergence of finite difference schemes of first order in 1-D

In particular, $u_m \longrightarrow u$ almost everywhere on $]a, b[\times]0, T[$, and Theorem 2.3.1 (Lax–Wendroff) gives us the first statement of the theorem. If the scheme is consistent with the entropy condition, we apply Theorem 2.3.4 in order to get the second statement. □

In order to treat the general case $u_0 \in L^1(\mathbb{R}) \cap L^\infty(\mathbb{R})$, we need the following theorem.

THEOREM 2.3.10 (Stability estimate) *Let $f \in C^2(\mathbb{R})$, $f'' > 0$, $u_0, v_0 \in L^\infty(\mathbb{R})$, $\|v_0\|_{L^\infty} \leq \|u_0\|_{L^\infty}$ and let u, v be the corresponding solution of (2.2.1),(2.2.2). Then for every $a, b \in \mathbb{R}$ such that $a < b$ and every $t > 0$ we have*

$$\int_a^b |u(x,t) - v(x,t)|\, dx \leq \int_{a-At}^{b+At} |u_0(x) - v_0(x)|\, dx .$$

where $A := \max\{ |f'(s)| \mid |s| \leq \|u_0\|_{L^\infty} \}$.

For the proof we refer to [218, Theorem 16.1].

COROLLARY 2.3.11 *Theorem 2.3.9 remains true if $u_0 \in L^1(\mathbb{R}) \cap L^\infty(\mathbb{R})$ and $f \in C^2(\mathbb{R})$ with $f'' > 0$.*

Proof Let $\varepsilon > 0$ be given and choose a, b and $v_0 \in C_0^\infty(]a, b[)$ such that $\|v_0 - u_0\|_{L^1} \leq \varepsilon/2T$. Let v denote the solution of (2.2.1) with respect to v_0 as constructed in the proof of Theorem 2.3.9. Then for sufficiently large l, m (see (2.3.15))

$$\int_0^T \int_a^b |v_l - v_m| \leq \frac{\varepsilon}{2},$$

and, letting $m \to \infty$,

$$\int_0^T \int_a^b |v_l - v| \leq \frac{\varepsilon}{2}.$$

Then if u denotes the exact solution of (2.2.1) with respect to u_0, we obtain

$$\int_0^T \int_a^b |v_l - u| \leq \int_0^T \int_a^b |v_l - v| + \int_0^T \int_a^b |v - u|$$

$$\leq \frac{\varepsilon}{2} + \int_0^T \int_{a-At}^{b+At} |v_0 - u_0|$$

$$\leq \frac{\varepsilon}{2} + \int_0^T \int_{\mathbb{R}} |v_0 - u_0|$$

$$\leq \frac{\varepsilon}{2} + \frac{\varepsilon}{2} = \varepsilon \,.$$

□

COROLLARY 2.3.12 (TVD scheme) *Theorem 2.3.9 remains true if we replace condition 3 by*

$$\mathrm{TV}(u_m(\cdot, t^{n+1})) \leq \mathrm{TV}(u_m(\cdot, t^n)) \leq \mathrm{TV}(u_0) < \infty \qquad (2.3.16)$$

for all $n \in \mathbb{N}$. Then the scheme is called a TVD scheme (total variation diminishing).

Now we shall show that TVD schemes preserve monotonicity.

LEMMA 2.3.13 *Let (u_i^n) be a grid function defined by a TVD scheme. Assume that u_i^n is a monotone (increasing/decreasing) grid function with respect to i. Then u_i^{n+1} is also a monotone (increasing/decreasing) grid function with respect to i.*

Proof Without restriction, let us assume that (u_i^n) is monotonically decreasing in i. Since we have $TV(u_i^n) < \infty$, the following limits exist:

$$u_L = \lim_{i \to -\infty} u_i^n, \quad u_R = \lim_{i \to \infty} u_i^n \,.$$

Furthermore, we have $u_L > u_R$. The scheme is defined by

$$u_i^{n+1} = u_i^n - \frac{\Delta t}{\Delta x}[g(u_i^n, u_{i+1}^n) - g(u_{i-1}^n, u_i^n)] \,.$$

2.3 Convergence of finite difference schemes of first order in 1-D

Since g is continuous, we also have

$$u_L = \lim_{i \to -\infty} u_i^{n+1}, \quad u_R = \lim_{i \to \infty} u_i^{n+1}.$$

Now let us suppose that u_i^{n+1} is not monotonically decreasing in i. Then there exist indices, i.e. with $i < l$, such that we have $(u_j := u_j^{n+1})$ $u_i < u_l$. Using this, we can derive for $r \in \mathbb{N}$

$$\mathrm{TV}_{[x_{-r}, x_r]}(u^{n+1}) \geq |u_{-r}^{n+1} - u_i^{n+1}| + |u_i^{n+1} - u_l^{n+1}| + |u_r^{n+1} - u_l^{n+1}|.$$

From the definition of u_R and u_L we obtain for $r \to \infty$

$$\mathrm{TV}(u^{n+1}) \geq |u_L - u_i^{n+1}| + |u_i^{n+1} - u_l^{n+1}| + |u_R - u_l^{n+1}|,$$

and on the other hand

$$\begin{aligned}\mathrm{TV}(u^n) = |u_L - u_R| &= u_L - u_R \\ &= u_L - u_i^{n+1} + u_i^{n+1} - u_l^{n+1} + u_l^{n+1} - u_R \\ &< |u_L - u_i| + |u_l - u_R| \leq TV(u^{n+1}),\end{aligned}$$

since u^n is assumed to be monotone. Here we have got the strict inequality because of $u_i < u_l$. But this is a contradiction, since we assumed that the scheme is TVD. □

Now we have to define numerical schemes in conservation form such that the conditions of Theorem 2.3.9 are satisfied. A class of schemes that has this property are the monotone schemes.

EXAMPLE 2.3.14 (Monotone schemes) Let us consider the simple equation (2.1.28) and the following discretizations using backward, forward and central difference quotients respectively:

$$u_j^{n+1} := u_j^n - \frac{\Delta t}{\Delta x} a(u_j^n - u_{j-1}^n), \tag{2.3.17}$$

$$u_j^{n+1} := u_j^n - \frac{\Delta t}{\Delta x} a(u_{j+1}^n - u_j^n), \tag{2.3.18}$$

$$u_j^{n+1} := u_j^n - \frac{\Delta t}{2\Delta x} a(u_{j+1}^n - u_{j-1}^n). \tag{2.3.19}$$

As in (2.1.28), we assume that $a > 0$ and $1 - a\Delta t/\Delta x > 0$. Then it can be seen very easily that in (2.3.17) the value u_j^{n+1} is a monotone function of u_j^n and u_{j-1}^n. In (2.3.18) and (2.3.19) u_j^{n+1} does not depend monotonically on u_{j+1}^n, u_j^n and u_{j+1}^n, u_{j-1}^n respectively. As we have already seen in Example 2.1.20, the schemes using forward (2.3.18) and central (2.3.19) difference quotients will not converge in general. But we shall prove that monotone schemes like (2.3.17) converge to the entropy solution. Therefore let us define the following class of numerical schemes.

DEFINITION 2.3.15 (Monotone schemes) *A numerical scheme is called a* monotone scheme *if it can be written in the form*

$$u_j^{n+1} := H(\ldots, u_{j-2}^n, u_{j-1}^n, u_j^n, u_{j+1}^n, \ldots),$$

with a function H that is monotonically nondecreasing in any argument.

An explicit scheme of the form (2.2.9) is monotone if

$$1 - \frac{\Delta t}{\Delta x}(\partial_1 g - \partial_2 g) > 0$$

and if

$$\begin{aligned} g = g(v, w) \text{ is monotonically increasing in } v \\ \text{and decreasing in } w. \end{aligned} \quad (2.3.20)$$

EXAMPLE 2.3.16 The Lax–Friedrichs scheme (see Example 2.2.6) and the Engquist–Osher scheme (see Example 2.2.7) are monotone in the sense of Definition 2.3.15.

EXAMPLE 2.3.17 (Godunov scheme [81]) The basic idea of the Godunov scheme is to compose the global solution by the exact solutions of local Riemann problems. For given initial values $u_0 \in L^1(\mathbb{R})$ we define u_i^0 as in (2.3.1). Now let us assume that we have already computed the approximation $(u_i^n)_{n \in \mathbb{N}}$ for the time t^n, where u_i^n is also constant on $]x_{i-1/2}, x_{i+1/2}[$ for all $i \in \mathbb{N}$.

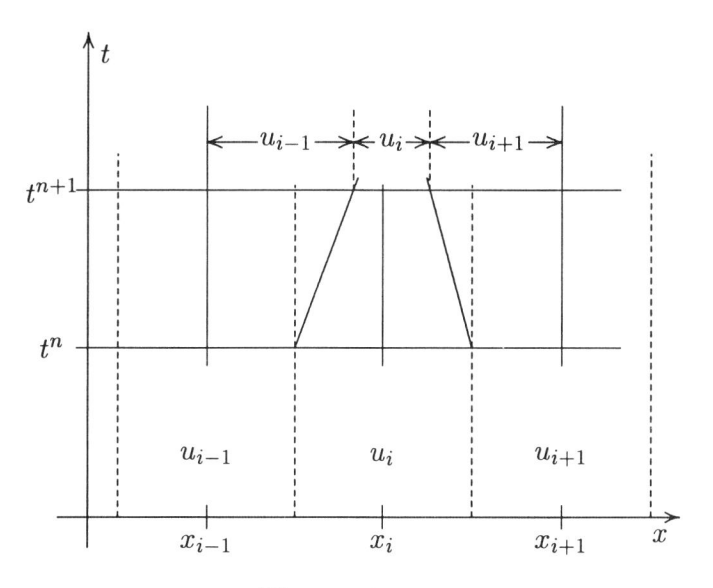

2.3 Convergence of finite difference schemes of first order in 1-D

Figure 2.3.1

On each cell $]x_{i-1}, x_i[$ for $i \in \mathbb{N}$ we determine the exact solution of the Riemann problem for

$$\partial_t u + \partial_x f(u) = 0 \quad \text{on } \mathbb{R} \times [t^n, t^{n+1}] \tag{2.3.21}$$

with respect to the initial condition

$$u(x, t^n) = \begin{cases} u_{i-1} & \text{if } x < x_{i-\frac{1}{2}}, \\ u_i & \text{if } x > x_{i-\frac{1}{2}}. \end{cases} \tag{2.3.22}$$

We denote this solution by $u(x, t; u_{i-1}, u_i)$. In order to ensure that the neighbouring solutions $u(x, t; u_{i-1}, u_i)$ and $u(x, t; u_i, u_{i+1})$ cannot influence each other, we have to assume that the shocks with

$$s_{i-\frac{1}{2}} = \frac{f(u_i) - f(u_{i-1})}{u_i - u_{i-1}} \quad \text{and} \quad s_{i+\frac{1}{2}} = \frac{f(u_{i+1}) - f(u_i)}{u_{i+1} - u_i}$$

must not intersect. This can be obtained if (see Figure 2.3.1)

$$|s_{i-\frac{1}{2}}|\frac{\Delta t}{\Delta x} \leq \frac{1}{2}, \quad |s_{i+\frac{1}{2}}|\frac{\Delta t}{\Delta x} \leq \frac{1}{2}$$

or

$$\frac{\Delta t}{\Delta x} \sup_{u \in \mathbb{R}} |f'(u)| \leq \frac{1}{2} . \tag{2.3.23}$$

The condition (2.3.23) is again the Courant, Friedrichs, Lewy condition or CFL condition (see Definition 2.2.9). If (2.3.23) is satisfied, the solution $u(x, t; u_{i-1}, u_i)$ of the local Riemann problem (2.3.21), (2.3.22) uniquely defines a function v on $\mathbb{R} \times [t^n, t^{n+1}]$ such that

$$v(x,t) := \begin{cases} u(x,t;,u_{i-1},u_i) & \text{if } x_{i-\frac{1}{2}} < x \leq x_i,\ t^n \leq t \leq t^{n+1}, \\ u(x,t;,u_i,u_{i+1}) & \text{if } x_i < x \leq x_{i+\frac{1}{2}},\ t^n \leq t \leq t^{n+1} . \end{cases}$$

As for the initial values, we have to ensure that the approximation u^{n+1} at time t^{n+1} is constant on $]x_{i-1/2}, x_{i+1/2}[$ for all $i \in \mathbb{N}$. Therefore we define

$$u_i^{n+1} := \frac{1}{\Delta x} \int_{x_{i-\frac{1}{2}}}^{x_{i+\frac{1}{2}}} v(x, t^{n+1})\, dx . \tag{2.3.24}$$

This means u_i^{n+1} is the mean value of v on $]x_{i-1/2}, x_{i+1/2}[$ and therefore contains parts of $u(x,t; u_{i-1}, u_i)$ and $u(x,t; u_i, u_{i+1})$. Since v is an exact solution on $]x_{i-1/2}, x_{i+1/2}[$, we get (see (2.2.6))

$$\int_{x_{i-\frac{1}{2}}}^{x_{i+\frac{1}{2}}} v(x, t^{n+1})\, dx = \int_{x_{i-\frac{1}{2}}}^{x_{i+\frac{1}{2}}} v(x, t^n)\, dx - \int_{t^n}^{t^{n+1}} f(v(x_{i+\frac{1}{2}}, s))\, ds$$

$$+ \int_{t^n}^{t^{n+1}} f(v(x_{i-\frac{1}{2}}, s))\, ds .$$

Using

$$v(x_{i+\frac{1}{2}}, t) = u(x_{i+\frac{1}{2}}, t; u_i, u_{i+1}) =: u_{i+\frac{1}{2}},$$
$$v(x_{i-\frac{1}{2}}, t) = u(x_{i-\frac{1}{2}}, t; u_{i-1}, u_i) =: u_{i-\frac{1}{2}},$$

2.3 Convergence of finite difference schemes of first order in 1-D

we obtain

$$u_i^{n+1} = u_i^n - \frac{\Delta t}{\Delta x}[f(u_{i+\frac{1}{2}}) - f(u_{i-\frac{1}{2}})], \qquad (2.3.25)$$

with the numerical flux

$$g(u_{i-1}, u_i) := f(u(x_{i-\frac{1}{2}}, t; u_{i-1}, u_i)) \quad \text{for all } i. \qquad (2.3.26)$$

This makes sense since $u(x_{i+1/2})$ is constant for $t^n < t \le t^{n+1}$. This scheme is called the Godunov scheme. If f is convex, i.e. $f'' > 0$, it can be shown that the scheme is monotone in the sense of Definition 2.3.15. In this case if $u_L > u_R$ the solution of the Riemann problem (2.3.21), (2.3.22) is (see (2.1.30))

$$u(x, t; u_L, u_R) = \begin{cases} u_L & \text{if } \xi < s, \\ u_R & \text{if } \xi > s, \end{cases}$$

for $s = \frac{f(u_R) - f(u_L)}{u_R - u_L}$ and $\xi = \frac{x - x_{i-1/2}}{t - t^n}$. If we have $u_L < u_R$ then (see (2.1.31))

$$u(x, t; u_L, u_R) = \begin{cases} u_L & \text{if } \xi \le f'(u_L), \\ a^{-1}(\xi) & \text{if } f'(u_L) < \xi < f'(u_R), \\ u_R & \text{if } \xi \ge f'(u_R), \end{cases}$$

where $a = f'$. Now we use these solutions of the local Riemann problems in (2.3.25) in order to compute the numerical flux $g(v, w)$. First let us consider the case where $v \ge w$. Let

$$s := \frac{f(v) - f(w)}{v - w}.$$

Then we have

$$u(x_{i-\frac{1}{2}}, t; v, w) = v \quad \text{if } 0 \le s \quad (\text{i.e. } f(v) \ge f(w)),$$

$$u(x_{i-\frac{1}{2}}, t; v, w) = w \quad \text{if } 0 > s \quad (\text{i.e. } f(v) \le f(w)).$$

70 Initial value problems for scalar conservation laws in 1-D

If $v < w$ then

$$u(x_{i-\frac{1}{2}}, t; v, w) = v \quad \text{if } f'(v) \geq 0 ,$$
$$u(x_{i-\frac{1}{2}}, t; v, w) = w \quad \text{if } f'(w) \leq 0 ,$$
$$u(x_{i-\frac{1}{2}}, t; v, w) = f'^{-1}(0) \quad \text{otherwise.}$$

This means that the numerical flux is given by

$$g(v, w) = \begin{cases} f(v) & \text{if } v \geq w, \ f(v) \geq f(w) , \\ f(w) & \text{if } v \geq w, \ f(v) \leq f(w) , \\ f(v) & \text{if } v < w, \ f'(v) \geq 0 , \\ f(w) & \text{if } v < w, \ f'(w) \leq 0 , \\ f(f'^{-1}(0)) & \text{otherwise.} \end{cases} \quad (2.3.27)$$

REMARK 2.3.18 *If f is convex, the numerical flux g for the Godunov scheme can also be written as*

$$g(v, w) := \begin{cases} \min_{v \leq s \leq w} f(s) & \text{if } v < w , \\ \max_{v \geq s \geq w} f(s) & \text{if } v \geq w . \end{cases}$$

Therefore we see that g is Lipschitz-continuous, consistent with the conservation law and, if $\Delta t / \Delta x$ is small enough, monotone in the sense of Definition 2.3.15.

We have tested the Godunov scheme (see Figures 2.3.2 and 2.3.3) for the same problem as described in Example 2.2.6. The results are shown in Figures 2.3.2 and 2.3.3.

Now we shall study the convergence of monotone schemes. It turns out that they define discrete solutions that will converge to a solution of the conservation law satisfying the entropy condition.

THEOREM 2.3.19 (Convergence of monotone schemes [89, 90]) *Assume that $u_0 \in L^1_{loc}(\mathbb{R})$, such that $TV(u_0) < \infty$. Consider the numerical scheme (2.2.10), which is assumed to be consistent with the conservation*

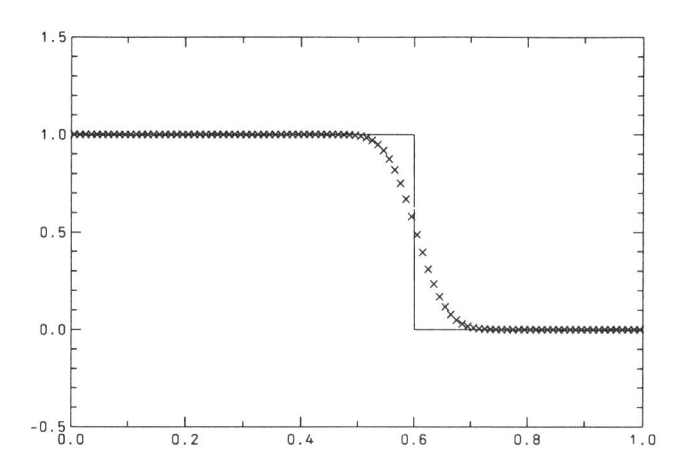

2.3 Convergence of finite difference schemes of first order in 1-D

Figure 2.3.2

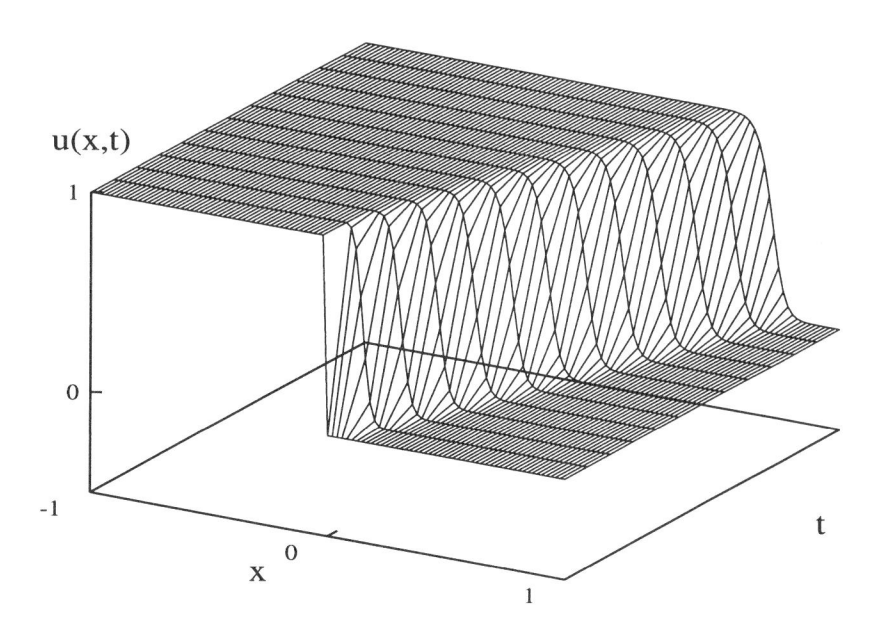

Figure 2.3.3 Godunov.

law, let $g \in C^1(\mathbb{R}^2)$ and let g satisfy (2.3.20). If L is the global Lipschitz constant of the numerical flux g, we assume

$$2(1-\theta)\frac{\Delta t}{\Delta x}L \le 1 \ . \tag{2.3.28}$$

Using the grid function u_i^n as defined in (2.2.10), we define $u_m \in L^\infty(\mathbb{R} \times \mathbb{R}^+)$ as in (2.3.3). Then if Δx and Δt converge to zero such that $\Delta t/\Delta x$ remains constant, the approximating functions u_m will converge to a function u in $L^1_{\text{loc}}(\mathbb{R} \times \mathbb{R}^+)$ such that u is a solution of (2.2.1), (2.2.2) in the distributional sense and u satisfies the entropy condition.

Proof We shall proof the theorem in several steps. Successively, we shall show that

$$\text{the scheme is TVD;} \tag{2.3.29}$$
$$\text{the approximating sequence } (u_m)_{m \in \mathbb{N}} \text{ is}$$
$$\text{uniformly bounded;} \tag{2.3.30}$$
$$\text{the approximating functions } u_m \text{ satisfy}$$
$$\text{the discrete entropy condition in Definition 2.3.3.} \tag{2.3.31}$$

Then the statement of the theorem follows from Theorem 2.3.9. Now let us start to prove (2.3.29). We define

$$\begin{aligned} L_i &:= L_i(u^{n+1}) := u_i^{n+1} + \lambda\theta(g_{i+\frac{1}{2}}^{n+1} - g_{i-\frac{1}{2}}^{n+1}) \ , \\ R_i &:= R_i(u^n) := u_i^n - \lambda(1-\theta)(g_{i+\frac{1}{2}}^n - g_{i-\frac{1}{2}}^n) \ , \end{aligned} \tag{2.3.32}$$

where $\lambda = \Delta t/\Delta x$. Then the scheme (2.2.10) can be written as

$$L(u^{n+1}) = R(u^n) \ . \tag{2.3.33}$$

We introduce the following notation:

$$\Delta_+ u_i^n = u_{i+1}^n - u_i^n \ ,$$
$$\Delta_- u_i^n = u_i^n - u_{i-1}^n \ ,$$
$$C_{i+\frac{1}{2}}^n := \lambda \frac{g_{i+\frac{1}{2}}^n - f(u_i^n)}{u_{i+1}^n - u_i^n} \ ,$$
$$D_{i-\frac{1}{2}}^n := \lambda \frac{g_{i-\frac{1}{2}}^n - f(u_i^n)}{u_i^n - u_{i-1}^n} \ .$$

2.3 Convergence of finite difference schemes of first order in 1-D

Then the operators L and R can be written in the form

$$L_i = u_i^{n+1} + \theta C_{i+\frac{1}{2}}^{n+1} \Delta_+ u_i^{n+1} - \theta D_{i-\frac{1}{2}}^{n+1} \Delta_- u_i^{n+1},$$

$$R_i = u_i^n - (1-\theta) C_{i+\frac{1}{2}}^n \Delta_+ u_i^n + (1-\theta) D_{i-\frac{1}{2}}^n \Delta_- u_i^n.$$

We should like to show that L_i increases while R_i decreases the total variation. For this purpose we need the following lemma. First let us fix some notation. We call $Z : L^1(\mathbb{R}) \to L^1(\mathbb{R})$ a TVD (total variation decreasing) operator if for $v \in L^1(\mathbb{R})$

$$TV(Zv) \leq TV(v).$$

Similarly, we call Z a TVI (total variation increasing) operator if

$$TV(Zv) \geq TV(v).$$

LEMMA 2.3.20 *Let Z be an operator of the form*

$$(Zv)_j := v_j + c_{j+\frac{1}{2}} \Delta_+ v_j - d_{j-\frac{1}{2}} \Delta_- v_j.$$

Then Z is a TVD operator if we have for all $j \in \mathbb{N}$: $c_{j+1/2} \geq 0$, $d_{j+1/2} \geq 0$, $c_{j+1/2} + d_{j+1/2} \leq 1$. This condition for the coefficients is equivalent to the monotonicity of $(Zv)_j$. Otherwise, if $-\infty < c \leq c_{j+1/2}, d_{j+1/2} \leq 0$ then Z is a TVI operator.

For the proof of this lemma we refer to 2.3.21 below. First let us continue the proof of Theorem 2.3.19. Consistency and monotonicity imply that

$$C_{j+\frac{1}{2}}^n = \lambda \frac{g(u_j^n, u_{j+1}^n) - f(u_j^n)}{u_{j+1}^n - u_j^n} \leq 0,$$

$$D_{j-\frac{1}{2}}^n = \lambda \frac{g(u_{j-1}^n, u_j^n) - f(u_j^n)}{u_j^n - u_{j-1}^n} \leq 0.$$
(2.3.34)

Thus by Lemma 2.3.20 we obtain (replace n by $n+1$ in (2.3.34)) that L is a TVI operator. In order to prove that R is a TVD operator, it remains to show that

$$-(1-\theta) C_{j+\frac{1}{2}}^n - (1-\theta) D_{j+\frac{1}{2}}^n \leq 1.$$

Again using consistency and monotonicity, we get

$$C^n_{j+\frac{1}{2}} + D^n_{j+\frac{1}{2}} = \lambda \frac{g^n(u^n_j, u^n_{j+1}) - f(u^n_j)}{u^n_{j+1} - u^n_j} + \lambda \frac{g^n(u^n_j, u^n_{j+1}) - f(u^n_{j+1})}{u^n_{j+1} - u^n_j}$$

$$\geq -2\lambda L,$$

and therefore

$$-(1-\theta)(C^n_{j+\frac{1}{2}} + D^n_{j+\frac{1}{2}}) \leq (1-\theta)2\lambda L \leq 1.$$

Hence Lemma 2.3.20 implies that R is a TVD operator. But then we can show that property (2.3.29) holds. This means that the operator $u^n \mapsto u^{n+1}$, where u^{n+1} is defined in (2.3.33), is a TVD operator. This follows now from

$$\operatorname{TV}(u^{n+1}) \leq \operatorname{TV}(Lu^{n+1}) = \operatorname{TV}(Ru^n) \leq \operatorname{TV}(u^n). \quad (2.3.35)$$

Now we are going to prove (2.3.30). We have to prove that the approximating solutions are uniformly bounded. By definition, we have

$$\operatorname{TV}(u^n) = \sum_j |u^n_{j+1} - u^n_j| \geq \sup_j u^n_j - \inf_j u^n_j,$$

and therefore

$$\sup_j u^n_j \leq \operatorname{TV}(u^n) + \inf_j u^n_j \leq \operatorname{TV}(u^0) + \inf_j u^n_j, \quad (2.3.36)$$

since the scheme is TVD. Furthermore, we have (we assume now that the scheme is explicit; the arguments can be used for an implicit one as well):

$$\sum_{j=-m}^{m} u^{n+1}_j - \sum_{j=-m}^{m} u^n_j = \lambda \sum_{j=-m}^{m} [g(u_{j-1}, u_j) - g(u_j, u_{j+1})]$$

$$= \lambda[g(u_{-m-1}, u_{-m}) - g(u_m, u_{m+1})].$$

Since $\lim_{m \to \infty} u_{\pm m}$ exists, we can estimate the mean value of all u^{n+1}_j, $j \in \mathbb{N}$:

$$\lim_{m \to \infty} \frac{1}{2m+1} \sum_{j=-m}^{m} u^{n+1}_j = \lim_{m \to \infty} \frac{1}{2m+1} \sum_{j=-m}^{m} u^n_j =: C_0.$$

2.3 Convergence of finite difference schemes of first order in 1-D

But this implies that $\inf_j u_j^n$ is bounded by C_0 from above:

$$\inf_j u_j^n \leq C_0 \leq \sup_j u_j^n .$$

Using (2.3.36), we obtain

$$\sup_j u_j^n \leq \mathrm{TV}(u^0) + \inf_j u_j^n \leq \mathrm{TV}(u^0) + C_0 ,$$
$$\inf_j u_j^n \geq -\mathrm{TV}(u^0) + \sup_j u_j^n \geq -\mathrm{TV}(u^0) + C_0 .$$

This proves (2.3.30). Finally let us show property (2.3.31); this means that the approximating solutions satisfy the discrete entropy condition (2.3.10). Because of Lemma 2.1.17, it is sufficient to consider the entropy function

$$U(u) := |u - c| \qquad (2.3.37)$$

for a fixed constant $c \in \mathbb{R}$ and

$$F(u) := [f(u) - f(c)]\,\mathrm{sign}(u - c) . \qquad (2.3.38)$$

Then we obtain

$$\begin{array}{llll} F'(u) = & f'(u) , & U'(u) = & 1 & \text{if } u > c, \\ F'(u) = & -f'(u) , & U'(u) = & -1 & \text{if } u < c. \end{array}$$

Since we have $F(c) = 0$, the function F is locally Lipschitz-continuous. For the numerical entropy flux we define

$$G(u, v)$$
$$:= g(\max\{u, c\}, \max\{v, c\}) - g(\min\{u, c\}, \min\{v, c\}). \quad (2.3.39)$$

G is locally Lipschitz-continuous because it is defined as a composition of Lipschitz-continuous functions. Furthermore, we have $G(u, u) = F(u)$. Now we have to verify (2.3.10) for U and G as defined in (2.3.37) and (2.3.39). For this purpose let us derive the following inequalities:

$$|u_i^n - c| - \lambda(1 - \theta)(G_{i+\frac{1}{2}}^n - G_{i-\frac{1}{2}}^n)$$
$$\geq |u_i^n - \lambda(1 - \theta)(g_{i+\frac{1}{2}}^n - g_{i-\frac{1}{2}}^n) - c| \qquad (2.3.40)$$

and

$$|u_i^{n+1} + \lambda\theta(g_{i+\frac{1}{2}}^{n+1} - g_{i-\frac{1}{2}}^{n+1}) - c|$$
$$\geq |u_i^{n+1} - c| + \lambda\theta(G_{i+\frac{1}{2}}^{n+1} - G_{i-\frac{1}{2}}^{n+1}). \tag{2.3.41}$$

The right-hand side of (2.3.40) and the left-hand side of (2.3.41) are the same because of (2.2.10). Thus (2.3.10) follows from (2.3.40) and (2.3.41). In order to show (2.3.40), define

$$H(u,v,w) := v - \lambda(1-\theta)[g(v,w) - g(u,v)] .$$

H is monotonically increasing with respect to all of its arguments. For u and w this follows from the corresponding properties of g. For v we obtain this because of the assumptions (2.3.28):

$$\partial_v H(u,v,w) = 1 - \lambda(1-\theta)(\partial_1 g - \partial_2 g) \geq 1 - \lambda(1-\theta)2L \geq 0 .$$

This can be used to prove (2.3.40):

$$|u_i^n - c| - \lambda(1-\theta)(G_{i+\frac{1}{2}}^n - G_{i-\frac{1}{2}}^n)$$
$$= \max\{u_i^n, c\} - \min\{u_i^n, c\}$$
$$\quad -\lambda(1-\theta)[g(\max\{u_i^n, c\}, \max\{u_{i+1}^n, c\})$$
$$\quad -g(\min\{u_i^n, c\}, \min\{u_{i+1}^n, c\})$$
$$\quad -g(\max\{u_{i-1}^n, c\}, \max\{u_i^n, c\}) + g(\min\{u_{i-1}^n, c\}, \min\{u_i^n, c\})]$$
$$= H(\max\{u_{i-1}^n, c\}, \max\{u_i^n, c\}, \max\{u_{i+1}^n, c\})$$
$$\quad - H(\min\{u_{i-1}^n, c\}, \min\{u_i^n, c\}, \min\{u_{i+1}^n, c\})$$
$$\geq \max\{H(u_{i-1}^n, u_i^n, u_{i+1}^n), H(c,c,c)\}$$
$$\quad - \min\{H(u_{i-1}^n, u_i^n, u_{i+1}^n), H(c,c,c)\}$$
$$= |H(u_{i-1}^n, u_i^n, u_{i+1}^n) - c| ,$$

since $H(c,c,c) = c$. This proves (2.3.40). In order to derive (2.3.41), we use

$$G_{i+\frac{1}{2}}^n = g(\max\{u_i^n, c\}, \max\{u_{i+1}^n, c\}) - g(\min\{u_i^n, c\}, \min\{u_{i+1}^n, c\})$$
$$\leq \operatorname{sign}(u_i^n - c)[g(u_i^n, u_{i+1}^n) - g(c,c)] . \tag{2.3.42}$$

2.3 Convergence of finite difference schemes of first order in 1-D

This follows from the monotonicity of g. Similarly, one can show that

$$-G^n_{i-\frac{1}{2}} \leq \text{sign}(u^n_i - c)[g(c,c) - g(u^n_{i-1}, u^n_i)] . \tag{2.3.43}$$

Then (2.3.42) and (2.3.43) imply that

$$\begin{aligned}
&|u^{n+1}_i - c| + \lambda\theta(G^{n+1}_{i+\frac{1}{2}} - G^{n+1}_{i-\frac{1}{2}}) \\
&\leq |u^{n+1}_i - c| + \lambda\theta\{\text{sign}(u^{n+1}_i - c)[g(u^{n+1}_i, u^{n+1}_{i+1}) - g(c,c)] \\
&\quad + \text{sign}(u^{n+1}_i - c)[g(c,c) - g(u^{n+1}_{i-1}, u^{n+1}_i)]\} \\
&= \text{sign}(u^{n+1}_i - c)\{u^{n+1}_i - c + \lambda\theta[g(u^{n+1}_i, u^{n+1}_{i+1}) - g(u^{n+1}_{i-1}, u^{n+1}_i)]\} \\
&\leq |u^{n+1}_i + \lambda\theta(g^{n+1}_{i+\frac{1}{2}} - g^{n+1}_{i-\frac{1}{2}}) - c| .
\end{aligned}$$

This proves the statement (2.3.41). Property (2.3.40) together with (2.3.41) implies (2.3.31) and finishes the proof of the convergence theorem for monotone schemes in conservation form. It remains to show the statement of Lemma 2.3.20.

2.3.21 (Proof of Lemma 2.3.20) Let us start with the TVD property. For any grid function we have

$$\begin{aligned}
(Zv)_{j+1} - (Zv)_j &= v_{j+1} + c_{j+\frac{3}{2}}\Delta_+ v_{j+1} - d_{j+\frac{1}{2}}\Delta_- v_{j+1} \\
&\quad - v_j - c_{j+\frac{1}{2}}\Delta_+ v_j + d_{j-\frac{1}{2}}\Delta_- v_j \\
&= (1 - c_{j+\frac{1}{2}} - d_{j+\frac{1}{2}})\Delta_+ v_j + c_{j+\frac{3}{2}}\Delta_+ v_{j+1} \\
&\quad + d_{j-\frac{1}{2}}\Delta_- v_j .
\end{aligned} \tag{2.3.44}$$

The coefficients of $\Delta_+ v_j$, $\Delta_+ v_{j+1}$ and $\Delta_- v_j$ are all non–negative, and therefore we have

$$\begin{aligned}
TV(Zv) &= \sum_j |(Zv)_{j+1} - (Zv)_j| \\
&\leq \sum_j [(1 - c_{j+\frac{1}{2}} - d_{j+\frac{1}{2}})|\Delta_+ v_j| + c_{j+\frac{3}{2}}|\Delta_+ v_{j+1}| \\
&\quad + d_{j-\frac{1}{2}}|\Delta_- v_j|] \\
&= \sum_j |\Delta_+ v_j| = TV(v) .
\end{aligned}$$

To obtain the last equality, we have to reorder the indices in the summation. This proves the TVD part in Lemma 2.3.20. For the TVI part we obtain from (2.3.44)

$$\Delta_+(Zv)_j - c_{j+\frac{3}{2}}\Delta_+v_{j+1} - d_{j-\frac{1}{2}}\Delta_-v_j = (1 - c_{j+\frac{1}{2}} - d_{j+\frac{1}{2}})\Delta_+v_j \ .$$

Since we have $c_{j+\frac{3}{2}}, d_{j-\frac{1}{2}} \leq 0$, we get

$$|\Delta_+(Zv)_j| - c_{j+\frac{3}{2}}|\Delta_+v_{j+1}| - d_{j-\frac{1}{2}}|\Delta_-v_j|$$
$$\geq (1 - c_{j+\frac{1}{2}} - d_{j+\frac{1}{2}})|\Delta_+v_j| \ ,$$

and if we sum up with respect to j,

$$\mathrm{TV}(Zv) - \sum_j (c_{j+\frac{1}{2}} + d_{j+\frac{1}{2}})|\Delta_+v_j|$$
$$= \sum_j |\Delta_+(Zv)_j| - \sum_j c_{j+\frac{3}{2}}|\Delta_+v_{j+1}| - \sum_j d_{j-\frac{1}{2}}|\Delta_-v_j|$$
$$\geq \sum_j (1 - c_{j+\frac{1}{2}} - d_{j+\frac{1}{2}})|\Delta_+v_j|$$
$$\geq \mathrm{TV}(v) - \sum_j (c_{j+\frac{1}{2}} + d_{j+\frac{1}{2}})|\Delta_+v_j| \ .$$

This implies that $\mathrm{TV}(Zv) \geq \mathrm{TV}(v)$ and proves the TVI part of Lemma 2.3.20. □

REMARK 2.3.22 *For linear problems $\partial_t u + a\partial_x u = 0$ and explicit schemes, (2.3.28) can be replaced by*

$$\frac{\Delta t}{\Delta x} a \leq 1 \ . \tag{2.3.45}$$

For the Proof see Example 2.4.8.

THEOREM 2.3.23 (Error estimate) *Assume that $u_0 \in L^1(\mathbb{R}) \cap L^\infty(\mathbb{R}) \cap BV(\mathbb{R})$ and that u is the entropy solution of (2.2.1), (2.2.2). Let u_h be given by (2.3.2) and (2.3.3) for a monotone numerical flux $g(u,v)$ and let $\Delta t/\Delta x$ be constant. Then for any $\sqrt{\Delta t} \leq t \leq T$ we have the following error estimate:*

$$\|u(\cdot,t) - u_h(\cdot,t)\|_{L^1(\mathbb{R})} \leq \|u(\cdot,0) - u_h(\cdot,0)\|_{L^1(\mathbb{R})} + ct\,TV(u_0)\sqrt{\Delta t} \ .$$

2.3 Convergence of finite difference schemes of first order in 1-D

Proof See [80, Theorem A1] and [124]. We shall not give the proof for this result in 1-D since we shall prove it in a more general context in 2-D on unstructured grids (see Section 3.4). □

Now we should like to mention some more conditions, which are sufficient for TVD. Explicit schemes of first order in conservation form can always be written in the form (see (2.2.9))

$$u_j^{n+1} = u_j^n - \frac{\Delta t}{\Delta x}(g_{j+\frac{1}{2}}^n - g_{j-\frac{1}{2}}^n) . \qquad (2.3.46)$$

Since we assume that the numerical flux is Lipschitz-continuous and consistent, i.e. $g(u,u) = f(u)$, the flux can be written in the form

$$g(u,v) = \begin{cases} \frac{1}{2}[g(u,u) + g(v,v)] - \frac{1}{2\lambda}(v-u)q(u,v) & \text{for } u \neq v, \\ f(u) & \text{for } u = v, \end{cases} \qquad (2.3.47)$$

where

$$q(u,v) = \lambda \frac{g(u,u) - 2g(u,v) + g(v,v)}{v - u} .$$

For a special grid function, this means that

$$g_{j+\frac{1}{2}} = \frac{1}{2}[f(u_j) + f(u_{j+1})] - \frac{1}{2\lambda}q_{j+\frac{1}{2}}\Delta_+ u_j , \qquad (2.3.48)$$

where

$$q_{j+\frac{1}{2}} := q(u_j, u_{j+1}) .$$

The expression $q_{j+\frac{1}{2}}$ can be interpreted as a numerical viscosity, which is necessary to approximate the entropy solution, similar to Theorem 2.1.7. Unfortunately, the numerical viscosity is responsible for smearing out the discontinuity. (See also Remark 2.2.6.)

Even for more general q one can show that the corresponding scheme is TVD. In the next theorem we shall give some sufficient conditions for q such that the corresponding scheme will become TVD.

THEOREM 2.3.24 (Sufficient condition for TVD [89, 90]) *Let the numerical flux be defined as in (2.3.48) with some function $q_{j+\frac{1}{2}} := q(u_j, u_{j+1})$ such that for a constant $C \in \mathbb{R}$*

$$\lambda \left| \frac{f_{j+1} - f_j}{u_{j+1} - u_j} \right| \leq q_{j+\frac{1}{2}} \leq \min\left\{\frac{1}{1-\theta}, C\right\}. \qquad (2.3.49)$$

Then the corresponding scheme as defined in (2.2.10) is TVD.

Proof Define $L_i(u^n)$ and $R_i(u^n)$ as in (2.3.32). Again we have to show that

$$u^n \longmapsto u^{n+1}, \qquad (2.3.50)$$

where u^{n+1} is defined by $L_i(u^{n+1}) = R_i(u^n)$, is a TVD operator. In the same way as in the preceding proof of Theorem 2.3.19, we shall show that L_i and R_i are TVI and TVD operators respectively. Then the same argument as in (2.3.35) proves the statement of the theorem. Before we shall show that L_i is a TVI operator, let us fix some notations:

$$\begin{aligned}
\nu_{j+\frac{1}{2}} &:= \lambda \frac{f_{j+1} - f_j}{u_{j+1} - u_j}, \\
C_{j+\frac{1}{2}} &:= \tfrac{1}{2}\theta(\nu_{j+\frac{1}{2}} - q_{j+\frac{1}{2}}), \\
D_{j+\frac{1}{2}} &:= -\tfrac{1}{2}\theta(\nu_{j+\frac{1}{2}} + q_{j+\frac{1}{2}}).
\end{aligned} \qquad (2.3.51)$$

It turns out that

$$\begin{aligned}
C_{j+\frac{1}{2}} &\leq \tfrac{1}{2}\theta(|\nu_{j+\frac{1}{2}}| - q_{j+\frac{1}{2}}) \leq 0, \\
C_{j+\frac{1}{2}} &\geq -C > -\infty, \quad D_{j+\frac{1}{2}} \geq -C > -\infty, \quad D_{j+\frac{1}{2}} \leq 0.
\end{aligned} \qquad (2.3.52)$$

The operator $L_j(u)$ can be written in the following form (we suppress n in u_j^n):

$$\begin{aligned}
L_j(u) &= u_j + \theta\lambda(g_{j+\frac{1}{2}} - g_{j-\frac{1}{2}}) \\
&= u_j + \theta\lambda\Bigg[\frac{1}{2}\left(\frac{f_{j+1} - f_j}{u_{j+1} - u_j} - \frac{1}{\lambda}q_{j+\frac{1}{2}}\right)\Delta_+ u_j \\
&\quad - \frac{1}{2}\left(\frac{-f_j + f_{j-1}}{u_j - u_{j-1}} - \frac{1}{\lambda}q_{j-\frac{1}{2}}\right)\Delta_+ u_{j-1}\Bigg]
\end{aligned}$$

2.3 Convergence of finite difference schemes of first order in 1-D

$$= u_j + \theta\left(\frac{1}{2}\nu_{j+\frac{1}{2}} - \frac{1}{2}q_{j+\frac{1}{2}}\right)\Delta_+ u_j$$

$$-\theta\left(-\frac{1}{2}\nu_{j-\frac{1}{2}} - \frac{1}{2}q_{j-\frac{1}{2}}\right)\Delta_+ u_{j-1}$$

$$= u_j + C_{j+\frac{1}{2}}\Delta_+ u_j - D_{j-\frac{1}{2}}\Delta_- u_j \,.$$

Then the properties (2.3.52) and Lemma 2.3.20 imply that L_j is a TVI operator. Now we are going to show that R_j is a TVD operator. As before, we have

$$R_j(u) = u_j - (1-\theta)\lambda\left(\frac{1}{2}\frac{f_{j+1}-f_j}{u_{j+1}-u_j}\Delta_+ u_j - \frac{1}{2\lambda}q_{j+\frac{1}{2}}\Delta_+ u_j\right.$$

$$\left.+ \frac{1}{2}\frac{f_j-f_{j-1}}{u_j-u_{j-1}}\Delta_+ u_{j-1} + \frac{1}{2\lambda}q_{j-1}\Delta_+ u_{j-1}\right)$$

$$= u_j + \left[-\frac{1-\theta}{2}(\nu_{j+\frac{1}{2}} - q_{j+\frac{1}{2}})\right]\Delta_+ u_j$$

$$-\left[\frac{1-\theta}{2}(\nu_{j-\frac{1}{2}} + q_{j-\frac{1}{2}})\right]\Delta_+ u_{j-1}$$

$$= u_j + C_{j+\frac{1}{2}}\Delta_+ u_j - D_{j-\frac{1}{2}}\Delta_+ u_{j-1} \,,$$

where

$$C_{j+\frac{1}{2}} = -\frac{1-\theta}{2}(\nu_{j+\frac{1}{2}} - q_{j+\frac{1}{2}}) \,,$$

$$D_{j-\frac{1}{2}} = \frac{1-\theta}{2}(\nu_{j-\frac{1}{2}} + q_{j-\frac{1}{2}}) \,.$$

Now it is easy to see that $C_{j+\frac{1}{2}}$ and $D_{j+\frac{1}{2}}$ satisfy the conditions of Lemma 2.3.20 concerning the TVD part:

$$C_{j+\frac{1}{2}} = -\frac{1-\theta}{2}(\nu_{j+\frac{1}{2}} - q_{j+\frac{1}{2}}) \geq \frac{1-\theta}{2}(q_{j+\frac{1}{2}} - |\nu_{j+\frac{1}{2}}|) \geq 0 \,,$$

$$D_{j+\frac{1}{2}} = \frac{1-\theta}{2}(\nu_{j+\frac{1}{2}} + q_{j+\frac{1}{2}}) \geq \frac{1-\theta}{2}(q_{j+\frac{1}{2}} - |\nu_{j+\frac{1}{2}}|) \geq 0 \,,$$

$$C_{j+\frac{1}{2}} + D_{j+\frac{1}{2}} = (1-\theta)q_{j+\frac{1}{2}} \leq 1 \,,$$

because of assumption (2.3.49).

Now Lemma 2.3.20 yields that $R_j(u)$ is a TVD operator, and therefore (2.3.50) is a TVD operator (see (2.3.35)). □

Next we show a simple corollary of Theorem 2.3.24

COROLLARY 2.3.25 (see [89, 90]) *Let $\nu_{j+1/2}$ be defined as in (2.3.51), let*

$$Q_\varepsilon(x) := \begin{cases} \varepsilon & \text{if } |x| < \varepsilon, \\ |x| & \text{if } |x| \geq \varepsilon, \end{cases}$$

and define $q_{j+1/2} := Q_\varepsilon(\nu_{j+1/2})$. If the conditions

$$\max_j |\nu_{j+\frac{1}{2}}| \leq \min\left\{\frac{1}{1-\theta}, C\right\} \quad \text{and} \quad 0 \leq \varepsilon < \min\left\{\frac{1}{1-\theta}, C\right\}$$

are satisfied then the corresponding scheme (2.3.46) with the numerical flux $g_{j+1/2}$ as defined in (2.3.47) is TVD.

Proof We have

$$q_{j+\frac{1}{2}} = Q_\varepsilon(\nu_{j+\frac{1}{2}}) = \begin{cases} \varepsilon & \text{if } |\nu_{j+\frac{1}{2}}| < \varepsilon, \\ |\nu_{j+\frac{1}{2}}| & \text{if } |\nu_{j+\frac{1}{2}}| \geq \varepsilon. \end{cases}$$

But in any case this expression is less than $\min\{1/(1-\theta), C\}$ by assumption and we can apply Theorem 2.3.24. □

REMARK 2.3.26 *Consider the following numerical flux*

$$g(u,v) = \frac{1}{2}[f(u) + f(v)] - \frac{1}{2\lambda}q(u,v)(v-u),$$

where $q(u,v) = |\lambda \frac{f(u)-f(v)}{u-v}|$. Then for sufficiently small Δt, q satisfies the conditions of Theorem 2.3.24 and therefore the corresponding scheme is TVD. But in general it will not approximate the entropy solution, as we shall see now. For f we assume that $f(u) = f(-u)$, $f'(0) = 0$ and $f''(0) > 0$.

2.3 Convergence of finite difference schemes of first order in 1-D

Now we use this numerical flux g to get an approximation for the following Riemann problem:

$$\partial_t u + \partial_x f(u) = 0 \quad \text{in } \mathbb{R} \times \mathbb{R}^+,$$
$$u(x,0) = \begin{cases} -1 & \text{if } x < 0, \\ 1 & \text{if } x \geq 0 \end{cases} \quad \text{in } \mathbb{R}.$$

Because of these special initial values, we always have $g(u_j, u_{j+1}) - g(u_{j-1}, u_j) = 0$. Consequently it will approximate the wrong function. The reason for this behaviour is a lack of numerical viscosity. As we have seen in Lemma 2.2.10, any scheme in conservation form can be considered as an approximation of at least second order of

$$\partial_t u + \partial_x f(u) = \Delta x \partial_x (b(u) \partial_x u),$$

where

$$b(u) = \tfrac{1}{2}[\partial_1 g(u,u) - \partial_2 g(u,u) - \lambda f'(u)^2].$$

Formally $b(u)$ has the physical meaning of a viscosity or diffusion. In this case we obtain for $b(u)$ if q is differentiable

$$b(u) = \tfrac{1}{4}\left[f'(u) + \tfrac{1}{\lambda}q(u,u) - f'(u) + \tfrac{1}{\lambda}q(u,u) - 2\lambda f'(u)^2\right]$$
$$= \frac{1}{2\lambda}q(u,u) - \frac{\lambda}{2}f'(u)^2 = \tfrac{1}{2}|f'(u)| - \frac{\lambda}{2}f'(u)^2.$$

For $u = 0$ the term $b(u)$ disappears and this means that in this case, there is not enough numerical viscosity to force the approximating solution to converge to the entropy solution. In practice, this problem can be solved as follows. Let

$$\nu(u,v) := \lambda \frac{f(u) - f(v)}{u - v} \quad \text{and} \quad q(u,v) = Q(\nu(u,v)),$$

where Q is defined by

$$Q(x) = \begin{cases} \dfrac{x^2}{4\varepsilon} + \varepsilon & \text{if } |x| < 2\varepsilon, \\ |x| & \text{if } |x| \geq 2\varepsilon. \end{cases}$$

84   Initial value problems for scalar conservation laws in 1-D

REMARK 2.3.27 *A numerical scheme in conservation form is called an E-scheme, if the numerical flux satisfies*

$$\text{sign}(u_{i+1} - u_i)[g(u_i, u_{i+1}) - f(u)] \leq 0$$

for all u between u_i and u_{i+1}. It can be shown that E-schemes have the TVD property and if one assumes a slightly stronger (CFL-like) condition, E-schemes will approximate the entropy solution. In particular, monotone schemes are E-schemes (see Tadmor [233] and Osher [173, Theorem 4.1]).

REMARK 2.3.28 *In [92] it is shown that the consistency for monotone schemes is of order one at most.*

REMARK 2.3.29 *In Sanders [208] the case of nonuniform meshes is treated. Sanders has proved that monotone schemes will also converge on a nonuniform spatial mesh. Additionally, the L^1 rate of convergence is estimated. The main idea of the proof is based on the Kuznetsov theory [124]. We shall use similar ideas in Section 3.4, where we shall prove error estimates for finite-volume schemes on triangular meshes.*

REMARK 2.3.30 *In general, discrete solutions given by a TVD scheme will not approximate the entropy solution (see Remark 2.3.26). For special schemes this has been done in [176] and also in [136].*

REMARK 2.3.31 *The existence problem for conservation laws in 1-D in Besov spaces is studied in [52].*

REMARK 2.3.32 *In [246] and [247] it is shown that for scalar conservation laws with convex flux function any sequence of approximate solutions, constructed by large-time-step schemes converges to the unique entropy solution if the Courant number is less than one. For special cases the Courant number can be chosen slightly larger than one.*

REMARK 2.3.33 *In [149] error estimates for the large-time-step method of Leveque have been derived.*

2.3 Convergence of finite difference schemes of first order in 1-D

REMARK 2.3.34 (Comparison of explicit and implicit schemes)
From (2.3.28) we obtain for explicit schemes (i.e. $\Theta = 0$, (2.3.28)) a stability condition

$$\frac{\Delta t}{\Delta x} L \leq \frac{1}{2} .$$

Actually this is a strong restriction on the size of the time step Δt. On the other hand, there is no restriction for implicit schemes (i.e. $\Theta = 1$, (2.3.28)). This means that even for arbitrary values of

$$\frac{\Delta t}{\Delta x} L$$

one will still get convergence. But in this case the solution will be less accurate. Even for the same value of $(\Delta t / \Delta x) L$ as in the explicit case the accuracy of the implicit schemes is bad. In order to see this, let us consider the simple linear transport equation in 1-D

$$\partial_t u + \partial_x u = 0 \quad \text{in } \mathbb{R} \times \mathbb{R}^+ , \tag{2.3.53}$$

with initial values

$$u(x, 0) = \begin{cases} 1 & \text{for } x < 0 , \\ 0 & \text{for } x \geq 0 . \end{cases} \tag{2.3.54}$$

For this problem the Engquist–Osher scheme reduces to

$$u_j^{n+1} = u_j^n - \frac{\Delta t}{\Delta x}(u_j^n - u_{j-1}^n) \tag{2.3.55}$$

in the explicit case and to

$$u_j^{n+1} = u_j^n - \frac{\Delta t}{\Delta x}(u_j^{n+1} - u_{j-1}^{n+1}) \tag{2.3.56}$$

in the implicit case. The equation (2.3.56) is equivalent to

$$u_j^{n+1} = \frac{1}{1+\lambda}(u_j^n + \lambda u_{j-1}^{n+1}) \tag{2.3.57}$$

86 Initial value problems for scalar conservation laws in 1-D

with $\lambda := \Delta t/\Delta x$. For the explicit case the CFL condition (see Remark 2.3.22)

$$\frac{\Delta t}{\Delta x} < 1$$

has to be satisfied, while there is no restriction in the implicit case. Now we have done some numerical experiments with (2.3.55) and (2.3.56) respectively. We have computed the discrete solutions u_h for (2.3.55) and (2.3.56) and have compared it with the exact solution u of the initial value problem (2.3.53) and (2.3.54). In Table 2.3.1 we have plotted the L^1-error

$$\|u - u_h\|_{L^1(]a,b[\times]0,T[)}$$

for different values of Δx and $\lambda := \Delta t/\Delta x$. All values refer to the time that is given by the implicit scheme with $\lambda = 10$ after one time step.

Δx	Expl. $\lambda = 0.9$	Impl. $\lambda = 0.9$	Impl. $\lambda = 2$	Impl. $\lambda = 10$	T
0.02	0.015341851	0.068472305	0.085722822	0.154217315	0.2
0.01	0.007670925	0.034236152	0.042861411	0.077108657	0.1
0.008	0.006136740	0.027388922	0.034289128	0.061686926	0.08
0.006	0.004602555	0.020541691	0.025716846	0.046265194	0.06
0.004	0.003068370	0.013694461	0.017144564	0.030843463	0.04
0.002	0.001534185	0.006847230	0.008572282	0.015421731	0.02
0.001	0.000767092	0.003423615	0.004286141	0.007710865	0.01
0.0001	0.000076709	0.000342361	0.000428614	0.000771086	0.001
0.0001	0.000251398	0.001098095	0.001379292	0.002624131	0.01

Table 2.3.1

It turns out that the explicit scheme with $\lambda = 0.9$ is much more accurate than the implicit one with the same value for λ. The results for the implicit scheme become worse if λ is increasing. In Figure 2.3.4 we have plotted the result for the explicit scheme with $\lambda = 0.9$ and in Figures 2.3.5 and 2.3.6 for the implicit schemes with $\lambda = 0.9$ and $\lambda = 2.0$ respectively.

REMARK 2.3.35 (Numbering of nodes for implicit schemes) *Again let us consider the linear transport equation (2.3.53) with initial values (2.3.54). We should like to point out that the implicit schemes depend*

	Upwind		Downwind		
Δx	L^1 error	CPU/s	L^1 error	CPU/s	Final time
0.02	0.068472	0.012	0.084472	0.271	0.2
0.01	0.034236	0.024	0.042235	0.557	0.1
0.008	0.027389	0.031	0.033789	0.691	0.08
0.006	0.0020542	0.042	0.025342	0.928	0.06
0.004	0.013694	0.140	0.016894	1.520	0.04
0.002	0.006847	0.540	0.008447	3.400	0.02
0.001	0.003424	1.400	0.004224	6.700	0.01
0.0001	0.000343	15.600	0.000422	67.900	0.001

Table 2.3.2 Implicit with $\lambda = 0.9$.

strongly on the numbering of the nodes of the underlying grid. In the first example the nodes have been numbered downwind ($x_j = j\Delta x$) and in the second one upwind ($x_j = (N-j)\Delta x$). In the downwind case the implicit scheme (2.3.56) reduces to (2.3.57) and we obtain the solution after one iteration step. In the upwind case one has to iterate much more in order to get the same L^1 error. The results for the L^1 error and the CPU-time for the downwind and upwind cases for different $\lambda := (\Delta t/\Delta x)$ and different Δx are shown in Tables 2.3.2 – 2.3.4. For comparison the corresponding results for an explicit scheme are given in Table 2.3.5. It is obvious that this problem also occurs for nonlinear and multidimensional problems. Concepts for getting optimal numbering of the nodes can be found in [23] and [60]. In [60] ordering effects for the stationary convection–diffusion equation

$$\Delta u + c\nabla u = f$$

on structured grids are investigated theoretically. Bey and Wittum [23] propose a new numbering strategy for $-\varepsilon\Delta u + c\nabla u = f$ with curl$c = 0$ on unstructured grids. Numerical experiments show that this algorithm has a robust behavior in a multigrid iteration.

2.4 Stability and consistency for linear initial value problems

In this section we should like to consider the stability of numerical schemes, which approximate solutions of linear initial value problems [221]. In partic-

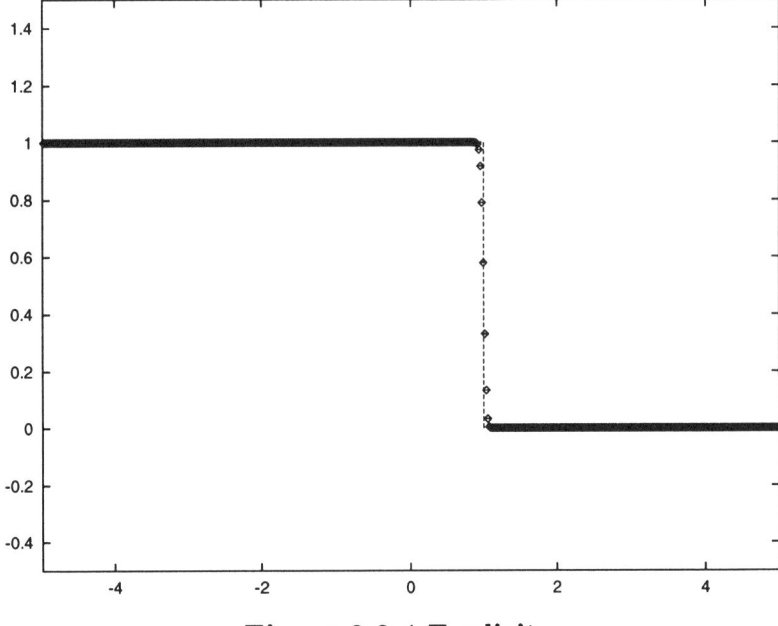

Figure 2.3.4 Explicit.

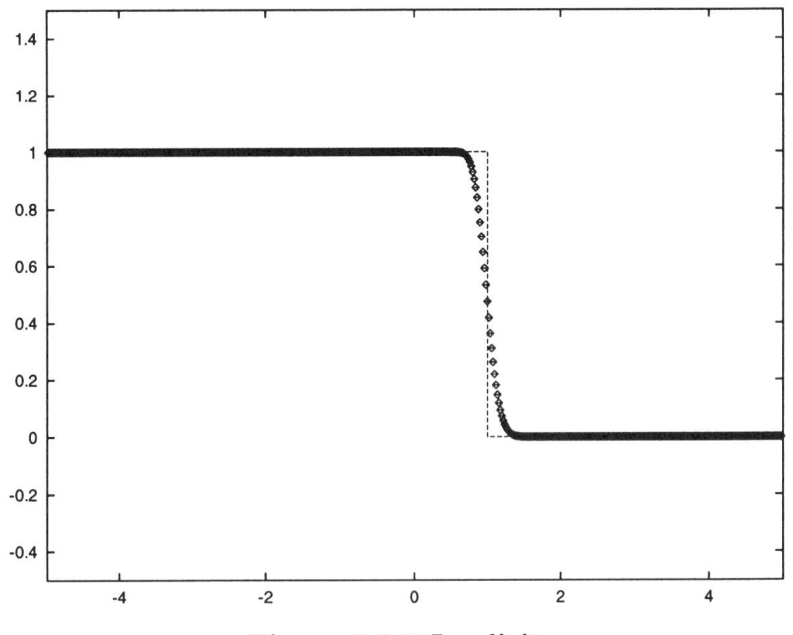

Figure 2.3.5 Implicit.

2.4 Stability and consistency for linear initial value problems

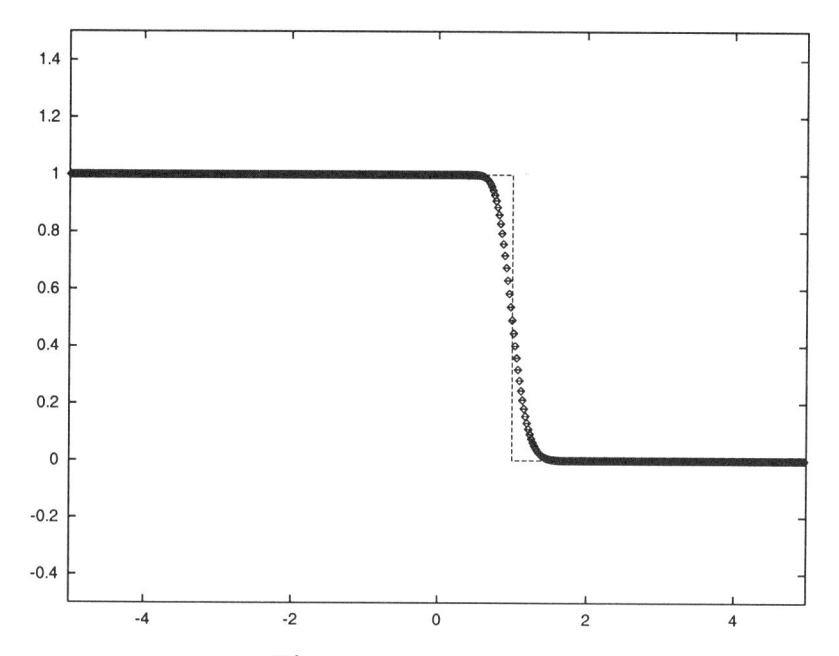

Figure 2.3.6 Implicit.

	Upwind		Downwind		
Δx	L^1 error	CPU/s	L^1 error	CPU/s	Final time
0.02	0.085723	0.006	0.085723	0.255	0.2
0.01	0.042861	0.011	0.042861	0.505	0.1
0.008	0.034289	0.014	0.034289	0.638	0.08
0.006	0.025717	0.031	0.025717	0.851	0.06
0.004	0.017145	0.031	0.017145	1.385	0.04
0.002	0.008572	0.160	0.008572	3.110	0.02
0.001	0.004286	0.570	0.004286	6.190	0.01
0.0001	0.000429	7.400	0.000429	62.000	0.001

Table 2.3.3 Implicit with $\lambda = 2$.

	Upwind		Downwind		
Δx	L^1 error	CPU/s	L^1 error	CPU/s	Final time
0.02	0.154217	0.0017	0.154217	0.246	0.2
0.01	0.077109	0.0033	0.077109	0.500	0.1
0.008	0.061687	0.0042	0.061687	0.632	0.08
0.006	0.046265	0.0050	0.046265	0.840	0.06
0.004	0.030843	0.0080	0.030843	1.379	0.04
0.002	0.015421	0.0190	0.015421	3.100	0.02
0.001	0.007711	0.0380	0.007711	6.190	0.01
0.0001	0.000771	1.4600	0.000771	61.900	0.001

Table 2.3.4 Implicit with $\lambda = 10$.

	Downwind/Upwind		
Δx	L^1 error	CPU/s	Final time
0.02	0.015342	0.017	0.2
0.01	0.007671	0.036	0.1
0.008	0.006137	0.045	0.08
0.006	0.004603	0.061	0.06
0.004	0.001534	0.220	0.04
0.001	0.000767	0.441	0.01
0.0001	0.000077	4.400	0.001

Table 2.3.5 Explicit with $\lambda = 0.9$.

2.4 Stability and consistency for linear initial value problems

ular, we derive the Lax theorem, which says that stability and consistency imply convergence in the case of linear equations. Furthermore, it turns out that the CFL condition as derived in §2.2 also implies stability in the sense as defined in this section. Let us consider

$$\partial_t u - P(\partial_x)u = 0 \quad \text{on } \mathbb{R} \times]0, T[,$$
$$u(x,0) = u_0(x) \quad \text{on } \mathbb{R} , \quad (2.4.1)$$

where P is a polynomial in ∂_x. Let us assume that the approximation of (2.4.1) is given by

$$L(S_+, S_-)u^{n+1} := Q(S_+, S_-)u^n , \quad u^0 = u_0 . \quad (2.4.2)$$

Here we have used the following notation:

$$S_+ f_j := f_{j+1} , \quad S_- f_j := f_{j-1} ,$$

and Q and L are a polynomials in S_+ and S_-. Notice that $I = S_+ S_-$.

EXAMPLE 2.4.1 Consider $\partial_t u + a \partial_x u = 0$ and the numerical scheme

$$u_j^{n+1} := u_j^n - a \frac{\Delta t}{\Delta x}(u_j^n - u_{j-1}^n) .$$

Then we have $P(\partial_x) = a\partial_x$ and

$$L(S_+, S_-) = I , \quad Q(S_+, S_-) = \left(1 - a\frac{\Delta t}{\Delta x}\right)I + a\frac{\Delta t}{\Delta x}S_- .$$

EXAMPLE 2.4.2 (Unstable schemes) Let us consider the exact solution u of

$$\partial_t u + \partial_x u = 0 \quad \text{in } \mathbb{R} \times \mathbb{R}^+ ,$$

$$u(x,0) = u_0(x) = \begin{cases} 1 & \text{if } x < 0 , \\ 0 & \text{if } x > 0 . \end{cases}$$

Now we use the Engquist–Osher scheme

$$u_j^{n+1} = u_j^n - \lambda(u_j^n - u_{j-1}^n) \text{ for } n \geq 0 ,$$

$$u_j^0 = \begin{cases} 1 & \text{if } x \leq 0, \\ 0 & \text{if } x > 0, \end{cases}$$

in order to compute a discrete solution. It turns out that we have for all $n \in \mathbb{N}$

$$u_n^n = \lambda^n ,$$
$$u_j^n = 0 \quad \text{for } j \geq n+1 . \tag{2.4.3}$$

This can be seen by induction. The statement (2.4.3) is true for $n = 0$ by definition. Let us assume that it holds for n. Then

$$u_{n+1}^{n+1} = u_{n+1}^n(1 - \lambda) + \lambda u_n^n$$
$$= \lambda \lambda^n = \lambda^{n+1} ,$$

and for $j \geq n+2$

$$u_j^{n+1} = u_j^n(1 - \lambda) + \lambda u_{j-1}^n = 0 .$$

This proves the statement (2.4.3). The exact solution is given by

$$u(x,t) = u_0(x - t) ,$$

i.e. u is just a shift of the initial value and stays between the values 0 and 1. But if $\lambda > 1$, (2.4.3) indicates that u_n^n can become large and unstable. Another example for an unstable scheme is given in Example 2.1.20 (b). Therefore it is necessary to introduce a stability condition.

DEFINITION 2.4.3 (Stability) *Let T be fixed. The numerical scheme (2.4.2) is called* stable *with respect to the norm $\|\cdot\|$ if there exist constants $c(T)$, β and τ such that*

$$\|u^n\| \leq c(T) e^{\beta n \Delta t / T} \|u^0\| \text{ for all } 0 \leq \Delta t < \tau \text{ and } n \leq \tfrac{T}{\Delta t} . \tag{2.4.4}$$

2.4 Stability and consistency for linear initial value problems

DEFINITION 2.4.4 (Truncation error, consistency) *We shall say that the numerical scheme (2.4.2) is consistent of order (q,p) with respect to the norm $\|\cdot\|$ if we have for any smooth solution v of the differential equation in (2.4.1)*

$$\|v(\cdot,t^{n+1}) - L^{-1}Qv(\cdot,t^n)\| = \mathcal{O}(\Delta t^{q+1}) + \mathcal{O}(\Delta x^p \Delta t).$$

DEFINITION 2.4.5 (Order of convergence) *The numerical scheme (2.4.2) is convergent of order (q,p) with respect to the norm $\|\cdot\|$ if*

$$\|v(\cdot,t^n) - u^n\| = \mathcal{O}(\Delta t^q) + \mathcal{O}(\Delta x^p) \quad \text{uniformly for all } n \in \mathbb{N},$$

where v is an exact solution of (2.4.1).

Now we can formulate the theorem of Lax, which gives us sufficient conditions for convergence.

THEOREM 2.4.6 (Lax theorem) *Let us assume that (2.4.1) and (2.4.2) are linear in u, that (2.4.2) is stable and consistent of order (q,p), and $\|u^0 - v(\cdot,0)\| = \mathcal{O}(\Delta x^p)$. Then (2.4.2) is convergent of order (q,p), i.e.*

$$\|u^n - v(t^n)\| = \mathcal{O}(\Delta x^p) + \mathcal{O}(\Delta t^q) \quad \text{uniformly for all } n \leq \tfrac{T}{\Delta t}.$$

Proof Let $t := n\Delta t \leq T$. The definition of Q and the stability of the scheme imply that

$$\|u^n\| = \|(L^{-1}Q)^n u^0\| \leq ce^{\beta n \Delta t/T}\|u^0\|.$$

Since the scheme is consistent of order (q,p), we obtain for the exact solution v

$$v^n := v(t^n) = L^{-1}Qv^{n-1} + \Delta t R,$$

where $\|R\| = \mathcal{O}(\Delta x^p) + \mathcal{O}(\Delta t^q)$. Then for $w^n := u^n - v^n$ we get

$$\begin{aligned}
w^n = u^n - v^n &= L^{-1}Q(u^{n-1}) - L^{-1}Q(v^{n-1}) - \Delta t R \\
&= L^{-1}Q(w^{n-1}) - \Delta t R \\
&= (L^{-1}Q)^2(w^{n-2}) - \Delta t L^{-1}QR - \Delta t R \\
&= (L^{-1}Q)^n(w^0) - \Delta t \sum_{j=0}^{n-1}(L^{-1}Q)^j R,
\end{aligned}$$

and therefore, since $w^0 = \mathcal{O}(\Delta x^p)$,

$$\|w^n\| \leq \Delta t \sum_{j=0}^{n-1} \|(L^{-1}Q)^j R\| + \mathcal{O}(\Delta x^p)$$

$$\leq \Delta t \|R\| \sum_{j=0}^{n-1} ce^{\beta j \Delta t/T} + \mathcal{O}(\Delta x^p)$$

$$\leq \Delta t \|R\| nce^{\beta n \Delta t/T} + \mathcal{O}(\Delta x^p) = \mathcal{O}(\Delta x^p) + \mathcal{O}(\Delta t^q) \ ,$$

where the constants may depend on T. This proves the convergence of the scheme. □

Now we should like to give a sufficient condition for stability. Before we can do this, we define:

$$\rho(\xi) := Q(e^{-i\xi}, e^{i\xi}) \tag{2.4.5}$$

for all $\xi \in \mathbb{R}$. The function ρ is called the symbol of Q. Then we are able to prove the following theorem.

THEOREM 2.4.7 (Stability condition) *A finite difference scheme $u^{n+1} = Q(S_+, S_-)u^n$ is stable with respect to the L^2 norm if and only if there exist constants C and τ such that*

$$|\rho(\xi)| \leq 1 + C\Delta t \tag{2.4.6}$$

for all $\xi \in \mathbb{R}$ and all $0 \leq \Delta t < \tau$.

Proof (see [221]) For this we need the discrete Fourier transform of $u := (u_j)_{j \in \mathbb{N}}$. It is defined as

$$\hat{u}(\xi) := \sum_j u_j e^{ij\xi}$$

for all $0 \leq \xi \leq 2\pi$ ($i := \sqrt{-1}$). Then it can be easily proved that

$$\widehat{S_+ u}(\xi) = e^{-i\xi} \hat{u}(\xi) \ , \quad \widehat{S_- u}(\xi) = e^{+i\xi} \hat{u}(\xi) \ .$$

2.4 Stability and consistency for linear initial value problems

This implies that

$$\hat{u}^{n+1} = \rho(\xi)\hat{u}^n .$$

Furthermore, we need the discrete L^2 norm:

$$\|u^n\|_2^2 := h \sum_j (u_j^n)^2 .$$

Since the $(e^{ij\xi})_{j \in \mathbb{N}}$ are an orthogonal system, we have the Parseval property

$$\|\hat{u}\|_2^2 = \int_0^{2\pi} |\hat{u}(\xi)|^2 \, d\xi = \int_0^{2\pi} \left| \sum_j u_j e^{ij\xi} \right|^2 d\xi$$

$$= 2\pi \sum_j u_j^2 = \frac{2\pi}{h} \|u\|_2^2 . \tag{2.4.7}$$

Let us assume that $|\rho(\xi)| \leq 1 + c\Delta t$. Then, using (2.4.7), we have

$$\|u^{n+1}\|_2^2 = \frac{h}{2\pi} \int_0^{2\pi} |\hat{u}^{n+1}(\xi)|^2 \, d\xi = \frac{h}{2\pi} \int_0^{2\pi} |\rho(\xi)|^2 \, |\hat{u}^n(\xi)|^2 \, d\xi$$

$$\leq \frac{h}{2\pi}(1 + c\Delta t)^2 \int_0^{2\pi} |\hat{u}^n(\xi)|^2 \, d\xi$$

$$\leq (1 + c\Delta t)^2 \|u^n\|_2^2 \leq (e^{c\Delta t})^2 \|u^n\|_2^2$$

$$\leq (e^{c\Delta t})^{2n} \|u^0\|_2^2 .$$

But this means that the scheme is stable.

Now we shall show that the scheme cannot be stable if the inequality (2.4.6) is not satisfied. Then for any $c > 0$ and $\tau > 0$ there exist a ξ and $0 \leq \Delta t < \tau$ such that

$$|\rho(\xi)| > 1 + c\Delta t , \tag{2.4.8}$$

and, in particular, there will be an interval I_c such that (2.4.8) holds for all $\xi \in I_c$. We consider the scheme $u^n = Qu^{n-1}$ for $1 \le n \le T/\Delta t$ with respect to initial values $(u_j^0)_j$ such that $\hat{u}^0 = 0$ on $\mathbb{R} \setminus I_c$. Thus

$$|\hat{u}^n(\xi)| = |\rho^n(\xi)\hat{u}^0(\xi)| > (1 + c\Delta t)^n |\hat{u}^0(\xi)| \quad \text{on } I_c ,$$

and, because of the Parseval property,

$$\|Q^n\| \|u_0\|_2 \ge \|Q^n u^0\|_2 = \|u^n\|_2 > (1 + c\Delta t)^n \|u^0\|_2 .$$

This means that

$$\|Q^n\| \ge (1 + c\Delta t)^n .$$

Now we assume that the scheme is stable,

$$\|Q^n\| \le C_0$$

for all $1 \le n \le T/\Delta t$. Therefore we have that for all c and all τ there exists a $\Delta t \in]0, \tau[$ such that for all $0 \le n \le T/\Delta t$

$$(1 + c\Delta t)^n \le \|Q^n\| \le C_0 .$$

Now let us choose $n = T/\Delta t$. Then we obtain

$$\left(1 + c\frac{T}{n}\right)^n \le C_0 .$$

Choose c sufficiently large such that

$$\frac{e^{cT}}{2} > C_0 .$$

Then choose τ sufficiently small such that for $n = T/\Delta t$, $\Delta t \in]0, \tau[$, $\left(1 + \frac{cT}{n}\right)^n > \frac{2}{3} e^{cT}$. But then we obtain a contradiction since

$$\frac{e^{cT}}{2} > \left(1 + \frac{cT}{n}\right)^n \ge \frac{2}{3} e^{cT} .$$

□

If one wants to have (see (2.4.4))

$$\|u^n\| \le c\|u^0\|$$

for all n then this condition is equivalent to

$$|\rho(\xi)| \le 1 \qquad (2.4.9)$$

for all ξ.

EXAMPLE 2.4.8 (Stability for the linear transport equation) Let us consider the linear transport equation $\partial_t u + a \partial_x u = 0$ and the following scheme ($\lambda := \Delta t / \Delta x$):

$$u_j^{n+1} = u_j^n - a\lambda[\Theta u_{j+1}^n + (1 - 2\Theta)u_j^n - (1 - \Theta)u_{j-1}^n] \qquad (2.4.10)$$

for some $\Theta \in [0,1]$. For $\Theta = 0$, $\Theta = \frac{1}{2}$ and $\Theta = 1$ we get the backward, central and forward difference respectively. In order to analyse the stability of this scheme, we write it in the form

$$\begin{aligned} u_j^{n+1} &= [(1 - a\lambda(1 - 2\Theta))I - a\lambda\Theta S_+ + a\lambda(1 - \Theta)S_-]u_j^n \\ &=: Q(S_+, S_-)u_j^n \ . \end{aligned}$$

Then for the symbol ρ of Q we get

$$\begin{aligned} \rho(\xi) &= 1 - a\lambda(1 - 2\Theta) - a\lambda\Theta e^{-i\xi} + a\lambda(1 - \Theta)e^{+i\xi} \\ &= [1 - a\lambda(1 - 2\Theta) + a\lambda(1 - 2\Theta)\cos\xi] + i(a\lambda \sin\xi) \end{aligned}$$

and

$$|\rho(\xi)|^2 = 1 - 2\lambda a(1 - \cos\xi)\{1 - 2\Theta - \lambda a[1 + 2\Theta(\Theta - 1)(1 - \cos\xi)]\} \ .$$

Let us assume that $a > 0$. Then it can be seen very easily that for $\Theta = 1$ (forward difference) and $\Theta = \frac{1}{2}$ (central difference) the scheme cannot be stable in the sense of (2.4.9). But for $\Theta = 0$ (backward difference) we have

$$|\rho(\xi)|^2 = 1 - 2\lambda a(1 - \cos\xi)(1 - \lambda a) \ .$$

Then the scheme is stable if $1 - \lambda a > 0$. This is just the well-known Courant–Friedrichs–Lewy (CFL) condition.

98   Initial value problems for scalar conservation laws in 1-D

2.5 Finite difference schemes of higher order for scalar equations in one space dimension

The main objection of this section is the construction of higher-order schemes for the conservation law

$$\partial_t u + \partial_x f(u) = 0 \ . \tag{2.5.1}$$

In the preceding sections we have seen that it is necessary to adapt the discretization to the flow direction. For the discretization of

$$\partial_t u + a \partial_x u = 0 \tag{2.5.2}$$

with $a > 0$, the backward difference quotient is stable while the forward and the central ones are unstable. These remarks refer to first-order discretizations. The situation for higher-order discretizations is much more complicated. Unfortunately the one-sided discretizations of higher order are also unstable. The same is true for the simplest scheme that is of second order in space and time. This is the so called Lax–Wendroff scheme, and will be constructed in the first part of this section. Then we develop the concept of higher-order schemes with limiters. For the linear transport equation (2.5.2) we make an ansatz and derive conditions for the free parameters under the constraint that the scheme should have the order of consistency two and it should be TVD (total variation diminishing). For the general nonlinear conservation law

$$\partial_t u + \partial_x f(u) = 0$$

we present the method of variable extrapolation. For proving the convergence of the discrete solutions that one get by this method, we refer to the corresponding results in the literature. We shall not discuss the proof in this section, since we shall prove a more general result for finite volume schemes on unstructured grids.

In order to get a first idea for constructing higher-order schemes we take a smooth solution u of (2.5.1) and expand it in its Taylor series around $(j\Delta x, n\Delta t)$. Using the notation $u_j^n := u(j\Delta x, n\Delta t)$, we obtain

$$u_j^{n+1} = u_j^n + \Delta t (\partial_t u)_j^n + \frac{\Delta t^2}{2} (\partial_t^2 u)_j^n + \mathcal{O}(\Delta t^3) \ . \tag{2.5.3}$$

Finite difference schemes of higher order for scalar equations in 1-D   99

In this chapter we always assume that a CFL condition is satisfied, i.e. $\Delta t = \lambda \Delta x$ and where λ is a fixed constant. Differentiating the differential equation (2.5.1) with respect to t, we derive

$$\partial_t^2 u = -\partial_t \partial_x f(u) = \partial_x(f'(u)\partial_x f(u)) \,. \tag{2.5.4}$$

Using (2.5.1) and (2.5.4) in (2.5.3), we obtain ($\lambda := \Delta t/\Delta x$)

$$\begin{aligned}
u_j^{n+1} &= u_j^n - \Delta t (\partial_x f(u))_j^n + \frac{\Delta t^2}{2}\partial_x(f'(u)\partial_x f(u))_j^n + \mathcal{O}(\Delta t^3) \\
&= u_j^n - \Delta t \left[\frac{f(u_{j+1}) - f(u_{j-1})}{2\Delta x} + \mathcal{O}(\Delta x^2) \right]_1 \\
&\quad + \frac{\Delta t^2}{2}\left[\frac{f'(u_{j+\frac{1}{2}})\partial_x f(u_{j+\frac{1}{2}}) - f'(u_{j-\frac{1}{2}})\partial_x f(u_{j-\frac{1}{2}})}{\Delta x} + \mathcal{O}(\Delta x^2) \right] \\
&\quad + \mathcal{O}(\Delta t^3) \\
&= u_j^n - \Delta t [\,]_1 + \frac{\Delta t^2}{2}\left[f'(u_{j+\frac{1}{2}})^2 \frac{u_{j+1} - u_j}{\Delta x^2} \right. \\
&\quad \left. - f'(u_{j-\frac{1}{2}})^2 \frac{u_j - u_{j-1}}{\Delta x^2} + \mathcal{O}(\Delta x^2) \right] + \mathcal{O}(\Delta t^3) \\
&= u_j^n - \frac{\lambda}{2}[f(u_{j+1}) - f(u_{j-1})] \\
&\quad + \frac{\lambda^2}{2}[f'(u_{j+\frac{1}{2}})^2(u_{j+1} - u_j) - f'(u_{j-\frac{1}{2}})^2(u_j - u_{j-1})] \\
&\quad + \mathcal{O}(\Delta x^3) \,. \tag{2.5.5}
\end{aligned}$$

If $f' = \text{const} = c$ then

$$u_j^{n+1} = u_j^n - \frac{\lambda c}{2}(u_{j+1} - u_{j-1}) + \frac{\lambda^2 c^2}{2}(u_{j+1} - 2u_j + u_{j-1}) \,.$$

Now we should like to express the identity (2.5.5) in the form

$$u_j^{n+1} = u_j^n - \frac{\Delta t}{\Delta x}[g(u_j^n, u_{j+1}^n) - g(u_{j-1}^n, u_j^n)] \,. \tag{2.5.6}$$

This can be done by using

$$\frac{f_{j+1} - f_j}{u_{j+1} - u_j} = f'(u_{j+\frac{1}{2}}) + \mathcal{O}(\Delta x^2) \,, \tag{2.5.7}$$

100   Initial value problems for scalar conservation laws in 1-D

and therefore

$$f'(u_{j+\frac{1}{2}})^2(u_{j+1} - u_j) = \frac{(f_{j+1} - f_j)^2}{u_{j+1} - u_j} + \mathcal{O}(\Delta x^3).$$

Using the notation

$$\Delta_+ u_j = u_{j+1} - u_j, \quad f_j := f(u_j), \quad \nu_{j+\frac{1}{2}} = \lambda \frac{f_{j+1} - f_j}{\Delta_+ u_j},$$

the numerical flux $g(u_j, u_{j+1})$ in (2.5.6) can be written as

$$g(u_j, u_{j+1}) := \frac{1}{2}\left[f_{j+1} + f_j - \frac{1}{\lambda}(\nu_{j+\frac{1}{2}})^2 \Delta_+ u_j\right]. \tag{2.5.8}$$

The numerical scheme (2.5.6) with the numerical flux (2.5.8) is called the Lax–Wendroff scheme. In (2.5.5) and (2.5.7) we have shown that the truncation error is of second order:

LEMMA 2.5.1 *The Lax–Wendroff scheme (2.5.6) with the numerical flux as defined in (2.5.8) is consistent of order two.*

Now we should like to analyse the stability of the Lax–Wendroff scheme in the linear case.

LEMMA 2.5.2 (Stability of the Lax–Wendroff scheme in the linear case) *Let us assume that $f'(u) = c$ and $|c\lambda| < 1$. Then the Lax–Wendroff scheme (2.5.8), (2.5.6) is stable in the sense of Definition 2.4.3.*

Proof Let S_+, S_- and ρ be defined as in §2.4. Then we get

$$\begin{aligned}
u_j^{n+1} &= u_j^n - c\frac{\lambda}{2}(S_+ - S_-)u_j^n + c^2\frac{\lambda^2}{2}(S_+ - 2I + S_-)u_j^n \\
&= \left[\left(-c\frac{\lambda}{2} + c^2\frac{\lambda^2}{2}\right)S_+ + \left(c^2\frac{\lambda^2}{2} + c\frac{\lambda}{2}\right)S_- + I(1 - c^2\lambda^2)\right]u_j^n \\
&= \left\{I[1 - (c\lambda)^2] + S_+\frac{c\lambda}{2}(-1 + c\lambda) + S_-\frac{c\lambda}{2}(c\lambda + 1)\right\}u_j^n \\
&=: Q(S_+, S_-)u_j^n. \tag{2.5.9}
\end{aligned}$$

Therefore

$$\rho(e^{-i\xi}, e^{i\xi}) = [1 - (c\lambda)^2] + e^{-i\xi}\frac{c\lambda}{2}(-1 + c\lambda) + e^{i\xi}\frac{c\lambda}{2}(c\lambda + 1)$$
$$= [1 - (c\lambda)^2] - c\lambda i\left(\frac{e^{-i\xi} - e^{i\xi}}{2i}\right) + (c\lambda)^2\left(\frac{e^{-i\xi} + e^{i\xi}}{2}\right)$$
$$= 1 - (c\lambda)^2 + c\lambda i \sin\xi + (c\lambda)^2 \cos\xi \ .$$

In order to show stability, we have to show that $|\rho| < 1$ (see Theorem 2.4.7). Using $|c|\lambda < 1$, we can derive

$$|\rho|^2 = [1 - (c\lambda)^2(1 - \cos\xi)]^2 + (c\lambda)^2 \sin^2\xi$$
$$= 1 - 2(c\lambda)^2(1 - \cos\xi) + (c\lambda)^4(1 - \cos\xi)^2 + (c\lambda)^2(1 - \cos^2\xi)$$
$$= 1 - (c\lambda)^2(2 - 2\cos\xi - 1 + \cos^2\xi) + (c\lambda)^4(1 - \cos\xi)^2$$
$$= 1 - (c\lambda)^2(1 - \cos\xi)^2 + (c\lambda)^4(1 - \cos\xi)^2 < 1 \ . \qquad \square$$

REMARK 2.5.3 *It is known that the standard Lax–Wendroff difference scheme produces approximations that in general do not converge to the entropy solution (see [221, page 301]). Majda and Osher [154] have modified this scheme such that it retains the desirable properties but the limit solution satisfies the entropy condition. In [155] it was shown that the Lax–Wendroff scheme is nonlinear unstable for $f(u) = \frac{1}{2}u^2$ at $u = 0$.*

REMARK 2.5.4 *In Lemma 2.5.1 we have shown that the Lax–Wendroff scheme is of second order. We have solved the same test problem as in Example 2.2.6 with the Lax–Wendroff scheme. It turns out that the shock is smeared out over fewer grid points than in the Lax–Friedrichs scheme or even in the Engquist–Osher or Godunov schemes. But the scheme produces oscillations near the discontinuity (see Figures 2.5.1 and 2.5.2). This is due to the fact that the Lax–Wendroff scheme does not have the TVD property.*

Let us consider again the Lax–Wendroff scheme for the linear conservation law

$$\partial_t u + a \partial_x u = 0 \quad \text{in } \mathbb{R} \times \mathbb{R}^+$$

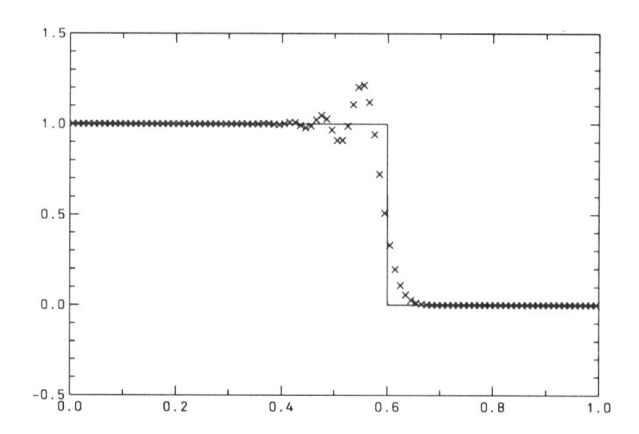

102 Initial value problems for scalar conservation laws in 1-D

Figure 2.5.1

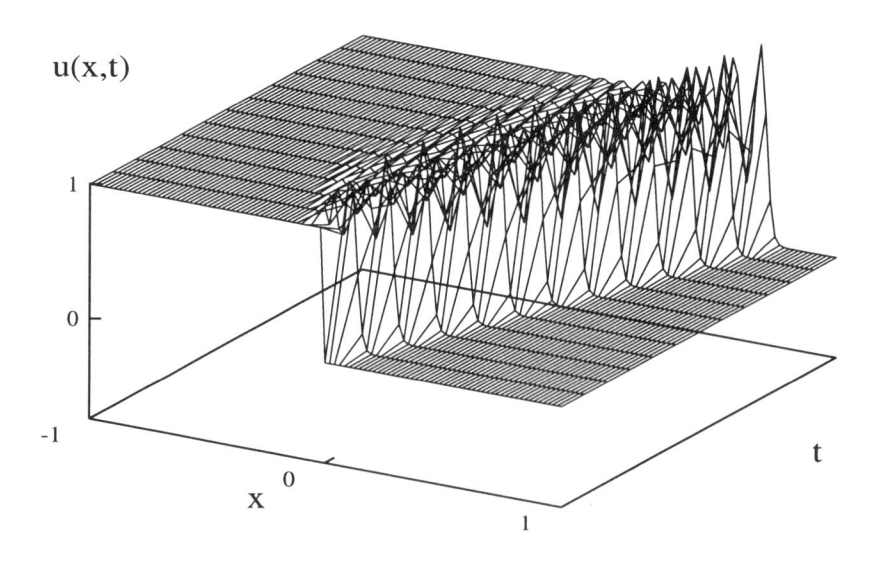

Figure 2.5.2

for some constant $a > 0$. The numerical flux for the Lax–Wendroff scheme is (see 2.5.8))

$$g_{j+\frac{1}{2}} = \frac{a}{2}[u_{j+1} + u_j - \lambda a(u_{j+1} - u_j)],$$

and the Lax–Wendroff scheme looks like $(u_j := u_j^n)$

$$\begin{aligned}
u_j^{n+1} &= u_j - \frac{\lambda a}{2}[u_{j+1} - u_{j-1} - \lambda a(u_{j+1} - 2u_j + u_{j-1})] \\
&= u_j - \lambda a(u_j - u_{j-1}) - \frac{\lambda a}{2}(u_{j+1} - 2u_j + u_{j-1})(1 - \lambda a) \\
&= u_j - \lambda a(u_j - u_{j-1}) - \frac{\lambda a}{2}(1 - \lambda a)\Delta_-\Delta_+ u_j .
\end{aligned} \quad (2.5.10)$$

This means that in this case the Lax–Wendroff method can be written as the sum of a monotone scheme of first order

$$R_1 u_j := u_j - \lambda a(u_j - u_{j-1})$$

and a second-order term

$$R_2 u_j := -\frac{\lambda a}{2}(1 - \lambda a)\Delta_-\Delta_+ u_j .$$

The operator $\Delta_-\Delta_+ u_j$ can be considered as an approximation of $\Delta x^2 \partial_x^2 u$. Since the CFL condition implies that $1 - \lambda a \geq 0$, $R_2 u_j$ has just the opposite sign of a diffusion term. We know already that the Lax–Wendroff scheme resolves the discontinuity only over a few grid points, but it produces oscillations near the discontinuity. On the other hand, monotone schemes resolve the discontinuity over more grid points, but without any oscillation. Therefore Sweby [230] had the idea of using the Lax–Wendroff scheme

$$u_j^{n+1} := R_1 u_j + R_2 u_j$$

in regions where the solution is smooth and to switch off $R_2 u_j$ in regions with strong oscillations

$$u_j^{n+1} := R_1 u_j .$$

The size of the quotient

$$r_j := \frac{\Delta_- u_j}{\Delta_+ u_j}$$

gives us a hint if the solution has strong oscillations. Sweby made the following ansatz (see (2.5.10):

$$u_j^{n+1} = u_j^n - \lambda a \Delta_- u_j - \frac{\lambda a}{2}(1 - \lambda a)\Delta_-(\varphi(r_j)\Delta_+ u_j) \quad (2.5.11)$$

where $\varphi \in C^0(\mathbb{R}^+)$ is the switch function. If $r_j < 0$ then $\Delta_- u_j$ and $\Delta_+ u_j$ have different signs, and this means that the solution oscillates and the switch function should be equal to zero. Then the scheme is monotone and the oscillations are damped out. If $\Delta_- u_j \leq 0$, $|\Delta_- u_j|$ is small and $\Delta_+ u_j \leq 0$, $|\Delta_+ u_j|$ is large then r_j is small and $\varphi(r_j)$ should be small. Thus near the upper part of the discontinuity the scheme is dominated by the monotone part $R_1 u_j$. Now if $\Delta_- u_j \leq 0$ and $|\Delta_- u_j|$ is large and $\Delta_+ u_j \leq 0$ and $|\Delta_+ u_j|$ is small then r_j is large. This occurs near the lower part of the discontinuity. In this case the Lax–Wendroff scheme makes no problem, and we keep the operator R_2. In order to obtain that the shock is resolved only over a few grid points, the switching function should suppress the numerical diffusion that is contained in the monotone part of the scheme (see Example 2.2.6). Since $R_2 u_j$ has the opposite sign to the numerical diffusion, this should be possible for a suitable choice of φ.

Now we shall derive sufficient conditions on φ such that the scheme (2.5.11) is TVD and of second order. To that end, we write the scheme (2.5.11) in a form for which we can apply Lemma 2.3.20. We obtain from (2.5.11) for $\tilde{\lambda} := \frac{1}{2}(1-\lambda a)\lambda a$ and $\Delta_- u_j \neq 0$

$$u_j^{n+1} = u_j^n - \lambda a \Delta_- u_j - \tilde{\lambda}\Delta_-(\varphi(r_j)\Delta_+ u_j)$$
$$= u_j^n - \left[\lambda a + \frac{\tilde{\lambda}\Delta_-(\varphi(r_j)\Delta_+ u_j)}{\Delta_- u_j}\right]\Delta_- u_j$$
$$= u_j^n + C_{j+\frac{1}{2}}\Delta_+ u_j - D_{j-\frac{1}{2}}\Delta_- u_j$$

for

$$C_{j+\frac{1}{2}} = 0 \quad \text{and} \quad D_{j-\frac{1}{2}} = \lambda a + \frac{\tilde{\lambda}\Delta_-(\varphi(r_j)\Delta_+ u_j)}{\Delta_- u_j}.$$

… Finite difference schemes of higher order for scalar equations in 1-D

If
$$0 \le D_{j-\frac{1}{2}} \le 1, \qquad (2.5.12)$$

Lemma 2.3.20 implies that the scheme (2.5.11) is TVD. Therefore we have to derive sufficient conditions for (2.5.12). We obtain

$$\frac{\Delta_-\varphi(r_j)\Delta_+ u_j}{\Delta_- u_j} = \frac{1}{\Delta_- u_j}[\varphi(r_j)(u_{j+1}-u_j) - \varphi(r_{j-1})(u_j - u_{j-1})]$$
$$= \frac{\varphi(r_j)}{r_j} - \varphi(r_{j-1}).$$

Let us assume that for $0 < r_k, r_{k-1}$

$$\left|\frac{\varphi(r_k)}{r_k} - \varphi(r_{k-1})\right| \le M \quad \text{for all } k, \qquad (2.5.13)$$
$$M \le 2 \qquad (2.5.14)$$

and

$$1 - \lambda a \ge 0 \quad \text{(CFL condition)}. \qquad (2.5.15)$$

Then we have for $D_{k-\frac{1}{2}}$

$$\lambda a - \tilde{\lambda}M \le \lambda a + \tilde{\lambda}\left[\frac{\varphi(r_k)}{r_k} - \varphi(r_{k-1})\right] = D_{k-\frac{1}{2}} \le \lambda a + \tilde{\lambda}M. \qquad (2.5.16)$$

Now (2.5.14) and (2.5.15) imply that

$$D_{k-\frac{1}{2}} \le \lambda a + \tilde{\lambda}M \le \lambda a + 2\tilde{\lambda}$$
$$= \lambda a + \lambda a(1 - \lambda a)$$
$$= 1 - (1 - \lambda a)^2 \le 1.$$

Furthermore,

$$D_{k-\frac{1}{2}} \ge \lambda a - \tilde{\lambda}M = \lambda a - \lambda a(1 - \lambda a)$$
$$= (\lambda a)^2 \ge 0.$$

Therefore we have shown th following lemma.

LEMMA 2.5.5 *Let* $1 - \lambda a \geq 0$ *and*

$$\varphi(r) = 0 \quad \text{if } r \leq 0, \tag{2.5.17}$$

$$0 \leq \max\left\{\frac{\varphi(r)}{r}, \varphi(r)\right\} \leq 2 \quad \text{for } 0 \leq r. \tag{2.5.18}$$

Then $D_{j-\frac{1}{2}}$ *satisfies (2.5.12), this means the scheme (2.5.11) has the TVD property. In particular, a sufficient condition for (2.5.18) is*

$$0 \leq \varphi(r) \leq 2r \quad \text{if } 0 < r \leq 1,$$
$$0 \leq \varphi(r) \leq 2 \quad \text{if } 1 \leq r.$$

This means that the graph of φ has to be in the shaded area of Figure 2.5.3.

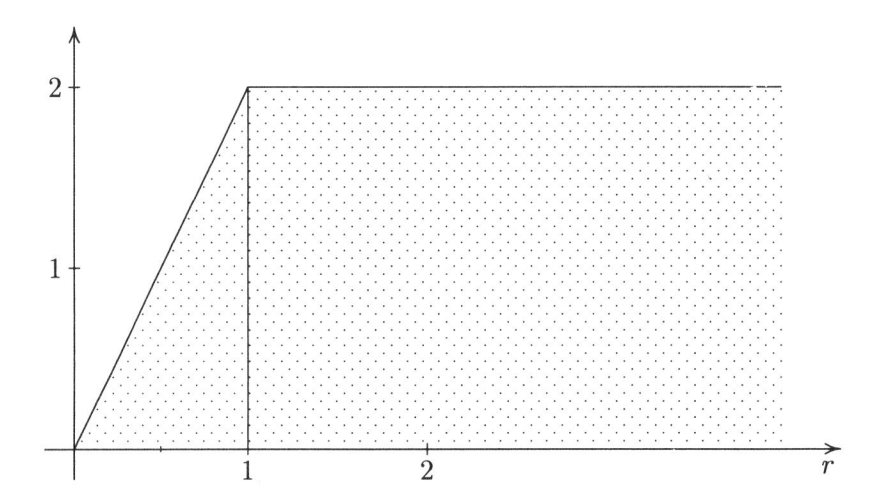

Figure 2.5.3

This condition is sufficient for the TVD property. Now we have to ensure that the scheme (2.5.11) is of second order. For proving this, we consider the following ansatz for φ:

$$\varphi(r) := [1 - \theta(r)]\varphi_{LW}(r) + \theta(r)\varphi_{WB}(r) \tag{2.5.19}$$

where $\varphi_{LW}(r) = 1$, $\varphi_{WB}(r) = r$ and $\theta \in C^{0,1}(\mathbb{R}, [0,1])$. The scheme (2.5.11) with $\varphi = \varphi_{LW}$ is just the Lax–Wendroff scheme, and (2.5.11) with $\varphi = \varphi_{WB}$ is known as the algorithm of Warming and Beam [248] (see Figure 2.5.4).

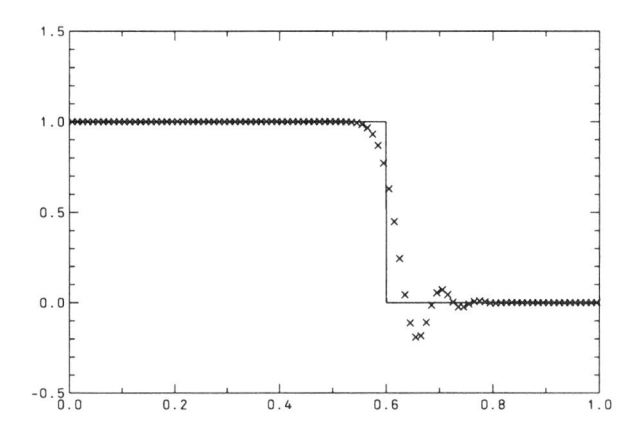

Figure 2.5.4

Although the Lax–Wendroff and the Warming and Beam schemes do not have the TVD property, the ansatz (2.5.19) for $\varphi(r)$ leads to second-order schemes and for special choices of $\theta(r)$ to schemes that have the TVD property.

LEMMA 2.5.6 Let

$$\varphi(r) := 1 - \theta(r) + \theta(r)r \qquad (2.5.20)$$

for $\theta \in C^{0,1}(\mathbb{R}, [0, 1])$. Then the corresponding scheme (2.5.11) is of second order where $u'(x) \neq 0$. A sufficient condition for (2.5.20) is that φ remains in the shaded region of Figure 2.5.5.

Proof Suppose that u is a smooth solution of $\partial_t u + a \partial_x u = 0$. Then we have

$$\begin{aligned}
u_j^{n+1} &- u_j^n + \lambda a \Delta_- u_j + \tilde{\lambda} \Delta_- (\varphi(r_j) \Delta_+ u_j) \\
&= [1 - \theta(r_j)](u_j^{n+1} - u_j^n + \lambda a \Delta_- u_j + \tilde{\lambda} \Delta_- \Delta_+ u_j) \\
&\quad + \theta(r_j)[u_j^{n+1} - u_j^n + \lambda a \Delta_- u_j + \tilde{\lambda} \Delta_- (r_j \Delta_+ u_j)] \\
&\quad + \tilde{\lambda}[\theta(r_j) - \theta(r_{j-1})]\Delta_+ u_{j-1}(r_{j-1} - 1) \\
&= \Delta t \mathcal{O}(\Delta x^2) .
\end{aligned}$$

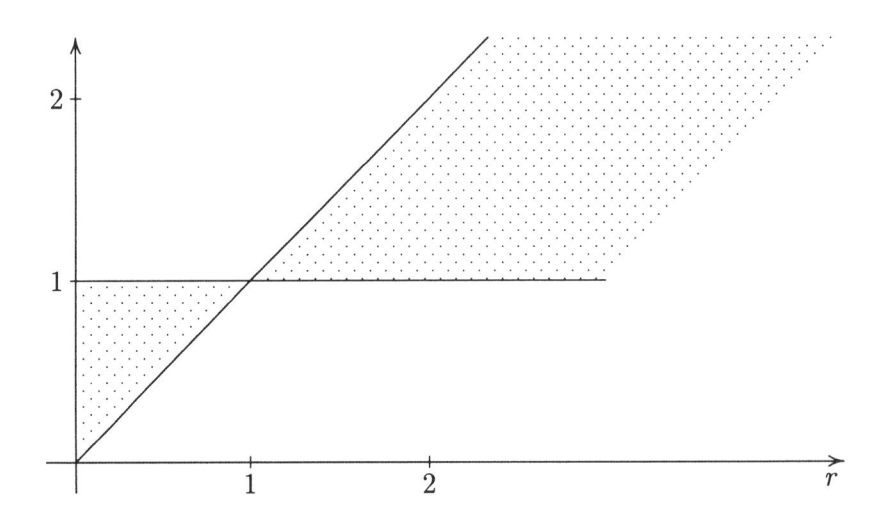

108   Initial value problems for scalar conservation laws in 1-D

Figure 2.5.5

This can be seen as follows. Since the Lax–Wendroff scheme (see Lemma 2.5.1) and the scheme of Warming and Beam (see Lemma 2.5.7) are of second order and since $\Delta_+ u_{j-1}(r_{j-1} - 1) = -\Delta_-\Delta_+ u_{j-1} = \mathcal{O}(\Delta x^2)$, we have

$$|\theta(r_j) - \theta(r_{j-1})| \, |\Delta_+ u_{j-1}(r_{j-1} - 1)|$$
$$\leq C\Delta x^2 |r_j - r_{j-1}|$$
$$= C\Delta x^2 \left| \frac{\Delta_- u_j}{\Delta_+ u_j} - \frac{\Delta_- u_{j-1}}{\Delta_+ u_{j-1}} \right|$$
$$= \frac{C\Delta x^2}{R} \left[\left(\Delta x u'_j - \frac{\Delta x^2}{2} u''_j \right)^2 \right.$$
$$\left. - \left(\Delta x u'_j - \frac{3}{2}\Delta x^2 u''_j \right)\left(\Delta x u'_j + \frac{\Delta x^2}{2} u''_j \right) + \mathcal{O}(\Delta x^4) \right]$$
$$= \frac{\mathcal{O}(\Delta x^3)}{u'^2_j + \mathcal{O}(\Delta x^2)},$$

where

$$R = \left(\Delta x u'_j + \frac{\Delta x^2}{2} u''_j \right)\left(\Delta x u'_j - \frac{\Delta x^2}{2} u''_j \right) + \mathcal{O}(\Delta x^4) \, .$$ □

LEMMA 2.5.7 *The scheme of Warming and Beam (i.e. (2.5.11) with $\varphi(r) := r$) is of second order.*

Proof This is left as an Exercise.

Now putting together Lemmas 2.5.5 and 2.5.6, we obtain a sufficient condition for the scheme (2.5.11) to be TVD and of second order.

THEOREM 2.5.8 *Let $\theta \in C^{0,1}(\mathbb{R}, [0,1])$ and φ be defined as in (2.5.20) for $r \geq 0$, suppose (2.5.15) and that φ satisfies (2.5.17), (2.5.18). Then the corresponding scheme (2.5.11) is TVD and for $r_j > 0$ of second order ($r_j < 0$ means local extrema).*

COROLLARY 2.5.9 *Let us suppose that φ satisfies*

$$\begin{aligned}
\varphi &= 0 & &\text{if } r \leq 0, \\
r &\leq \varphi(r) \leq \min\{2r, 1\} & &\text{if } 0 \leq r \leq 1, \\
1 &\leq \varphi(r) \leq \min\{2, r\} & &\text{if } 1 \leq r.
\end{aligned}$$

Then the corresponding scheme (2.5.11) is TVD and for $r \geq 0$ of second order. This means the graph of φ has to be in the shaded region of Figure 2.5.6.

EXAMPLE 2.5.10 (Limiters) There are several well-known limiter functions φ that satisfy the conditions in Corollary 2.5.9. They are defined as follows for $r > 0$:

$$\varphi_1(r) := \max\{0, \min\{r, 1\}\} \qquad \text{(minmod)},$$

$$\varphi_2(r) := \max\{0, \min\{2r, 1\}, \min\{r, 2\}\} \quad \text{(superbee of Roe)},$$

$$\varphi_3(r) := \frac{|r|+r}{1+|r|} \qquad \text{(van Leer)},$$

$$\varphi_4(r) := \frac{r^2+r}{1+r^2} \qquad \text{(van Albada)},$$

$$\varphi_5(r) := \max\{0, \min\{r, \beta\}\}, \; 1 \leq \beta \leq 2 \text{ (Chakravarthy and Osher)}.$$

110 Initial value problems for scalar conservation laws in 1-D

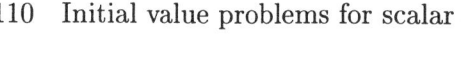

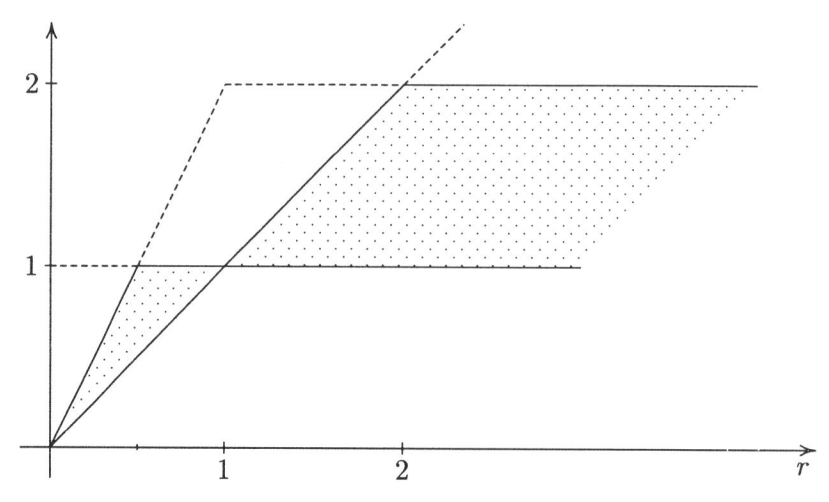

Figure 2.5.6

REMARK 2.5.11 *For generalizing the above ideas to nonlinear conservation laws we refer to [230, Theorem 5.18].*

EXAMPLE 2.5.12 Here we have solved the same test problem as in Example 2.2.6 with the scheme (2.5.11) (see Figure (2.5.7)) and the minmod limiter φ_1 (see Figure 2.5.8). There are no oscillations, and the shock is smeared out over nearly the same number of grid points as in the Lax–Wendroff scheme (see Figures 2.5.1 and 2.5.2).

REMARK 2.5.13 *It can be shown that the order of consistency of semidiscrete schemes in local extrema is at most one. For the special TVB schemes (i.e. schemes for which the TV norm remains bounded) it can be shown that they are uniformly of higher order (see [215]).*

REMARK 2.5.14 *In [83] Goodman and Leveque define TVD schemes in 2-D, and they prove that they have at most first order consistency.*

There is another technique to construct second order schemes, the so-called variable extrapolation.

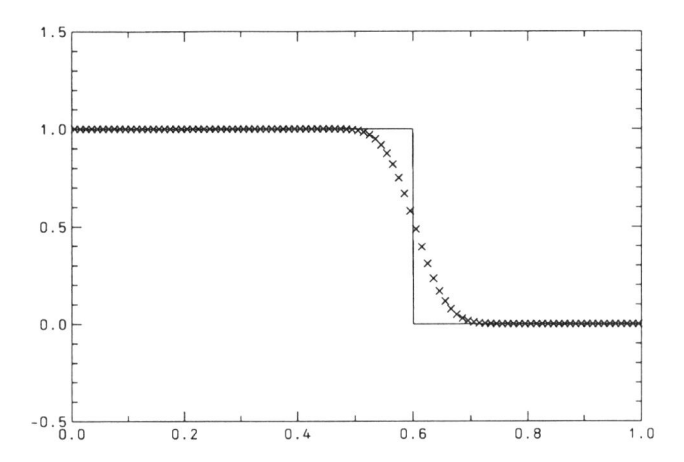

Figure 2.5.7 Godunov.

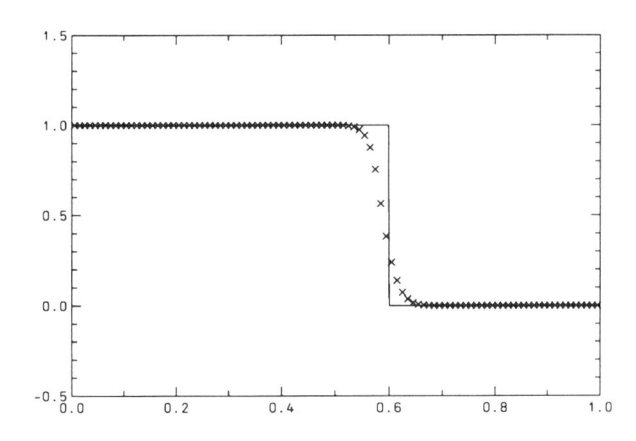

Figure 2.5.8 Higher order: Sweby with minmod limiter.

112 Initial value problems for scalar conservation laws in 1-D

LEMMA 2.5.15 (Variable extrapolation) *Let us consider a scheme in conservation form*

$$u_j^{n+1} = u_j^n - \frac{\Delta t}{\Delta x}[g(u_j^n, u_{j+1}^n) - g(u_{j-1}^n, u_j^n)]$$

for the initial value problem

$$\begin{aligned}\partial_t u + \partial_x f(u) &= 0 &\text{in } \mathbb{R} \times \mathbb{R}^+, \\ u(x,0) &= u_0(x) &\text{in } \mathbb{R}.\end{aligned}$$

We assume that g is a consistent numerical flux that is sufficiently smooth. Then let (see Figure 2.5.9)

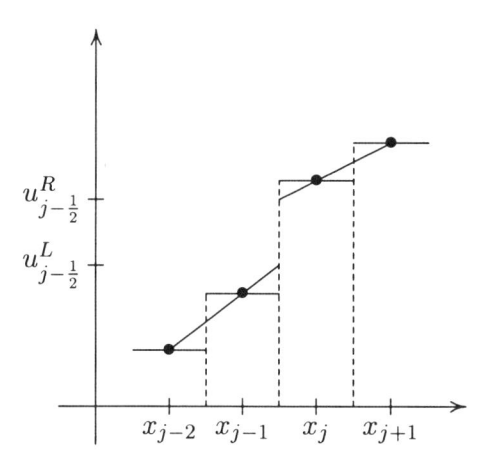

Figure 2.5.9

$$\begin{aligned} u_{j-\frac{1}{2}}^L &:= u_{j-1} + \tfrac{1}{2}(u_{j-1} - u_{j-2}), \\ u_{j-\frac{1}{2}}^R &:= u_j - \tfrac{1}{2}(u_{j+1} - u_j). \end{aligned} \quad (2.5.21)$$

Then the numerical scheme

$$u_j^{n+1} = u_j^n - \frac{\Delta t}{\Delta x}[g(u_{j+\frac{1}{2}}^L, u_{j+\frac{1}{2}}^R) - g(u_{j-\frac{1}{2}}^L, u_{j-\frac{1}{2}}^R)] \quad (2.5.22)$$

is consistent of order 2 with respect to x.

Finite difference schemes of higher order for scalar equations in 1-D

Proof In order to derive sufficient conditions for the variable extrapolation to be consistent of order 2 with respect to x, we prove a more general result. Let $v, w, \tilde{v}$ and $\tilde{w}$ be extrapolation values of u_j. The conditions (2.5.24), (2.5.25), (2.5.27), (2.5.28) and (2.5.29) which we shall assume for $v, w, \tilde{v}, \tilde{w}$ will be justified below. We have to show that

$$g(\tilde{v}, \tilde{w}) - g(v, w) = \Delta x \partial_x f(u_j) + \mathcal{O}(\Delta x^3) \ .$$

By Taylor series expansion in (v_0, w_0), we obtain ($h := \Delta x$)

$$\begin{aligned} g(\tilde{v}, \tilde{w}) - g(v, w) &= (\tilde{v} - v)[\partial_1 g + \tfrac{1}{2}(\tilde{v} + v - 2v_0)\partial_1^2 g \\ &\quad + (\tilde{w} - w_0)\partial_1 \partial_2 g] + (\tilde{w} - w)[\partial_2 g \\ &\quad + \tfrac{1}{2}(\tilde{w} + w - 2w_0)\partial_2^2 g + (v - v_0)\partial_1 \partial_2 g] \\ &\quad + \mathcal{O}(h^3) \ , \end{aligned} \quad (2.5.23)$$

where we have assumed that

$$\begin{aligned} \tilde{v} - v_0 &= \mathcal{O}(h) \ , \quad \tilde{w} - w_0 = \mathcal{O}(h) \ , \\ v - v_0 &= \mathcal{O}(h) \ , \quad w - w_0 = \mathcal{O}(h) \ . \end{aligned} \quad (2.5.24)$$

Using

$$\tilde{v} - v_0 + v - v_0 = \mathcal{O}(h^2) \ , \quad (2.5.25)$$

(2.5.23), $g(v, v) = f(v)$ and $v_0 = w_0$, we get

$$\begin{aligned} f(\tilde{v}) - f(v) &= g(\tilde{v}, \tilde{v}) - g(v, v) \\ &= (\tilde{v} - v)[(\partial_1 g + \partial_2 g) + \frac{1}{2}(\tilde{v} + v - 2v_0)(\partial_1^2 g + \partial_2^2 g) \\ &\quad + [(\tilde{v} - v_0) + (v - v_0)]\partial_1 \partial_2 g] + \mathcal{O}(h^3) \\ &= (\tilde{v} - v)(\partial_1 g + \partial_2 g) + \mathcal{O}(h^3) \ . \end{aligned} \quad (2.5.26)$$

On the other hand, using (2.5.23), (2.5.25),

$$\tilde{w} + w - 2w_0 = \mathcal{O}(h^2) \ , \quad (2.5.27)$$

$$\tilde{w} - w - (\tilde{v} - v) = \mathcal{O}(h^3) \quad (2.5.28)$$

114 Initial value problems for scalar conservation laws in 1-D

and

$$\left.\begin{array}{c}\tilde{v}-v_0\\ \tilde{w}-v_0\end{array}\right\} = h\partial_x u_j + \mathcal{O}(h^2)\,,$$

$$\left.\begin{array}{c}v-v_0\\ w-v_0\end{array}\right\} = -h\partial_x u_j + \mathcal{O}(h^2)\,,$$

(2.5.29)

we get

$$\begin{aligned}g(\tilde{v},\tilde{w}) &- g(v,w)\\ &= (\tilde{v}-v)\partial_1 g(v_0,v_0) + (\tilde{w}-w)\partial_2 g(v_0,v_0) + \mathcal{O}(h^3)\\ &= (\tilde{v}-v)[\partial_1 g(v_0,v_0) + \partial_2 g(v_0,v_0)]\\ &\quad + (\tilde{w}-w-(\tilde{v}-v))\partial_2 g(v_0,v_0) + \mathcal{O}(h^3)\\ &= (\tilde{v}-v)(\partial_1 g + \partial_2 g) + \mathcal{O}(h^3)\,.\end{aligned}$$

Therefore, under the assumptions of (2.5.24), (2.5.25), (2.5.27), (2.5.28) and (2.5.29), we obtain

$$\begin{aligned}g(\tilde{v},\tilde{w}) - g(v,w) &= f(\tilde{v}) - f(v) + \mathcal{O}(h^3)\\ &= hf'(v_0)\partial_x u_j + \mathcal{O}(h^3)\,.\end{aligned}$$

(2.5.30)

Now we assume that u is a smooth function. Let $v := u^L_{j-\frac{1}{2}}$ and $w := u^R_{j-\frac{1}{2}}$ be defined as in (2.5.21) and $\tilde{v} := u^L_{j+\frac{1}{2}}$ and $\tilde{w} := u^R_{j+\frac{1}{2}}$, $w_0 = v_0 := u_j$. Then the assumptions (2.5.24), (2.5.25), (2.5.27), (2.5.28) and (2.5.29) are satisfied. □

REMARK 2.5.16 *In order to avoid oscillation, this scheme has to be used with limiter functions as defined in Example 2.5.10. Then, instead of (2.5.21), one has to use*

$$\begin{aligned}u^L_{j-\frac{1}{2}} &= u_{j-1} + \tfrac{1}{2}\varphi(r_{j-1})(u_{j-1} - u_{j-2})\,,\\ u^R_{j-\frac{1}{2}} &= u_j - \tfrac{1}{2}\varphi(r_j)(u_{j+1} - u_j)\,.\end{aligned}$$

(2.5.31)

LEMMA 2.5.17

(a) Define

$$S_j = \begin{cases} 0 & \text{if } \Delta_+ u_j \Delta_- u_j < 0, \\ \text{sign}(\Delta_+ u_j) \min\left\{\dfrac{|\Delta_+ u_j|}{\Delta x}, \dfrac{|\Delta_- u_j|}{\Delta x}\right\} & \text{otherwise,} \end{cases} \qquad (2.5.32)$$

and let $u^L_{j-\frac{1}{2}}$ and $u^R_{j-\frac{1}{2}}$ be as in (2.5.31) with $\varphi = \varphi_1 = $ minmod limiter, $L_j(x) = u_j + S_j(x - x_j)$ Then we have

$$u^R_{j-\frac{1}{2}} = u_j + S_j(x_{j-\frac{1}{2}} - x_j) = L_j(x_{j-1/2}),$$
$$u^L_{j-\frac{1}{2}} = u_{j-1} + S_{j-1}(x_{j-\frac{1}{2}} - x_{j-1}) = L_{j-1}(x_{j-1/2}).$$

(b) If $u^L_{j-\frac{1}{2}}$ and $u^R_{j-\frac{1}{2}}$ are defined as in (2.5.31) with $\varphi = \varphi_2 = $ superbee limiter then one has to replace "min" in (2.5.32) by max.

Proof Let us show the statement a) for $u^R_{j-\frac{1}{2}}$.

1st case. Let $\Delta_+ u_j \Delta_- u_j < 0$. Then $S_j = 0$ and

$$\varphi_1(r) = \varphi_1\left(\dfrac{\Delta_- u_j}{\Delta_+ u_j}\right) = \max\{0 \,.\, \min\{r\,,1\}\} = 0\,.$$

2nd case. Here sign $\Delta_- u_j = $ sign $\Delta_+ u_j > 0$ and $\Delta_- u_j > \Delta_+ u_j$. We have

$$u_j + S_j(x_{j-\frac{1}{2}} - x_j) = u_j + \dfrac{\Delta_+ u_j}{\Delta x}\left(-\dfrac{\Delta x}{2}\right) = u_j - \dfrac{\Delta_+ u_j}{2}$$
$$= u_j - \varphi_1(r_j)\dfrac{\Delta_+ u_j}{2}$$
$$= u^R_{j-\frac{1}{2}}\,.$$

3rd case. Here sign $\Delta_- u_j = $ sign $\Delta_+ u_j > 0$ and $\Delta_- u_j \leq \Delta_+ u_j$. Then we have

$$u_j + S_j(x_{j-\frac{1}{2}} - x_j) = u_j + \dfrac{\Delta_- u_j}{\Delta x}\left(-\dfrac{\Delta x}{2}\right)$$
$$= u_j - \dfrac{1}{2}\Delta_- u_j = u_j - \dfrac{1}{2}\dfrac{\Delta_- u_j}{\Delta_+ u_j}\Delta_+ u_j$$
$$= u_j - \dfrac{1}{2}\varphi_1(r_j)\Delta_+ u_j = u^R_{j-\frac{1}{2}}\,.$$

The other cases and (b) can be treated similarly. □

DEFINITION 2.5.18 *Schemes as defined in (2.5.22) with $u^R_{j-\frac{1}{2}}$ and $u^L_{j-\frac{1}{2}}$ as in Lemma 2.5.17 (a) and (b) are called MUSCL type schemes (Monotone Upstream–Centred Schemes for Conservation Laws; see [129], [130], [131], [132] and [133]).*

REMARK 2.5.19 *In [174] and [176] the convergence of a higher-order scheme to the uniquely defined entropy solution is proved. Furthermore, in a more general context in n-D on unstructured grids the convergence to the entropy solution for finite volume schemes has been proved in [121]. The details will be given in §§3.5 and 3.6.*

A slightly different extrapolation method is given in the following lemma.

LEMMA 2.5.20 *Let g and f be as in Lemma 2.5.15 and let*

$$u^L_{j+\frac{1}{2}} = u_j + \frac{1}{4}(u_{j+1} - u_{j-1}) , +\frac{\kappa}{4}(u_{j+1} - 2u_j + u_{j-1})$$
$$u^R_{j-\frac{1}{2}} = u_j - \frac{1}{4}(u_{j+1} - u_{j-1}) + \frac{\kappa}{4}(u_{j+1} - 2u_j + u_{j-1}) .$$

Then the scheme

$$u^{n+1}_j = u^n_j - \frac{\Delta t}{\Delta x}[g(u^L_{j+\frac{1}{2}}, u^R_{j+\frac{1}{2}}) - g(u^L_{j-\frac{1}{2}}, u^R_{j-\frac{1}{2}})]$$

is consistent of order 2 with respect to Δx. For $\kappa = \frac{1}{2}$ this scheme is known as the QUICK scheme (see [135]). In this case it is of third order.

Proof Choose

$$v := u^L_{j-\frac{1}{2}}, \; w := u^R_{j-\frac{1}{2}}, \; \tilde{v} := u^L_{j+\frac{1}{2}}, \; \tilde{w} := u^R_{j+\frac{1}{2}}, \; v_0 := w_0 := u_j.$$

Then we have to verify (2.5.24), (2.5.25), (2.5.27), (2.5.28) and (2.5.29) for v, w, $\tilde{v}$, $\tilde{w}$, v_0 and w_0. □

REMARK 2.5.21 *As mentioned in Lemma 2.5.6, higher-order TVD schemes have the disadvantage that they degenerate locally to first-order accuracy (i.e. truncation error) at critical points with $u'(x) = 0$. In [215] TVB schemes (i.e. total variation bounded) are developed that are uniformly of higher order in space, in particular at critical points. But since these schemes use a fixed stencil, they smear out discontinuities more than Engquist–Osher TVD or ENO schemes (see §3.7). Several numerical examples confirm this property. On the other hand, for smooth solutions they have the expected order of convergence.*

REMARK 2.5.22 *If one uses schemes that are higher-order in space, it is necessary to take care of a higher-order time discretization as well. In [215, Example 1 case (ii)] an Engquist–Osher TVD discretization is applied to the one-dimensional Burgers equation such that the exact solution is a rarefaction wave. This scheme is of first order in time and third order in space. But the scheme converges to the wrong solution, namely to an entropy violating expansion shock. If the time discretization is also of third order, the scheme converges to the correct solution. For explicit schemes the CFL condition is necessary for getting a stable scheme. But this means a strong restriction to the size of the timestep. In [138] and [139], a generalized method that is still stable for Courant numbers much larger than 1 is developed.*

2.6 Streamline diffusion method

We consider the initial value problem for the Burgers equation

$$\begin{aligned}\partial_t u + u \partial_x u &= 0 \quad \text{in } \mathbb{R} \times \mathbb{R}^+ , \\ u(\cdot, 0) &= u_0 \quad \text{in } \mathbb{R} .\end{aligned} \qquad (2.6.1)$$

Let us assume for this section that u_0 has compact support. Because of (2.1.33), $u(\cdot, t)$ also has compact support for any finite time t. Up to now we have studied difference methods for treating (2.6.1). In this section we discuss finite element methods for discretizing (2.6.1). Unfortunately the standard Galerkin finite element method is not stable. This can be seen very easily in the following example.

EXAMPLE 2.6.1 (Standard Galerkin finite element method for convection-dominated diffusion equation) We know that the solution of (2.1.12) converges to the solution of (2.1.13) as $\varepsilon \to 0$. Now if (2.1.13) reduce to a simple stationary linear case in 1-D we obtain

118 Initial value problems for scalar conservation laws in 1-D

$$-\varepsilon u'' + u' = 0 \ .$$

Therefore let us consider the boundary value problem

$$\begin{aligned}-\varepsilon u'' + u' = 0 \quad &\text{in }]0,1[\ , \\ u(0) = 1 \ , \ u(1) = 0 \ , &\end{aligned} \qquad (2.6.2)$$

with $0 < \varepsilon << 1$ (see [109]). The exact solution of (2.6.2) is given by

$$u(x) = \alpha(1 - e^{-(1-x)/\varepsilon}) \qquad (2.6.3)$$

where $\alpha := (1 - e^{-1/\varepsilon})^{-1}$. For small ε the solution u is nearly equal to 1 beyond a small layer in a neighbourhood of $x = 1$, where u decays very rapidly to zero. Let us apply the standard Galerkin finite element method on a uniform grid with mesh size h. Then we obtain (see [109]) the following system of linear equations:

$$-\frac{\varepsilon}{h^2}(u_{i-1} - 2u_i + u_{i+1}) + \frac{1}{2h}(u_{i+1} - u_{i-1}) = 0 \qquad (2.6.4)$$

for $i = 1, \ldots, N-1$,

$$u_0 = 1 \ , u_N = 0 \ ,$$

where u_i is assumed to be an approximation of the exact solution in the grid point ih. We consider the case where $0 < \varepsilon << h$, since this is the most interesting one in fluid dynamics. Then for a heuristic argument we can neglect the first term with the coefficient ε/h^2. For simplicity we assume that N is an odd number. Then u_i is close to 0 if i is an even number and close to 1 for odd numbers. This means that the approximate solution given by (2.6.4) oscillates very strongly and is not close to the exact solution given by (2.6.3).

Therefore the standard Galerkin finite element method is not stable for conservation laws (see [109]), and we need a damping term for stabilization. Let us explain the idea for the stationary problem

$$\beta \nabla u + bu = f \quad \text{in } \Omega :=]0,1[\times]0,1[\ ,$$

where $\beta = \binom{\beta_1}{\beta_2} \in \mathbb{R}^2$ and $b \in \mathbb{R}$.

2.6 Streamline diffusion method

Instead of the usual finite element ansatz

$$\int_\Omega (\beta \nabla u_h v + b u_h v) = \int_\Omega f v$$

for u_h and for all $v \in S \subset \overset{\circ}{H}{}^{1,2}(\Omega)$, where S consists of the piecewise linear ansatz functions with respect to a given triangulation, we consider

$$\int_\Omega \beta \nabla u_h (v + h\beta \nabla v) + \int_\Omega b u_h (v + h\beta \nabla v) = \int_\Omega f(v + h\beta \nabla v) .$$

The additional term $h\beta \nabla v$ stabilizes the scheme, since, after integration by parts, we have formally

$$-h \int_\Omega (\beta_1^2 \partial_x^2 u_h + 2\beta_1 \beta_2 \partial_x \partial_y u_h + \beta_2^2 \partial_y^2 u_h) v + \int_\Omega \beta \nabla u_h v (1 - hb)$$
$$+ \int_\Omega b u_h v = \int_\Omega f(v + h\beta \nabla v) .$$

The second-order term scaled with h works similarly to the numerical diffusion for the finite difference methods (see (2.2.13)).

For the formulation of the discrete time-dependent problem we need the following notation:

$$S_n := \mathbb{R} \times]t^n, t^{n+1}[,$$
$$T_h^n : \text{triangulation of } S_n$$
$$P_k(K) \text{ polynomials on } K \text{ of degree } \leq k, \text{ for } K \in T_h^n,$$
$$V_h^n := \{v \in H^1(S_n) \mid v|_K \in P_k(K), K \in T_h^n\} .$$

Now the weak formulation of the discrete problem is the following. Find u_h such that $u_h|_{S_n} \in V_h^n$ and for $n = 0, 1, 2, \ldots$

$$\int_{S_n} (\partial_t u_h + u_h \partial_x u_h)[v + h(\partial_t v + u_h \partial_x v)]$$
$$+ \int_\mathbb{R} (u_{h+}^n - u_{h-}^n) v_+^n = 0 \qquad (2.6.5)$$

for all $v \in V_h^n$ and

$$u_+^0 = u_0 .$$

Here we have used the notation

$$v_\pm^n = \lim_{s \to 0\pm} v(t^n + s) .$$

REMARK 2.6.2 *Let us consider the following formal arguments to motivate the streamline diffusion method and to interpret it in the context of least square methods. The initial boundary value problem for the linear advection equation is given by* $(a > 0)$

$$Au := \partial_t u + a \partial_x u = f \quad \text{for } (x,t) \in \Omega :=]0,1[\,\times\,]0,T[\,, \qquad (2.6.6)$$
$$u(x,0) = u_0(x), \quad x \in]0,1[\,, \qquad (2.6.7)$$
$$u(0,t) = 0, \quad t \in]0,1[\,. \qquad (2.6.8)$$

We assume that f, u_0 *and* u *are sufficiently smooth. Writing this problem as a least square problem, we obtain the following: Minimize the functional*

$$J(u) := \|Au - f\|_{L^2(\Omega)}^2$$

in the set of all functions u *satisfying (2.6.7) and (2.6.8). For* $\varphi \in C_0^\infty(\Omega)$, *the sum* $u + \varepsilon \varphi$ *is still an admissible function, and we obtain*

$$0 = \frac{d}{d\varepsilon} J(u + \varepsilon\varphi)|_{\varepsilon=0} = \frac{d}{d\varepsilon} \int_\Omega [A(u + \varepsilon\varphi) - f]^2|_{\varepsilon=0}$$

$$= 2 \int_\Omega [A(u + \varepsilon\varphi) - f] A\varphi|_{\varepsilon=0} = 2 \int_\Omega [A^t A(u + \varepsilon\varphi) - A^t f]\varphi|_{\varepsilon=0}$$

$$= 2 \int_\Omega (A^t A u - A^t f)\varphi , \qquad (2.6.9)$$

where A^t *denotes the formal adjoint w.r.t.* $(\cdot,\cdot)_{L^2(\Omega)}$. *This means that* u *has to satisfy*

$$(-\partial_t - a\partial_x)(\partial_t u + a\partial_x u) = (-\partial_t - a\partial_x)f =: \tilde{f}$$

2.6 Streamline diffusion method

or

$$-\partial_t^2 u - 2a\partial_t\partial_x u - a^2\partial_x^2 u = \tilde{f} \ .$$

From (2.6.9) we can also obtain a weak formulation:

$$\int_\Omega (\partial_t u + a\partial_x u)(\partial_t\varphi + a\partial_x\varphi) = \int_\Omega f(\partial_t\varphi + a\partial_x\varphi) \ . \tag{2.6.10}$$

The standard Galerkin ansatz gives

$$\int_\Omega (\partial_t u + a\partial_x u)\varphi = \int_\Omega f\varphi \ . \tag{2.6.11}$$

Now if we multiply (2.6.10) by h and add the equation (2.6.11), we get

$$\int_\Omega (\partial_t u + a\partial_x u)[\varphi + h(\partial_t\varphi + a\partial_x\varphi)] = \int_\Omega f[\varphi + h(\partial_t\varphi + a\partial_x\varphi)] \ . \tag{2.6.12}$$

This corresponds to (2.6.5).

Now we shall study the convergence properties of u_h. Let $\Omega \subset\subset \mathbb{R}\times]0,T[$, such that $\mathrm{supp}(u) \subset \Omega$ for all $t \in]0,T[$.

THEOREM 2.6.3 (Lax–Wendroff theorem for finite elements) *If $\|u_h\|_{L^\infty(\Omega)} \leq C$ and $u_h \to u$, $h \to 0$ a.e. in Ω then u is an entropy solution of (2.6.1).*

Proof See [111, Theorem 3.1.] □

THEOREM 2.6.4 (Convergence of the streamline diffusion method) *Let $\|u_h\|_{L^\infty(\Omega)} \leq C$ uniformly in h. Then we have for a subsequence*

$$u_h \to u \quad \text{for } h \to 0 \quad \text{a.e. in } \Omega \ ,$$

and u is the entropy solution of (2.6.1).

REMARK 2.6.5 *In order to show that u_h converges to the entropy solution in the proof of Theorem 2.6.3, it is sufficient to consider only one strictly convex entropy pair, i.e. $U(u) = \frac{1}{2}u^2$ and therefore $F(u) = \frac{1}{3}u^3$ (see Remark 30 in [237]). This is a motivation to consider these terms for proving convergence $u_h \to u$.*

Now we shall prove Theorem 2.6.4 in several steps. First we collect the results and at the end of this section we shall give the proofs.

LEMMA 2.6.6 *Let u_h be a solution of (2.6.5) and assume that*

$$\|u_h\|_{L^\infty(\Omega)} \leq \text{const} \qquad (2.6.13)$$

uniformly in h. Then we have

$$\partial_t\left(\frac{u_h^2}{2}\right) + \partial_x\left(\frac{u_h^3}{3}\right) = F_h + G_h,$$

where

$$F_h + G_h \text{ bounded in } [H^{1,\infty}(\Omega)]^*, \qquad (2.6.14)$$
$$G_h \text{ bounded in } C^0(\overline{\Omega})^*, \qquad (2.6.15)$$
$$\text{there exists } F \in [H^{1,2}(\Omega)]^* \text{ such that}$$
$$F_h \to F \text{ in } [\mathring{H}^{1,2}(\Omega)]^*. \qquad (2.6.16)$$

Proof See 2.6.12.

LEMMA 2.6.7 (Theorem of Murat) *Let $g^\varepsilon := g_1^\varepsilon + g_2^\varepsilon$ be a sequence of distributions such that*

$$g^\varepsilon \text{ is bounded in } [H^{1,\infty}(\Omega)]^* \qquad (2.6.17)$$
$$g_1^\varepsilon \text{ is bounded in } M(\Omega) := [C^0(\overline{\Omega})]^* \qquad (2.6.18)$$
$$g_2^\varepsilon \to g_2 \text{ in } [\mathring{H}^{1,2}(\Omega)]^* \text{ (i.e. compact in } H^{-1,2}(\Omega)). \qquad (2.6.19)$$

Then for a subsequence there exists $g_1 \in H^{-1,2}_{\text{loc}}(\Omega)$ such that

$$g^\varepsilon \to g_1 + g_2 \text{ in } H^{-1,2}_{\text{loc}}(\Omega).$$

Proof See 2.6.14.

COROLLARY 2.6.8 *Let us assume (2.6.13) of Lemma 2.6.6. Then*

$$\partial_t\left(\frac{u_h^2}{2}\right) + \partial_x\left(\frac{u_h^3}{3}\right) \quad \text{is compact in } H_{\text{loc}}^{-1,2}(\Omega).$$

Proof This follows from Lemma 2.6.6 and Lemma 2.6.7. □

COROLLARY 2.6.9 *Suppose that $\|u_h\|_{L^\infty} \leq C$ uniformly in h. Then*

$$\partial_t u_h + \partial_x\left(\frac{u_h^2}{2}\right) \quad \text{is compact in } H_{\text{loc}}^{-1,2}(\Omega).$$

Proof This is similar to the above.

COROLLARY 2.6.10 (Compactness theorem of Tartar) *Let u_h be a solution of (2.6.5). Assume that*

$$\partial_t u_h + \partial_x\left(\frac{u_h^2}{2}\right) \quad \text{is compact in } H_{\text{loc}}^{-1,2}(\Omega),$$

and that

$$\partial_t\left(\frac{u_h^2}{2}\right) + \partial_x\left(\frac{u_h^3}{3}\right) \quad \text{is compact in } H_{\text{loc}}^{-1,2}(\Omega).$$

Then we have

$$u_h \to u \quad \text{strongly in } L^p(\Omega) \text{ for all } p \leq \infty.$$

Proof See 2.6.15.

2.6.11 (Proof of Theorem 2.6.4) Corollary 2.6.8 implies that

$$\partial_t\left(\frac{u_h^2}{2}\right) + \partial_x\left(\frac{u_h^3}{3}\right) \quad \text{is compact in } H_{\text{loc}}^{-1,2}(\Omega).$$

Therefore Corollary 2.6.10 implies that $u_h \to u$ strongly in $L^p(\Omega)$ for all $p \leq \infty$ and we have $u_h \to u$ a.e. in Ω. Since u_h is uniformly bounded in $L^\infty(\Omega)$, this result together with Theorem 2.6.3 proves the statement of Theorem 2.6.4. □

Now we shall prove Lemmas 2.6.6 and 2.6.7 and Corollary 2.6.10.

2.6.12 (Proof of Lemma 2.6.6) In the first step of the proof we shall use the differential equation to write

$$\partial_t\left(\frac{u_h^2}{2}\right) + \partial_x\left(\frac{u_h^3}{3}\right)$$

as a sum of terms (see (2.6.24)), which can be easily estimated in order to get (2.6.14), (2.6.15) and (2.6.16). Let $\varphi \in C_0^\infty(\Omega)$ and use $v := \pi_h(u_h\varphi)$ as a test function in (2.6.5), where π_h denotes the standard interpolation operator. Then we obtain, using $A(v) := \partial_t v + u_h \partial_x v$,

$$0 = \int_{S_n} (\partial_t u_h + u_h \partial_x u_h)[v + h(\partial_t v + u_h \partial_x v)] + \int_{\mathbb{R}} (u_{h+}^n - u_{h-}^n)v_+^n$$

$$= \int_{S_n} A(u_h)[v + hA(v)] + \int_{\mathbb{R}} (u_{h+}^n - u_{h-}^n)v_+^n = 0 \ . \qquad (2.6.20)$$

Let $w := u_h\varphi$ and obtain

$$\int_{S_n} A(u_h)w + \int_{\mathbb{R}} (u_{h+}^n - u_{h-}^n)w_+^n + h\int_{S_n} A(u_h)^2\varphi$$

$$= \int_{S_n} A(u_h)(w-v) + \int_{\mathbb{R}} (u_{h+}^n - u_{h-}^n)(w_+^n - v_+^n)$$

$$+ h\int_{S_n} A(u_h)^2\varphi - h\int_{S_n} A(u_h)A(v)$$

$$= \int_{S_n} A(u_h)(w-v) + \int_{\mathbb{R}} (u_{h+}^n - u_{h-}^n)(w_+^n - v_+^n) \qquad (2.6.21)$$

$$+ h\int_{S_n} A(u_h)[A(w) - A(v)]$$

$$- h\int_{S_n} A(u_h)[A(w) - \varphi A(u_h)] =: R \ .$$

2.6 Streamline diffusion method

Since

$$A(w) - \varphi A(u_h) = \partial_t w + u_h \partial_x w - \varphi \partial_t u_h - \varphi u_h \partial_x u_h$$
$$= \varphi \partial_t u_h + \partial_t \varphi u_h + u_h^2 \partial_x \varphi + u_h \varphi \partial_x u_h$$
$$\quad - \varphi \partial_t u_h - \varphi u_h \partial_x u_h$$
$$= u_h \partial_t \varphi + u_h^2 \partial_x \varphi ,$$

we can continue

$$R = \underbrace{\int_{S_n} A(u_h)(w-v)}_{=:F_1} + \underbrace{\int_{\mathrm{I\!R}} (u_{h+}^n - u_{h-}^n)(w_+^n - v_+^n)}_{=:F_2}$$
$$+ \underbrace{h \int_{S_n} A(u_h)[\partial_t w - \partial_t v + u_h(\partial_x w - \partial_x v)]}_{=:F_3}$$
$$\underbrace{- h \int_{S_n} A(u_h)(u_h \partial_t \varphi + u_h^2 \partial_x \varphi)}_{=:F_4}$$
$$= F_1 + F_2 + F_3 + F_4 .$$

Furthermore, we have by integration by parts

$$\int_{S_n} A(u_h) w = \int_{S_n} (\partial_t u_h + u_h \partial_x u_h)(u_h \varphi) \tag{2.6.22}$$
$$= - \int_{S_n} u_h \partial_t (u_h \varphi) + \int_{\mathrm{I\!R}} u_h^2 \varphi \Big|_{t^n}^{t^{n+1}}$$
$$\quad - \int_{S_n} \frac{u_h^2}{2} \partial_x (u_h \varphi)$$
$$= -\frac{1}{2} \int_{S_n} u_h^2 \partial_t \varphi + \frac{1}{2} \int_{\mathrm{I\!R}} [(u_h^2)_-^{n+1} \varphi^{n+1} - (u_h^2)_+^n \varphi^n] - \int_{S_n} \frac{u_h^3}{3} \partial_x \varphi ,$$

since

$$\int_{S_n} u_h \partial_t (u_h \varphi) = \int_{S_n} u_h^2 \partial_t \varphi + \int_{S_n} u_h \varphi \partial_t u_h$$

$$= \int_{S_n} u_h^2 \partial_t \varphi - \int_{S_n} \partial_t(u_h\varphi) u_h$$
$$+ \int_{\mathbb{R}} [(u_h^2)_-^{n+1} \varphi^{n+1} - (u_h^2)_+^n \varphi^n],$$

and therefore

$$\int_{S_n} u_h \partial_t(u_h\varphi) = \frac{1}{2} \int_{S_n} u_h^2 \partial_t \varphi + \frac{1}{2} \int_{\mathbb{R}} [\ldots].$$

Similarly, we have

$$\int_{S_n} u^2 \partial_x(\varphi u) = \int_{S_n} u^2(\partial_x \varphi \, u + \varphi \partial_x u) = \int_{S_n} (u^3 \partial_x \varphi + u^2 \partial_x u \varphi)$$

$$\int_{S_n} \frac{u^3}{3} \partial_x \varphi = -\int_{S_n} u^2 \partial_x u \varphi,$$

and therefore

$$\int_{S_n} u^2 \partial_x(\varphi u) = \int_{S_n} \left(u^3 \partial_x \varphi - \frac{u^3}{3} \partial_x \varphi \right) = \frac{2}{3} \int_{S_n} u^3 \partial_x \varphi$$

and

$$\frac{1}{2} \int_{S_n} u^2 \partial_x(\varphi u) = \frac{1}{3} \int_{S_n} u^3 \partial_x \varphi.$$

Now we sum up in (2.6.22):

$$\sum_n \int_{S_n} A(u_h) w = -\frac{1}{2} \int_\Omega u_h^2 \partial_t \varphi - \frac{1}{3} \int_\Omega u_h^3 \partial_x \varphi$$
$$+ \frac{1}{2} \sum_n \int_{\mathbb{R}} [(u_h^2)_-^{n+1} \varphi^{n+1} - (u_h^2)_+^n \varphi^n].$$

Rearranging gives

$$-\frac{1}{2}\int_\Omega u_h^2 \partial_t \varphi - \frac{1}{3}\int_\Omega u_h^3 \partial_x \varphi$$
$$= \sum_n \int_{S_n} A(u_h) w - \frac{1}{2}\sum_n \int_{\mathbb{R}} [\ldots] \qquad (2.6.23)$$

and, using (2.6.21),

$$= -\sum_n \int_{\mathbb{R}} (u_{h+}^n - u_{h-}^n) w_+^n - h \int_\Omega A(u_h)^2 \varphi$$
$$+ \sum_n (F_1 + F_2 + F_3 + F_4) - \frac{1}{2}\sum_n \int_{\mathbb{R}} [\ldots]$$
$$=: -h \int_\Omega A(u_h)^2 \varphi + \sum_n (F_1 + F_2 + F_3 + F_4) - R .$$

For R we obtain

$$R = \sum_n \int_{\mathbb{R}} (u_{h+}^n - u_{h-}^n)(u_h \varphi)_+^n + \frac{1}{2}\sum_n \int_{\mathbb{R}} [(u_h^2)_-^{n+1}\varphi^{n+1} - (u_h^2)_+^n \varphi^n]$$
$$= \sum_n \int_{\mathbb{R}} \left[(u_{h+}^n)^2 \varphi^n - u_{h-}^n u_{h+}^n \varphi^n + \frac{1}{2}(u_{h-}^n)^2 \varphi^n - \frac{1}{2}(u_{h+}^n)^2 \varphi^n \right]$$
$$= \frac{1}{2}\sum_n \int_{\mathbb{R}} (u_{h+}^n - u_{h-}^n)^2 \varphi^n .$$

Then (2.6.23) implies

$$-\frac{1}{2}\int_\Omega u_h^2 \partial_t \varphi - \frac{1}{3}\int_\Omega u_h^3 \partial_x \varphi$$
$$= -h \int_\Omega A(u_h)^2 \varphi + \sum_n (F_1 + F_2 + F_3 + F_4)$$
$$- \frac{1}{2}\sum_n \int_{\mathbb{R}} (u_{h+}^n - u_{h-}^n)^2 \varphi^n . \qquad (2.6.24)$$

Now we need

LEMMA 2.6.13 *We have*

$$h \int_\Omega A(u_h)^2 \leq \text{const} \qquad (2.6.25)$$

$$\left|\sum_n F_i\right| \leq c\sqrt{h}\|\varphi\|_{H^1(\Omega)}, \qquad i = 1,\ldots,4, \qquad (2.6.26)$$

$$\sum_n \int_{\mathbb{R}} (u_{h+}^n - u_{h-}^n)^2 \varphi^n \leq c\|\varphi\|_{C^0(\Omega)}. \qquad (2.6.27)$$

Proof The statements (2.6.25) and (2.6.27) follow from (2.6.5) with $v = u_h$ (see [111]). The assertion (2.6.26) can be obtained using interpolation estimates (see Johnson and Szepessy [111, §2]). □

Now we continue with (2.6.24). Let $\varphi \in C_0^\infty(\Omega)$ and

$$G_h(\varphi) := -h\int_\Omega A(u_h)^2 \varphi - \frac{1}{2}\sum_n \int_{\mathbb{R}} (u_{h+}^n - u_{h-}^n)^2 \varphi,$$

$$F_h(\varphi) := \sum_n (F_1 + \ldots + F_4).$$

Then (2.6.24) implies that

$$-\frac{1}{2}\int_\Omega u_h^2 \partial_t \varphi - \frac{1}{3}\int_\Omega u_h^3 \partial_x \varphi = F_h(\varphi) + G_h(\varphi)$$

for all $\varphi \in C_0^\infty(\Omega)$,

or

$$\partial_t \left(\frac{u_h^2}{2}\right) + \partial_x \left(\frac{u_h^3}{3}\right) = F_h + G_h$$

in the distributional sense. Now we have to show that $F_h + G_h$ satisfies the assumptions of Lemma 2.6.7 (theorem of Murat):

$$F_h + G_h \text{ bounded in } [H^{1,\infty}(\Omega)]^*, \qquad (2.6.28)$$
$$G_h \text{ bounded in } C^0(\overline{\Omega})^*, \qquad (2.6.29)$$
$$F_h \to F \text{ in } [\mathring{H}^{1,2}(\Omega)]^* \text{ with } F = 0. \qquad (2.6.30)$$

Let $\varphi \in H^{1,\infty}(\Omega)$. Then we have $\varphi \in C^{0,1}$ and

$$(F_h + G_h)(\varphi) \le c\|\varphi\|_{C^{0,1}} \le c\|\varphi\|_{H^{1,\infty}}.$$

This implies (2.6.28). Furthermore, (2.6.29) follows immediately from (2.6.25) and (2.6.27) and (2.6.30) from (2.6.26). This finishes the proof of Lemma 2.6.6. □

2.6.14 (Proof of Lemma 2.6.7 (Murat)) First let us show that

$$C^0(\overline{\Omega})^* \hookrightarrow [H^{1,p}(\Omega)]^* =: H^{-1,q}(\Omega) \tag{2.6.31}$$

is compact for all $q < N/(N-1)$ where $\Omega \subset\subset \mathbb{R}^N$ and $1/p + 1/q = 1$.

Proof The embedding $H^{1,p}(\Omega) \hookrightarrow C^0(\overline{\Omega})$ is compact if and only if $1 - N/p > 0$ or $1 > N/p$ or $p > N$. This is equivalent to $q < -N/(1-N) = N/(N-1)$. This proves (2.6.31). Now let us show Lemma 2.6.7 of Murat. Consider

$$\begin{aligned}-\Delta v_1^\varepsilon &= g_1^\varepsilon &&\text{in } \Omega, \\ v_1^\varepsilon &= 0 &&\text{on } \partial\Omega,\end{aligned} \tag{2.6.32}$$

$$\begin{aligned}-\Delta v_2^\varepsilon &= g_2^\varepsilon &&\text{in } \Omega, \\ v_2^\varepsilon &= 0 &&\text{on } \partial\Omega.\end{aligned} \tag{2.6.33}$$

By assumption and (2.6.31), we have $g_1^\varepsilon \in H^{-1,q}(\Omega)$, $q < N/(N-1) \le 2$ and $g_2^\varepsilon \in H^{-1,2}(\Omega)$. Using interior L^p estimates from [5] and [6], we obtain for $\Omega' \subset\subset \Omega$

$$\|v_1^\varepsilon\|_{H^{1,q}(\Omega')} \le c\|g_1^\varepsilon\|_{H^{-1,q}(\Omega')},$$
$$\|v_2^\varepsilon\|_{H^{1,2}(\Omega')} \le c\|g_2^\varepsilon\|_{H^{-1,2}(\Omega')}.$$

Since (2.6.32) and (2.6.33) are linear, we get similar estimates for the differences

$$\|v_1^\varepsilon - v_1^{\varepsilon'}\|_{H^{1,q}(\Omega')} \le c\|g_1^\varepsilon - g_1^{\varepsilon'}\|_{H^{-1,q}(\Omega')},$$
$$\|v_2^\varepsilon - v_2^{\varepsilon'}\|_{H^{1,2}(\Omega')} \le c\|g_2^\varepsilon - g_2^{\varepsilon'}\|_{H^{-1,2}(\Omega')}.$$

Furthermore,

$$\|v_1^\varepsilon\|_{H^{1,r}(\Omega')} \le \text{const} \quad \text{for } r < \infty. \tag{2.6.34}$$

130 Initial value problems for scalar conservation laws in 1-D

This follows from
$$-\Delta v_1^\varepsilon = g_1^\varepsilon \in H^{-1,\infty}(\Omega).$$

Now we can show that
$$\|v_1^\varepsilon - v_1^{\varepsilon'}\|_{H^{1,2}(\Omega')} \to 0 \quad \text{as } |\varepsilon - \varepsilon'| \to 0. \tag{2.6.35}$$

Proof
$$\|v_1^\varepsilon - v_1^{\varepsilon'}\|_{H^{1,2}(\Omega')}^2 = \int_{\Omega'} |\nabla v_1^\varepsilon - \nabla v_1^{\varepsilon'}|^2 + \int_{\Omega'} |v_1^\varepsilon - v_1^{\varepsilon'}|^2$$
$$\leq \left(\int_{\Omega'} |\nabla v_1^\varepsilon - \nabla v_1^{\varepsilon'}|^\alpha\right)^{1/\alpha} \left(\int_{\Omega'} |\nabla v_1^\varepsilon - \nabla v_1^{\varepsilon'}|^\beta\right)^{1/\beta}$$
$$+ \left(\int_{\Omega'} |v_1^\varepsilon - v_1^{\varepsilon'}|^\alpha\right)^{1/\alpha} \left(\int_{\Omega'} |v_1^\varepsilon - v_1^{\varepsilon'}|^p\right)^{1/\beta}$$
$$\leq c\|v_1^\varepsilon - v_1^{\varepsilon'}\|_{H^{1,q}(\Omega')} \|v_1^\varepsilon - v_1^{\varepsilon'}\|_{H^{1,p}(\Omega')}$$
$$\leq C\|g_1^\varepsilon - g_1^{\varepsilon'}\|_{H^{-1,q}} (\|v_1^\varepsilon\|_{H^{1,p}} + \|v_1^{\varepsilon'}\|_{H^{1,p}}) \tag{2.6.36}$$

Let $\alpha := q$ and $\beta := p$. Then we have $p \geq 2$. Because of (2.6.18) and (2.6.31), the first factor in (2.6.36) converges to 0 and the second is bounded (see (2.6.34)). This proves (2.6.35).

Now we can show the statement of the lemma of Murat. For $v^\varepsilon := v_1^\varepsilon + v_2^\varepsilon$ we have
$$\|g^\varepsilon - g^{\varepsilon'}\|_{H^{-1,2}(\Omega')} = \|\Delta(v^\varepsilon - v^{\varepsilon'})\|_{H^{-1,2}(\Omega')}$$
$$\leq \|v^\varepsilon - v^{\varepsilon'}\|_{H^{1,2}(\Omega')}.$$

This implies that g_ε is compact in $H^{-1}(\Omega')$ and the proof of Lemma 2.6.7 is finished. □

2.6.15 (Proof of Corollary 2.6.10) Corollaries 2.6.8 and 2.6.9 imply

$$\partial_t u_h + \partial_x \left(\frac{u_h^2}{2}\right) \text{ is compact in } H^{-1}_{\text{loc}}(\Omega),$$

$$\partial_t \frac{u_h^2}{2} + \partial_x \left(\frac{u_h^3}{3}\right) \text{ is compact in } H^{-1}_{\text{loc}}(\Omega).$$

2.6 Streamline diffusion method

Let

$$v_h := \left(u_h, \frac{u_h^2}{2}\right),$$

$$w_h := \left(-\frac{u_h^3}{3}, \frac{u_h^2}{2}\right).$$

Then we have

$$\text{div}\, v_h \text{ is compact in } H_{\text{loc}}^{-1}(\Omega)$$

$$\text{curl}\, w_h = \partial_t\left(\frac{u_h^2}{2}\right) + \partial_x\left(\frac{u_h^3}{3}\right) \text{ is compact in } H_{\text{loc}}^{-1}(\Omega).$$

Since $\|u_h\|_{L^\infty} \leq \text{const}$,

$$v_h \rightharpoonup v \quad \text{in } L^2(\Omega),$$
$$w_h \rightharpoonup w \quad \text{in } L^2(\Omega).$$

Then we can apply the following lemma.

LEMMA 2.6.16 (div–curl lemma) *Let* $\text{curl}\, w_h$ *and* $\text{div}\, v_h$ *be compact in* $H_{\text{loc}}^{-1}(\Omega)$ *and*

$$v_h \rightharpoonup v \quad \text{in } L^2(\Omega),$$
$$w_h \rightharpoonup w \quad \text{in } L^2(\Omega).$$

Then $v_h w_h \to vw$ *in the sense of distributions.*

Proof See Example 3,[236] and Lemma 2.6.20.

Now using this lemma, we obtain that u_h^4 converges to $\tilde{u}$ in $L_{\text{loc}}^1(\Omega)$, and therefore a.e. in Ω' to $\tilde{u}$. Since $\|u_h\|_{L^\infty(\Omega)}$ is bounded and $u_h \rightharpoonup u$ weak-$*$, it turns out that $u^4 = \tilde{u}$ a.e. in $\Omega' \subset\subset \Omega$. But this implies that $u_h \to u$ a.e. in Ω' and $u_h \to u$ in $L_{\text{loc}}^1(\Omega)$. Now if we choose Ω and Ω' such that $\text{supp}\, u \subset \Omega' \subset\subset \Omega$ for all $t \in]0, T[$, we obtain the statement of Corollary 2.6.10.

$\square$

REMARK 2.6.17 *Beyond this special application, the div–curl lemma is a powerful tool in existence theory for conservation laws. Consult for example [54], [235], [236] and [245].*

Now we are going to prove Lemma 2.6.16. First we consider a more general situation as in Lemma 2.6.16. Let $\{u^n\} \subset [L^2(\Omega)]^p, \Omega \subset \mathbb{R}^N$ denote a sequence with the properties

$$u^n \rightharpoonup u \quad \text{in} \quad [L^2(\Omega)]^p$$

$$\text{and} \quad A_i u^n := \sum_{j=1}^{p} \sum_{k=1}^{N} a_{ijk} \frac{\partial u_j^n}{\partial x_k} \quad \text{belongs to a} \qquad (2.6.37)$$

compact subset of $H_{\text{loc}}^{-1}(\Omega)$, $i = 1, \ldots, q$ for some $a_{ijk} \in \mathbb{R}$.

We define the manifold V by

$$V := \left\{ (\lambda, \xi) \in \mathbb{R}^p \times \mathbb{R}^N \setminus \{0\} \Big| \sum_{j=1}^{p} \sum_{k=1}^{N} a_{ijk} \lambda_j \xi_k = 0, i = 1, \ldots, q \right\}$$

and the set Λ as the projection of V onto the first coordinate λ:

$$\Lambda := \left\{ \lambda \in \mathbb{R}^p \Big| \exists \xi \in \mathbb{R}^N \setminus \{0\} : \sum_{j=1}^{p} \sum_{k=1}^{N} a_{ijk} \lambda_j \xi_k = 0, i = 1, \ldots, q \right\}.$$

Now we can show the following theorem, which goes back to Murat [162].

THEOREM 2.6.18 *Let $Q : \mathbb{R}^p \to \mathbb{R}$ be defined by $Q(x) := x^T A x$, $A \in \mathbb{R}^{p \times p}$ and $A = (a_{ij})$ symmetric. The sequence $\{u^n\} \subset [L^2(\Omega)]^p$ may satisfy condition (2.6.37).*

(a) If $Q(\lambda) \geq 0$ for all $\lambda \in \Lambda$ then

$$\liminf_{n \to \infty} \int_{\Omega} \Phi^2(x) Q(u^n(x)) \, dx$$

$$\geq \int_{\Omega} \Phi^2(x) Q(u(x)) \, dx \quad \text{for all } \Phi \in C_0^{\infty}(\Omega) .$$

2.6 Streamline diffusion method 133

(b) If $Q(\lambda) = 0$ for all $\lambda \in \Lambda$ then

$$\lim_{n\to\infty} \int_\Omega \Phi^2(x) Q(u^n(x))\, dx$$
$$= \int_\Omega \Phi^2(x) Q(u(x))\, dx \quad \text{for all } \Phi \in C_0^\infty(\Omega).$$

Proof (a) Define the sequence $\{w^n\}$ by

$$w^n(x) = \begin{cases} u^n(x)\Phi(x) : x \in \Omega, \\ 0 \qquad\qquad : x \in \mathbb{R}^N\setminus\Omega \end{cases}$$

for $\Phi \in C_0^\infty(\Omega)$ and w respectively. Using $\Phi^2 Q(u^n) = Q(w^n)$ for $x \in \Omega$, we have to show

$$\liminf_{n\to\infty} \int_{\mathbb{R}^N} Q(w^n(x))\, dx \geq \int_{\mathbb{R}^N} Q(w(x))\, dx.$$

Let $\tilde{Q} : \mathbb{C}^p \to \mathbb{R}, z \mapsto \bar{z}^T A z$ be the extension of Q to the complex plane. Application of the Plancherel formula – note that $\{w^n\} \subset [L^2(\Omega)]^p$ – yields

$$\int_{\mathbb{R}^N} \tilde{Q}(w^n(x))\, dx = \sum_{i,j=1}^p a_{ij} \int_{\mathbb{R}^N} w_i^n(x)\overline{w_j^n(x)}\, dx$$
$$= \sum_{i,j=1}^p a_{ij} \int_{\mathbb{R}^N} \hat{w}_i^n(\xi)\overline{\hat{w}_j^n(\xi)}\, d\xi \qquad (2.6.38)$$
$$= \int_{\mathbb{R}^N} \tilde{Q}(\hat{w}^n(\xi))\, d\xi . \qquad (2.6.39)$$

where $\hat{w}_i^n(\xi) = (2\pi)^{N/2} \int_{\mathbb{R}^N} e^{ix\cdot\xi} w_j^n(x)\, dx$ is the Fourier transform of w_j^n ($j = 1, \ldots, p$). Splitting up the integral in (2.6.39), we get for $R > 0$

$$\int_{\mathbb{R}^N} \tilde{Q}(\hat{w}^n(\xi))\, d\xi = \int_{|\xi|\leq R} \tilde{Q}(\hat{w}^n(\xi))\, d\xi + \int_{|\xi|>R} \tilde{Q}(\hat{w}^n(\xi))\, d\xi$$
$$=: A_n + B_n .$$

Let us begin with the term A_n. Because of $e^{-i(\cdot)\xi} \in L^2(K), K = \operatorname{supp} \Phi$ and $w^n \rightharpoonup w$ in $[L^2(\Omega)]^p$, we conclude that

$$\hat{w}^n(\xi) \to \hat{w}(\xi) \qquad (2.6.40)$$

pointwise for every $\xi \in \mathbb{R}^N$. Continuity of $\tilde{Q}$ guarantees $\tilde{Q}(\hat{w}^n(\xi)) \to \tilde{Q}(\hat{w}(\xi))$ pointwise for every $\xi \in \mathbb{R}^N$. Furthermore the Cauchy–Schwarz inequality gives

$$|\hat{w}_j^n(\xi)| \leq C \|w_j^n\|_{L^2(\Omega)} \quad (j = 1, \ldots, p).$$

Using the boundedness in $L^2(\Omega)$ of the weakly convergent sequence $\{w^n\} \subset L^2(\Omega)$ and again the continuity of $\tilde{Q}$, we arrive at

$$|\tilde{Q}(\hat{w}(\xi))| \leq C \quad \text{for all } \xi \in \mathbb{R}^N. \tag{2.6.41}$$

Then (2.6.40) and (2.6.41) enable us to apply Lebesgue's dominated convergence theorem on bounded domains:

$$A_n \overset{n \to \infty}{\longrightarrow} \int_{|\xi| \leq R} \tilde{Q}(\hat{w}(\xi)) \, d\xi. \tag{2.6.42}$$

Note that by the Plancherel formula

$$\int_{|\xi| \leq R} \tilde{Q}(\hat{w}(\xi)) \, d\xi = \int_{\mathbb{R}^N} \tilde{Q}(\hat{w}(\xi)) \, d\xi - \int_{|\xi| > R} \tilde{Q}(\hat{w}(\xi)) \, d\xi$$

$$= \int_{\mathbb{R}^N} \tilde{Q}(w(x)) \, dx - \int_{|\xi| > R} \tilde{Q}(\hat{w}(\xi)) \, d\xi. \tag{2.6.43}$$

Turning to the term B_n, Lemma 2.6.22 tells us that for all $\alpha > 0$ there exists a constant $C_\alpha > 0$ with

$$\tilde{Q}(\hat{w}^n(\xi)) \geq -\alpha |\hat{w}^n(\xi)|^2 - C_\alpha \left(\sum_{i=1}^{q} \left| \sum_{j=1}^{p} \sum_{k=1}^{N} a_{ijk} \frac{\xi_k}{|\xi|} \hat{w}_j^n(\xi) \right|^2 \right).$$

Integration with respect to ξ implies by the uniform bound $\|\hat{w}^n\|_{L^2} \leq M$ that

$$B_n \geq -\alpha \int_{|\xi| > R} |\hat{w}(\xi)|^2 \, d\xi - C_\alpha \int_{|\xi| > R} \sum_{i=1}^{q} \left| \sum_{j=1}^{p} \sum_{k=1}^{N} a_{ijk} \frac{|\xi_k|}{|\xi|} \hat{w}_j^n(\xi) \right|^2 d\xi$$

$$\geq -\alpha M - C_\alpha \int_{|\xi| > R} \sum_{i=1}^{q} \left| \sum_{j=1}^{p} \sum_{k=1}^{N} \frac{a_{ijk} \hat{w}_j^n(\xi)}{1 + |\xi|} \xi_k \right|^2 \left| \frac{1 + |\xi|}{|\xi|} \right|^2 d\xi.$$

Passing to the limit as $n \to \infty$ for R and α fixed, one obtains with Lemma 2.6.23

$$\liminf B_n \geq -\alpha M$$
$$-C_\alpha \sum_{i=1}^{q} \int_{|\xi|>R} \underbrace{\left|\sum_{j=1}^{p}\sum_{k=1}^{N} \frac{a_{ijk}\hat{w}_j(\xi)}{1+|\xi|}\xi_k\right|^2 \left|\frac{1+|\xi|}{|\xi|}\right|^2 d\xi}_{= I(\xi)} \quad (2.6.44)$$

Now (2.6.42), (2.6.43) and (2.6.44) imply

$$\liminf_{n\to\infty} \int_{\mathbb{R}^N} Q(w^n(\xi))\,d\xi = \liminf_{n\to\infty} \int_{\mathbb{R}^N} \tilde{Q}(w^n(\xi))\,d\xi$$
$$= \liminf_{n\to\infty} A_n + \liminf_{n\to\infty} B_n$$
$$\geq \int_{\mathbb{R}^N} \tilde{Q}(w(x))\,dx$$
$$- \int_{|\xi|>R} \tilde{Q}(\hat{w}(\xi))\,d\xi$$
$$- \alpha M - C_\alpha \sum_{i=1}^{q} \int_{|\xi|>R} I(\xi)\,d\xi\ .$$

Letting R be sufficiently large, we get – since $\alpha > 0$ was arbitrary – the statement of the theorem by $I \in L^1(\mathbb{R}^N)$.

(b) For $Q(\lambda) \leq 0$ for all $\lambda \in \Lambda$ we get in the same way as in (a)

$$\limsup_{n\to\infty} \int_\Omega \Phi^2(x)Q(u^n(x))\,dx$$
$$\leq \int_\Omega \Phi^2(x)Q(u(x))\,dx \quad \text{for all } \Phi \in C_0^\infty.$$

Then for $Q(\lambda) = 0$ for all $\lambda \in \Lambda$ we obtain b). □

REMARK 2.6.19 Recall that for $w \in H^1(\mathbb{R}^N, \mathbb{R}^N), N \geq 2, \operatorname{curl} w : \mathbb{R}^N \to \mathbb{R}^{N(N-1)/2}$ is defined by

$$(\operatorname{curl} w)_{1 \leq j < k \leq N} := (-1)^{j+k} \left(\frac{\partial w_j}{\partial x_k} - \frac{\partial w_k}{\partial x_j} \right).$$

Equipped with Theorem 2.6.18 we can formulate and prove the following lemma, which implies the statement of the div–curl Lemma 2.6.16.

LEMMA 2.6.20 Let $\{v^n\}, \{w^n\} \subset [L^2(\Omega)]^N$ be weakly convergent sequences to v and w respectively so that $\{\operatorname{div}(v^n)\}$ and $\{\operatorname{curl}(w^n)\}$ belong to compact subsets of $H_{\text{loc}}^{-1}(\Omega)$ and $[H_{\text{loc}}^{-1}(\Omega)]^{N(N-1)/2}$ respectively. Then

$$\sum_{i=1}^{N} v_i^n w_i^n \to \sum_{i=1}^{N} v_i w_i$$

in the distributional sense.

Proof First we have to show that the conditions (2.6.37) are satisfied. Let $u^n := \binom{v^n}{w^n}$ and for $i = 1$ define

$$a_{ijk} = a_{1jk} = \begin{cases} 0 & \text{if } j > N \\ \delta_{jk} & \text{otherwise .} \end{cases}$$

Then we have

$$\sum_{j=1}^{p} \sum_{k=1}^{N} a_{ijk} \partial_k u_j^n = \sum_{j=1}^{N} \sum_{k=1}^{N} a_{ijk} \partial_k u_j^n$$

$$= \sum_{k=1}^{N} \partial_k u_k^n = \operatorname{div} v^n .$$

Choose $i = 2$ and

$$a_{2jk} = \begin{cases} 1 & \text{if } j = s, \, k = 1 , \\ -1 & \text{if } j = r, \, k = 2 , \\ 0 & \text{otherwise ,} \end{cases}$$

where $r := N+1, s := N+2$. Then

$$\sum_{j=1}^{p}\sum_{k=1}^{N} a_{2jk}\partial_k u_j^n = \partial_1 u_s^n - \partial_2 u_r^n = \partial_1 w_2^n - \partial_2 w_1^n \ .$$

Similarly for $i = 3,\ldots, N(N-1)/2+1$ we get the other components of curl w^n. Therefore the conditions (2.6.37) are satisfied. For this choice of the a_{ijk} we obtain for $\lambda = (\lambda^1, \lambda^2) \subset \mathbb{R}^{2N}$

$$\sum_{j=1}^{p}\sum_{k=1}^{N} a_{1jk}\lambda_j\xi_k = \sum_{j=1}^{N}\sum_{k=1}^{N} \delta_{jk}\lambda_j\xi_k = \sum_{k=1}^{N} \lambda_k\xi_k = \sum_{k=1}^{N} \lambda_k^1\xi_k \ ,$$

$$\sum_{j=1}^{p}\sum_{k=1}^{N} a_{2jk}\lambda_j\xi_k = \xi_1\lambda_s - \xi_2\lambda_r = \xi_1\lambda_{N+2} - \xi_2\lambda_{N+1} = \xi_1\lambda_2^2 - \xi_2\lambda_1^2 \ ,$$

and similar expressions for $i = 3,\ldots, N$. Therefore in this case we get for V and Λ

$$V = \Big\{((\lambda^1, \lambda^2), \xi) \subset \mathbb{R}^{2N} \times \mathbb{R}^N\setminus\{0\}\Big| \sum_{j=1}^{N}\xi_j\lambda_j^1 = 0,$$

$$\xi_k\lambda_j^2 - \xi_j\lambda_k^2 = 0, 1 \leq j < k \leq N\Big\}$$

$$= \Big\{((\lambda^1, \lambda^2), \xi) \in \mathbb{R}^{2N} \times \mathbb{R}^N\setminus\{0\}\Big|\lambda^1 \perp \xi \text{ and } \lambda^2\|\xi\Big\}$$

$$\Lambda = \Big\{(\lambda^1, \lambda^2) \in \mathbb{R}^{2N}\Big|\lambda_1 \perp \lambda_2\Big\} \ .$$

Then define $A = \frac{1}{2}\begin{pmatrix} 0 & I \\ I & 0 \end{pmatrix}$ and for $\lambda = (\lambda^1, \lambda^2) \in \mathbb{R}^{2N}$

$$Q(\lambda) := \lambda^T A \lambda = \lambda^1\lambda^2 \ ,$$

and we obtain

$$Q(\lambda) = 0 \quad \text{for all } \lambda \in \Lambda \ .$$

Then Theorem 2.6.18 (b) gives the statement. $\square$

REMARK 2.6.21 *It can be shown that the scalar product is merely the only nonlinear function f such that $w^n \rightharpoonup w$ implies $f(w^n) \rightharpoonup f(w)$ (see [162]).*

LEMMA 2.6.22 *Let $Q(\lambda) \geq 0$ for all $\lambda \in \Lambda$. Then for each $\alpha > 0$ there exists a constant $C_\alpha > 0$ with*

$$\tilde{Q}(z) \geq -\alpha|z|^2 - C_\alpha \left(\sum_{i=1}^{q} \left| \sum_{j=1}^{p} \sum_{k=1}^{N} a_{ijk} \xi_k z_j \right|^2 \right)$$

for all $z \in \mathbb{C}^p$ and $\xi \in \mathbb{R}^N$, $|\xi| = 1$.

Proof We give an indirect proof. Assume that there exists some $\alpha_0 > 0$ so that for all $n \in \mathbb{N}$ there exist sequences $\{z^n\} \subset \mathbb{C}^p$ and $\{\xi^n\} \subset \mathbb{R}^N$ with $|\xi^n| = 1$ that fulfil

$$\tilde{Q}(z^n) < -\alpha_0 |z^n|^2 - n \left(\sum_{i=1}^{q} \left| \sum_{j=1}^{p} \sum_{k=1}^{N} a_{ijk} \xi_k^n z_j^n \right|^2 \right). \qquad (2.6.45)$$

Without loss of generality, we can divide (2.6.45) by $|z^n|^2$ and assume $|z^n| = 1$. Together with $|\xi^n| = 1$, we get some subsequences – also denoted by $\{z^n\}$, $\{\xi^n\}$ – that satisfy

$$z^n \to \tilde{z}, \quad \xi^n \to \tilde{\xi}.$$

The inequality (2.6.45) now implies

$$\sum_{i=1}^{q} \left| \sum_{j=1}^{p} \sum_{k=1}^{N} a_{ijk} \xi_k^n z_j^n \right|^2 < -\frac{\tilde{Q}(z^n)}{n} - \frac{\alpha_0}{n}$$

$$< -\frac{\tilde{Q}(z^n)}{n}.$$

So we have for $n \to \infty$, by the boundedness of $\tilde{Q}(z^n)$,

$$\sum \sum a_{ijk} \tilde{\xi}_k \tilde{z}_j = 0. \qquad (2.6.46)$$

Because of $\tilde{\xi} \neq 0$, this means by definition $\text{Re}\,\tilde{z}$ and $\text{Im}\,\tilde{z} \in \Lambda$. Using $Q(\lambda) \geq 0$ for all $\lambda \in \Lambda$, we get

$$\tilde{Q}(\tilde{z}) = \text{Re}\,\tilde{z}\, A\, \text{Re}\,\tilde{z} + \text{Im}\,\tilde{z}\, A\, \text{Im}\,\tilde{z} \geq 0\,.$$

But (2.6.45) implies $\tilde{Q}(z^n) \leq -\alpha_0 < 0$, and therefore $\tilde{Q}(\tilde{z}) < 0$. This is a contradiction. □

LEMMA 2.6.23 *Let $\{u^n\} \subset [L^2(\Omega)]^p$, supp $u^n \subset K$, $K \subset \Omega \subset \mathbb{R}^N$ compact, be a sequence satisfying conditions (2.6.37). Then there exists a subsequence of $\{u^n\}$ with*

$$\frac{1}{1+|\xi|}\tilde{A}_i\hat{u}^n(\xi) \to \frac{1}{1+|\xi|}\tilde{A}_i\hat{u}(\xi) \quad in\ [L^2(\mathbb{R}^N)]^p\,,$$

where

$$\tilde{A}_i\hat{u}^n(\xi) := \sum_{j=1}^{p}\sum_{k=1}^{N} a_{ijk}\xi_k\hat{u}_j^n\,.$$

Proof Because of the compactness of K and (2.6.37) there exists a subsequence of $\{A_i u^n\}$ ($i = 1, \ldots, q$) that converges strongly in $H^{-1}(\Omega)$. Linearity and continuity of $A_i : [L^2(\Omega)]^p \to H^{-1}(\Omega)$ gives with $u^n \rightharpoonup u$ in $[L^2(\Omega)]^p$

$$A_i u^n \to A_i u \quad \text{in } H^{-1}(\Omega) \text{ for the strongly convergent subsequence}\,.$$

The Fourier transform is an isometric mapping with respect to L^2. So we get analogously

$$\tilde{A}_i\hat{u}^n \to \tilde{A}_i\hat{u} \quad \text{in } H^{-1}(\mathbb{R}^N)\,.$$

By definition of $H^{-1}(\mathbb{R}^N)$, this means

$$\int_{\mathbb{R}^N} (1+|\xi|^2)^{-1}[A_i(\hat{u}^n - \hat{u})]^2\, d\xi \to 0\,.$$

Using $\sqrt{1+|\xi|^2} \leq 1+|\xi|$, we get the statement of the lemma. □

This finishes the proof of the convergence of the streamline diffusion method as defined in (2.6.5) for getting approximating solutions of (2.6.1).

REMARK 2.6.24 *A similar method on unstructured grids in 2-D or 3-D, the streamline diffusion shock-capturing finite element method, is described in [112] and [226]. For this scheme an L^∞ estimate can be proved, and also the convergence to the entropy solution. As far as we know, an L^∞ estimate for the streamline diffusion method (2.6.5) is still an open problem.*

3 Initial value problems for scalar conservation laws in 2-D

3.1 Dimensional splitting schemes for scalar conservation laws in 2-D

In this section we extend the schemes presented in §§2.2 and 2.5 to n-D on structured (or cartesian) grids. The details and proofs will be given for 2-D. The basic idea is to split the whole partial differential equation in several one-dimensional operators in the x- and y-directions respectively. Then in the first step the operator in the x-direction is solved with respect to the given initial values. The new values obtained by this step are used as initial values for the remaining operator in the second step. We define this algorithm more precisely and give the proof for the convergence to the entropy solution. As before the main idea for the proof consists in BV estimates for the discrete solution. The same ideas can be used in 3-D and n-D.

We shall consider the following scalar equation in 2-D:

$$\partial_t u + \partial_x f_1(u) + \partial_y f_2(u) = 0 \quad \text{in } \mathbb{R}^2 \times \mathbb{R}^+ , \tag{3.1.1}$$

$$u(\cdot, 0) = u_0 \quad \text{in } \mathbb{R}^2 \tag{3.1.2}$$

for given functions f_1 and f_2 and u_0. We always assume that f_1 and f_2 are sufficiently smooth. In a similar manner as in §2.1, we define a weak solution of (3.1.1), (3.1.2).

DEFINITION 3.1.1 (Weak solution) Let $u_0 \in L^\infty(\mathbb{R}^2)$. Then u is called a weak solution of (3.1.1), (3.1.2) if and only if $u \in L^1_{\text{loc}}(\mathbb{R}^2 \times \mathbb{R}^+)$ and if

$$\int_{\mathbb{R}^2}\int_{\mathbb{R}^+} [u\partial_t\varphi + f_1(u)\partial_x\varphi + f_2(u)\partial_y\varphi] + \int_{\mathbb{R}^2} u_0\varphi(\cdot,0) = 0$$

for all $\varphi \in C_0^\infty(\mathbb{R}^2 \times [0,\infty[)$.

As in the 1-D case, the weak solution is not uniquely defined. Therefore we add an entropy condition in order to select the solution that has a physical meaning. The entropy condition can be derived by using the viscosity limit as before (see [123]).

DEFINITION 3.1.2 (Entropy solution) A weak solution of (3.1.1), (3.1.2) is called an entropy solution if we have for all $\varphi \in C_0^\infty(\mathbb{R}^2 \times \mathbb{R}^+)$, $\varphi \geq 0$ and for all $k \in \mathbb{R}$

$$\int_{\mathbb{R}^2}\int_{\mathbb{R}^+} \Big\{\partial_t\varphi |u-k| + \partial_x\varphi \operatorname{sign}(u-k)[f_1(u)-f_1(k)] $$
$$+\partial_y\varphi \operatorname{sign}(u-k)[f_2(u)-f_2(k)]\Big\} \geq 0 \,.$$

This condition is also known as the Kruzkov entropy condition. Kruzkov [123] has shown that every entropy solution can be considered as a viscosity limit.

THEOREM 3.1.3 (Existence and uniqueness of an entropy solution in 2-D) Let $u_0 \in L^1(\mathbb{R}^2) \cap L^\infty(\mathbb{R}^2)$. Then there exists one and only one entropy solution u of (3.1.1), (3.1.2) and $u \in C^0([0,T], L^1(\mathbb{R}^2)) \cap L^\infty([0,T] \times \mathbb{R}^2)$.

For the proof we refer to Kruzkov [123]. We omit the details of this proof, since we shall outline the proof for the corresponding initial boundary value problem (see Theorem 6.2.10 for the details of the proof).

REMARK 3.1.4 *The existence problem in the BV spaces in many dimensions is also studied in [240].*

In this section we shall discuss the "dimensional splitting method" for solving (3.1.1), (3.1.2) numerically. This method is also called the "fractional step method" (see [254]) and can be described as follows. Instead of (3.1.1), (3.1.2), we consider the following one-dimensional problems:

$$\partial_t v + \partial_x f_1(v) = 0 \qquad v(\cdot, 0) = u_0 \qquad \text{on } \mathbb{R} \times [0, \Delta t], \qquad (3.1.3)$$
$$\partial_t w + \partial_y f_2(w) = 0 \qquad w(\cdot, 0) = v(\cdot, \Delta t) \quad \text{on } \mathbb{R} \times [0, \Delta t]. \qquad (3.1.4)$$

The functions v and w are assumed to be the uniquely defined entropy solution of (3.1.3), (3.1.4). Let us denote the corresponding solution operators of (3.1.1), (3.1.2), (3.1.3) and (3.1.4) by $S(t)$, $S^x(t)$ and $S^y(t)$; this means that

$$u(t) = S(t)u_0, \quad v(t) = S^x(t)u_0, \quad w(t) = S^y(t)v(\Delta t).$$

The basic idea of the dimensional splitting method is to approximate $S(T)u_0$ for $T := n\Delta t$ by

$$S(T)u_0 \approx [S^x(\Delta t)S^y(\Delta t)]^n u_0. \qquad (3.1.5)$$

The idea is due to Godunov [81]. This semidiscrete method is of first order in time. Strang [227] has improved this method using

$$S(T)u_0 \approx \left[S^y\left(\frac{\Delta t}{2}\right)S^x(\Delta t)S^y\left(\frac{\Delta t}{2}\right)\right]^n u_0, \qquad (3.1.6)$$

which is now of second order in time. It can also be shown that both methods converge if $n = T/\Delta t$ tends to infinity for fixed T.

EXAMPLE 3.1.5 (Dimensional splitting for linear equations)
Consider for $a, b \in \mathbb{R}$

$$\begin{aligned}\partial_t u + a\partial_x u + b\partial_y u &= 0 & \text{in } \mathbb{R}^2 \times \mathbb{R}^+, \\ u(x, y, 0) &= u_0(x, y) & \text{in } \mathbb{R}^2.\end{aligned} \qquad (3.1.7)$$

Then in the first step we solve

$$\partial_t v + a\partial_x v = 0, \qquad v(x,y,0) = u_0(x,y)$$

and obtain $v(x,y,t) = u_0(x-at, y)$, and in the second step we solve

$$\partial_t w + b\partial_y w = 0, \qquad w(x,y,0) = v(x,y,\Delta t) = u_0(x - a\Delta t, y)$$

and get

$$w(x,y,t) = w(x, y - bt, 0) = u_0(x - a\Delta t, y - bt).$$

Of course w solves (3.1.7).

THEOREM 3.1.6 (Convergence for the semidiscrete dimensional splitting scheme) *Assume that $u_0 \in L^1(\mathbb{R}^2) \cap L^\infty(\mathbb{R}^2)$ and $T > 0$ is fixed. Then we have*

$$\max_{0 \le t \le T} \|S(t)u_0 - [S^x(\Delta t) S^y(\Delta t)]^n u_0\|_{L^1(\mathbb{R}^2)} \longrightarrow 0,$$

$$\max_{0 \le t \le T} \left\|S(t)u_0 - \left[S^y\left(\frac{\Delta t}{2}\right) S^x(\Delta t) S^y\left(\frac{\Delta t}{2}\right)\right]^n u_0\right\|_{L^1(\mathbb{R}^2)} \longrightarrow 0$$

if $n = t/\Delta t \to \infty$.

Proof See Crandall and Majda [46, Theorem 1].

Since we are interested in numerical schemes for approximating entropy solutions of (3.1.1), (3.1.2), we shall study the problem if we replace S^x and S^y in (3.1.5) and (3.1.6) by some suitable 1-D difference operators. It turns out that in general we cannot expect convergence to the entropy solution of (3.1.1). For instance, if we use the corresponding 1-D Lax–Wendroff difference operators instead of S^x and S^y, it can be shown (see [46, §5]), that in the limit $n \to \infty$ the resulting solution will not satisfy the entropy condition. But for a large class of 1-D schemes the convergence to the entropy solution can be proved. For monotone schemes in 1-D this has been done by Crandall and Majda [45].

First let us describe the algorithm, and later on we shall show the precise result of Crandall and Majda. We start with initial value u_0 such that

$$u_0 \in L^1(\mathbb{R}^2) \cap L^\infty(\mathbb{R}^2) . \tag{3.1.8}$$

In order to start the algorithm, we have to define piecewise-constant initial values $U_{j,k}^0$. Let

$$U_{j,k}^0 := \frac{1}{\Delta x \Delta y} \int_{R_{j,k}} u_0(x,y)\, dx\, dy , \tag{3.1.9}$$

where

$$R_{j,k} := [(j-\tfrac{1}{2})\Delta x, (j+\tfrac{1}{2})\Delta x] \times [(k-\tfrac{1}{2})\Delta y, (k+\tfrac{1}{2})\Delta y] . \tag{3.1.10}$$

Let $g_1, g_2 : \mathbb{R}^2 \longrightarrow \mathbb{R}$ be the numerical fluxes of monotone schemes in 1-D, which are Lipschitz-continuous and consistent. Then define

$$U_{j,k}^{n+\frac{1}{2}} := U_{j,k}^n - \lambda_y[g_2(U_{j,k}^n, U_{j,k+1}^n) - g_2(U_{j,k-1}^n, U_{j,k}^n)] , \tag{3.1.11}$$

$$U_{j,k}^{n+1} := U_{j,k}^{n+\frac{1}{2}} - \lambda_x[g_1(U_{j,k}^{n+\frac{1}{2}}, U_{j+1,k}^{n+\frac{1}{2}}) - g_1(U_{j-1,k}^{n+\frac{1}{2}}, U_{j,k}^{n+\frac{1}{2}})] , \tag{3.1.12}$$

where

$$\lambda_x := \frac{\Delta t}{\Delta x} , \quad \lambda_y := \frac{\Delta t}{\Delta y} .$$

Now we shall extend $U_{j,k}^n$ to a function U_h that is defined on all of $\mathbb{R}^2$. Let $\chi_{j,k}^n$ be the characteristic function of $R_{j,k} \times [n\Delta t, (n+1)\Delta t]$, $h = \Delta t$ and

$$U_h(x,y,t) := \sum_{n=0}^{\infty} \sum_{j,k=-\infty}^{\infty} U_{j,k}^n \chi_{j,k}^n(x,y,t) , \tag{3.1.13}$$

where $t = n\Delta t$. Now we can give the precise statement concerning the convergence of the splitting scheme.

THEOREM 3.1.7 (Convergence of monotone schemes in 2-D) *Let $T > 0$ and $u_0 \in L^1(\mathbb{R}^2) \cap L^\infty(\mathbb{R}^2)$. Let $U_{j,k}^0$, $U_{j,k}^n$ and U_h be defined as in (3.1.9),(3.1.11), (3.1.12) and (3.1.13) respectively. The 1-D schemes in (3.1.11) and (3.1.12) are assumed to be monotone and the numerical fluxes g_1, g_2 to be Lipschitz-continuous and consistent. Then there exists a $u \in C^0([0,T], L^1(\mathbb{R}^2)) \cap L^\infty([0,T] \times \mathbb{R}^2)$ such that for $\Delta t \longrightarrow 0$ with fixed λ_x and λ_y, sufficiently small, we have*

$$\sup_{0 \le t \le T} \|u(t) - U_h(t)\|_{L^1(\mathbb{R}^2)} \longrightarrow 0$$

for each $T > 0$ and u is an entropy solution of (3.1.1), (3.1.2).

In the following we shall prove this result in several steps (see 3.1.17). First let us define monotonicity in 2-D.

DEFINITION 3.1.8 (Monotonicity in 2-D) *A difference scheme in $2 - D$*

$$U_{j,k}^{n+1} := H(U_{j-q,k-p}^n, \ldots, U_{j+q,k+p}^n) =: \vec{H}(U^n)_{j,k} \tag{3.1.14}$$

is called locally monotone *on $[a, b] \subset \mathbb{R}$ if and only if H is monotonically nondecreasing with respect to all of its arguments. The scheme is said to be in conservation form if it can be written in the form (see (3.1.11) and (3.1.12))*

$$\sum_{j,k} U_{j,k}^{n+1} = \sum_{j,k} U_{j,k}^n \tag{3.1.15}$$

REMARK 3.1.9 *The scheme defined in Theorem 3.1.7 is monotone in the sense of Definition 3.1.8 (as a composition of monotone schemes) if λ_x and λ_y are sufficiently small, and a scheme in conservation form.*

DEFINITION 3.1.10 (Functions of bounded variation) *We say that $f \in L^1(\mathbb{R}^2)$ is of bounded variation in $\mathbb{R}^2$, i.e. $f \in BV(\mathbb{R}^2)$, if and only if*

$$\|f\|_{BV(\mathbb{R}^2)} := \sup_{h \in \mathbb{R}\setminus\{0\}} \frac{1}{h} \int_{\mathbb{R}^2} |f(x+h, y) - f(x, y)|\, dx\, dy$$

$$+ \sup_{h \in \mathbb{R}\setminus\{0\}} \frac{1}{h} \int_{\mathbb{R}^2} |f(x, y+h) - f(x, y)|\, dx\, dy$$

is bounded. For piecewise-constant functions this definition corresponds to Definition 2.3.5.

DEFINITION 3.1.11 (Discrete L^1 and BV norms) Let $\Delta := \{(i\Delta x, j\Delta y)|\ i \in \mathbb{Z}, j \in \mathbb{Z}\}$ and let U be a grid function on Δ. Then we define

$$\|U\|_{L^1(\Delta)} := \sum_{i,j} |U_{i,j}|\, \Delta x\, \Delta y,$$

$$\|U\|_{BV(\Delta)} := \frac{1}{\Delta x}\|\Delta_+^x U\|_{L^1(\Delta)} + \frac{1}{\Delta y}\|\Delta_+^y U\|_{L^1(\Delta)}.$$

LEMMA 3.1.12 (Compactness in L^1) (a) Let $H \subset L^1(\mathbb{R}^2)$ be bounded and assume that

$$\lim_{\substack{h\to 0 \\ s\to 0}} \sup_{f\in H} \int_{\mathbb{R}^2} |f(x+h, y+s) - f(x,y)| = 0.$$

Then the set H is pre-compact in $L^1_{\text{loc}}(\mathbb{R}^2)$. (b) Bounded sets in $L^1(\mathbb{R}^2) \cap BV(\mathbb{R}^2)$ are pre-compact in $L^1_{\text{loc}}(\mathbb{R}^2)$.

Proof See [77].

LEMMA 3.1.13 (Theorem of Crandall and Tartar) Let $\Omega \subset \mathbb{R}^2$ be a measurable set and let $C \subset L^1(\Omega)$ be such that $f, g \in C$ implies $\max\{f, g\} \in C$. Let $T : C \longrightarrow L^1(\Omega)$ satisfy

$$\int_\Omega T(f) = \int_\Omega f \quad \text{for all } f \in C. \tag{3.1.16}$$

Then the following properties are equivalent:

(a) $f, g \in C$ and $f \leq g$ implies $T(f) \leq T(g)$ a.e.,

(b) $\displaystyle\int_\Omega (T(f) - T(g))^+ \leq \int_\Omega (f-g)^+$ for $f, g \in C$,

(c) $\displaystyle\int_\Omega |T(f) - T(g)| \leq \int_\Omega |f-g|$ for $f, g \in C$.

Proof First, we prove that (a) implies (b). By assumption, for $f, g \in C$ we have $h := \max\{f, g\} = g + (f - g)^+ \in C$ and the monotonicity of T implies $T(h) - T(f) \geq 0$ and therefore

$$T(f) - T(g) \leq T(h) - T(g) \quad \text{and} \quad [T(f) - T(g)]^+ \leq T(h) - T(g).$$

Then (b) follows from

$$\int_\Omega [T(f) - T(g)]^+ \leq \int_\Omega T(h) - T(g) = \int_\Omega h - g = \int_\Omega (f - g)^+.$$

(b) implies (c), since

$$\int_\Omega |T(f) - T(g)| = \int_\Omega [T(f) - T(g)]^+ + \int_\Omega [T(g) - T(f)]^+$$

$$\leq \int_\Omega (f - g)^+ + \int_\Omega (g - f)^+ = \int_\Omega |f - g|.$$

Now let us prove that (c) implies (a). We assume that $f, g \in C$, $f \leq g$ and that (c) holds. Then, using $2s^+ = |s| + s$, we obtain

$$2\int_\Omega [T(f) - T(g)]^+ = \int_\Omega |T(f) - T(g)| + \int_\Omega [T(f) - T(g)]$$

$$\leq \int_\Omega |f - g| + \int_\Omega (f - g) = 0,$$

since $f \leq g$. □

LEMMA 3.1.14 *Let $U_{j,k}, V_{j,k}$ be grid functions in $\mathbb{R}^2$ with step size Δx and Δy such that $a \leq U_{j,k}, V_{j,k} \leq b$ for all $j, k \in \mathbb{Z}$. Let G be a monotone operator on $[a, b]$ as in Definition 3.1.8. We assume that the corresponding scheme is consistent and in conservation form (see Definition 3.1.8). Then we have the following properties:*

(a) $\|G(U) - G(V)\|_{L^1(\Delta)} \leq \|U - V\|_{L^1(\Delta)}$ *if* $U, V \in L^1(\Delta)$;

(b) $\|G(U)\|_{BV(\Delta)} \leq \|U\|_{BV(\Delta)}$ *if* $U \in L^1(\Delta)$ *and* $U \in BV(\Delta)$;

(c) $\min_{l,m} U_{l,m} \leq G(U)_{j,k} \leq \max_{l,m} U_{l,m}$ for all $j, k \in \mathbb{N}$, where the $\min_{l,m}, \max_{l,m}$ is taken for $j - q \leq l \leq j + q$, $k - p \leq m \leq k + p$;

(d) $U \leq V \Rightarrow G(U) \leq G(V)$ pointwise for all grid points.

Proof (d) This is nothing else than the monotonicity of G, which has been assumed.

(c) Let us assume that j, k, p and q are fixed and defined as in Definition 3.1.8. Set

$$c := \max\{U_{l,m} | j - q \leq l \leq j + q, \ k - p \leq m \leq k + p\},$$

and let V be the constant grid function $V_{r,s} := c$ for all $r, s \in \mathbb{N}$. The conservation property implies $G(V) = V$, and since we have $V_{l,m} \geq U_{l,m}$ for all l, m as before,

$$\max_{l,m} U_{l,m} = c = V_{j,k} = G(V)_{j,k} \geq G(U)_{j,k}.$$

This can be done for all j and k. Similarly, one can prove the estimate from below.

(a) We apply Lemma 3.1.13 with $\Omega = \Delta$ and the discrete measure assigning the mass $\Delta x \Delta y$ to each point. Let

$$C := \{U \in L^1(\Delta) | a \leq U_{j,k} \leq b \text{ for all } j, k\}$$

and $T(U) := G(U)$. From part (c) we know that

$$\|G(U)\|_{L^1(\Delta)} = \sum_{j,k} |G(U)|_{j,k} \Delta x \Delta y$$

$$\leq \sum_{j,k} (\max\{|U_{l,m}||j - q \leq l \leq j + q,$$

$$k - p \leq m \leq k + p\}) \Delta x \Delta y$$

$$\leq \sum_{j,k} \sum_{\substack{l=j-q,\ldots,j+q \\ m=k-p,\ldots,k+p}} |U_{l,m}| \Delta x \Delta y.$$

In this sum each term $U_{j,k}$ occurs $r := (1+2p)(1+2q)$ times. Therefore it can be estimated by

$$\leq r \sum_{j,k} |U_{j,k}| \Delta x \Delta y$$

$$\leq r \|U\|_{L^1(\Delta)} . \qquad (3.1.17)$$

This means that $G : C \longrightarrow L^1(\Delta)$. Now we shall show that (3.1.16) is true. The conservation form yields

$$\int_\Delta G(U) = \sum_{i,k} G(U)_{j,k} \Delta x \Delta y = \sum_{i,k} U_{j,k} \Delta x \Delta y = \int_\Delta U . \qquad (3.1.18)$$

Therefore the assumptions of Lemma 3.1.13 are satisfied. Since G is monotone, (a) of Lemma 3.1.13 is true and therefore (c) of Lemma 3.1.13. But this is just the statement that we have to prove.

(b) We shall show that (a) implies (b). Set $(\tau U)_{j,k} := U_{j+1,k}$. By the definition of G, we have $\tau G(U) = G(\tau U)$. Using (a), we can show

$$\|\Delta_+^x G(U)\|_{L^1(\Delta)} = \|\tau G(U) - G(U)\|_{L^1(\Delta)}$$
$$= \|G(\tau U) - G(U)\|_{L^1(\Delta)} \leq \|\tau U - U\|_{L^1(\Delta)}$$
$$= \|\Delta_+^x U\|_{L^1(\Delta)} .$$

A similar estimate holds for $\Delta_+^y U$. Now using the definition of $\|U\|_{BV(\Delta)}$ (see Definition 3.1.11) we have proved (b). □

LEMMA 3.1.15 *Let us assume that the assumptions of Lemma 3.1.14 are satisfied and let U_h be defined as in (3.1.13). Let $u_0 \in L^1(\mathbb{R}^2)$, $a \leq u_0 \leq b$ and U^0 as in (3.1.9). Then there exists a constant C independent of U^0 and $\Delta t, \Delta x$ such that we have for all $0 \leq t_1, t_2$*

$$\int_{\mathbb{R}^2} |U_h(\cdot, t_2) - U_h(\cdot, t_1)| \leq C(|t_1 - t_2| + \Delta t) \|U^0\|_{BV(\mathbb{R}^2)} . \qquad (3.1.19)$$

Furthermore there is a constant C_1 such that

$$\frac{1}{\Delta t} \|G(U) - U\|_{L^1(\Delta)} \leq C_1 \|U\|_{BV(\Delta)} . \qquad (3.1.20)$$

3.1 Dimensional splitting schemes in 2-D

Proof Let $n > m$, $n\Delta t \leq t_2 \leq (n+1)\Delta t$, $m\Delta t \leq t_1 < (m+1)\Delta t$ and $U^n := U_h(\cdot, t_2)$, $U^m := U_h(\cdot, t_1)$. Then, using Lemma 3.1.14, we obtain

$$\begin{aligned}\|U^n - U^m\|_{L^1(\mathbb{R}^2)} &= \|G^n(U^0) - G^m(U^0)\|_{L^1(\Delta)} \\ &\leq \|G^{n-1}(U^0) - G^{m-1}(U^0)\|_{L^1(\Delta)} \\ &\leq \|G^{n-m}(U^0) - U^0\|_{L^1(\Delta)} \\ &\leq \|G^{n-m}(U^0) - G^{n-m-1}(U^0)\|_{L^1(\Delta)} + \\ &\quad \cdots + \|G(U^0) - U^0\|_{L^1(\Delta)} \\ &\leq (n-m)\|G(U^0) - U^0\|_{L^1(\Delta)} \\ &\leq \left(\frac{t_2 - t_1}{\Delta t} + 1\right)\|G(U^0) - U^0\|_{L^1(\Delta)}.\end{aligned}$$

Applying the estimate (3.1.20), we obtain (3.1.19). Therefore it remains to show (3.1.20). Since the scheme (3.1.14) is assumed to be in conservation form, it can be written in the form (3.1.11), (3.1.12) with $U^* = U^{n+1/2}$:

$$U^{n+1}_{j,k} = (G_1(U^*))_{j,k} = U^*_{j,k} - \lambda_x [g_1(U^*_{j,k}, U^*_{j+1,k}) - g_1(U^*_{j-1,k}, U^*_{j,k})],$$

(see [45, Proposition 3.5]). Then we have $(U^* := U^{n+1/2})$

$$\begin{aligned}\frac{1}{\Delta x \Delta y}&\|G_1(U^*) - U^*\|_{L^1(\Delta)} \\ &= \sum_{j,k} |(G(U^*))_{j,k} - U^*_{j,k}| \\ &\leq \sum_{j,k} \lambda_x |g_1(U^*_{j,k}, U^*_{j+1,k}) - g_1(U^*_{j-1,k}, U^*_{j,k})| \\ &\leq \sum_{j,k} \Big[\lambda_x |g_1(U^*_{j,k}, U^*_{j+1,k}) - g_1(U^*_{j,k}, U^*_{j,k})| \\ &\qquad + \lambda_x |g_1(U^*_{j,k}, U^*_{j,k}) - g_1(U^*_{j-1,k}, U^*_{j,k})|\Big] \\ &\leq L \sum_{j,k} \lambda_x |U^*_{j+1,k} - U^*_{j,k}| + \lambda_x |U^*_{j,k} - U^*_{j-1,k}| \\ &= \Delta t L \sum_{j,k} \frac{1}{\Delta x}|\Delta^x_+ U^*_{j,k}| + \frac{1}{\Delta x}|\Delta^x_+ U^*_{j-1,k}|\end{aligned}$$

$$\leq \frac{c\Delta t L}{\Delta x \Delta y}\left(\frac{1}{\Delta x}\|\Delta_+^x U^n\|_{L^1(\Delta)} + \frac{1}{\Delta y}\|\Delta_+^y U^n\|_{L^1(\Delta)}\right)$$

$$\leq \frac{c\Delta t L}{\Delta x \Delta y}\|U^n\|_{BV(\Delta)},$$

where L estimates the Lipschitz constant of g_1 and g_2. □

The following lemma ensures that a convergent sequence of grid functions as defined in (3.1.13) will converge to the entropy solution.

LEMMA 3.1.16 *Assume that (3.1.11) and (3.1.12) define a scheme in conservation form that is consistent and monotone on $[a,b]$. Let $U_{j,k}^0 \in [a,b]$ and let U_h ($h = \Delta t$) be the approximating sequence as defined in (3.1.13). Furthermore, assume that there exists an $U \in L^\infty(\mathbb{R}^2 \times [0,T])$ such that $U_h \longrightarrow U$ a.e. Then U is an entropy solution of (3.1.1) and (3.1.2).*

Similarly as in the Lax–Wendroff Theorem 2.3.1, it can be shown that U is a weak solution of the conservation law. For the statement concerning the entropy we refer to [45, Proposition 4.1]. Now let us prove Theorem 3.1.7.

3.1.17 (Proof of Theorem 3.1.7) Let us assume that u_0 has compact support. The more general case can be treated by approximation similar to that in Corollary 2.3.11. Lemma 3.1.14 (b) and (3.1.18) imply that

$$\|U_h(\cdot,t)\|_{L^1(\mathbb{R}^2)} \leq c\|U^0\|_{L^1(\mathbb{R}^2)},$$
$$\|U_h(\cdot,t)\|_{BV(\mathbb{R}^2)} \leq \|U^0\|_{BV(\mathbb{R}^2)}.$$

Therefore the set $\{U_h(\cdot,t)|0 \leq h \leq 1\}$ is bounded in $L^1(\mathbb{R}^2) \cap BV(\mathbb{R}^2)$ for any $0 \leq t \leq T$ and, using Lemma 3.1.12, pre-compact in $L^1_{\text{loc}}(\mathbb{R}^2)$. Lemma 3.1.15 implies that

$$\|U_h(\cdot,t_1) - U_h(\cdot,t_2)\|_{L^1(\mathbb{R}^2)} \leq C(|t_1 - t_2| + \Delta t)\|U^0\|_{BV(\mathbb{R}^2)}. \quad (3.1.21)$$

But now we can apply the ideas of the proof of the Arzelá–Ascoli theorem as in Theorem 2.3.9. We obtain a subsequence $h \to 0$ and a function $U \in C^0([0,\infty[; L^1(\mathbb{R}^2))$ such that

$$\lim_{h \to 0} \max_{0 \leq t \leq T} \|U_h(\cdot,t) - U(\cdot,t)\|_{L^1(\mathbb{R}^2)} = 0, \quad (3.1.22)$$

and consequently

$$\|U(\cdot,t_1) - U(\cdot,t_2)\|_{L^1(\mathbb{R}^2)} \leq C|t_1 - t_2| \cdot \|U^0\|_{BV(\mathbb{R}^2)} \ . \tag{3.1.23}$$

Passing to another subsequence, we have the pointwise convergence a.e. of U_h to U on $\mathbb{R}^2 \times [0,T[$. Using Lemma 3.1.14 (c) and the assumption $U_0 \in L^\infty(\mathbb{R}^2)$, we obtain that $U \in L^\infty(\mathbb{R}^2 \times [0,T])$. Therefore Lemma 3.1.16 implies that the limit U is the entropy solution of (3.1.1). Because of (3.1.19) and (3.1.23), the initial condition (3.1.2) is satisfied in the sense of $C^0([0,T[, L^1(\mathbb{R}^2))$, since

$$\int_{\mathbb{R}^2} |U(x,\cdot) - u_0(x)| \ dx \leq \int_{\mathbb{R}^2} |U(x,\cdot) - U_h(x,\cdot)| \ dx$$
$$+ \int_{\mathbb{R}^2} |U_h(x,\cdot) - u_0(x)| \ dx \ .$$

The first term converges to zero because of (3.1.22) and the second because of (3.1.21). This finishes the proof of Theorem 3.1.7. □

The result of Theorem 3.1.7 can be generalized to inhomogeneous equations in N space dimensions of the form

$$\partial_t U + \sum_{i=1}^N \partial_i f_i(U) = F(t,x) \ ,$$
$$U(x,0) = U_0(x) \ . \tag{3.1.24}$$

Then the algorithm looks like

$$U_{j,k}^{n+1} := G(U_{j-q,k-r}^n, \ldots, U_{j+q+1,k+s+1}^n) + \Delta t F_{j,k}^n \ . \tag{3.1.25}$$

The convergence of this scheme can be proved in the case of the following theorem.

THEOREM 3.1.18 (Dimensional splitting for inhomogeneous equations) Let $U_0 \in L^1(\mathbb{R}^2) \cap L^\infty(\mathbb{R}^2)$, $F \in L^\infty(\mathbb{R}^2 \times]0,T[) \cap L^1(\mathbb{R}^2 \times]0,T[)$ and $a, b \in \mathbb{R}$ such that

$$a + T\|F\|_{L^\infty} \leq U_0 \leq b - T\|F\|_{L^\infty} \quad \text{a.e.},$$

$$F_{j,k}^n := \frac{1}{\Delta t \Delta x \Delta y} \int_{n\Delta t}^{(n+1)\Delta t} \int_{R_{j,k}} F,$$

where $R_{j,k}$ is defined in (3.1.10). Let $U_{j,k}^0$, $U_{j,k}^n$ and $U_h b$ be defined as in (3.1.9), (3.1.25) and (3.1.13) respectively. The scheme is assumed to be in conservation form, consistent, monotone and Lipschitz-continuous. Then we have

$$a \leq U_h \leq b \quad \text{for all } n, j, k, \Delta t n \leq T,$$

and U_h converges in $L^1(\mathbb{R}^2)$ uniformly on compact subsets of $[0,T[$ to the entropy solution of (3.1.1).

For the proof we refer to [45, Theorem 6.1].

REMARK 3.1.19 (Higher-order dimensional splitting schemes)
If we use 1-D higher-order methods as in §2.5 instead of the monotone schemes in (3.1.11) and (3.1.12), convergence can be shown if we use slightly modified limiters of those defined in Example 5.10. For more details we refer to [41] and [42]. We shall show the convergence to the entropy solution in a more general situation for finite volume methods of higher order on unstructured grids in many dimensions (see §3.5).

REMARK 3.1.20 Another method for approximating the solution of conservation laws has been proposed by Dafermos [47], where the flux function is approximated by a polygon, i.e. a continuous, piecewise-linear function. In [98] and [99] this approach has been developed into a numerical method in 1-D. In [41] and [42] the convergence for the generalization to 2-D as a fractional step method was shown.

REMARK 3.1.21 (Error estimates) *Kuznetsov [124] has shown error estimates of the form*

$$\|u(t) - u_h(t)\|_{L^1(\mathbb{R}^m)} \leq c h^{1/2}$$

for dimensional splitting schemes on uniform cartesian grids. We shall use this technique later for error estimates for finite volume methods on unstructured grids (see §3.4).

3.2 Finite volume schemes in 2-D

Up to now we have considered finite difference methods on structured grids. Now in this section we shall consider finite volume discretizations on unstructured grids for hyperbolic conservation laws in two space dimensions

$$\partial_t u + \partial_x f_1(u) + \partial_y f_2(u) = 0 \quad \text{in } \mathbb{R}^2 \times \mathbb{R}^+ , \qquad (3.2.1)$$

with initial values

$$u(x, y, 0) = u_0(x, y) \quad \text{in } \mathbb{R}^2 , \qquad (3.2.2)$$

where $u : \mathbb{R}^2 \times \mathbb{R}^+ \to \mathbb{R}$, $f_1, f_2 : \mathbb{R} \to \mathbb{R}$ and $u_0 \in L^\infty(\mathbb{R}^2)$. We assume that f_1 and f_2 are sufficiently smooth, $f_1(0) = 0$ and $f_2(0) = 0$. Essentially there are three different numerical methods that are used in general for solving (3.2.1), (3.2.2). There are the finite difference methods on cartesian grids in the form of dimensional splitting as described in § 3.1 (or fractional steps) [40, 41, 42, 45], and the finite volume [29, 119, 160, 70] and finite element [111, 231, 101, 107] methods on arbitrary meshes. In order to treat complex geometries and to do local mesh refinement, it is preferable to use unstructured grids consisting of triangles, quadrangles or more general polygons. Concerning the geometry, it is easier to adapt the unstructured cells to the boundary of the domain. Concerning the adaptivity, mesh refinement can be done really locally without any non conforming nodes. On a cartesian grid local mesh refinement always has some global effects or one has to use unstructured grids locally to avoid non conforming nodes. If one decides to work on unstructured grids then one can use finite elements as in § 2.6 or finite volume methods. One always has to take care of the upwinding or the numerical damping that is necessary for the discretization of conservation

156 Initial value problems for scalar conservation laws in 2-D

laws. Since it is easier to combine the upwinding with finite volume discretization than with finite element methods, we study in this section the finite volume methods in more detail.

In the elliptic case the situation is much easier, since in general upwinding is not necessary (see [85], [93] and [94]).

First we need some notations and basic assumptions.

DEFINITION 3.2.1 *Let a k-polygon be a closed polygon with k vertices. The set*

$$T := \{T_i \mid T_i \text{ is a } k\text{-polygon for } i \in I \subseteq \mathbb{N}\},$$

(where $I \subseteq \mathbb{N}$ is an index set) is called an unstructured grid *of $\Omega \subset \mathbb{R}^2$ if the following two properties are satisfied.*

1. $\Omega = \bigcup_{i \in I} T_i$.

2. *For two different T_i and T_j we have $T_j \cap T_i = \emptyset$ or $T_j \cap T_i = $ a common vertex of T_i and T_j or $T_j \cap T_i = $ a common edge of T_i and T_j.*

NOTATION 3.2.2 *Let $(T_j)_j$ denote an unstructured grid of the $\mathbb{R}^2$. In particular, we shall use the following notation:*

- T_j : the jth cell of the unstructured grid;

- $|T_j|$: area of T_j;

- w_j: centre of gravity of T_j;

- T_{jl}, $l = 1, \ldots, k$: neighbouring cells of T_j;

- $\alpha(j, l)$: global number of the l-th neighbouring cell T_{jl} of T_j such that $T_{jl} = T_{\alpha(j,l)}$;

- w_{jl}, $l = 1, \ldots, k$: centres of gravity of T_{jl}, $w_{jl} = w_{\alpha(j,l)}$;

- u_j^n: approximation of the exact solution u on T_j at time $n\Delta t$; the function u_j^n is assumed to be constant on T_j;

- u_{jl}^n: approximation of the exact solution u on T_{jl} at time $n\Delta t$; $u_{jl}^n = u_{\alpha(j,l)}^n$;

- $u_j := u_j^n$;
- S_{jl}: lth edge of T_j;
- z_{jl}: midpoint of the lth edge of the cell j;
- ν_{jl}: outer normal to S_{jl} of length $|S_{jl}|$;
- $n_{jl} := \nu_{jl}/|\nu_{jl}|$; we denote the x- and y-coordinates of n_{jl} respectively, by n_{jlx} and n_{jly};
- $h := \sup_{j \in I} \operatorname{diam}(T_j)$;
- $u^n(z) := u_j^n$ if $z \in T_j$;
- $u_h(z,t) := u_j^n$ if $z \in T_j$ and $t^n \leq t < t^{n+1}$, $n = 0, 1, \ldots$

Figure 3.2.1 illustrates the notation for unstructured grids of quadrangles ($k = 4$) and triangles ($k = 3$), respectively. Also the dual cells (see Definition 3.7.6 and Example 7.1.5) of a cell-vertex mesh can be considered as an unstructured cell-centred grid. In Figure 3.2.2 this situation is pointed out for hexagons as dual cells.

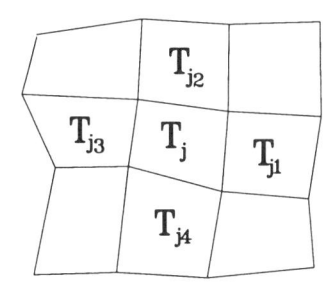

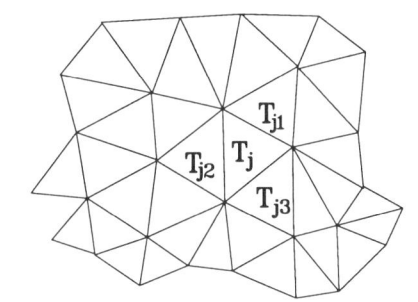

Figure 3.2.1

NOTATION 3.2.3 Simultaneously we use a global and a local numbering. For any j and l there exists a unique number m such that for $i := \alpha(j, l)$ we have

$$\alpha(i, m) = j.$$

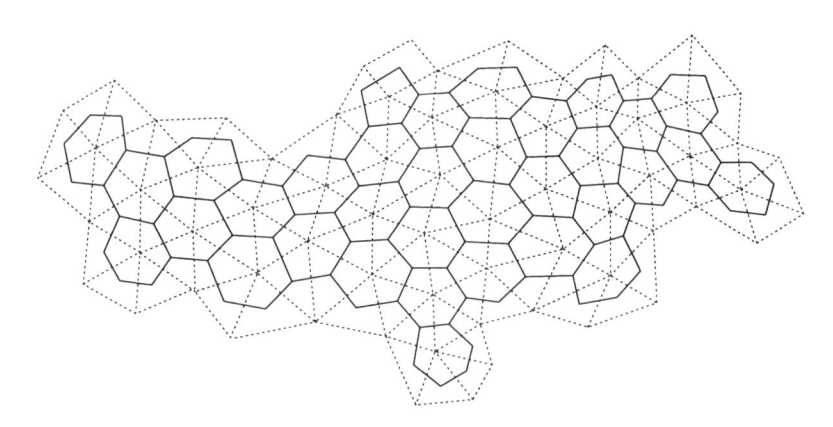

Figure 3.2.2

As we use the notation $\nu_{j,\alpha(j,l)} := \nu_{jl}$, it turns out that $\nu_{im} = -\nu_{jl}$.

ASSUMPTION 3.2.4 We assume that there exist constants $c_1, c_2 \geq 0$ such that

$$0 < c_1 \leq \frac{\Delta t}{h} \leq c_2 \qquad (3.2.3)$$

as $\Delta t, h \to 0$. Moreover, we assume that there exists a constant $c_v > 0$ such that

$$\sup_j \frac{h^2}{|T_j|} \leq c_v . \qquad (3.2.4)$$

This means that the smallest angle in the triangulation remains bounded from below uniformly in h.

3.2.5 (Idea of finite volume methods) Let φ be in $C^\infty(\mathbb{R}^2, \mathbb{R}^2)$. Then on a single triangle T_j we have (n is the outer unit normal)

$$\operatorname{div} \varphi = \frac{1}{|T_j|} \int_{T_j} \operatorname{div} \varphi + \mathcal{O}(h)$$

$$= \frac{1}{|T_j|} \int_{\partial T_j} n\varphi + \mathcal{O}(h) = \frac{1}{|T_j|} \sum_{l=1}^{3} \int_{S_{jl}} n_{jl}\varphi + \mathcal{O}(h)$$

$$= \frac{1}{|T_j|} \sum_{l=1}^{3} n_{jl}\,\varphi(z_{jl})|S_{jl}| + \mathcal{O}(h)$$

$$= \frac{1}{|T_j|} \sum_{l=1}^{3} \nu_{jl}\,\varphi(z_{jl}) + \mathcal{O}(h) \ .$$

Therefore

$$\operatorname{div} \varphi|_{T_j} = \frac{1}{|T_j|} \sum_{l=1}^{3} \nu_{jl}\,\varphi(z_{jl}) + \mathcal{O}(h) \ . \tag{3.2.5}$$

This formula gives us the basic idea for finite volume methods. Since for piecewise-constant φ the term $\varphi(z_{jl})$ makes no sense we replace it in this case by some weighted meanvalue $g_{jl}(\varphi(w_j), \varphi(w_{jl}))$ of the adjacent values $\varphi(w_j), \varphi(w_{jl})$, where g_{jl} is similar to the numerical fluxes that we know already from the one dimensional schemes. This means that we consider

$$\frac{1}{|T_j|} \sum_{l=1}^{3} g_{jl}(\varphi(w_j), \varphi(w_{jl}))$$

instead of (3.2.5).

DEFINITION 3.2.6 (Finite volume scheme) *For given initial values $u_0 \in L^\infty(\mathbb{R}^2)$ let u_j^n be defined by the following numerical scheme:*

$$u_j^0 := \frac{1}{|T_j|} \int_{T_j} u_0 \ , \tag{3.2.6}$$

$$u_j^{n+1} := u_j^n - \frac{\Delta t}{|T_j|} \sum_{l=1}^{k} g_{jl}(u_j^n, u_{jl}^n) \ , \tag{3.2.7}$$

where for g_{jl}, $l = 1, \ldots, k$, we assume that for any $R > 0$ and for all $u, v, u', v' \in B_R(0)$ we have

$$|g_{jl}(u,v) - g_{jl}(u',v')| \leq c(R)\,h\,(|u-u'| + |v-v'|) \ , \tag{3.2.8}$$
$$g_{j,\alpha(j,l)}(u,v) = -g_{\alpha(j,l),j}(v,u) \ , \tag{3.2.9}$$
$$g_{j,\alpha(j,l)}(u,u) = \nu_{j,\alpha(j,l)}f(u) \ , \tag{3.2.10}$$

where

$$f(u) := \begin{pmatrix} f_1(u) \\ f_2(u) \end{pmatrix}$$

and where we have used the notation $g_{j,\alpha(j,l)} := g_{jl}$. The condition (3.2.8) is a local Lipschitz condition, (3.2.9) is the conservation property and (3.2.10) consistency.

Now we shall consider some examples for numerical schemes in 2-D that satisfy the conditions given in Definition 3.2.6.

EXAMPLE 3.2.7 Consider (3.2.1), (3.2.2) as a scalar conservation law in 2-D. Let

$$c_{jl}(u) := n_{jl} f(u),$$

and

$$\begin{aligned} c_{jl}^+(u) &:= c_{jl}(0) + \int_0^u \max\{c'_{jl}(s), 0\} ds, \\ c_{jl}^-(u) &:= \int_0^u \min\{c'_{jl}(s), 0\} ds. \end{aligned} \quad (3.2.11)$$

(In 1-D this leads us just to the Engquist–Osher scheme – see [62] and Example 2.2.7.) Now let

$$g_{jl}(u,v) := |S_{jl}| \left[c_{jl}^+(u) + c_{jl}^-(v) \right]. \quad (3.2.12)$$

Then g_{jl} satisfies the conditions (3.2.8)–(3.2.10).

Proof The condition (3.2.8) is obvious. Let us prove (3.2.9). For $i := \alpha(j,l)$ we have

$$\begin{aligned} g_{jl}(u,v) = g_{ji}(u,v) &= [\nu_{ji} f(u)]^+ + [\nu_{ji} f(v)]^- \\ &= [-\nu_{ij} f(u)]^+ + [-\nu_{ij} f(v)]^- \\ &= |S_{ij}| \left[(-c_{ij})^+(u) + (-c_{ij})^-(v)\right] \\ &= -|S_{ij}| \left[c_{ij}^+(v) + c_{ij}^-(u)\right] = -g_{ij}(v,u). \end{aligned}$$

Here we have used $[-c_{ij}(u)]^+ = -c_{ij}(u)^- - c_{ij}(0)$ and $[-c_{ij}(u)]^- = -c_{ij}(u)^+ + c_{ij}(0)$.

The property (3.2.10) follows from

$$g_{jl}(u,u) = |S_{jl}| c_{jl}(u) = \nu_{jl} f(u).$$ □

In the following lemma we shall study the pointwise consistency of finite volume schemes in 2-D. Similar situations have been considered in [201] and [202]. It turns out that in general for an unstructured grid one cannot expect pointwise consistency.

LEMMA 3.2.8 *Let $(T_j)_j$ be an unstructured grid satisfying Definition 3.2.1, Notation 3.2.3 and Assumption 3.2.4, and let $g_{jl}(u,v)$ be the numerical flux for the numerical scheme (3.2.7), satisfying (3.2.8)–(3.2.10). Then in general the scheme (3.2.7) is not pointwise-consistent. This means that if u is a smooth solution of (3.2.1), u^n need not necessarily satisfy*

$$\frac{u_j^{n+1} - u_j^n}{\Delta t} + \frac{1}{|T_j|} \sum_{l=1}^{k} g_{jl}(u_j, u_{jl}) = \mathcal{O}(h). \tag{3.2.13}$$

Proof A counterexample can be given for the linear convection equation

$$\partial_t u + a\partial_x u + b\partial_y u = 0 \quad \text{in} \quad \mathbb{R}^2 \times \mathbb{R}^+. \tag{3.2.14}$$

In this case the scheme defined in Example 3.2.7 reduces to the scheme where the numerical flux can be written in the following form:

$$g_{jl}(u_j, u_{jl}) = q_{jl}^+ u_j + q_{jl}^- u_{jl}$$

$$= \frac{q_{jl}}{2} [u_j + u_{jl} + \text{sgn}(q_{jl})(u_j - u_{jl})],$$

where we use the notation

$$q_{jl} := \nu_{jl} \begin{pmatrix} a \\ b \end{pmatrix} \tag{3.2.15}$$

162 Initial value problems for scalar conservation laws in 2-D

Let u be a smooth solution of (3.2.14) and $u_j^n = u(w_j, t^n)$. Then we have

$$u_j^{n+1} = u_j^n + \partial_t u_j^n \Delta t + \mathcal{O}(\Delta t^2),$$

$$u_{ji} := u_{ji}^n = u_j^n + \nabla u_j^n (w_{ji} - w_j) + \mathcal{O}(h^2),$$

and therefore

$$Q := \frac{u_j^{n+1} - u_j^n}{\Delta t} + \frac{1}{|T_j|} \sum_{l=1}^{k} g_{jl}(u_j, u_{jl})$$

$$= \partial_t u_j^n + \mathcal{O}(\Delta t) + \frac{1}{2|T_j|} \sum_{l=1}^{k} q_{jl} \Big\{ 2u_j + \nabla u_j (w_{jl} - w_j) + \mathcal{O}(h^2)$$

$$+ \frac{|q_{jl}|}{q_{jl}} [-\nabla u_j (w_{jl} - w_j)] \Big\}$$

$$= \partial_t u_j^n + \mathcal{O}(\Delta t) + \frac{u_j}{|T_j|} \sum_{l=1}^{k} q_{jl}$$

$$+ \frac{1}{2|T_j|} \sum_{l=1}^{k} q_{jl} \nabla u_j (w_{jl} - w_j) \left(1 - \frac{|q_{jl}|}{q_{jl}}\right) + \mathcal{O}(h). \qquad (3.2.16)$$

The last term results from $q_{jl} = \mathcal{O}(h)$ (see (3.2.15) and (3.2.4)). Since

$$\sum_l q_{jl} = \binom{a}{b} \sum_l \nu_{jl} = 0, \qquad (3.2.17)$$

the first sum in (3.2.16) cancels and we get

$$Q = \partial_t u_j^n + \frac{1}{2|T_j|} \sum_{l=1}^{k} q_{jl} \nabla u_j (w_{jl} - w_j) \left(1 - \frac{|q_{jl}|}{q_{jl}}\right) + \mathcal{O}(h) \quad (3.2.18)$$

$$= \partial_t u_j^n + \frac{1}{|T_j|} \sum_{q_{jl} < 0} q_{jl} \nabla u_j (w_{jl} - w_j) + \mathcal{O}(h). \qquad (3.2.19)$$

Using the conservation law (3.2.14), we obtain

$$Q = -\nabla u_j \binom{a}{b} + \frac{1}{|T_j|} \sum_{q_{jl} < 0} q_{jl} \nabla u_j (w_{jl} - w_j) + \mathcal{O}(h) \qquad (3.2.20)$$

$$= \nabla u_j \left[\sum_{q_{jl}<0} \frac{q_{jl}(w_{jl} - w_j)}{|T_j|} - \begin{pmatrix} a \\ b \end{pmatrix} \right] + \mathcal{O}(h). \tag{3.2.21}$$

Now, because of (3.2.15), q_{jl} has the same sign as $\cos\varphi$, where φ is the angle between the vectors ν_{jl} and $\binom{a}{b}$. Using this fact, we can easily construct the element T_j and the neighbouring cell T_{j1} (see Figure 3.2.3) with any given number of edges k in such a way that the side S_{j1} is parallel to the x-axis and the angles $\alpha_{12}, \alpha_{1k} < \frac{1}{2}\pi - \varepsilon$.

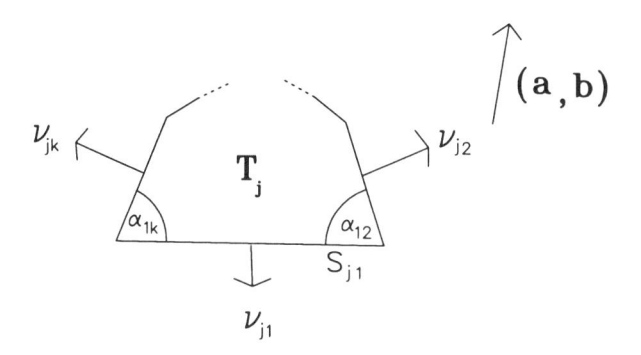

Figure 3.2.3

Then there exist $a \neq 0$, b such that

$$q_{jl} > 0, \quad l = 2, \ldots, k, \quad q_{j1} < 0. \tag{3.2.22}$$

Furthermore, we assume that $(w_{j1} - w_j)_x = 0$ (x-coordinate of $w_{j1} - w_j$), and therefore (3.2.21) reduces to

$$Q = \partial_x u_j(-a) + \partial_y u_j \left[\frac{q_{j1}(w_{j1} - w_j)_y}{|T_j|} - b \right] + \mathcal{O}(h). \tag{3.2.23}$$

As $a \neq 0$, it turns out that (3.2.23) cannot be $\mathcal{O}(h)$. This proves the lemma. □

REMARK 3.2.9 *We emphasize that (3.2.23) holds in particular for any equal-sided triangle one of whose edges is parallel to the x-axis.*

3.3 Convergence of finite volume schemes in 2-D on unstructured grids

In this section we present the proof given in [120] for the convergence of upwind finite volume schemes. The basic idea consists in the concept of measure valued solution, first developed by DiPerna [53, 54]. By the maximum principle, it is easy to get an L^∞ bound for the sequence (u_h) of discrete solutions that converges weak-$*$. That is not enough to get a weak solution in the limit $h \to 0$ since the non linearity cannot be controlled. But it is still possible to get a measure-valued solution ν in the limit. In order to obtain a weak solution, one has to show that ν is a point measure in every point $(x,t) \in \mathbb{R}^2 \times \mathbb{R}^+$. This can be done by analysing the entropy condition very carefully. In § 3.4 another proof of the convergence of the finite volume method will be given that will not use the theory of measure-valued solutions. But the result in § 3.5 concerning higher-order schemes is also based on this theory. Therefore the reader who is only interested in the convergence result for first-order schemes can skip this section and continue with § 3.4.

For the following notation we refer to [53] and [64]. By $K \subset\subset \mathbb{R}^2$ we denote a compact set in $\mathbb{R}^2$.

DEFINITION 3.3.1 (Measure-valued solution) *Let*

$$\text{Prob}(\mathbb{R}) := \{\mu \mid \mu \text{ is a probability measure on } \mathbb{R}\} \ .$$

Then a Young measure ν, i.e. is a measurable map

$$\nu : \mathbb{R}^2 \times [0,T] \to \text{Prob}(\mathbb{R})$$

is a measure-valued solution *to the conservation law (3.2.1) if we have*

$$\partial_t \langle \nu, \text{id} \rangle + \partial_x \langle \nu, f_1 \rangle + \partial_y \langle \nu, f_2 \rangle = 0$$

in the distributional sense. Here we have used the notation

$$\langle \nu, f \rangle (x,y,t) := \int_{\mathbb{R}} f(\lambda) \, d\nu_{x,y,t}(\lambda) \ .$$

3.3 Convergence of finite volume schemes in 2-D on unstructured grids

REMARK 3.3.2 *If $\nu(x,y,t) = \delta_{u(x,y,t)}$, where δ_u concentrated in u is the Dirac distribution and u the exact weak solution, then Definition 3.3.1 reduces to the definition of weak solutions.*

THEOREM 3.3.3 (Tartar's theorem) *Let $(u_h)_{h>0}$ denote a family of functions mapping $\mathbb{R}^m \to \mathbb{R}^n$ that is bounded in $L^\infty(\mathbb{R}^m, \mathbb{R}^n)$ with $\|u_h\|_{L^\infty} \le M$. Then there is a subsequence $(u_h)_h$ and a Young measure ν with support in a ball of radius M a.e. (x,t) such that for all continuous g, the weak-* limit of $g(u_h)$ exists and*

$$g(u_h) \overset{*}{\rightharpoonup} \langle \nu, g \rangle.$$

Proof See [236] and [235].

DEFINITION 3.3.4 (Consistency with the entropy pair) *Let $F_1, F_2 \in C^1(\mathbb{R})$, $U \in C^2(\mathbb{R})$ be defined such that U is convex and*

$$U'(s) f_1'(s) = F_1'(s), \quad U'(s) f_2'(s) = F_2'(s)$$

for all $s \in \mathbb{R}$. Then (U, F_1, F_2) is called an entropy pair *for the conservation law (3.2.1). Now a Young measure ν is said to be consistent with the entropy pair (U, F_1, F_2) if we have*

$$\partial_t \langle \nu, U \rangle + \partial_x \langle \nu, F_1 \rangle + \partial_y \langle \nu, F_2 \rangle \le 0 \tag{3.3.1}$$

in the distributional sense. If (3.3.1) is satisfied for all entropy pairs (U, F_1, F_2) and if ν is a measure-valued solution of (3.2.1) then ν is called the entropy measure-valued *or* admissible measure-valued *solution.*

The most important tool for proving the convergence will be the following Theorem of DiPerna [53].

THEOREM 3.3.5 *Let us assume that $u_0 \in L^1(\mathbb{R}^2) \cap L^\infty(\mathbb{R}^2)$ and that there is a Young measure ν satisfying the following properties:*

(a) *the function $(x, y, t) \to \langle \nu_{x,y,t}, |\text{id}| \rangle$ is in $L^\infty([0,T], L^1(\mathbb{R}^2))$;*

(b) *ν is a measure-valued solution to the conservation law (3.2.1), (3.2.2);*

(c) *ν is consistent with all entropies (U, F_1, F_2);*

(d) ν assumes the initial values u_0 in the following sense:

$$\lim_{t\to 0^+,\, t>0} \frac{1}{t} \int_0^t \int_{\mathbb{R}^2} \langle \nu_{x,y,s}, |\mathrm{id} - u_0(x,y)| \rangle \, dx\, dy\, ds = 0 \,.$$

Then the Young measure reduces to a Dirac measure; that is,

$$\nu_{x,y,t} = \delta_{u(x,y,t)} \quad \text{a.e. in } \mathbb{R}^2 \times [0,T]\,,$$

where $u \in L^\infty(\mathbb{R}^2 \times [0,T])$ is the unique entropy solution of the problem (3.2.1), (3.2.2) in the sense of Kruzkov [123]. (For existence see Theorem 3.1.3.)

COROLLARY 3.3.6 (Strong convergence) Let u_h be defined as in Notation 3.2.2, (3.2.6) and (3.2.7) and let $\|u_h\|_{L^\infty(\mathbb{R}^2 \times [0,T])}$ be uniformly bounded in h. Then the sequence u_h converges to u strongly in L^1_{loc} if and only if the Young measure ν reduces at almost all points (x,y,t) to the Dirac measure concentrated in $u(x,y,t)$.

For application of these results in the context of conservation laws it is useful to apply the following sufficient condition for the property (d) of Theorem 3.3.5.

THEOREM 3.3.7 Assume that $u_0 \in L^\infty(\mathbb{R}^2) \cap L^1(\mathbb{R}^2)$ and that $\nu : \mathbb{R}^2 \times [0,T] \to \mathrm{Prob}(\mathbb{R})$ is a Young measure. Furthermore, assume that the conditions (a) and (b) of Theorem 3.3.5 are valid and that

$$\lim_{t\to 0^+,\, t>0} \frac{1}{t} \int_0^t \int_{\mathbb{R}^2} \langle \nu_{x,y,s}, \mathrm{id}\rangle \varphi(x,y) \, dx\, dy\, ds$$

$$= \int_{\mathbb{R}^2} u_0(x,y) \varphi(x,y) \, dx\, dy \qquad (3.3.2)$$

for all $\varphi \in C_0^1(\mathbb{R}^2)$. Additionally we suppose

$$\lim_{t\to 0^+,\, t>0} \frac{1}{t} \int_0^t \int_{\mathbb{R}^2} \langle \nu_{x,y,s}, U\rangle \, dx\, dy\, ds \leq \int_{\mathbb{R}^2} U(u_0(x,y)) \, dx\, dy \qquad (3.3.3)$$

3.3 Convergence of finite volume schemes in 2-D on unstructured grids

for **one** *strictly convex continuous function* $U : \mathbb{R} \to \mathbb{R}$ *such that* $U(0) = 0$. *Then* ν *satisfies the condition* (d) *of Theorem 3.3.5; that is,*

$$\lim_{t \to 0, \ t > 0} \frac{1}{t} \int_0^t \int_{\mathbb{R}^2} \langle \nu_{x,y,s}, |\mathrm{id} - u_0(x,y)|\rangle \, dx \, dy \, ds = 0.$$

For the proofs of Theorems 3.3.5 and 3.3.7 see [53, §6] and [42, Theorem 2.2].

Now we have to show that the finite volume scheme (3.2.6), (3.2.7) will define a subsequence that converges to a measure-valued solution of (3.2.1) in the limit $h \to 0$, and this limit will satisfy the conditions (a)–(d) of Theorem 3.3.5. In 3.3.8 – 3.3.16 we shall derive sufficient conditions for (a)–(d) in Theorem 3.3.5, and we shall summarize these results in Theorem 3.3.17. Let us start with the following proposition.

PROPOSITION 3.3.8 *For given initial values* $u_0 \in L^\infty(\mathbb{R}^2) \cap L^1(\mathbb{R}^2)$ *let* $M := \sup_{x,y} |u_0(x,y)|$, u_j^n *be defined by the numerical scheme (3.2.6), (3.2.7) such that* $c_1 \leq (\Delta t)/h \leq c_2$ *and*

$$\|u_h\|_{L^\infty(\mathbb{R}^2 \times \mathbb{R}^+)} \leq M \quad \text{uniformly in } h. \tag{3.3.4}$$

Furthermore, assume that there exists $\beta \in [0, 1[$ *such that for all* $K \subset\subset \mathbb{R}^2$

$$\Delta t^\beta h \sum_n \sum_{j, T_j \cap K \neq \emptyset} \sum_{l=1}^k |g_{jl}(u_j, u_{jl}) - \nu_{jl} f(u_j)| \leq C_2(M, K) \tag{3.3.5}$$

uniformly in $\Delta t, h$. *Then there exists a subsequence* $(u_h)_h$ *and a Young measure* $\nu \ (= \nu_{x,y,t})$ *such that*

$$a(u_h) \rightharpoonup \int_\mathbb{R} a(\lambda) d\nu_{x,y,t}(\lambda) \quad \text{weak-* in } L^\infty$$

for all $a \in C^0(\mathbb{R})$ *and that* $\nu_{x,y,t}$ *is a measure-valued solution of*

$$\partial_t u + \partial_x f_1(u) + \partial_y f_2(u) = 0 \quad \text{in } \mathbb{R}^2 \times \mathbb{R}^+. \tag{3.3.6}$$

In particular, we have

$$\int_{\mathbb{R}^2 \times \mathbb{R}^+} u \partial_t \varphi + \int_{\mathbb{R}^2 \times \mathbb{R}^+} \overline{f} \nabla \varphi + \int_{\mathbb{R}^2} u_0 \varphi(\cdot, 0) = 0 \qquad (3.3.7)$$

for all $\varphi \in C_0^\infty(\mathbb{R}^2 \times \mathbb{R}^+)$, where $u = \int_{\mathbb{R}} d\nu_{x,y,t}$, $\overline{f}(x,y,t) := \int_{\mathbb{R}} f(\lambda) \, d\nu_{x,y,t}(\lambda)$. This means that ν satisfies the condition (b) of Theorem 3.3.5.

REMARK 3.3.9 *A sufficient criterion for (3.3.4) is given in Theorem 4.8 in [73], see also Proposition 3.3.14 here.*

Proof of Proposition 3.3.8 In this proof the summation with respect to j is always taken over $\{j \mid T_j \cap K \neq \emptyset\}$. To prove this proposition, we have to multiply (3.2.7) by $|T_j| \varphi(w_j, t^n)$, where $\varphi \in C_0^\infty(\mathbb{R}^2 \times \mathbb{R}_0^+)$ and $\mathrm{supp}(\varphi) \subset K$. We obtain after summation

$$\sum_n \sum_j |T_j|(u_j^{n+1} - u_j^n)\varphi(w_j, t^n)$$
$$= -\Delta t \sum_n \sum_j \sum_l g_{jl}(u_j^n, u_{jl}^n)\varphi(w_j, t^n) \ . \qquad (3.3.8)$$

Using summation by parts [73], we get

$$\sum_n \sum_j |T_j|(u_j^{n+1} - u_j^n)\varphi(w_j, t^n)$$
$$= - \int_{\mathbb{R}^2 \times \mathbb{R}^+} u_h \partial_t \varphi - \int_{\mathbb{R}^2} u_0 \varphi(\cdot, 0) + \mathcal{O}(h) \ . \qquad (3.3.9)$$

For the right-hand side in (3.3.8) we obtain, using (3.2.9), (3.2.5), (cf. Lemma 4.4 and 4.5 of [73]) and the notation $\varphi_{jl}^n := \varphi(z_{jl}, t^n)$ and $\varphi_j^n := \varphi(w_j, t^n)$,

$$-\Delta t \sum_n \sum_j \sum_l g_{jl}(u_j^n, u_{jl}^n)\varphi_j^n$$
$$= -\Delta t \sum_n \sum_j \sum_l g_{jl}(u_j^n, u_{jl}^n)(\varphi_j^n - \varphi_{jl}^n)$$
$$= -\Delta t \sum_n \sum_j \sum_l [g_{jl}(u_j^n, u_{jl}^n) - \nu_{jl} f(u_j)](\varphi_j^n - \varphi_{jl}^n) \qquad (3.3.10)$$

3.3 Convergence of finite volume schemes in 2-D on unstructured grids

$$+\Delta t \sum_n \sum_j \sum_l \nu_{jl} f(u_j) \varphi_{jl}^n$$

$$= -R + \int_{\mathbb{R}^2 \times \mathbb{R}^+} f(u_h) \nabla \varphi + \mathcal{O}(h),$$

where

$$R := \Delta t \sum_n \sum_j \sum_l \left[g_{jl}(u_j^n, u_{jl}^n) - \nu_{jl} f(u_j) \right] (\varphi_j^n - \varphi_{jl}^n). \quad (3.3.11)$$

Since u_h is uniformly bounded, there exists $u \in L^\infty(\mathbb{R}^2 \times \mathbb{R}^+)$ such that

$$u_h \rightharpoonup u \quad \text{weak-}* \quad \text{in } L^\infty,$$

and, since f is continuous, we obtain (see [235] and Theorem 3.3.3)

$$f(u_h) \rightharpoonup \overline{f} \quad \text{weak-}* \quad \text{in } L^\infty,$$

where

$$\overline{f}(x, y, t) := \int_{\mathbb{R}} f(\lambda) \, d\nu_{x,y,t}(\lambda).$$

Therefore if

$$R \to 0 \quad \text{as } h, \Delta t \to 0, \quad (3.3.12)$$

we obtain

$$\int_{\mathbb{R}^2 \times \mathbb{R}^+} u \partial_t \varphi + \int_{\mathbb{R}^2 \times \mathbb{R}^+} \overline{f} \nabla \varphi + \int_{\mathbb{R}^2} u_0 \varphi(\cdot, 0) = 0, \quad (3.3.13)$$

which proves Proposition 3.3.8. It remains to show (3.3.12). Using the assumption (3.3.5), we get

$$|R| = \Delta t \left| \sum_{n,j,l} [g_{jl} - \nu_{jl} f(u_j)](\varphi_j^n - \varphi_{jl}^n) \right|$$

$$\leq c(\varphi) \Delta t^{1-\beta} h \Delta t^\beta \sum_{n,j,l} |g_{jl} - \nu_{jl} f(u_j)| \quad (3.3.14)$$

$$\leq c(\varphi, M) \Delta t^{1-\beta} \to 0 \quad \text{if } \beta < 1,$$

where $\operatorname{supp}(\varphi) \subset K$. This finishes the proof of Proposition 3.3.8. □

Now let us verify the condition (c) in Theorem 3.3.5. In particular, we shall show that the finite volume scheme (3.2.6), (3.2.7) will define a measure-valued solution of (3.2.1) in the limit $h \to 0$, and this limit is consistent with all entropies (U, F_1, F_2).

PROPOSITION 3.3.10 *Let us assume the conditions of Proposition 3.3.8 and additionally that there is an entropy pair (U, F_1, F_2) and a numerical entropy flux $G_{jl} \in C^{0,1}(\mathbb{R}^2, \mathbb{R})$ for $j \in \mathbb{N}$, $l = 1, \ldots, k$ such that U is convex and G_{jl} satisfies the following conditions. We assume that for any $R > 0$ and for all $u, v, u', v' \in B_R(0)$ we have*

$$|G_{jl}(u,v) - G_{jl}(u',v')| \leq c(R)\, h\, (|u-u'| + |v-v'|),\quad (3.3.15)$$
$$G_{j,\alpha(j,l)}(u,v) = -G_{\alpha(j,l),j}(v,u),\quad (3.3.16)$$
$$G_{j,\alpha(j,l)}(u,u) = \nu_{j,\alpha(j,l)} F(u),\quad (3.3.17)$$

where

$$F(u) := \begin{pmatrix} F_1(u) \\ F_2(u) \end{pmatrix}.$$

Furthermore, we assume

$$U(u_j^{n+1}) - U(u_j^n) \leq -\frac{\Delta t}{|T_j|} \sum_{l=1}^{k} G_{jl}(u_j^n, u_{jl}^n) \quad (3.3.18)$$

and that there exists a $\beta \in [0,1[$ such that for all $K \subset\subset \mathbb{R}^2$ we have

$$\Delta t^\beta h \sum_n \sum_{j, T_j \cap K \neq \emptyset} \sum_{l=1}^{k} |G_{jl}(u_j, u_{jl}) - \nu_{jl} F(u_j)| \leq C_3(K) \quad (3.3.19)$$

uniformly in $\Delta t, h$. Then the measure-valued solution ν as obtained in Proposition 3.3.8 is consistent with the entropy pair (U, F_1, F_2); that is,

$$\partial_t \langle \nu, U \rangle + \partial_x \langle \nu, F_1 \rangle + \partial_y \langle \nu, F_2 \rangle \leq 0 \quad (3.3.20)$$

in the distributional sense.

3.3 Convergence of finite volume schemes in 2-D on unstructured grids

COROLLARY 3.3.11 *Suppose that the same assumptions as in Proposition 3.3.10 hold. Additionally assume that there exists $\beta \in [0,1[$ such that (3.3.18) and (3.3.19) are valid for **all** entropies U and all corresponding entropy fluxes G_{jl}. Then the condition (c) of Theorem 3.3.5 is satisfied.*

Proof of Proposition 3.3.10 In this proof the summation with respect to j is always taken over $\{j \,|\, T_j \cap K \neq \emptyset\}$. Let $\varphi \in C_0^\infty(\mathbb{R}^2 \times \mathbb{R}^+)$, $\varphi \geq 0$, $\mathrm{supp}(\varphi) \subset K$. Then we multiply (3.3.18) by $(|T_j|/\Delta t)\varphi_j^n$ and sum over n and j. Then, using the same ideas as in the proof of Proposition 3.3.8, we obtain

$$- \int_{\mathbb{R}^2 \times \mathbb{R}^+} U(u_h)\partial_t \varphi - \int_{\mathbb{R}^2 \times \mathbb{R}^+} F(u_h)\nabla \varphi + R + \mathcal{O}(h) \leq 0, \quad (3.3.21)$$

where

$$R := \Delta t \sum_n \sum_j \sum_l \left[G_{jl}(u_j^n, u_{jl}^n) - \nu_{jl} F(u_j) \right] (\varphi_j^n - \varphi_{jl}^n). \quad (3.3.22)$$

Now if $h \to 0$ then (3.3.21) and (3.3.19) imply

$$- \int_{\mathbb{R}^2 \times \mathbb{R}^+} \overline{U} \partial_t \varphi - \int_{\mathbb{R}^2 \times \mathbb{R}^+} \overline{F} \nabla \varphi \leq 0, \quad (3.3.23)$$

where

$$\overline{U}(x,y,t) := \int_\mathbb{R} U(\lambda)\, d\nu_{x,y,t}(\lambda) = \langle \nu, U \rangle(x,y,t),$$

$$\overline{F}_1(x,y,t) := \int_\mathbb{R} F_1(\lambda)\, d\nu_{x,y,t}(\lambda) = \langle \nu, F_1 \rangle(x,y,t),$$

$$\overline{F}_2(x,y,t) := \int_\mathbb{R} F_2(\lambda)\, d\nu_{x,y,t}(\lambda) = \langle \nu, F_2 \rangle(x,y,t).$$

This completes the proof of Proposition 3.3.10. □

Before we can verify the property (d) in Theorem 3.3.5, we need some auxiliary results.

PROPOSITION 3.3.12 *Let the assumption of Proposition 3.3.8 be valid. Then the measure-valued solution ν of Proposition 3.3.8 satisfies (3.3.2) of Theorem 3.3.7; that is,*

$$\lim_{t \to 0, \, t>0} \frac{1}{t} \int_0^t \int_{\mathbb{R}^2} \langle \nu_{x,y,s}, \mathrm{id} \rangle v(x,y) \, dx \, dy \, ds$$
$$= \int_{\mathbb{R}^2} u_0(x,y) v(x,y) \, dx \, dy \qquad (3.3.24)$$

for all $v \in C_0^1(\mathbb{R}^2)$.

Proof (See [231]) To prove (3.3.24), we have to show that for all $v \in C_0^1(\mathbb{R}^2, \mathbb{R})$

$$\lim_{t \to 0, \, t>0} \frac{1}{t} \int_0^t \int_{\mathbb{R}^2} u(x,y,s) \, v(x,y) \, dx \, dy \, ds$$
$$= \int_{\mathbb{R}^2} u_0(x,y) \, v(x,y) \, dx \, dy \, . \qquad (3.3.25)$$

since $u_h \rightharpoonup u$ weak-$*$ and

$$u(x,y,s) = \langle \nu_{x,y,s}, \mathrm{id} \rangle = \int_{\mathbb{R}} \lambda \, d\nu_{x,y,s}(\lambda) \, . \qquad (3.3.26)$$

Now let $\psi \in C_0^\infty([0,T[)$, $\psi(0) = 0$ and $\varphi(x,y,t) := \psi(t) \, v(x,y)$. Then we obtain for any φ of this form (cf. (3.3.7))

$$\int_0^T \int_{\mathbb{R}^2} (u \partial_t \varphi + \langle \nu, f_1 \rangle \partial_x \varphi + \langle \nu, f_2 \rangle \partial_y \varphi) = 0 \, . \qquad (3.3.27)$$

This implies that

$$\int_0^T \int_{\mathbb{R}^2} (uv \partial_t \psi + \langle \nu, f_1 \rangle \psi \partial_x v + \langle \nu, f_2 \rangle \psi \partial_y v) = 0 \, . \qquad (3.3.28)$$

3.3 Convergence of finite volume schemes in 2-D on unstructured grids

Using the notation

$$A(t) := \int_{\mathbb{R}^2} u(\cdot, t) v ,$$

$$B(t) := \int_{\mathbb{R}^2} \langle \nu_{x,t}, f_1 \rangle \partial_x v \, dx ,$$

$$C(t) := \int_{\mathbb{R}^2} \langle \nu_{x,t}, f_2 \rangle \partial_y v ,$$

we obtain

$$\int_0^T A \partial_t \psi + \int_0^T B \psi + \int_0^T C \psi = 0 . \tag{3.3.29}$$

Since

$$\int_0^T |B(t)| \, dt = \int_0^T \left| \int_{\mathbb{R}^2} \langle \nu, f_1 \rangle \partial_x v \right|$$

$$\leq \int_0^T \int_{\mathbb{R}^2} |\partial_x v| \int_{\mathbb{R}} |f_1(\lambda)| \, d\nu(\lambda) \leq c(\mathrm{supp}(v)) \tag{3.3.30}$$

we have $B \in L^1([0,T])$ and similarly $C \in L^1([0,T])$. The last inequality follows from the definition of $\langle \nu, f_1 \rangle$ as a weak limit of $f(u_h)$. Therefore we get $\partial_t A \in L^1([0,T])$ and

$$A(0) := \lim_{t \to 0 , \, t > 0} A(t)$$

exists. Now instead of ψ we choose the sequence of smooth functions ψ_n such that $\psi_n(0) = 1$, $|\psi_n| \leq 1$, $\psi_n \to 0$ a.e. We obtain

$$\int_0^T A \partial_t \psi_n + \int_0^T B \psi_n + \int_0^T C \psi_n + \int_{\mathbb{R}^2} u_0 v = 0 . \tag{3.3.31}$$

Then

$$\int_0^T A\partial_t \psi_n = -\int_0^T \partial_t A\psi_n + A\psi_n \Big|_0^T$$

$$= -\int_0^T \partial_t A\psi_n - A(0)\psi_n(0) \to -A(0),$$

since $\psi_n \to 0$ a.e. and $|\psi_n| \le 1$. For the same reason, we obtain

$$\int_0^T B\psi_n \to 0, \qquad \int_0^T C\psi_n \to 0,$$

and therefore (3.3.31) implies

$$A(0) = \lim_{t \to 0,\, t>0} \int_{\mathbb{R}^2} u(\cdot,t)v = \int_{\mathbb{R}^2} u_0 v.$$

This proves Proposition 3.3.12. □

Now we shall recall the discrete maximum principle and the L^1 estimate for the scheme (3.2.7). To do this, we need the notion of monotonicity of the scheme (3.2.7).

DEFINITION 3.3.13 (Monotonicity [73]) *The scheme (3.2.7) is called* monotone *if the function*

$$H(u_j, u_{j1}, \ldots, u_{jk}) := u_j - \frac{\Delta t}{|T_j|} \sum_{l=1}^k g_{jl}(u_j, u_{jl}) \qquad (3.3.32)$$

is monotonically non decreasing in any argument $u_j, u_{j1}, \ldots, u_{jk}$.

PROPOSITION 3.3.14 *Let* $u_0 \in L^\infty(\mathbb{R}^2) \cap L^1(\mathbb{R}^2)$ *with* $\sup_{x,y} |u_0(x,y)| \le M$.

3.3 Convergence of finite volume schemes in 2-D on unstructured grids

(a) Let u_j^n be defined by the monotone numerical scheme (3.2.6), (3.2.7). Then we have for all $n \in \mathbb{N}$

$$\|u^n\|_{L^1(\mathbb{R}^2)} \leq \|u^0\|_{L^1(\mathbb{R}^2)} \;. \tag{3.3.33}$$

(b) Instead of the monotonicity, we assume that

$$\frac{g_{jl}(u_j, u_{jl}) - g_{jl}(u_j, u_j)}{u_j - u_{jl}} \geq 0 \quad \text{for } l = 0, 1, \ldots, k \;. \tag{3.3.34}$$

(In particular, (3.3.34) is automatically fulfilled for any monotone scheme.)

Choose Δt such that

$$\frac{\Delta t}{h} \leq \frac{1}{k \, c(M) \, c_v} \;, \tag{3.3.35}$$

where $c(M)$ is the constant of Lipschitz continuity of g_{jl} and G_{jl} (see (3.2.8) and (3.3.15)) and c_v is defined in (3.2.4). Then we have for all $n \in \mathbb{N}$

$$\|u^n\|_{L^\infty(\mathbb{R}^2)} \leq \|u^0\|_{L^\infty(\mathbb{R}^2)} \;. \tag{3.3.36}$$

Proof (a) The monotonicity of the scheme (3.2.7) and the property $H(0,0,\ldots,0) = 0$ imply that

$$[H(u^n)]^+ \leq H((u^n)^+) \quad \text{and} \quad [H(u^n)]^- \geq H((u^n)^-) \;.$$

Here we use the notation $a^+ := \max\{a, 0\}$ and $a^- := \min\{a, 0\}$. Then we have

$$\begin{aligned}
|u_j^{n+1}| &= |H(u^n)| = [H(u^n)]^+ - [H(u^n)]^- \\
&\leq H((u^n)^+) - H((u^n)^-) \\
&= |u_j^n| - \frac{\Delta t}{|T_j|} \sum_{l=1}^k g_{jl}((u_j^n)^+, (u_{jl}^n)^+) \\
&\quad + \frac{\Delta t}{|T_j|} \sum_{l=1}^k g_{jl}((u_j^n)^-, (u_{jl}^n)^-) \;.
\end{aligned} \tag{3.3.37}$$

176 Initial value problems for scalar conservation laws in 2-D

Now the properties of g_{jl} imply that for $v = u^0$ or $v = (u^0)^+$ or $v = (u^0)^-$ we have

$$\sum_{l=1}^{k} |g_{jl}(v_j, v_{jl})| = \sum_{l=1}^{k} |g_{jl}(v_j, v_{jl}) - g_{jl}(0,0)|$$

$$\leq c(M)h \left(k|v_j| + \sum_{l=1}^{k} |v_{jl}| \right)$$

$$\leq c(M)h \left(k|u_j^0| + \sum_{l=1}^{k} |u_{jl}^0| \right).$$

So, since $u_0 \in L^1(\mathbb{R}^2)$, $\sum_j \sum_{l=1}^{k} g_{jl}(v_j, v_{jl})$ is absolutely convergent for $v = u^0$ or $v = (u^0)^+$ or $v = (u^0)^-$, and we have by the conservation property (3.2.9) of g_{jl} that

$$\sum_j \sum_{l=1}^{k} g_{jl}(u_j^0, u_{jl}^0) = 0 \ , \quad \sum_j \sum_{l=1}^{k} g_{jl}((u_j^0)^+, (u_{jl}^0)^+) = 0 \ , \qquad (3.3.38)$$

$$\sum_j \sum_{l=1}^{k} g_{jl}((u_j^0)^-, (u_{jl}^0)^-) = 0 \ .$$

Then, multiplying (3.3.37) by $|T_j|$ and summing over all j, we get

$$\|u^1\|_{L^1(\mathbb{R}^2)} \leq \|u^0\|_{L^1(\mathbb{R}^2)} \ , \qquad (3.3.39)$$

and the proof is finished by induction.

(b) The L^∞ estimate can be obtained as follows. The definition of the algorithm (3.2.7) implies

$$u_j^{n+1} = u_j^n - \frac{\Delta t}{|T_j|} \sum_{l=1}^{k} g_{jl}(u_j^n, u_{jl}^n)$$

$$= u_j^n - \frac{\Delta t}{|T_j|} \sum_{l=1}^{k} [g_{jl}(u_j^n, u_{jl}^n) - g_{jl}(u_j^n, u_j^n)]$$

$$= u_j^n - \frac{\Delta t}{|T_j|} \sum_{l=1}^{k} K_{jl}^n (u_j^n - u_{jl}^n) \ ,$$

3.3 Convergence of finite volume schemes in 2-D on unstructured grids

where

$$K_{jl}^n := \frac{g_{jl}(u_j, u_{jl}) - g_{jl}(u_j, u_j)}{u_j - u_{jl}} \quad \text{for } u_{jl} \neq u_j \tag{3.3.40}$$

and

$$K_{jl}^n := 0 \quad \text{for } u_{jl} = u_j. \tag{3.3.41}$$

Then we can continue:

$$u_j^{n+1} = u_j^n \left(1 - \frac{\Delta t}{|T_j|} \sum_{l=1}^{k} K_{jl}^n \right) + \frac{\Delta t}{|T_j|} \sum_{l=1}^{k} K_{jl}^n u_{jl}^n.$$

Let us assume $n = 0$. Then, by the assumption (3.2.8),

$$|K_{jl}^0| \leq c(M)h,$$

and therefore

$$\left| \frac{\Delta t}{|T_j|} \sum_{l=1}^{k} K_{jl}^0 \right| \leq k \frac{\Delta t}{\min_j |T_j|} c(M)h \leq 1.$$

Altogether, we get

$$|u_j^1| \leq |u_j^0| \left(1 - \frac{\Delta t}{|T_j|} \sum_{l=1}^{k} K_{jl}^0 \right) + \frac{\Delta t}{|T_j|} \sum_{l=1}^{k} K_{jl}^0 |u_{jl}^0| \tag{3.3.42}$$

$$\leq \| u^0 \|_{L^\infty(\mathbb{R}^2)} \left(1 - \frac{\Delta t}{|T_j|} \sum_{l=1}^{k} K_{jl}^0 \right) + \frac{\Delta t}{|T_j|} \sum_{l=1}^{k} K_{jl}^0 \| u^0 \|_{L^\infty(\mathbb{R}^2)}$$

$$\leq \| u^0 \|_{L^\infty(\mathbb{R}^2)} \leq M.$$

Then we have $\| u^1 \|_{L^\infty(\mathbb{R}^2)} \leq M$, and the proof can be finished by induction.

$\square$

REMARK 3.3.15 $\sum_j \sum_{l=1}^{k} g_{jl}(u_j^n, u_{jl}^n) = 0$ for all $n \in \mathbb{N}$ is a consequence of the proof of Proposition 3.3.14, and the same can be proved for G_{jl}.

178 Initial value problems for scalar conservation laws in 2-D

PROPOSITION 3.3.16 *Let the assumptions of Propositions 3.3.8 and 3.3.10 be valid and let $U(u) = \frac{1}{2}u^2$. Then the measure-valued solution ν of Proposition 3.3.8 satisfies (3.3.3) of Theorem 3.3.7; that is,*

$$\lim_{t \to 0,\ t > 0} \frac{1}{t} \int_0^t \int_{\mathbb{R}^2} \langle \nu_{x,y,s}, U \rangle \, dx\, dy\, ds \leq \int_{\mathbb{R}^2} U(u_0(x,y))\, dx\, dy \,. \quad (3.3.43)$$

Proof The properties of G_{jl} as defined in (3.3.15)–(3.3.17) imply that we have, as in (3.3.38),

$$\sum_j \sum_l G_{jl}(u_j, u_{jl}) = 0 \,.$$

Therefore we obtain from (3.3.18)

$$\sum_j [U(u_j^{n+1}) - U(u_j^n)] |T_j| \leq 0 \,,$$

and, summing this up with respect to n:

$$\sum_j [U(u_j^n) - U(u_j^0)][|T_j|] \leq 0 \,.$$

Of course the value of the sum $\sum_j U(u_j^n)$ is finite uniformly in n, since we have, thanks to (3.3.33) and (3.3.36),

$$\|U(u^n)\|_{L^1} \leq \tfrac{1}{2}\|u^n\|_{L^1}\|u^n\|_{L^\infty} \leq \tfrac{1}{2}\|u^0\|_{L^1}\|u^0\|_{L^\infty} \leq \text{const} \,.$$

Let $t_1 := n_1 \Delta t$ and $t_2 = n_2 \Delta t$. Then we get

$$\frac{\Delta t}{t_2 - t_1} \sum_{n=n_1}^{n_2-1} \sum_j U(u_j^n) |T_j| \leq \sum_j U(u_j^0)|T_j| \,,$$

or

$$\frac{1}{t_2 - t_1} \int_{t_1}^{t_2} \int_{\mathbb{R}^2} U(u_h) \leq \int_{\mathbb{R}^2} U(u_0) \,.$$

3.3 Convergence of finite volume schemes in 2-D on unstructured grids

Let $\varphi_l \in L^1(\mathbb{R}^2 \times [0,T])$ such that $\varphi_l \leq 1$ and $\varphi_l \to 1$ a.e. as $l \to \infty$. Then

$$U(u_h)\varphi_l \to U(u_h) \text{ a.e.},$$

and, since $\varphi_l \leq 1$, the sequence also converges in $L^1(\mathbb{R}^2 \times [0,T])$. For each $\varepsilon > 0$ there exists an l_0 such that for all $l \geq l_0$

$$\frac{1}{t_2 - t_1} \int_{t_1}^{t_2} \int_{\mathbb{R}^2} U(u_h)\varphi_l \leq \int_{\mathbb{R}^2} U(u_0) + \varepsilon.$$

Now as $h \to 0$ we get

$$\frac{1}{t_2 - t_1} \int_{t_1}^{t_2} \int_{\mathbb{R}^2} \langle \nu, U \rangle \varphi_l \leq \int_{\mathbb{R}^2} U(u_0) + \varepsilon$$

for all $l \geq l_0$. Then $l \to \infty$ implies

$$\frac{1}{t_2 - t_1} \int_{t_1}^{t_2} \int_{\mathbb{R}^2} \langle \nu, U \rangle \leq \int_{\mathbb{R}^2} U(u_0) + \varepsilon.$$

Since this is true for every $\varepsilon > 0$, we obtain

$$\frac{1}{t_2 - t_1} \int_{t_1}^{t_2} \int_{\mathbb{R}^2} \langle \nu, U \rangle \leq \int_{\mathbb{R}^2} U(u_0).$$

Now we show that

$$(x, y, t) \mapsto \langle \nu_{x,y,t}, a \rangle \in L^\infty(\mathbb{R}^2 \times [0,T]) \tag{3.3.44}$$

for all functions $a \in C^0(\mathbb{R})$: Since $\nu_{x,y,t}$ is a probability measure with a compact support K, we have

$$|\langle \nu_{x,y,t}, a \rangle| = \left| \int_{\mathbb{R}} a(\lambda) \, d\nu_{x,y,t}(\lambda) \right|$$

$$= \left| \int_K a(\lambda) \, d\nu_{x,y,t}(\lambda) \right|$$

$$\leq \nu_{x,y,t}(K) \sup_K |a(\lambda)| \leq \text{const},$$

independently of (x,y,t). This implies (3.3.44). Therefore we have that the function

$$t \mapsto \frac{1}{t_2 - t} \int_t^{t_2} \int_{\mathbb{R}^2} \langle \nu, U \rangle$$

is Lipschitz-continuous and admits a trace at $t = 0$. So we can pass to the limit $t \to 0$ to obtain

$$\frac{1}{t} \int_0^t \int_{\mathbb{R}^2} \langle \nu, U \rangle \leq \int_{\mathbb{R}^2} U(u_0) \ .$$

Then the limit $t \to 0$ proves the proposition. □

Now let us summarize the results of this section in order to get a general convergence result for the finite volume scheme (3.2.6), (3.2.7).

THEOREM 3.3.17 *(Let us assume that Conditions 3.2.1 – 3.2.4 are satisfied. Let $u_0 \in L^\infty(\mathbb{R}^2) \cap L^1(\mathbb{R}^2)$ and let u_j^n be defined by the numerical scheme (3.2.6), (3.2.7) such that for g_{jl} the properties (3.2.8)–(3.2.10) and for G_{jl} the properties (3.3.15)–(3.3.17) are valid. Furthermore, let us assume that there exists a constant M such that the following conditions are satisfied:*

$$\|u_h\|_{L^\infty(\mathbb{R}^2 \times \mathbb{R}^+)} \leq M \quad \text{uniformly in } h, \tag{3.3.45}$$
$$\|u_h(\cdot, t)\|_{L^1(\mathbb{R}^2)} \leq M \quad \text{uniformly in } h \text{ and } t. \tag{3.3.46}$$

Moreover, let there exist $\beta \in [0, 1[$ such that for all $K \subset\subset \mathbb{R}^2$

$$\Delta t^\beta h \sum_n \sum_{j, T_j \cap K \neq \emptyset} \sum_{l=1}^k |g_{jl}(u_j, u_{jl}) - \nu_{jl} f(u_j)| \leq C_2(K, M) \tag{3.3.47}$$

uniformly in $\Delta t, h$,

$$\Delta t^\beta h \sum_n \sum_{j, T_j \cap K \neq \emptyset} \sum_{l=1}^k |G_{jl}(u_j, u_{jl}) - \nu_{jl} F(u_j)| \leq C_3(K, M) \tag{3.3.48}$$

3.3 Convergence of finite volume schemes in 2-D on unstructured grids

uniformly in $\Delta t, h$, *and*

$$U(u_j^{n+1}) - U(u_j^n) \leq -\frac{\Delta t}{|T_j|} \sum_{l=1}^{k} G_{jl}(u_j^n, u_{jl}^n) \tag{3.3.49}$$

for all entropies (U, F_1, F_2) *and* G_{jl} *satisfying the conditions of Proposition 3.3.10. Then*

$$u_h \rightharpoonup u \quad weak\text{-}* \quad in \ L^\infty, \quad u \in L^\infty(\mathbb{R}^2 \times \mathbb{R}^+), \tag{3.3.50}$$

and u *is the uniquely determined solution in the sense of Kruzkov of the initial value problem (3.2.1), (3.2.2).*

Proof We have to verify the conditions (a)–(d) of Theorem 3.3.5. From (3.3.46) it follows that

$$\sup_{[0,T]} \int_{\mathbb{R}^2} |u_h(x, y, t)| \, dx \, dy \leq M \ .$$

Then $h \to 0$ gives us (a). Now (b) follows from Proposition 3.3.8, and (c) from Proposition 3.3.10 and Corollary 3.3.11. Propositions 3.3.16 and 3.3.12 imply that the assumptions of Theorem 3.3.7 are satisfied. Then from Theorem 3.3.7 we obtain the property (d) of Theorem 3.3.5. □

Now we describe a class of upwind finite volume schemes for which the conditions of Theorem 3.3.17 are fulfilled. In particular, we show that the Engquist–Osher scheme and the Lax–Friedrichs scheme with suitably chosen numerical entropy fluxes belong to that class. This yields the convergence result for these schemes (see Theorem 3.3.28).

ASSUMPTION 3.3.18 Using the notation and assumptions introduced in 3.2.1, we shall assume in the sequel that the numerical flux $g_{jl} \in C^1(\mathbb{R}^2)$ satisfies the conditions (3.2.8)–(3.2.10) and the numerical entropy flux $G_{jl} \in C^1(\mathbb{R}^2)$ satisfies the conditions (3.3.15)–(3.3.17). Additionally, we assume the monotonicity of the scheme (3.2.7) (see Definition 3.3.13 and (3.3.34)) and $u_0 \in L^1(\mathbb{R}^2) \cup L^\infty(\mathbb{R}^2)$.

EXAMPLE 3.3.19 (Engquist–Osher scheme and Lax–Friedrichs scheme) Let

$$c_{jl}(u) := n_{jl} f(u) ,$$

and define c_{jl}^+ and c_{jl}^- as in Example 3.2.7. Then the Engquist–Osher numerical flux is given by (see [62])

$$g_{jl}^{EO}(u,v) := |S_{jl}| \left[c_{jl}^+(u) + c_{jl}^-(v) \right] .$$

The Lax–Friedrichs numerical flux is given explicitly by (see e.g. [29])

$$g_{jl}^{LF}(u,v) := \frac{1}{2}[\nu_{jl} f(u) + \nu_{jl} f(v)] - \frac{|S_{jl}|}{2\lambda_{jl}}(v - u) ,$$

where λ_{jl} are arbitrarily chosen constants satisfying

$$\lambda_{jl} = \lambda_{lj} > \tilde{c} > 0 ,$$
$$\lambda_{jl} \sup_u (n_{jl} f'(u)) \leq 1 .$$

Then g_{jl}^{EO} and g_{jl}^{LF} satisfy the Conditions (3.2.8)–(3.2.10). Moreover, if Δt is chosen such that in the case of the Engquist–Osher scheme we have

$$\sup_j \frac{\Delta t}{|T_j|} \sum_{l=1}^{k} \max\{\nu_{jl} f'(u_j), 0\} \leq 1$$

while in the case of the Lax–Friedrichs scheme

$$\sup_j \frac{\Delta t}{|T_j|} \sum_{l=1}^{k} \left(\frac{1}{2} \nu_{jl} f'(u_j) + \frac{|S_{jl}|}{2\lambda_{jl}} \right) \leq 1$$

then both schemes are monotone.

As we have seen in Proposition 3.3.14, the first two Conditions (3.3.45) and (3.3.46) of Theorem 3.3.17 are satisfied for monotone schemes with properly chosen Δt. Next we shall show that the condition (3.3.47) implies the condition (3.3.48) for suitably chosen numerical entropy fluxes.

We recall that the Assumptions 3.3.18 are supposed to be satisfied throughout this section.

3.3 Convergence of finite volume schemes in 2-D on unstructured grids

PROPOSITION 3.3.20 *Assume that for the numerical flux g_{jl} there exists $\beta \in [0, 1[$ such that for all $K \subset\subset \mathbb{R}^2$*

$$\Delta t^\beta h \sum_n \sum_{j, T_j \cap K \neq \emptyset} \sum_{l=1}^k |g_{jl}(u_j, u_{jl}) - \nu_{jl} f(u_j)| \leq C_2(K, M) \quad (3.3.51)$$

uniformly in $\Delta t, h$, where $M := \|u_0\|_{L^\infty(\mathbb{R}^2)}$. Let G_{jl} be the numerical entropy flux corresponding to the entropy (U, F_1, F_2) such that additionally the compatibility condition

$$\frac{\partial G_{jl}}{\partial v}(u, v) = U'(v) \frac{\partial g_{jl}}{\partial v}(u, v) \quad (3.3.52)$$

is fulfilled for all $u, v \in \mathbb{R}$. Moreover we suppose that

$$g_{jl} \text{ is monotonically non increasing in the second variable} \quad (3.3.53)$$

(which is fulfilled by monotone schemes). Then for any $K \subset\subset \mathbb{R}^2$ we have

$$\Delta t^\beta h \sum_n \sum_{j, T_j \cap K \neq \emptyset} \sum_{l=1}^k |G_{jl}(u_j, u_{jl}) - \nu_{jl} F(u_j)|$$
$$\leq C_2(K, M) c(U', M) \quad (3.3.54)$$

uniformly in $\Delta t, h$.

Proof We have

$$|G_{jl}(u_j, u_{jl}) - \nu_{jl} F(u_j)|$$

$$= \left| \int_{u_j}^{u_{jl}} \frac{\partial G_{jl}}{\partial s}(u_j, s) \, ds \right|$$

$$= \left| \int_{u_j}^{u_{jl}} U'(s) \frac{\partial g_{jl}}{\partial s}(u_j, s) \, ds \right|$$

$$\leq c(U', M) \int_{\min(u_j, u_{jl})}^{\max(u_j, u_{jl})} \left[-\frac{\partial g_{jl}}{\partial s}(u_j, s) \right] ds$$

$$= c(U', M) |g_{jl}(u_j, u_{jl}) - g_{jl}(u_j, u_j)| ,$$

which proves our proposition. $\square$

REMARK 3.3.21 *The compatibility condition (3.3.52) can be viewed as an analogy of the well-known entropy compatibility condition*

$$F'_j(s) = U'(s)f'_j(s), \qquad j = 1, 2$$

(cf. Definition 3.3.4). The condition (3.3.52) turns out to be crucial in obtaining global control of the entropy dissipation, as we shall see later. In the following example we show that for the Engquist–Osher and the Lax–Friedrichs schemes the numerical entropy flux can be defined in such a way that the conditions (3.3.15)–(3.3.17) and the compatibility condition (3.3.52) are all fulfilled.

EXAMPLE 3.3.22 (Entropy fluxes for the E-O and L-F schemes)
Without loss of generality, we assume that $F(0) = 0$. We choose the numerical entropy flux in the case of the Engquist–Osher scheme as (see examples 3.3.19 and 3.2.7)

$$G_{jl}^{EO}(u,v) := |S_{jl}| \int_0^u U'(s) \max\{c'_{jl}(s), 0\}\, ds$$

$$+ |S_{jl}| \int_0^v U'(s) \min\{c'_{jl}(s), 0\}\, ds\ ,$$

and in the case of the Lax–Friedrichs scheme as

$$G_{jl}^{LF}(u,v) := \frac{1}{2}[\nu_{jl}F(u) + \nu_{jl}F(v)] - \frac{1}{2\lambda_{jl}}[U(v) - U(u)]\ .$$

Then these numerical entropy fluxes satisfy the conditions (3.3.15)–(3.3.17) and the compatibility condition (3.3.52).

Next we shall consider the condition (3.3.47) of Theorem 3.3.17.

PROPOSITION 3.3.23 *Let $u_0 \in L^1(\mathbb{R}^2) \cap L^\infty(\mathbb{R}^2)$, $\|u_0\|_{L^\infty(\mathbb{R}^2)} \leq M$ and let u_j^n be defined by the numerical scheme (3.2.7) with numerical fluxes g_{jl} and G_{jl} chosen such that the conditions (3.3.52) and (3.3.53) (of Proposition 3.3.20) are fulfilled. Moreover, choose Δt such that the CFL-like condition*

$$\frac{\Delta t}{h} \leq \frac{1}{(k+2)\, c(M)\, c_v} \tag{3.3.55}$$

3.3 Convergence of finite volume schemes in 2-D on unstructured grids

holds. Here $c(M)$ is the constant of Lipschitz continuity of g_{jl} and G_{jl} (see (3.2.8) and (3.3.15)) and c_v is defined in (3.2.4). Then we have for all $\beta \geq \frac{1}{2}$ (see also Proposition 3.3.10) and for all $K \subset\subset \mathbb{R}^2$

$$\Delta t^\beta h \sum_n \sum_{j, T_j \cap K \neq \emptyset} \sum_{l=1}^k |g_{jl}(u_j, u_{jl}) - \nu_{jl} f(u_j)| \leq C_2(K, M) \qquad (3.3.56)$$

uniformly in $\Delta t, h$.

This proposition will now be proved in two lemmas.

LEMMA 3.3.24 *Under the assumptions of Proposition 3.3.23, we have*

$$\frac{(u_j^{n+1})^2 - (u_j^n)^2}{2} + \frac{\Delta t}{|T_j|} \sum_{l=1}^k G_{jl}(u_j, u_{jl})$$
$$+ \left(\frac{\Delta t}{|T_j|}\right)^2 \sum_{l=1}^k [g_{jl}(u_j, u_{jl}) - g_{jl}(u_j, u_j)]^2 \leq 0 , \qquad (3.3.57)$$

where G_{jl} is the numerical entropy flux corresponding to the entropy $U(u) = \frac{1}{2} u^2$.

Proof We set $\lambda := \Delta t / |T_j|$ and for all $t \in [-M, M]$

$$\Delta_2 G(t) := G_{jl}(u_j, t) - G_{jl}(u_j, u_j) ,$$
$$\Delta_2 g(t) := g_{jl}(u_j, t) - g_{jl}(u_j, u_j) ,$$
$$\partial_2 G(t) := \frac{\partial G_{jl}}{\partial t}(u_j, t) ,$$
$$\partial_2 g(t) := \frac{\partial g_{jl}}{\partial t}(u_j, t) .$$

Note that $\sum_{l=1}^k g_{jl}(u_j, u_{jl}) = \sum_{l=1}^k \Delta_2 g(u_{jl})$, since

$$\sum_{l=1}^k g_{jl}(u_j, u_j) = f(u_j) \sum_{l=1}^k \nu_{jl} = 0 ,$$

and the same holds for G_{jl}. Then we can write using (3.2.7)

$$\frac{(u_j^{n+1})^2 - (u_j^n)^2}{2} = u_j^n(u_j^{n+1} - u_j^n) + \frac{1}{2}(u_j^{n+1} - u_j^n)^2$$

$$= u_j(-\lambda)\sum_{l=1}^{k}\Delta_2 g(u_{jl}) + \frac{1}{2}\lambda^2\left[\sum_{l=1}^{k}\Delta_2 g(u_{jl})\right]^2$$

$$\leq -\lambda u_j\sum_{l=1}^{k}\Delta_2 g(u_{jl}) + \frac{1}{2}k\lambda^2\sum_{l=1}^{k}[\Delta_2 g(u_{jl})]^2 .$$

Hence the left-hand side of (3.3.57) is less then or equal to

$$\lambda\sum_{l=1}^{k}\left\{-u_j\Delta_2 g(u_{jl}) + \Delta_2 G(u_{jl}) + \lambda\left(\frac{k+2}{2}\right)[\Delta_2 g(u_{jl})]^2\right\} . \quad (3.3.58)$$

Setting $a := u_j$ and

$$p(t) := -a\Delta_2 g(t) + \Delta_2 G(t) + \lambda\left(\frac{k+2}{2}\right)[\Delta_2 g(t)]^2 ,$$

the proof will be completed by showing that

$$p(t) \leq 0 \quad \text{for all } t \in [-M, M] . \quad (3.3.59)$$

Now

$$p'(t) = -a\partial_2 g(t) + \partial_2 G(t) + \lambda(k+2)[\Delta_2 g(t)][\partial_2 g(t)] . \quad (3.3.60)$$

Using (3.3.52) with $U(u) = \frac{1}{2}u^2$ and $\Delta_2 g(t) = \partial_2 g(\xi)(t-a)$ for ξ between a and t, we have

$$p'(t) = (t-a)\partial_2 g(t)[1 + \lambda(k+2)\partial_2 g(\xi)] . \quad (3.3.61)$$

Now since $\partial_2 g(t) \leq 0$ for all $t \in\,]-M, M[$ (cf. (3.3.53)), we are finished provided the term $1 + \lambda(k+2)\partial_2 g(\xi)$ in (3.3.61) is nonnegative. In fact $p'(t) \geq 0$ for $t \leq a$ and $p'(t) \leq 0$ for $t \geq a$, which, together with $p(a) = 0$, gives us (3.3.59) (see Figure 3.3.1).

But $\partial_2 g(\xi) \geq -c(M)h$ by Lipschitz continuity (3.3.15), and consequently

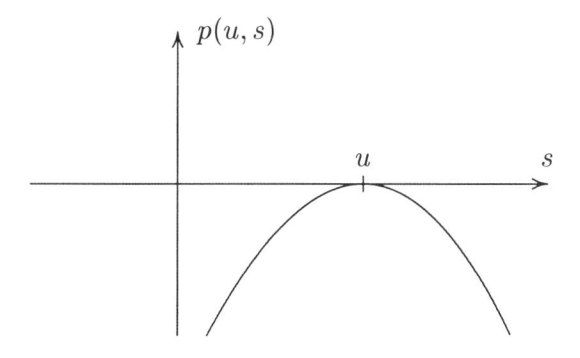

3.3 Convergence of finite volume schemes in 2-D on unstructured grids

Figure 3.3.1

$$1 + \lambda(k+2)\partial_2 g(\xi) \geq 1 - \lambda(k+2)hc(M) \quad (3.3.62)$$
$$\geq 1 - \frac{\Delta t}{h} c_v(k+2)c(M) \geq 0$$

because of (3.3.55). □

LEMMA 3.3.25 *Under the assumptions of Proposition 3.3.23, we have*

$$\sum_n \sum_j \sum_{l=1}^k [g_{jl}(u_j, u_{jl}) - \nu_{jl} f(u_j)]^2 \leq c(u_0) . \quad (3.3.63)$$

Proof Multiplying (3.3.57) by $|T_j|$ and summing over j, we get

$$\tfrac{1}{2}\|u^{n+1}\|^2_{L^2(\mathbb{R}^2)} - \tfrac{1}{2}\|u^n\|^2_{L^2(\mathbb{R}^2)}$$
$$+ \sum_j \frac{(\Delta t)^2}{|T_j|} \sum_{l=1}^k [g_{jl}(u_j, u_{jl}) - \nu_{jl} f(u_j)]^2 \leq 0 , \quad (3.3.64)$$

since $\sum_j \sum_{l=1}^k G_{jl}(u_j, u_{jl}) = 0$ (see Remark 3.3.15). Note that the L^2 norms on the left-hand side of (3.3.64) are finite, since (see Proposition 3.3.14)

$$\|u^n\|^2_{L^2(\mathbb{R}^2)} \leq \|u^n\|_{L^\infty(\mathbb{R}^2)} \|u^n\|_{L^1(\mathbb{R}^2)} \leq M\|u_0\|_{L^1(\mathbb{R}^2)} .$$

Now from the trivial property of unstructured grids $|T_j| \leq ch^2$ it follows that we can replace $|T_j|$ by ch^2 in the third term of (3.3.64) with the inequality still holding. Since the resulting coefficient is bounded

$$0 < c_1^2 \leq \frac{1}{c}\left(\frac{\Delta t}{h}\right)^2 \leq c_2^2$$

(see (3.2.3)), (3.3.64) yields, after summing over n from 0 to N,

$$c_1^2 \sum_{n=0}^{N}\sum_{j}\sum_{l=1}^{k}[g_{jl}(u_j,u_{jl}) - \nu_{jl}f(u_j)]^2$$
$$\leq \tfrac{1}{2}\|u^0\|_{L^2(\mathbb{R}^2)}^2 - \tfrac{1}{2}\|u^{N+1}\|_{L^2(\mathbb{R}^2)}^2$$
$$\leq \tfrac{1}{2}\|u^0\|_{L^2(\mathbb{R}^2)}^2 \leq \tfrac{1}{2}\|u^0\|_{L^\infty(\mathbb{R}^2)}\|u^0\|_{L^1(\mathbb{R}^2)} =: c(u_0) \ .$$

This gives us the desired result, since the right-hand side does not depend on N. $\square$

Now we are ready to prove Proposition 3.3.23.

Proof of Proposition 3.3.23 In this proof the summation with respect to j is always taken over $\{j \,|\, T_j \cap K \neq \emptyset\}$. Using the Cauchy-Schwarz inequality and Lemma 3.3.25, we get

$$\Delta t^\beta h \sum_n \sum_j \sum_{l=1}^{k} |g_{jl}(u_j,u_{jl}) - \nu_{jl}f(u_j)| = \sum_n \sum_j \sum_{l=1}^{k} |\Delta_2 g(u_{jl})| \Delta t^\beta h$$
$$\leq \left(\sum_n \sum_j \sum_{l=1}^{k} [\Delta_2 g(u_{jl})]^2\right)^{\frac{1}{2}} \left(\sum_n \sum_j \sum_{l=1}^{k} \Delta t^{2\beta} h^2\right)^{\frac{1}{2}}$$
$$\leq c(u_0, K) \left(\sum_n \sum_j \sum_{l=1}^{k} \Delta t^{2\beta} h^2\right)^{\frac{1}{2}}$$
$$\leq c(u_0, K)(\tilde{c}\Delta t^{2\beta - 1})^{\frac{1}{2}} \ .$$

The last term is bounded as $\Delta t \to 0$ provided $\beta \geq \tfrac{1}{2}$. $\square$

Next, we shall focus on global control of the entropy dissipation (3.3.49) for all entropies (U, F_1, F_2) and corresponding G_{jl}. We shall see that our compatibility condition (3.3.52) is again the crucial point in proving (3.3.49).

3.3 Convergence of finite volume schemes in 2-D on unstructured grids

PROPOSITION 3.3.26 *Under the assumptions of Proposition 3.3.23, we have*

$$U(u_j^{n+1}) - U(u_j^n) \leq -\frac{\Delta t}{|T_j|} \sum_{l=1}^{k} G_{jl}(u_j^n, u_{jl}^n) \quad (3.3.65)$$

for all convex entropies $U \in C^2(\mathbb{R})$.

Proof Throughout this proof we shall use the same notation as in the proof of Lemma 3.3.24. We begin by rewriting the scheme (3.2.7):

$$u_j^{n+1} = \frac{1}{k} \sum_{l=1}^{k} \left[u_j^n - \lambda k \Delta_2 g(u_{jl}) \right] . \quad (3.3.66)$$

The main step in the proof is the following mean-value property of convex functions:

$$U\left(\frac{a_1 + \ldots + a_m}{m}\right) \leq \frac{1}{m}[U(a_1) + \ldots + U(a_m)] , \quad (3.3.67)$$

which can be easily proved by induction. Applying this to (3.3.66), we obtain

$$U(u_j^{n+1}) \leq \frac{1}{k} \sum_{l=1}^{k} U(u_j^n - \lambda k \Delta_2 g(u_{jl})) . \quad (3.3.68)$$

Hence the proof will be completed if we prove the following inequality:

$$U(u_j^n - \lambda k \Delta_2 g(u_{jl})) \leq U(u_j^n) - \lambda k \Delta_2 G(u_{jl}) . \quad (3.3.69)$$

Now, setting $a := u_j^n$ and defining for all $t \in [-M, M]$

$$p(t) := U(a - \lambda k \Delta_2 g(t)) - U(a) + \lambda k \Delta_2 G(t) ,$$

we have (similarly to the proof of Lemma 3.3.26)

$$\begin{aligned} p'(t) &= U'(a - \lambda k \Delta_2 g(t))\left[-\lambda k \partial_2 g(t)\right] + \lambda k \partial_2 G(t) \\ &= \lambda k\, \partial_2 g(t)[U'(t) - U'(a - \lambda k \Delta_2 g(t))] \\ &= \lambda k\, \partial_2 g(t)\, U''(\eta)\left[t - a + \lambda k \partial_2 g(\tilde{\eta})(t - a)\right] \end{aligned}$$

for some η between t and $a - \lambda k \Delta_2 g(t)$ and $\tilde{\eta}$ between a and t. Finally,

$$p'(t) = \lambda k\, \partial_2 g(t)\, U''(\eta)\, (t-a)\, [1 + \lambda k \partial_2 g(\tilde{\eta})]\ .$$

Since $\partial_2 g(t) \leq 0$ for all $t \in\ ]-M, M[$ (the monotonicity of g_{jl} - see (3.3.53)) and $U'' \geq 0$ (the convexity of U), the proof can be finished (as in Lemma 3.3.24) by showing that $1 + \lambda k \partial_2 g(\tilde{\eta}) \geq 0$. But this holds true because of the CFL-like condition (3.3.55) (cf. (3.3.62) and the end of the proof of Lemma 3.3.24). □

REMARK 3.3.27 *Note that the CFL-like condition (3.3.55) that was used at the end of the proof is a bit stronger than the CFL-like condition (3.3.35).*

Now we can summarize the results of this section in the following theorem.

THEOREM 3.3.28 *Let us assume that the conditions from 3.2.1 - 3.2.4 are satisfied. Let $u_0 \in L^\infty(\mathbb{R}^2) \cap L^1(\mathbb{R}^2)$ and let u_j^n be defined by the numerical scheme (3.2.6), (3.2.7) such that for g_{jl} the properties (3.2.8)–(3.2.10) and for G_{jl} the properties (3.3.15)–(3.3.17) are valid. Let the scheme (3.2.6), (3.2.7) be monotone in the sense of Definition 3.3.13. Moreover, let the compatibility condition*

$$\frac{\partial G_{jl}}{\partial v}(u,v) = U'(v) \frac{\partial g_{jl}}{\partial v}(u,v) \qquad (3.3.70)$$

be satisfied for all $u, v \in \mathbb{R}$ and let

$$\frac{\Delta t}{h} \leq \frac{1}{(k+2)\, c(M)\, c_v}\ . \qquad (3.3.71)$$

Then all the conditions of Theorem 3.3.17 are satisfied, and consequently

$$u_h \rightharpoonup u \quad \text{weak-*}, \quad u \in L^\infty(\mathbb{R}^2 \times \mathbb{R}^+)\ , \qquad (3.3.72)$$

and u is the uniquely determined solution in the sense of Kruzkov of the initial value problem (3.2.1), (3.2.2).

Proof This is a trivial consequence of Propositions 3.3.14, 3.3.20, 3.3.23 and 3.3.26. □

REMARK 3.3.29 *Because of corollary 3.3.6, the convergence in (3.3.72) is in $L^1_{\text{loc}}(\mathbb{R}^2 \times \mathbb{R})$ (see [53]).*

EXAMPLE 3.3.30 (Engquist–Osher and Lax–Friedrichs schemes) *As a consequence of Theorem 3.3.28, we have the convergence proof for the Engquist–Osher and Lax–Friedrichs schemes with numerical entropy fluxes chosen as in Example 3.3.22. Note that the convergence result obtained in [29] can be applied to the Lax–Friedrichs scheme but not to the Engquist–Osher one, since the latter does not satisfy the condition (2.37) of Theorem 2.3 in [29].*

REMARK 3.3.31 *Convergence of finite volume schemes for general multi-dimensional conservation laws with boundary conditions are studied in [21]. They also use the theory of measure-valued solution and apply it to bounded domains.*

REMARK 3.3.32 *Using also the theory of measure-valued solutions, in [231] the convergence of the streamline diffusion shock-capturing method has been proved, including error estimates for the linear case. This method is an extension of the streamline diffusion method with additional artificial viscosity terms, which are weighted in a suitable way.*

3.4 Error estimates for first-order schemes

Now we should like to show how to get error estimates for finite volume methods on unstructured grids in 2-D. For error estimates in 1-D we refer to [234]. Kuznetsov [124] has shown for first-order equations in 2-D that monotone fractional schemes on uniform cartesian grids satisfy

$$\|u(t) - u_h(t)\|_{L^1(\mathbb{R}^n)} = \mathcal{O}(h^{1/2}).$$

Sanders [208] has generalized the result of Kuznetsov to nonuniform structured grids in 1-D. Cockburn *et al.* [31, 32, 33] have generalized the previous results to first-order quasi-monotone schemes on cartesian grids and have proved

$$\|u(t) - u_h(t)\|_{L^1(\mathbb{R}^n)} = \mathcal{O}(h^{\frac{1}{2}\alpha})$$

192 Initial value problems for scalar conservation laws in 2-D

for some $\alpha \in]0,1]$. In this section we shall consider error estimates for monotone finite volume schemes on unstructured grids. This means (see Proposition 3.3.14) that we can assume an L^∞ bound for u_h. Therefore we always assume that g_{jl} is monotone in this section. Now we shall describe a method on the basis of the Kuznetsov results [124] for getting error estimates. The method that we shall describe in this section is due to Vila [238] and Cockburn [34]. We still assume 3.2.1 – 3.2.4.

LEMMA 3.4.1 *Let $g = g_{jl}$ be the monotone numerical flux as defined in Definition 3.2.6 with $g(u,u) = \nu_{jl} f(u) =: f_{jl}$, and let G, H_c and u^* be defined by*

$$G(u,v) := g(\max\{u,c\}, \max\{v,c\}) - g(\min\{u,c\}, \min\{v,c\})$$
$$H_c(u) := [f_{jl}(u) - f_{jl}(c)]\operatorname{sign}(u-c)$$

and

$$u^* := u - \lambda[g(u,v) - f_{jl}(u)] . \tag{3.4.1}$$

Let Δt be sufficiently small such that (2.3.28) is satisfied. Then we obtain

$$|u^* - c| + \lambda G(u,v) \leq |u - c| + \lambda H_c(u) .$$

Proof See (2.3.40) with

$$\Theta = 0, \quad u_i = u, \quad u_{i+1} = v, \quad u_{i-1} = u .$$

□

LEMMA 3.4.2 *Let*

$$u^* := u - \lambda[g_{jl}(u,v) - g_{jl}(u,u)] ,$$

where g_{jl} is a monotone numerical flux satisfying (3.2.8)–(3.2.10), let F be the entropy flux corresponding to $U(u) := u^2$, and let G_{jl} be the corresponding numerical entropy flux satisfying (3.3.15)–(3.3.17) and (3.3.52). Then if λ and ε satisfy

$$1 - (1+\varepsilon)\lambda \sup_{|u|,|v| \leq c} |\partial_2 g_{jl}(u,v)| \geq 0 , \tag{3.4.2}$$

3.4 Error estimates for first-order schemes

there exists a constant $c \in \mathbb{R}^+$ such that for all u with

$$\|u\|_{L^\infty(\mathbb{R}^2 \times \mathbb{R}^+)} \leq c$$

we have

$$(u^*)^2 - u^2 + \lambda[G_{jl}(u,v) - G_{jl}(u,u)] \\ + \varepsilon[g_{jl}(u,v) - g_{jl}(u,u)]^2 \lambda^2 \leq 0 . \quad (3.4.3)$$

Proof The following arguments are similar as in the proof of Lemma 3.3.24. We have $(u^*)^2 - u^2 = 2u(u^* - u) + (u^* - u)^2$, and therefore, using (3.4.1), the left-hand side LS in (3.4.3) can be written as

$$LS = 2u(-\lambda)[g_{jl}(u,v) - g_{jl}(u,u)] \\ + \lambda^2[g_{jl}(u,v) - g_{jl}(u,u)]^2(1+\varepsilon) + \lambda[G_{jl}(u,v) - G_{jl}(u,u)] \\ =: \lambda p(u,v) .$$

Now it turns out that

$$p(u,u) = 0$$

and

$$\partial_2 p(u,v) = -2u\partial_2 g_{jl}(u,v) + 2\lambda(g_{jl}(u,v)$$

$$-g_{jl}(u,u))\partial_2 g_{jl}(u,v)(1+\varepsilon) + \underbrace{U'(v)}_{2v}\partial_2 g_{jl}(u,v)$$

$$= \partial_2 G_{jl}(u,v)$$

$$= \partial_2 g_{jl}(u,v)(2(v-u) + (1+\varepsilon)2\lambda \underbrace{(g_{jl}(u,v) - g_{jl}(u,u))}_{= \partial_2 g_{jl}(u,\xi)(v-u)}$$

$$= 2(v-u)\partial_2 g_{jl}(u,v) \underbrace{(1 + (1+\varepsilon)\lambda \partial_2 g_{jl}(u,\xi))}_{\geq 0 \text{ if } \lambda \text{ is sufficiently small}}$$

194 Initial value problems for scalar conservation laws in 2-D

for a suitable ξ between u and v. This holds since g_{jl} is monotone. Therefore $p(u,v) \leq 0$ for all $v \in \mathbb{R}$ (see Figure 3.3.1). □

LEMMA 3.4.3 *Let $u_0, v_0 \in L^1(\mathbb{R}^2) \cap L^\infty(\mathbb{R}^2)$ and let u_h, v_h be the corresponding solutions of the finite volume schemes (3.2.6), (3.2.7). Then we have*

$$\|u_h(\cdot, t+\tau) - v_h(\cdot, t+\tau)\|_{L^1(\mathbb{R}^2)} \leq \|u_h(\cdot, t) - v_h(\cdot, t)\|_{L^1(\mathbb{R}^2)}$$

for any t (L^1 contraction), and if

$$h \sum_j \sum_{l=1}^{3} |u_j^0 - u_{jl}^0| \leq c_0 ,$$

there exists a constant c_1 such that we have uniformly for all $t, \tau > 0$ with $\tau > \Delta t$, $t + \tau \leq T$

$$\|u_h(\cdot, t+\tau) - u_h(\cdot, t)\|_{L^1(\mathbb{R}^2)} \leq c_1 \tau$$

Proof For Q defined as

$$(Q(u))_j := u_j - \frac{\Delta t}{|T_j|} \sum_{l=1}^{3} g_{jl}(u_j^n, u_{jl}^n)$$

we have, due to monotonicity and the CFL-condition

$$u \leq v \quad \text{implies} \quad Qu \leq Qv .$$

Using Lemma 3.1.13, this is equivalent to

$$\int_{\mathbb{R}^2} |Q(u^n) - Q(v^n)| \leq \int_{\mathbb{R}^2} |u^n - v^n| ,$$

or

$$\|u_h(\cdot, t^{n+1}) - v_h(\cdot, t^{n+1})\|_{L^1(\mathbb{R}^2)} \leq \|u_h(\cdot, t^n) - v_h(\cdot, t^n)\|_{L^1(\mathbb{R}^2)} .$$

3.4 Error estimates for first-order schemes 195

Since u_h is piecewise-constant in t, we have the first statement. For the same reason and because of (3.2.8), we have

$$\|u_h(\cdot, t^{n+1}) - u_h(\cdot, t^n)\|_{L^1(\mathbb{R}^2)} \leq \|u_h(\cdot, t^n) - u_h(\cdot, t^{n-1})\|_{L^1(\mathbb{R}^2)}$$
$$\leq \ldots \leq \|u_h(\cdot, t^1) - u_h(\cdot, t^0)\|_{L^1(\mathbb{R}^2)}$$
$$= \sum_j |T_j| |u_h(w_j, t^1) - u_h(w_j, t^0)|$$
$$\leq \Delta t \sum_j \sum_l |g_{jl}(u_j^0, u_{jl}^0) - g_{jl}(u_j^0, u_j^0)|$$
$$\leq c\Delta t h \sum_j \sum_l |u_{jl}^0 - u_j^0| \leq c\Delta t .$$

$\square$

LEMMA 3.4.4 *Assume that $u_0 \in L^2(\mathbb{R}^2) \cap L^\infty(\mathbb{R}^2)$, g_{jl} satisfies the assumptions in Lemma 3.4.2, G_{jl} satisfies (3.3.15)–(3.3.17), and Δt is sufficiently small. Then*

$$\frac{1}{2} \sum_n \sum_j \sum_{l=1}^3 |u_j - u_{jl}|^2 Q_{jl}^2(u_j, u_{jl}) \leq C \|u_0\|_{L^2(\mathbb{R}^2)}^2 ,$$

where

$$Q_{jl}(u, v) = \frac{g_{jl}(u, u) - 2g_{jl}(u, v) + g_{jl}(v, v)}{v - u} .$$

Proof Using $u_j := u_j^n$

$$u_{jl}^* := u_j - \frac{\Delta t}{|T_j|} \frac{[g_{jl}(u_j, u_{jl}) - g_{jl}(u_j, u_j)]}{|S_{jl}|} |\partial T_j| ,$$

we have

$$u_j^{n+1} = u_j - \frac{\Delta t}{|T_j|} \sum_{l=1}^3 [g_{jl}(u_j, u_{jl}) - g_{jl}(u_j, u_j)]$$
$$= \frac{1}{|\partial T_j|} \sum_l u_{jl}^* |S_{jl}| .$$

The convexity of the function $u \to u^2$ implies that

$$|u_j^{n+1}|^2 \leq \sum_l \frac{|S_{jl}|}{|\partial T_j|}(u_{jl}^*)^2 \ .$$

Therefore, using lemma 3.4.2 with $\lambda = \lambda_{jl} = \Delta t |\partial T_j|/|T_j|\|S_{jl}|$,

$$|u_j^{n+1}|^2 - |u_j^n|^2 + \frac{\Delta t}{|T_j|}\sum_l G_{jl}(u_j, u_{jl})$$

$$\leq \sum_l \frac{|S_{jl}|}{|\partial T_j|}(u_{jl}^*)^2 - \sum_l \frac{|S_{jl}|}{|\partial T_j|}(u_j^n)^2 + \frac{\Delta t}{|T_j|}\sum_l G_{jl}(u_j, u_{jl})$$

$$= \sum_l \frac{|S_{jl}|}{|\partial T_j|}[(u_{jl}^*)^2 - (u_j^n)^2] + \sum_l \frac{\Delta t}{|T_j|}[G_{jl}(u_j, u_{jl}) - G_{jl}(u_j, u_j)]$$

$$\leq -\varepsilon \sum_l \frac{|S_{jl}|}{|\partial T_j|}\lambda^2 |u_j - u_{jl}|^2 C_{jl}(u_j, u_{jl})^2 \ ,$$

where

$$C_{jl}(u, v) := \frac{g_{jl}(u, u) - g_{jl}(u, v)}{v - u} \ .$$

Multiplying by $|T_j|$ and summing over all T_j, we obtain

$$\sum_j |u_j^{n+1}|^2|T_j| + c\varepsilon \sum_{j,l} \lambda^2 |T_j| |u_j - u_{jl}|^2 C_{jl}(u_j, u_{jl})^2$$

$$+ \Delta t \sum_{j,l} G_{jl}(u_j, u_{jl}) \leq \sum_j |u_j^n|^2 |T_j| \ .$$

Summing over $n = 1$ to $n = N$, we obtain for $Q_{jl} = Q_{jl}(u_j, u_{jl})$

$$\sum_j |u_j^{N+1}|^2|T_j| + \frac{c\varepsilon}{2} \sum_n \sum_{\text{edges}} |u_j - u_{jl}|^2 Q_{jl}^2 \leq \sum_j |u_j^0|^2 |T_j| \ ,$$

and therefore

$$\frac{1}{2}\sum_n \sum_{\text{edges}} |u_j^n - u_{jl}^n|^2 Q_{jl}^2 \leq \frac{c}{\varepsilon}\|u^0\|^2_{L^2(\mathbb{R}^2)} \leq \frac{c}{\varepsilon}\|u_0\|^2_{L^2(\mathbb{R}^2)} \ .$$

Here we have used the fact that

$$C_{jl}(u,v)^2 + C_{lj}(v,u)^2 \geq \tfrac{1}{2}(C_{jl}(u,v) - C_{lj}(v,u))^2 = \tfrac{1}{2}Q_{jl}(u,v)^2 \ .$$

□

3.4 Error estimates for first-order schemes

NOTATION 3.4.5 For any $r : \mathbb{R} \to \mathbb{R}$ let

$$H_r(u) = [f(u) - f(r)]\,\mathrm{sgn}(u - r),$$

let u_h be the discrete solution of (3.2.6), (3.2.7) and $\omega \in C_0^\infty(\mathbb{R}^2 \times \mathbb{R})$, $g \in L^1_{\mathrm{loc}}(\mathbb{R}^2 \times \mathbb{R})$. Then for $\Omega \subset \mathbb{R}^2$ bounded define

$$\mathcal{H}(g, u_h, \omega)(y, s) := -\int_{\Omega \times [t_0, t_1]} \Big[|u_h(x,t) - g(y,s)|\,\partial_t \omega(\xi, \chi)$$

$$+ H_{g(y,s)}(u_h(x,t))\nabla_x \omega(\xi, \chi)\Big]\,dx\,dt,$$

$$\mathcal{H}_0(g, u_h, \omega)(y, s) := \int_\Omega |u_h(x, t_0) - g(y, s)|\,\omega(\xi, t_0 - s)\,dx$$

$$\mathcal{H}_1(g, u_h, \omega)(y, s) := \int_\Omega |u_h(x, t_1) - g(y, s)|\,\omega(\xi, t_1 - s)\,dx \qquad (3.4.4)$$

$$\mathcal{H}_g(\omega)(y, s) := \int_{\Omega \times [t_0, t_1]} H_{g(y,s)}(0) \nabla_x \omega(\xi, \chi)\,dx\,dt,$$

where $\xi = x - y$ and $\chi = t - s$.

THEOREM 3.4.6 (Kuznetsov [124]) *Assume that the following hold.*

(1) u_h *is given by the finite volume scheme (3.2.6), (3.2.7).*

(2) There are $C_1, C_2 \in \mathbb{R}^+$ *and a function* $\varphi_1(\tau, h)$ *such that for all* $t \in \mathbb{R}^+$

$$\|u_h(\cdot, t)\|_{L^1(\mathbb{R}^2)} \leq C_1,$$

$$\|u_h(\cdot, t + \tau) - u_h(\cdot, t)\|_{L^1(\mathbb{R}^2)} \leq \varphi_1(\tau, h) C_2,$$

where φ_1 *is monotonicaly increasing in* τ.

(3) $u_0 \in L^\infty(\mathbb{R}^2) \cap L^2(\mathbb{R}^2) \cap BV(\mathbb{R}^2)$, *and* u_0 *has compact support,*

$$h \sum_j \sum_l |u^0_{jl} - u^0_j| \leq c.$$

(4) For all $\Omega \subset \mathbb{R}^2$, Ω open and bounded, for all $g \in L^\infty([t_0, t_1], L^1(\mathbb{R}^2) \cap L^\infty(\mathbb{R}^2))$, for all $\omega \in C_0^\infty(\mathbb{R}^2 \times \mathbb{R})$ such that $\omega(x, t) = \omega(-x, t) = \omega(x, -t)$ there exists $C_3(t_0, t_1) \in \mathbb{R}^+$ such that

$$\int_{\mathbb{R}^2 \times [t_0, t_1]} (\mathcal{H} + \mathcal{H}_g - \mathcal{H}_0 + \mathcal{H}_1)(g, u_h, \omega)(y, s)\, dy\, ds$$

$$\leq C_3(t_0, t_1) \sqrt{|\Omega|} \varphi_2(h) \|\omega\|_{1,1}$$

where $\|\cdot\|_{1,1} = \|\cdot\|_{H^{1,1}(\mathbb{R}^2 \times \mathbb{R})}$. (Notice that Ω is contained in the definition of $\mathcal{H}, \mathcal{H}_g, \mathcal{H}_0, \mathcal{H}_1$).

Then there exists a positive constant $C(u^0, T)$, such that for any $\varepsilon > 0$

$$\|u(\cdot, t_1) - u_h(\cdot, t_1)\|_{L^1(\Omega)} \leq \|u(\cdot, t_0) - u_h(\cdot, t_0)\|_{L^1(\Omega)} \tag{3.4.5}$$

$$+C\left(\varepsilon + \varphi_1(\varepsilon, h)\right) + \frac{C\varphi_2(h)}{\varepsilon}. \tag{3.4.6}$$

Proof See 3.4.11.

Now we have to show that the finite volume scheme (3.2.6), (3.2.7) satisfies the conditions (1)-(4) in Theorem 3.4.6. This follows from the following theorem.

THEOREM 3.4.7 *Assume conditions (1)–(3) of Theorem 3.4.6, Lemma 3.4.1 and that Δt is sufficiently small. Then the condition (4) of Theorem 3.4.6 holds with the right-hand side*

$$\ldots \leq C_4(u^0) |\Omega|^{\frac{1}{2}} h^{\frac{1}{2}} (t_1 - t_0)^{\frac{1}{2}} \|\omega\|_{1,1}.$$

Proof See 3.4.10.

3.4 Error estimates for first-order schemes

THEOREM 3.4.8 (A priori error estimate) *Assume conditions (1) and (3) of Theorem 3.4.6 and that g_{jl} and G_{jl} satisfy (3.2.8)–(3.2.10) and (3.3.15)–(3.3.17) respectively, g_{jl} is monotone and*

$$\partial_2 G_{jl}(u,v) = U'(v)\partial_2 g_{jl}(u,v) \tag{3.4.7}$$

for all convex functions U. Furthermore, let Δt be sufficiently small. Then there exists a constant C independent of h such that

$$\|u(\cdot,t_1) - u_h(\cdot,t_1)\|_{L^1(\Omega)} \leq \|u(\cdot,t_0) - u_h(\cdot,t_0)\|_{L^1(\Omega)} + Ch^{\frac{1}{4}}. \tag{3.4.8}$$

Proof Proposition 3.3.14 and Lemma 3.4.3 imply condition (2) of Theorem 3.4.6 such that the assumptions of Theorem 3.4.7 are satisfied. Then Theorem 3.4.7 implies (4) of Theorem 3.4.6. Therefore

$$\|u(\cdot,t_1) - u_h(\cdot,t_1)\|_{L^1(\Omega)} \leq \|u(\cdot,t_0) - u_h(\cdot,t_0)\|_{L^1(\Omega)} + C\left(\varepsilon + \frac{h^{\frac{1}{2}}}{\varepsilon}\right). \tag{3.4.9}$$

If we choose $\varepsilon = h^{\frac{1}{4}}$, we obtain (3.4.8). □

REMARK 3.4.9 *The Engquist–Osher and the Lax–Friedrichs scheme as described in 3.3.22 satisfy condition (3.4.7), this means (3.4.9) holds for these schemes.*

3.4.10 (Proof of Theorem 3.4.7) Let $t_0 = t^0 < t^1 < \ldots < t^{N-1} < t^N = t_1$. Since $u_h|_T = u_T^n$ on $]t^n, t^{n+1}[$ and using the notation ($\xi := x-y$, $\chi := t-s$)

$$\omega_{T,n}(y,s) := \frac{1}{|T|} \int_T \omega(\xi, t^n - s)\, dx$$

$$\omega_{e,n}(y,s) := \frac{1}{\Delta t|e|} \int_e \int_{t^n}^{t^{n+1}} \omega(\cdot - y, \chi)\, dt\, d\sigma,$$

where e denotes an oriented edge of the triangulation. The orientation of e should not depend on T. We obtain

$$\mathcal{H}(g, u_h, \omega)(y, s) = -\sum_T \sum_{n=0}^{N-1} \int_T \int_{t^n}^{t^{n+1}} \Big[|u_T^n - g(y,s)| \partial_t \omega(\xi, \chi)$$
$$+ H_{g(y,s)}(u_T^n) \nabla_x \omega(\xi, \chi) \Big] dx\, dt$$

$$= \sum_{T,n} \Big\{ -|u_T^n - g(y,s)| |T| [\omega_{T,n+1}(y,s) - \omega_{T,n}(y,s)]$$

$$- H_{g(y,s)}(u_T^n) \int_{\partial T} \int_{t^n}^{t^{n+1}} n(\sigma) \omega(\cdot - y, \chi)\, d\sigma\, dt \Big\}$$

$$= \sum_{T,n} \Big\{ -|u_T^n - g(y,s)| |T| [\omega_{T,n+1}(y,s) - \omega_{T,n}(y,s)]$$

$$- H_{g(y,s)}(u_T^n) \sum_{e \in \partial T} \Delta t |e| \omega_{e,n}(y,s) n_{Te} \Big\}, \quad (3.4.10)$$

where the sum over e is taken over all e of the boundary ∂T of T. Now T always refers to the triangle T and e to the eth edge of triangle T. The value u_{Te} corresponds to the eth-neighbouring triangle of triangle T and ν_{Te} is the scaled outer normal on the eth edge in triangle T. We get

$$\sum_T \sum_{n=0}^{N-1} |u_T^n - g(y,s)| |T| [\omega_{T,n}(y,s) - \omega_{T,n+1}(y,s)]$$
$$= \sum_T |u_T^0 - g(y,s)| |T| \omega_{T,0} - \sum_T |u_T^N - g(y,s)| |T| \omega_{T,N}$$
$$+ \sum_T \sum_{n=1}^{N} [|u_T^n - g(y,s)| - |u_T^{n-1} - g(y,s)|] |T| \omega_{T,n}. \quad (3.4.11)$$

For the terms containing u_0 and u^N we obtain

$$\sum_T |u_T^0 - g(y,s)| |T| \omega_{T,0}$$

$$= \int_\Omega |u_h^0 - g(y,s)|\omega(\xi, t^0 - s)\, dx = \mathcal{H}_0(g, u_h, \omega)(y,s) \qquad (3.4.12)$$

$$\sum_T |u_T^N - g(y,s)|\,|T|\omega_{T,N}$$

$$= \int_\Omega |u_h^N - g(y,s)|\omega(\xi, t^N - s)\, dx = \mathcal{H}_1(g, u_h, \omega)(y,s).$$

Now using (3.4.4) and (3.4.10)–(3.4.12), $\mathcal{H}(g, u_h, \omega)$ can be written in the following form:

$$\mathcal{H} = \mathcal{H}_0 - \mathcal{H}_1 - \mathcal{H}_g(\omega) + \sum_T \sum_{n=0}^{N-1} (|u_T^{n+1} - g| - |u_T^n - g|)|T|\omega_{T,n+1}$$

$$+ \sum_{T,n,e} \Delta t\, \omega_{e,n}[G_{Te}(u_T^n, u_{Te}^n) - H_g(u_T^n)\nu_{Te}]_1$$

$$- \sum_{T,n,e} \Delta t\, \omega_{e,n}[G_{Te}(u_T^n, u_{Te}^n) - H_g(0)\nu_{Te}]_2$$

$$- \sum_{T,n,e} \Delta t\, \omega_{e,n} H_g(0)\nu_{Te} + \mathcal{H}_g(\omega),$$

where $G_{Te}(u,v)$ is defined as in Lemma 3.4.1. The notation $[\ ]_1, [\ ]_2$ is used below to refer to the corresponding expressions.

By definition of $H_g(\omega)$ (see (3.4.4)), the sum of the last two terms is equal to zero. In order to rewrite the differences $u_T^{n+1} - g$, we use the notation

$$u_T^{n+1,e} := u_T^n - \frac{\Delta t|\partial T|}{|e||T|}[g_{Te}(u_T^n, u_{Te}^n) - f(u_T^n)\nu_{Te}] \qquad (3.4.13)$$

and obtain using the definition of a finite volume scheme (see (3.2.7))

$$u_T^{n+1} = u_T^n - \sum_e \frac{\Delta t|\partial T|}{|e||T|}[g_{Te}(u_T^n, u_{Te}^n) - f(u_T^n)\nu_{Te}]\frac{|e|}{|\partial T|}$$

$$= \sum_e \frac{|e|}{|\partial T|} u_T^{n+1,e}, \qquad (3.4.14)$$

and therefore

$$|u_T^{n+1} - g| - |u_T^n - g| \leq \sum_e \frac{|e|}{|\partial T|}[|u_T^{n+1,e} - g| - |u_T^n - g|]_3 \ .$$

Using this estimate, we get

$$[\mathcal{H} - \mathcal{H}_0 + \mathcal{H}_1 + \mathcal{H}_g(\omega)](y,s)$$

$$\leq \sum_{T,n,e} \frac{|e|}{|\partial T|}[\ ]_3 |T| \omega_{T,n+1} + \sum_{T,n,e} \Delta t \omega_{e,n}[\ ]_1 - \sum_{T,n,e} \Delta t \omega_{e,n}[\ ]_2$$

$$= \sum_{T,n,e} \frac{|e||T|}{|\partial T|} \left\{ [\ ]_3 (\omega_{T,n+1} - \omega_{e,n}) + \omega_{e,n}[\ ]_3 + \frac{\Delta t |\partial T|}{|e||T|} \omega_{e,n}[\ ]_1 \right\}$$

$$- \sum_{T,n,e} \Delta t \omega_{e,n}[\ ]_2 \ . \tag{3.4.15}$$

Lemma 3.4.1 implies that

$$[\ ]_3 + \frac{\Delta t |\partial T|}{|e||T|}[\ ]_1 \leq 0 \ . \tag{3.4.16}$$

Since u^0 has compact support, and if Ω is large, then

$$\sum_{T,n,e} \Delta t \omega_{e,n}[\ ]_2 = 0 \ . \tag{3.4.17}$$

Now if we integrate (3.4.15) with respect to $\int_{\mathbb{R}^2} \int_{t_0}^{t_1} \ldots ds \, dy$ and we use (3.4.16), we obtain

$$\int (\mathcal{H} - \mathcal{H}_0 + \mathcal{H}_1 + \mathcal{H}_g) \, dy \, ds$$

$$\leq \sum_{T,n,e} \frac{|e||T|}{|\partial T|}|u^{n+1,e} - u_T^n| \int_{\mathbb{R}^2 \times [t_0,t_1]} |\omega_{T,n+1} - \omega_{e,n}| \ . \tag{3.4.18}$$

Now let us prove the theorem for a special class of functions ω as defined in the following lines. The general case can then be obtained by an approximation argument. Let

$$\zeta \in C_0^\infty(\mathbb{R}^2, \mathbb{R}^+), \quad \theta \in C_0^\infty(\mathbb{R}^1, \mathbb{R}^+)$$

3.4 Error estimates for first-order schemes

such that

$$\zeta(x)\zeta(-x), \quad \operatorname{supp}\zeta \subset [-1,1]^2, \quad \int_{\mathbb{R}^2} \zeta(x)\,dx = 1\,,$$

$$\theta(t) = \theta(-t), \quad \operatorname{supp}\theta \subset [-1,1], \quad \int_{\mathbb{R}^1} \theta(t)\,dt = 1\,,$$

$$w(x,t) := \zeta(x)\theta(t)\,.$$

In order to estimate

$$R := \int_{\mathbb{R}^2 \times [t_0, t_1]} |w_{T,n+1} - w_{e,n}|\, dx\, dt\,,$$

consider

$$R = \int_{\mathbb{R}^2} \int_{t_0}^{t_1} \left| \frac{1}{|T|} \int_T w(x-y, t^{n+1} - s)\, dx \right.$$

$$\left. - \frac{1}{\Delta t |e|} \int_e \int_{t^n}^{t^{n+1}} w(\cdot - y, t - s)\, d\sigma\, dt \right|\, ds\, dy$$

$$= \left[\int_{\mathbb{R}^2} \int_{t_0}^{t_1} \left| \frac{1}{|T|} \int_T \zeta(x-y)\theta(t^{n+1}-s)\,dx \right. \right.$$

$$\left. - \frac{1}{\Delta t|e|} \int_e \int_{t^n}^{t^{n+1}} \zeta(\cdot - y)\,\theta(t-s)\,d\sigma\, dt \right|\, ds\, dy \right]$$

$$= \left[\int_{\mathbb{R}^2} \int_{t_0}^{t_1} \left| \frac{1}{|T|} \int_T \zeta(x-y)\, dx \cdot \frac{1}{\Delta t} \int_{t^n}^{t^{n+1}} \theta\left(t^{n+1} - s\right) dt \right. \right.$$

$$\left. - \frac{1}{\Delta t|e|} \int_e \zeta(\cdot - y)\, d\sigma \int_{t^n}^{t^{n+1}} \theta(t-s)\, dt \right|\, ds\, dy \right]$$

$$= \int_{\mathbb{R}^2} \int_{t_0}^{t_1} \left[\left| \frac{1}{|T|} \int_T \zeta(x-y)\, dx \right. \right.$$

Initial value problems for scalar conservation laws in 2-D

$$\left[\frac{1}{\Delta t}\int_{t^n}^{t^{n+1}}\left\{\theta\left(t^{n+1}-s\right)-\theta\left(t-s\right)\right\}dt\right]_4$$

$$+\frac{1}{\Delta t}\int_{t^n}^{t^{n+1}}\theta\left(t-s\right)dt$$

$$\left[\frac{1}{|T|}\int_T \zeta\left(x-y\right)dx - \frac{1}{|e|}\int_e \zeta\left(\cdot-y\right)d\sigma\right]_5\bigg]\,dsdy\,. \qquad (3.4.19)$$

Let us first estimate

$$\left[\frac{1}{|T|}\int_T \zeta\left(x-y\right)dx - \frac{1}{|e|}\int_e \zeta\left(\cdot-y\right)d\sigma\right]_5.$$

For simplicity, let us consider the case when $T =:\,]0, a[\,\times\,]0,b[$ is a rectangle and e is part of $\{(x,0)|x \in \mathbb{R}\}$.

$$\left|\frac{1}{|T|}\int_T v(x)\,dx - \frac{1}{|e|}\int_e v(x_1,0)\,dx_1\right|$$

$$= \left|\frac{1}{ab}\int_0^a\!\!\int_0^b v(x)\,dx_2\,dx_1 - \frac{1}{ab}\int_0^a\!\!\int_0^b v(x_1,0)\,dx_2\,dx_1\right|$$

$$\leq \frac{1}{ab}\int_0^a\!\!\int_0^b |(v(x_1,x_2) - v(x_1,0))|\,dx_2\,dx_1$$

$$\leq \frac{1}{ab}\int_0^a\!\!\int_0^b\!\!\int_0^1 |\partial_2 v(x_1,\tau x_2)x_2|\,d\tau\,dx_2\,dx_1$$

$$\leq \frac{1}{ab}\int_0^a\!\!\int_0^b\!\!\int_0^{x_2} |\partial_2 v(x_1,z)|\,dz\,dx_2\,dx_1$$

$$\leq \frac{1}{ab}b\|\partial_2 v(\cdot)\|_{L^1(T)}\,.$$

3.4 Error estimates for first-order schemes

Similarly, it can be shown that

$$\frac{1}{|T|} \int_T \zeta(x-y)\,dx - \frac{1}{|e|} \int_e \zeta(\cdot - y)\,d\sigma$$

$$\leq \frac{ch}{|T|} \int_T |\zeta'(x-y)|\,dx.$$

Therefore the second term in (3.4.19) can be estimated by

$$\frac{1}{\Delta t}\frac{h}{|T|} \int_{t^n}^{t^{n+1}} \int_{t_0}^{t_1} \theta(t-s)\,ds\,dt \int_T \int_{\mathbb{R}^2} |\zeta'(x-y)|\,dy\,dx$$

$$\leq \frac{h}{\Delta t|T|}\Delta t \int_{\mathbb{R}} \theta(\tau)\,d\tau |T| \int_{\mathbb{R}^2} |\zeta'(z)|\,dz$$

$$\leq h\|\theta\|_{L^1}\|\zeta'\|_{L^1} \leq ch\|\omega\|_{1,1}.$$

To estimate the term $[\ ]_4$ in (3.4.19), consider

$$\left|\frac{1}{\Delta t} \int_{t_0}^{t_1} \int_{t_n}^{t^{n+1}} \int_t^{t^{n+1}} \frac{d}{d\tau}\theta(\tau-s)\,d\tau\,dt\,ds\right|$$

$$\leq \frac{1}{\Delta t} \int_{t_0}^{t_1} \int_{t_n}^{t^{n+1}} \int_t^{t^{n+1}} |\theta'(\tau-s)|\,d\tau\,dt\,ds$$

$$\leq \int_{t_n}^{t^{n+1}} \int_{t_0}^{t_1} |\theta'(\tau-s)|\,ds\,d\tau$$

$$\leq \int_{t_n}^{t^{n+1}} \int_{t_0-s}^{t_1-s} |\theta'(\sigma)|\,d\sigma$$

$$\leq \Delta t \int_{\mathbb{R}} |\theta'(\sigma)|\,d\sigma.$$

Altogether, we obtain

$$R \leq C(h+\Delta t)\|\omega\|_{1,1}. \qquad (3.4.20)$$

Therefore (3.4.18) and (3.4.20) imply that

$$\int_{\mathbb{R}^n \times [t_0, t_1]} (\mathcal{H} - \mathcal{H}_0 + \mathcal{H}_1 + \mathcal{H}_g)$$

$$\leq C\|\omega\|_{1,1}(h + \Delta t) \sum_{T,n,e} \frac{|e||T|}{|\partial T|} |u_T^{n+1,e} - u_T^n|.$$

By definition, we have

$$|u_T^{n+1,e} - u_T^n| = \left|\frac{\Delta t |\partial T|}{|e||T|} [g_{Te}(u_T, u_{Te}) - f(u_T^n)\nu_{Te}]\right|$$

$$= \frac{\Delta t |\partial T|}{|e||T|} \underbrace{\left|\frac{g_{Te}(u_T, u_{Te}) - f(u_T^n)\nu_{Te}}{u_{Te} - u_T}\right|}_{=C_{Te}(\nu_{Te}, u_T, u_{Te})} |u_{Te} - u_T|.$$

Therefore we have for $Q_{Te}(n, u, v) := C_{Te}(n, u, v) + C_{Te}(-n, v, u)$

$$\int_{\mathbb{R}^2 \times [t_0, t_1]} (\mathcal{H} - \mathcal{H}_0 + \mathcal{H}_1 + \mathcal{H}_g)$$

$$\leq C(h + \Delta t)\|\omega\|_{1,1} \sum_{T,n,e} \frac{|T||e|}{|\partial T|} \frac{|\partial T|}{|T||e|} \Delta t C_{Te}(\nu_{Te}, u_T, u_{Te})|u_{Te} - u_T|$$

$$\leq C(h + \Delta t)\|\omega\|_{1,1} \sum_{n,e} \Delta t Q_{Te}(\nu_{Te}, u_T, u_{Te})|u_{Te} - u_T|,$$

where the sum is now taken over all time steps and all edges e of the triangulations. Now u_T and u_{Te} refer to the two triangles having e as a joint edge. Using Lemma 3.4.4, (3.2.3) and (3.2.4), we can continue:

$$\leq C\frac{(h + \Delta t)}{\sqrt{h}}\|\omega\|_{1,1} \left(\sum_{n,e} \Delta t^2 h\right)^{\frac{1}{2}} \left[\sum_{n,e} Q_{Te}^2(u_T, u_{Te})|u_{Te} - u_T|^2\right]^{\frac{1}{2}}$$

$$\leq C\sqrt{|\Omega|}\frac{h + \Delta t}{\sqrt{h}}\|\omega\|_{1,1}\sqrt{t_1 - t_0}\|u_0\|_{L^2}.$$

This finishes the proof of Theorem 3.4.7. □

3.4 Error estimates for first-order schemes

3.4.11 (Proof of Theorem 3.4.6) Let $I := [t_0, t_1]$. The Kruzkov entropy inequality (see [123] and Definition 3.1.2) implies that we have for all $c \in \mathbb{R}$ and for all test functions $\varphi \in C_0^\infty(\mathbb{R}^2 \times \mathbb{R})$, $\varphi \geq 0$.

$$-\int_{\mathbb{R}^2 \times I} (|u-c| \partial_s \varphi + H_c(u) \nabla_y \varphi) \, dy \, ds$$

$$+ \int_{\mathbb{R}^2} |u(t_1) - c| \varphi(y, t_1) \, dy \qquad (3.4.21)$$

$$- \int_{\mathbb{R}^2} |u - c| \varphi(y, t_0) \, dy \leq 0 .$$

For $\Omega \subset \mathbb{R}^2$ and $(x, t) \in \Omega \times I$ choose

$$c = u_h(x, t), \quad \varphi(y, s) = \omega(x - y, t - s)$$

and integrate (3.4.21) over $\Omega \times I$. Then we obtain

$$\int_{\mathbb{R}^2 \times \Omega \times I^2} |u(y, s) - u_h(x, t)| \, \partial_t \omega \, dy \, dx \, dt \, ds$$

$$+ \int_{\mathbb{R}^2 \times \Omega \times I^2} H_{u_h(x,t)}(u(y, s)) \nabla_x \omega(x - y, t - s) \, dy \, dx \, dt \, ds$$

$$\leq - \int_{\mathbb{R}^2 \times \Omega \times I} |u(y, t_1) - u_h(x, t)| \, \omega(x - y, t_1 - t) \, dy \, dx \, dt$$

$$+ \int_{\mathbb{R}^2 \times \Omega \times I} |u(y, t_0) - u_h(x, t)| \, \omega(x - y, t_0 - t) \, dy \, dx \, dt . \qquad (3.4.22)$$

Using

$$S_1(\tau) := \int_{\mathbb{R}^2 \times \Omega \times I} |u(y, \tau) - u_h(x, t)| \, \omega(x - y, \tau - t) \, dy \, dx \, dt, \qquad (3.4.23)$$

$$S_2(\tau) := \int_{\mathbb{R}^2 \times \Omega \times I} |u_h(x, \tau) - u(y, t)| \, \omega(x - y, \tau - t) \, dy \, dx \, dt, \qquad (3.4.24)$$

(3.4.22) can be written as

$$\int_{\mathbb{R}^2\times\Omega\times I^2} |u(y,s) - u_h(x,t)| \, \partial_t w(x-y, t-s) \, dy \, dx \, dt \, ds$$

$$+ \int_{\mathbb{R}^2\times\Omega\times I^2} H_{u_h(x,t)}(u(y,s)) \nabla_x w(x-y, t-s) \, dy \, dx \, dt \, ds \qquad (3.4.25)$$

$$\leq -S_1(t_1) + S_1(t_0) \, .$$

Using the notation (3.4.4) with $g = u$, we obtain

$$-\int \mathcal{H} \leq -S_1(t_1) + S_1(t_0) \, .$$

If $J(\tau) := S_1(\tau) + S_2(\tau)$, we obtain

$$J(t_1) = S_1(t_1) + S_2(t_1) = S_1(t_1) + \int \mathcal{H}_1$$

$$\leq \int \mathcal{H} + S_2(t_0) + S_1(t_0) + \int \mathcal{H}_1 - \int \mathcal{H}_0$$

$$= J(t_0) + \int [\mathcal{H} + \mathcal{H}_u - \mathcal{H}_0 + \mathcal{H}_1]_1 - \int \mathcal{H}_u \, . \qquad (3.4.26)$$

Now choose $w(x,t) = \zeta_\varepsilon(x)\theta_\varepsilon(t)$, where

$$\zeta_\varepsilon(x) := \frac{1}{\varepsilon^2}\zeta\left(\frac{x}{\varepsilon}\right), \quad \theta_\varepsilon(t) := \frac{1}{\varepsilon}\theta\left(\frac{t}{\varepsilon}\right) \, .$$

Then we obtain for $J(t_0)$

$$J(t_0) = S_1(t_0) + S_2(t_0)$$

$$= \int_{\mathbb{R}^2\times\Omega\times I} |u(y,t_0) - u_h(x,t)|\zeta_\varepsilon(\xi)\theta_\varepsilon(t-t_0) \, dy \, dx \, dt$$

$$+ \int_{\mathbb{R}^2\times\Omega\times I} |u_h(x,t_0) - u(y,t)|\zeta_\varepsilon(\xi)\theta_\varepsilon(t_0-t) \, dy \, dx \, dt$$

$$= \int_{t_0}^{t_1} \theta_\varepsilon(t-t_0) \int_{\mathbb{R}^2\times\Omega} \Big(|u(y,t_0) - u_h(x,t)|$$

$$+ |u_h(x,t_0) - u(y,t)|\Big)_1 \zeta_\varepsilon(\xi) \, dy \, dx \, dt \, .$$

3.4 Error estimates for first-order schemes

The term $(\ )_1$ can be estimated as follows:

$$\overbrace{|u(y,\tau) - u_h(x,t)|}^{\tilde{J}_2(\tau)} + \overbrace{|u_h(x,\tau) - u(y,t)|}^{\tilde{J}_1(\tau)}$$
$$\leq 2\overbrace{|u(y,\tau) - u(x,\tau)|}^{\tilde{J}_3} + 2\overbrace{|u(x,\tau) - u_h(x,\tau)|}^{\tilde{J}_4}$$
$$+ \overbrace{|u_h(x,\tau) - u_h(x,t)|} + \overbrace{|u(y,\tau) - u(y,t)|} \ .$$

Let

$$J_i(\tau) = \int_{t_0}^{t_1} \theta_\varepsilon(t-\tau) \int_{\mathbb{R}^2 \times \Omega} \tilde{J}_i(\tau)(x,y,t)\zeta_\varepsilon(\xi) \, dy \, dx \, dt \ .$$

Then we have

$$J(t_0) \leq 2 J_1(t_0) + 2 J_2(t_0) + J_3(t_0) + J_4(t_0) \tag{3.4.27}$$

and the following estimates are valid:

$$J_1(t_i) = \|u(.,t_i) - u_h(.,t_i)\|_{L^1(\Omega)} ,$$
$$J_2(t_i) \leq c\epsilon \|u\|_{BV(\mathbb{R}^2 \times]0,T[)} ,$$
$$J_3(t_i) \leq \sup_{0 \leq |\tau| \leq \epsilon} \int_\Omega |u_h(x, t_i - \tau) - u_h(x, t_i)| \, dx , \tag{3.4.28}$$
$$J_4(t_i) \leq c\epsilon$$

$$\int_{\mathbb{R}^2 \times \Omega \times I^2} |H_u(0)\nabla_x \omega(x-y, t-s)| \, dy \, dx \, dt \, ds \leq c\epsilon \ .$$

Similarly, it can be shown that

$$J(t_1) \geq 2 J_1(t_1) - 2 J_2(t_1) - J_3(t_1) - J_4(t_1) \ . \tag{3.4.29}$$

Let us prove these estimates for J_1, J_3 and $\int \mathcal{H}_u$. For the first two,

$$J_1(t_i) = \int_{t_0}^{t_1} \theta_\epsilon(t - t_i) \, dt \int_{\Omega \times \mathbb{R}^2} |u(x,t_i) - u_h(x,t_i)|\zeta_\epsilon(x-y) \, dx \, dy$$
$$= \int_\Omega |u(x,t_i) - u_h(x,t_i)| \, dx$$

and

$$J_3(t_i) = \int_{t_0}^{t_1} \theta_\epsilon(t-t_i) \int_{\Omega \times \mathbb{R}^2} |u_h(x,t) - u_h(x,t_i)| \zeta_\epsilon(x-y) \, dx \, dy \, dt$$

$$= \int_{t_0}^{t_1} \theta_\epsilon(t-t_i) \int_{\Omega} |u_h(x,t) - u_h(x,t_i)| \, dx \, dt$$

$$\leq \sup_{0 \leq |\tau| \leq \epsilon} \int_{\Omega} |u_h(x,t_i-\tau) - u_h(x,t_i)| \, dx \, ,$$

since $\operatorname{supp} \theta_\epsilon \subset [-\epsilon, \epsilon]$. The estimates for J_2 and J_4 and for the integral in (3.4.28) can be proved similarly.

Let us now prove the fifth estimate in (3.4.28). For this, we have to estimate

$$\int \mathcal{H}_u \, .$$

We obtain from the definition of $\mathcal{H}_u$ (see Notation 3.4.5)

$$\int \mathcal{H}_u = \int_{\mathbb{R}^2 \times I} \mathcal{H}_{u(y,s)} \, dy \, ds$$

$$= \int_{\mathbb{R}^2 \times \Omega \times I^2} [f(u(y,s)) - f(0)] \operatorname{sgn} u(y,s) \, \nabla_x \omega(x-y, t-s) \, dy \, dx \, dt \, ds$$

$$= \int_{\mathbb{R}^2 \times \partial\Omega \times I^2} n(\sigma)[f(u(y,s)) - f(0)]$$

$$\cdot \operatorname{sgn} u(y,s) \, \omega(x(\sigma) - y, t-s) \, dy \, dx(\sigma) \, dt \, ds$$

$$= -\int_{\mathbb{R}^2 \times \partial\Omega \times I^2} n(\sigma)[f(u(x(\sigma) - z, s)) - f(0)]$$

$$\cdot \operatorname{sgn} u(x(\sigma) - z, s) \, \omega(z, t-s) \, dy \, dx(\sigma) \, dt \, ds \, . \qquad (3.4.30)$$

3.4 Error estimates for first-order schemes

Since ω is symmetric, we also obtain

$$\int \mathcal{H}_u = \int_{\mathbb{R}^2 \times \Omega \times I^2} -[f(u(y,s)) - f(0)]$$

$$\cdot \operatorname{sgn} u(y,s) \; \nabla_x \omega(y-x, t-s) \, dy \, dx \, dt \, ds$$

$$= \int_{\mathbb{R}^2 \times \partial\Omega \times I^2} n(\sigma)[f(u(x(\sigma)+z,s)) - f(0)]$$

$$\cdot \operatorname{sgn} u(x(\sigma)+z, s) \; \omega(z, t-s) \, dz \, dx(\sigma) \, dt \, ds \; . \quad (3.4.31)$$

Adding (3.4.30) and (3.4.31), we obtain

$$\int \mathcal{H}_u = \frac{1}{2} \int_{\mathbb{R}^2 \times \partial\Omega \times I^2} n(\sigma)[f(u(x(\sigma)+z,s)) - f(u(x(\sigma)-z,s))]$$

$$\cdot \operatorname{sgn} u(x(\sigma)+z, s) \; \omega(z, t-s) \, dz \, dx(\sigma) \, dt \, ds \; ,$$

and, since f is Lipschitz-continuous,

$$\left| \int \mathcal{H}_u \right| \leq c(\|u\|_{L^\infty}) \int_{\mathbb{R}^2 \times \partial\Omega \times I^2} |u(x(\sigma)+z,s) - u(x(\sigma)-z,s)|$$

$$\cdot |\omega(z, t-s)| \, dz \, dx(\sigma) \, dt \, ds$$

$$\leq c \int_{\mathbb{R}^2 \times \partial\Omega \times I} |u(x(\sigma)+z,s) - u(x(\sigma)-z,s)| \zeta_\varepsilon(z) \, dz \, dx(\sigma) \, ds$$

$$\leq c \sup_{|z| \leq \varepsilon} \int_{\partial\Omega \times I} |u(x(\sigma)+z,s) - u(x(\sigma)-z,s)| \, dx(\sigma) \, ds \int_{\mathbb{R}^2} \zeta_\varepsilon(z) \, dz \; .$$

This term disappears if Ω is sufficiently large, since u_0 and therefore u have compact support. For the last conclusion we refer to [80, Theorem 3.1, (3.14) in Chapter 2]. This finishes the proof of the fifth estimate in (3.4.28).

Now using (3.4.26)–(3.4.29), the error estimate (3.4.5) can be obtained as follows:

$$\|u(.,t_1) - u_h(.,t_1)\|_{L^1(\Omega)}$$
$$\leq \|u(.,t_0) - u_h(.,t_0)\|_{L^1(\Omega)} + c\varepsilon$$
$$+ \sum_{i=1}^{2} \sup_{0 \leq \tau \leq \varepsilon} \int_{\Omega} |u_h(x, t_i - \tau) - u_h(x, t_i)| dx - \frac{1}{2} \int \mathcal{H}_u + \frac{1}{2} \int [\,]_1.$$

Now the assumptions of Theorem 3.4.6 imply that

$$\sup_{0 \leq \tau \leq \varepsilon} \int_{\Omega} |u_h(x, t_i - \tau) - u_h(x, t_i)| dx \leq \varphi_1(\varepsilon, h) C_2,$$

$$\int [\,]_1 \leq C_3 \sqrt{|\Omega|} \varphi_2(h) \|\omega\|_{1,1},$$

and therefore we obtain

$$\|u(\cdot,t_1) - u_h(\cdot,t_1)\|_{L^1(\Omega)}$$
$$\leq \|u(\cdot,t_0) - u_h(\cdot,t_0)\|_{L^1(\Omega)} + C(\varepsilon + \varphi_1(\varepsilon, h)) + \frac{C\varphi_2(h)}{\varepsilon},$$

since $\|\omega\|_{1,1} \leq c/\varepsilon$. □

REMARK 3.4.12 *Noelle and [170] obtains error estimates for grids with thin triangles, also nonconvex fluxfunctions and general E–schemes.*

3.5 Higher-order finite volume schemes for scalar equations

Now we shall show the convergence of finite volume schemes of higher-order on unstructured grids consisting of triangles. On more general grids with arbitrary cells the proof of the convergence is still an open question. For the definition of the higher-order schemes on arbitrary cells we refer to [250]. First we shall present the theoretical background for convergence proofs and after that we shall discuss some special limiters and reconstructions on general unstructured triangular meshes and also some numerical experiments.

3.5 Higher-order finite volume schemes for scalar equations

The proof of the convergence theorem will be given in Section 3.6 (see [121]). The main idea for higher-order finite volume schemes is to use functions that are piecewise-linear on each triangle instead of the piecewise-constant functions. More precisely, we shall obtain the piecewise-linear functions from piecewise-constant ones by some reconstruction technique as described in the following definition.

DEFINITION 3.5.1 (Linear reconstruction) *Let $\mathcal{T}_h := \{T_j \mid j \in I\}$ be a family of triangulations as in Section, 3.2, where $h := \sup_{j \in I} \operatorname{diam}(T_j)$ and let v_h be a piecewise-constant function on $\mathbb{R}^2$ such that*

$$v_h(x) = v_j \quad \text{if} \quad x \in T_j. \tag{3.5.1}$$

Then the map

$$L: \mathbb{R}^I \to L^\infty(\mathbb{R}^2), \quad v_h \to L(v_h) \tag{3.5.2}$$

is called a (piecewise-) linear reconstruction if $L(v_h)$ is linear on T_j for all $j \in I$. Usually we use the notation

$$L_j(x) := L(v_h)|_{T_j}(x). \tag{3.5.3}$$

Since $L(v_h)$ may be discontinuous on the edges of the triangles, we define for the midpoints z_{jl} of the edges of the triangles

$$L_j(z_{jl}) := \lim_{x \to z_{jl}, x \in T_j} L(v_h)(x). \tag{3.5.4}$$

Figure 3.5.1 illustrates the situation of this definition. On the triangles T_j and T_l we have piecewise-constant values v_j and v_l and piecewise-linear functions L_j and L_l respectively.

EXAMPLE 3.5.2 (Reconstruction of Durlovsky, Engquist, Osher)
Let w_j, w_{j1} and w_{j2} denote the centres of gravity of the marked triangles in the first diagram of Figure 3.5.2 and let S_1 denote the plane through the points (w_j, u_j), (w_{j1}, u_{j1}) and (w_{j2}, u_{j2}) associated with the black points in Figure 3.5.2. As before, the index j refers to the central triangle. The linear functions S_2 and S_3 are defined similarly.

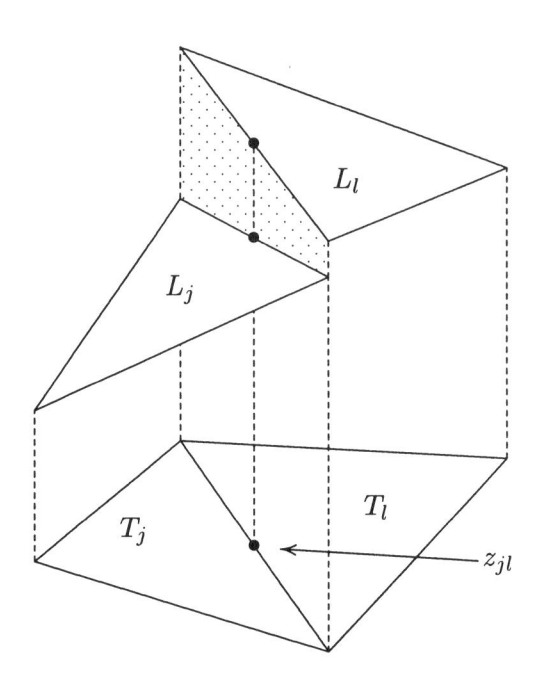

214   Initial value problems for scalar conservation laws in 2-D

Figure 3.5.1

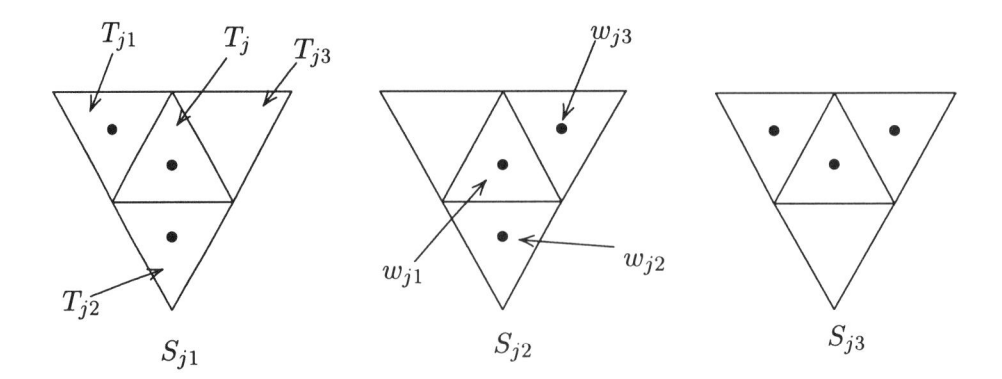

Figure 3.5.2

3.5 Higher-order finite volume schemes for scalar equations

The min–mod limiter defined in [58] can be expressed in our notation by (see also Lemma 2.5.17)

$$L_j^n(x) := u_j^n + \sigma \nabla S_i \cdot (x - w_j),$$

where i is defined by

$$|\nabla S_i| < |\nabla S_k| \quad \forall k \in \{1, 2, 3\}, k \neq i,$$

and

$$\sigma = \begin{cases} 0 & \text{if } u_j^n > u_{jl}^n \;\; \forall l \text{ or } u_j^n < u_{jl}^n \;\; \forall l, \\ 1 & \text{otherwise}. \end{cases}$$

This mean that $\sigma = 1$ except for local extrema, where $\sigma = 0$. The second limiter in [58] can be expressed as

$$L_j^n(x) := u_j^n + \sigma \nabla S_i \cdot (x - w_j),$$

with σ as above and where i is defined by

$$\tau_i |\nabla S_i| > \tau_k |\nabla S_k| \quad \forall k \neq i$$

and

$$\tau_i = \begin{cases} 1 & \text{if } \nabla S_i \cdot (z_{jl} - w_j) \in I(0, u_{jl} - u_j) \;\; \forall l, \\ 0 & \text{otherwise}. \end{cases}$$

Now we are ready to define the higher-order finite volume schemes.

DEFINITION 3.5.3 (Higher-order finite volume scheme) *Assume that $u_0 \in L^1 \cap L^\infty(\mathbb{R}^2)$ with compact support and define*

$$u_j^0 := \frac{1}{|T_j|} \int_{T_j} u_0(x). \tag{3.5.5}$$

Assume that u_j^n and the corresponding linear reconstruction $L(u^n)$ are already defined. Let g_{jl} be a monotone numerical flux satisfying (3.2.8) – (3.2.10). Remember that a first order finite volume scheme is given by

(3.2.7). Then the higher-order (in space) finite volume scheme is defined by

$$u_j^{n+1} := u_j^n - \frac{\Delta t}{|T_j|} \sum_{l=1}^{3} g_{jl}(L_j^n(z_{jl}), L_{jl}^n(z_{jl})) ,\qquad (3.5.6)$$

where $L_{jl}^n(x)$ is the linear reconstruction evaluated on the neighbouring triangle T_{jl}.

The higher-order finite volume scheme in space and time is defined by the TVD Runge–Kutta method of second order concerning the time discretization [30]:

$$u_j^{n+\frac{1}{2}} := u_j^n - \frac{\Delta t}{|T_j|} \sum_{l=1}^{3} g_{jl}(L_j^n(z_{jl}), L_{jl}^n(z_{jl})) \qquad (3.5.7)$$

$$u_j^{n+1} := \frac{1}{2}\left[u_j^n + u_j^{n+\frac{1}{2}} - \frac{\Delta t}{|T_j|} \sum_{l=1}^{3} g_{jl}(L_j^{n+\frac{1}{2}}(z_{jl}), L_{jl}^{n+\frac{1}{2}}(z_{jl}))\right]$$

where the linear functions $L_j^{n+\frac{1}{2}}$ and $L_{jl}^{n+\frac{1}{2}}$ are defined with the intermediate values $u_j^{n+\frac{1}{2}}$.

In order to show the convergence of the higher-order finite volume scheme, the linear reconstructions have to satisfy suitable conditions. Therefore we have to define admissible reconstructions.

DEFINITION 3.5.4 (Admissible linear reconstructions) *Let v_h and $L(v_h)$ be as in Definition 3.5.1 and let z_{jl} be the midpoint of the joint edge S_{jl} of the triangles T_j and T_{jl}. Then a reconstruction $L_j, j \in I$, is called admissible if $L_j(w_j) = v_j$ and if there exist constants $C_1 \in \mathbb{R}$ and $\alpha \in]\frac{1}{2}, 1]$ such that for all $j \in I$*

$$|L_j(z_{jl}) - v_j| \leq C_1 h^\alpha , \qquad (3.5.8)$$

$$[L_j(z_{jl}) - v_j](v_j - v_{jl}) \leq C_1 h^{2\alpha} , \qquad (3.5.9)$$

where $h := \max_j \operatorname{diam} T_j$.

3.5 Higher-order finite volume schemes for scalar equations

The condition (3.5.8) implies that the slope of the L_j cannot be too large. The condition (3.5.9) ensures that if $L_j(z_{jl}) - v_j$ and $v_j - v_{jl}$ have the same signs then one of the terms should be sufficiently small. Otherwise, if they have different signs, (3.5.9) gives no additional restriction. Therefore this condition forces the sign of the gradient of the reconstruction in the direction of $z_{jl} - w_j$ to have the same sign as $v_{jl} - v_j$.

Furthermore, in addition to (3.3.52), we need a similar compatibility condition for the second argument of the numerical flux g_{jl}.

ASSUMPTION 3.5.5 (Compatibility condition) *Assume (3.2.8)–(3.2.10) and require that the numerical flux g_{jl} admits a numerical entropy flux G_{jl} satisfying (3.3.52) and*

$$\frac{\partial G_{ij}}{\partial u}(u,v) = U'(u)\frac{\partial g_{ij}}{\partial u}(u,v) \qquad (3.5.10)$$

almost everywhere.

Now we can state the following convergence theorem.

THEOREM 3.5.6 (Convergence of higher-order finite volume schemes [121, 250]) *Let $\mathcal{T}_h := \{T_j \mid j \in I\}$ be a family of triangulations where $h := \max_{j \in I} \operatorname{diam}(T_j)$. Assume initial data $u_0 \in L^1(\mathbb{R}^2) \cap L^\infty(\mathbb{R}^2)$ with compact support and define*

$$u_j^0 := \frac{1}{|T_j|}\int_{T_j} u_0(x)\, dx \ . \qquad (3.5.11)$$

Suppose that g_{jl} are numerical fluxes satisfying the assumptions (3.2.8)–(3.2.10) with corresponding numerical entropy fluxes G_{jl} satisfying the assumptions (3.3.15)–(3.3.17), (3.3.52) and (3.5.10). Assume that a sequence of approximate solutions of (3.2.1) and (3.2.2) is given by Definition 3.5.3 with

$$u_h(x,t) := u_j^{n+1} \quad \text{for } x \in T_j \text{ and } t^n \leq t < t^{n+1} \qquad (3.5.12)$$

such that there is a constant $M < \infty$ independent of h such that

$$\|u_h\|_{L^\infty(\mathbb{R}^2\times[0,T])} \leq M \ . \qquad (3.5.13)$$

Furthermore, suppose that the following conditions are satisfied. There exist constants $\alpha \in]\frac{1}{2}, 1]$ and C_1 such that for any n there is a linear reconstruction $L^n := L(u^n)$ that is admissible uniformly in n, i.e.

$$|L_j^n(z_{jl}) - u_j^n| \leq C_1 h^\alpha , \qquad (3.5.14)$$

$$[L_j^n(z_{jl}) - u_j^n](u_j^n - u_{jl}^n) \leq C_1 h^{2\alpha} \qquad (3.5.15)$$

for all n and j, where C_1 and α do not depend on n and j. Additionally, we need the CFL condition

$$\sup_j \frac{\Delta t}{|T_j|} C_g(M) \sum_{l=1}^{3} |S_{jl}| \leq 1 \qquad (3.5.16)$$

where the constant $C_g(M)$ is defined in (3.2.8). Then for any given $T > 0$, $u_h \in L^1 \cap L^\infty(\mathbb{R}^2 \times [0,T])$, and as $h \to 0$ the sequence $(u_h)_h$ that converges to the solution of (3.2.1), (3.2.2) strongly in $L^1_{\text{loc}}(\mathbb{R}^2 \times [0,T])$ satisfying the condition in Definition 3.1.2.

For more details and the proof of this theorem we refer to Section 3.6 (see also [72] and [250]). Sufficient conditions for (3.5.13) can be found in [249]. The case $u_0 \in L^1 \cap L^\infty(\mathbb{R}^2)$ without compact support is also treated in [121].

EXAMPLE 3.5.7 (Admissible reconstruction, limiter function ansatz C) The following admissible reconstruction was developed by Wierse in [250] according to the conditions that ensure convergence. It turns out that the order of convergence for approximating smooth solutions is nearly two if one uses this limiter function with a special reconstruction. Unfortunately this has so far been shown only in numerical experiments and not theoretically. But the experimental order of convergence as well as the L^1 error are better than for similar schemes known in the literature, in particular for first-order schemes. Now let us explain the ideas from [72]. Determine

$$u_j^{max} := \max\{u_j^n, u_{jl}^n, u_{jl+1}^n, u_{jl-1}^n\} ,$$
$$u_j^{min} := \min\{u_j^n, u_{jl}^n, u_{jl+1}^n, u_{jl-1}^n\} .$$

3.5 Higher-order finite volume schemes for scalar equations 219

Construct for each triangle a linear reconstruction function of the form $\tilde{L}_j^n(x) = u_j^n + S_j^n \cdot (x - w_j)$ with some $S_j^n \in \mathbb{R}$ (for example with the approach in [18], [58], [68], [69] or [79], or see Example 3.5.8). Let $R(x) = S_j^n \cdot (x - w_j)$ and define an admissible reconstruction $L_j^n(x)$ in the following way. If

$$R(z_{jl}) \cdot (u_j^n - u_{jl}^n) \leq 0 \text{ for all } l \quad \text{then} \quad L_j^n(z_{jl}) := u_j^n + \alpha \cdot R(z_{jl}),$$

with $\alpha = \min_l(\alpha_l)$ and α_l maximal such that

(a) $\qquad \alpha_l |R(z_{jl})| \leq C_1 h^\alpha, \quad l = 1, 2, 3\ ;$

(b) $\qquad u_j^{\min} \leq u_j^n + \alpha_l R(z_{jl}) \leq u_j^{\max}.$ \hfill (3.5.17)

If this is not the case, but only for one l

$$R(z_{jl}) \cdot (u_j^n - u_{jl}^n) > 0$$

and

$$R(z_{jl-1}) R(z_{jl+1}) \leq 0,$$

then

$$L_j^n(z_{jl}) := u_j^n,$$
$$L_j^n(z_{jl+1}) := u_j^n + \alpha \cdot \mathrm{sign}(R(z_{ij+1})) \cdot \mathrm{minvalue},$$
$$L_j^n(z_{jl-1}) := u_j^n + \alpha \cdot \mathrm{sign}(R(z_{ij-1})) \cdot \mathrm{minvalue}$$

with minvalue $= \min(|R(z_{jl+1})|, |R(z_{jl-1})|)$ and α such that

(a) $\qquad \alpha \cdot \mathrm{minvalue} \leq C_1 h^\alpha,$

(b) $\qquad u_j^{\min} \leq u_j^n + \alpha \cdot \mathrm{minvalue} \leq u_j^{\max}.$ \hfill (3.5.18)

For every other case set

$$L_j^n(x) := u_j^n \quad \text{for all } j.$$

EXAMPLE 3.5.8 (Linear reconstructions, see [68] and [250]) Determine values $\tilde{u}_{jl}^n$, $l = 1, 2, 3$, at the vertices P_{jl} of the triangle T_j by averaging the piecewise-constant values on the triangles. This can be done for example with

$$\tilde{u}_{jl}^n = \frac{\sum_{k \in I, P_{jl} \in T_k} (u_k^n / r_k)}{\sum_{k \in I, P_{jl} \in T_k} (1/r_k)}; \qquad r_k = |P_{jl} - w_k|.$$

Let $\tilde{u}_{jl+1}^n$ and $\tilde{u}_{jl+2}^n$ be the values at the remaining vertices of T_j. Define S_j^n such that

$$S_j^n \cdot (z_{jl} - w_j) := \frac{\frac{1}{2}(\tilde{u}_{jl+1}^n + \tilde{u}_{jl+2}^n) - \tilde{u}_{jl}^n}{3} \quad \text{for } l = 1, 2, 3.$$

For later reference we denote the combination of this way of reconstruction and limiter functions as in ansatz C as scheme C.

REMARK 3.5.9 *It can be shown (see [250]) that the scheme described in Examples 3.5.7 and 3.5.8 satisfies the conditions (3.5.13)–(3.5.15) of Theorem 3.5.6. Also, the second limiter in Example 3.5.2 satisfies the conditions in Theorem 3.5.6 except for (3.5.14).*

So far, the higher-order property of the schemes for nonlinear problems as defined in Definition 3.5.3 has not been proved theoretically, e.g. by an error estimate of the form

$$\|u(\cdot, t) - u_h(\cdot, t)\|_{L^1(\mathbb{R}^2)} \leq c h^\alpha \qquad (3.5.19)$$

with α sufficiently large. Here u denotes the exact and u_h the discrete solution. For first-order schemes as defined in (3.2.6), (3.2.7) an error estimate of this form (3.5.19) has been proved with $\alpha = \frac{1}{4}$. An example of LeVeque [141] shows that at most $\mathcal{O}(h^{\frac{1}{2}})$ can be expected. For a higher-order scheme Szepessy [231] gets at least $\mathcal{O}(h^{\frac{3}{2}})$ for the streamline–diffusion–shock–capturing method with finite elements applied to linear conservation laws. For the schemes as defined in Definition 3.5.3 it turns out that we can obtain error estimates (3.5.19) with $\alpha \in [1.5, 2]$, at least for numerical experiments in cases where the solution is smooth. We shall show this in the following part of this section (see [250]). We consider 2-D test examples

3.5 Higher-order finite volume schemes for scalar equations

where the exact solution is known explicitly, so that the numerical error and the experimental order of convergence (EOC) can be calculated. The EOC is defined in the following way. On the triangulation $\mathcal{T}_h := \{T_j \mid j \in I\}$ we assume that there is an asymptotic expansion

$$u_h = u + u_1 h^\alpha + \ldots \qquad (3.5.20)$$

where α has to be determined. Then on a coarser grid $\mathcal{T}_{2h}$ we have of course

$$u_{2h} = u + u_1 (2h)^\alpha + \ldots \qquad (3.5.21)$$

Therefore, as a first approximation, we get

$$\text{EOC} := \alpha = \ln\left(\frac{\|u - u_h\|_{L^1}}{\|u - u_{2h}\|_{L^1}}\right) \Big/ \ln\left(\tfrac{1}{2}\right). \qquad (3.5.22)$$

We approximate the L^1 norm at time t by $\|u - u_h\|_{L^1} \approx \sum_{i \in I} |T_i| |u(x_i,t) - u_h(x_i,t)|$. The grid with maximum length $\tfrac{1}{2}h$ of an edge has been constructed by dividing the triangles into four in the grid with maximum length h. For the scalar conservation law we use the CFL condition

$$\sup_i \frac{\Delta t^n}{|T_i|} \sum_{j=1}^3 \max\{\nu_{ij} f'(u_i^n), 0\} \leq \text{CFL} \qquad (\text{CFL} < 1).$$

As a test problem, we consider the initial value problem

$$\partial_t u + \partial_x u^2 + \partial_y u^2 = f \quad \text{in}[0,1]^2$$
$$u(x,0) = u_0. \qquad (3.5.23)$$

The initial values u_0 and the right-hand side f are chosen such that in the first example the solution is smooth over the whole time interval and in the second example the solution has a discontinuity. As numerical flux g_{jl} the flux of Engquist and Osher (see Example 3.3.19) is used.

Since there is a dependence of the consistency order of the scheme on the form of the grid, the calculations are made on different grids. The macro triangulations are given in Figures 3.5.3 and 3.5.4.

We shall use the following notation for the solution calculated with different schemes:

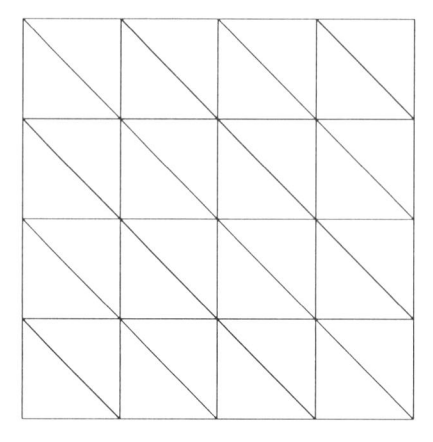

Figure 3.5.3 Grid A.

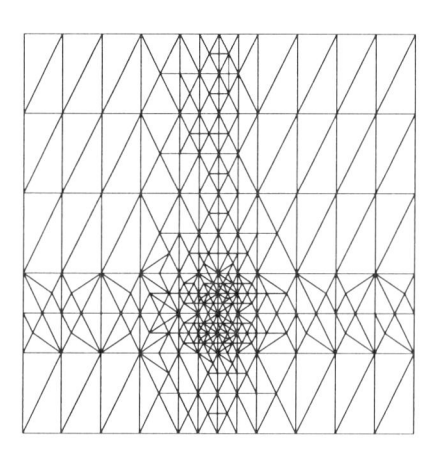

Figure 3.5.4 Grid B.

3.5 Higher-order finite volume schemes for scalar equations

u_h: solution of the first order scheme as defined in (3.2.6), (3.2.7);

u_h^C: solution of the higher-order scheme in (3.5.6) or (3.5.7) with the limiter function of Example 3.5.7, where the term $S_i^n \cdot (z_{ij} - x_i)$ is constructed as described in Example 3.5.8;

u_h^S: solution of the higher-order scheme as described in Example 3.5.2 with the second limiter function.

EXAMPLE 3.5.10 (Approximation of a smooth solution) We choose

$$u_0(x,y) = \tfrac{1}{5}\sin(2\pi x)\sin(2\pi y)$$

and

$$\begin{aligned} f(x,y,t) = \tfrac{2}{5}\pi\Big\{ &-\cos[2\pi(x-t)] \cdot \sin[2\pi(y-t)] \\ &-\sin[2\pi(x-t)] \cos[2\pi(y-t)] \\ &+\cos[2\pi(x-t)] \sin[2\pi(y-t)]\, 2u(x,y,t) \\ &+\sin[2\pi(x-t)] \cos[2\pi(y-t)]\, 2u(x,y,t)\Big) \end{aligned}$$

with

$$u(x,y,t) = \tfrac{1}{5}\sin[2\pi(x-t)]\sin[2\pi(y-t)].$$

The exact solution of problem (3.5.23) is given by u. As higher-order scheme the scheme in (3.5.7) is used. The term

$$\frac{\Delta t}{|T_i|} \int_{T_i} f(x,y,t)$$

evaluated with a quadrature formula of degree three is added on the right-hand side of (3.5.6) or (3.5.7) respectively. To avoid boundary effects, periodic boundary conditions have been used. The result are presented in the following tables. For different schemes on different grids (see column **h**) the L^1 error and the EOC are shown.

Results on grid A (CFL $= 0.5, T = 0.3$)

h	$\|u - u_h\|_{L^1}$	EOC	$\|u - u_h^C\|_{L^1}$	EOC	$\|u - u_h^S\|_{L^1}$	EOC
0.35	0.0635		0.0284		0.0261	
0.18	0.0299	1.09	0.0082	1.79	0.0118	1.14
0.09	0.0148	1.01	0.0020	2.07	0.0046	1.35
0.04	0.0075	0.98	0.0005	2.04	0.0012	1.92
0.02	0.0038	0.99	0.0001	2.01	0.0003	2.03

Results on grid B (CFL $= 0.5, T = 0.3$)

h	$\|u - u_h\|_{L^1}$	EOC	$\|u - u_h^C\|_{L^1}$	EOC	$\|u - u_h^S\|_{L^1}$	EOC
0.22	0.0219		0.0065		0.0085	
0.11	0.0075	1.54	0.0020	1.70	0.0033	1.36
0.06	0.0031	1.28	0.0006	1.62	0.0013	1.37
0.03	0.0015	1.03	0.0002	1.55	0.0005	1.50

We see that the schemes with ansatz C and ansatz S are schemes of higher-order.

It turns out that the L^1 error for u_h^C is much better than for u_h^S and also the EOC (except for one case $h = 0.02$ on grid A).

EXAMPLE 3.5.11 (Approximation of a solution with a discontinuity) For this example we choose $f = 0$ and

$$u_0(x, y) = \begin{cases} 2 & \text{if } \frac{1}{2}(x+y) - 0.5 < 0, \\ 1 & \text{if } \frac{1}{2}(x+y) - 0.5 > 0. \end{cases}$$

Then the exact solution can be determined as

$$u(x, y, t) = \begin{cases} 2 & \text{if } \frac{1}{2}x + y - 0.5 < 3t, \\ 1 & \text{if } \frac{1}{2}x + y - 0.5 > 3t. \end{cases}$$

In this case (3.5.6) is used as the higher-order scheme. Numerical experiments show that the use of the scheme (3.5.7) yields no further improvement in the L^1 error.

Results on grid A ($CFL = 0.9, T = 0.1$)

h	$\|u - u_h\|_{L^1}$	EOC	$\|u - u_h^C\|_{L^1}$	EOC	$\|u - u_h^S\|_{L^1}$	EOC
0.35	0.0310		0.0223		0.0364	
0.18	0.0180	0.78	0.0122	0.70	0.0142	1.36
0.09	0.0087	1.04	0.0041	1.57	0.0060	1.24
0.04	0.0046	0.93	0.0025	0.70	0.0026	1.21
0.02	0.0022	1.06	0.0011	1.25	0.0014	0.94

Results on grid B ($CFL = 0.9, T = 0.1$)

h	$\|u - u_h\|_{L^1}$	EOC	$\|u - u_h^C\|_{L^1}$	EOC	$\|u - u_h^S\|_{L^1}$	EOC
0.22	0.0551		0.034		0.0356	
0.11	0.0366	0.59	0.0187	0.85	0.0200	0.83
0.06	0.0224	0.71	0.0096	0.96	0.0105	0.92
0.03	0.0131	0.78	0.0049	0.98	0.0054	0.95

The use of a higher-order method leads to an improvement in the L^1 error compared with that of the first order scheme. The higher-order schemes produce no oscillations at the discontinuities. The best results are obtained with ansatz C.

3.6 Proof of convergence of higher-order finite volume schemes for scalar equations

In this section we prove Theorem 3.5.6, i.e. the convergence of the finite volume scheme as described in (3.5.6) (see [121], [34] and [29]). In the first step we prove a cell entropy inequality that implies the convergence to an admissible measure-valued solution. This is the main part of the convergence proof. The key tools for obtaining this entropy inequality are (3.3.52) and (3.5.10). In the following we always assume the conditions of Theorem 3.5.6. For further results we refer to [34], [29], [101] and [107].

THEOREM 3.6.1 (Cell entropy inequality) *Let (U, F) be a convex entropy pair with $U \in C^2$ and assume that G_{jl} satisfies Assumption 3.5.5. Suppose that the CFL condition (3.5.16) holds. Then there are constants C*

226 Initial value problems for scalar conservation laws in 2-D

and $h_0 > 0$ such that for all $h \leq h_0$ the following cell entropy inequality holds

$$U(u_j^{n+1}) - U(u_j^n) + \frac{\Delta t}{|T_j|} \sum_l G_{jl} \leq C\|U''\|_{L^\infty(B_M(0))} h^{2\alpha}, \tag{3.6.1}$$

where α is defined in Theorem 3.5.6 and

$$G_{jl} := G_{jl}(L_j(z_{jl}), L_{jl}(z_{jl})), \tag{3.6.2}$$

M is given in (3.5.13) and $B_M(0)$ is the ball of radius M in $\mathbb{R}$.

The proof of this theorem is the central piece of this section. It relies on a careful analysis of the entropy dissipation and on the properties (3.5.14) and (3.5.15). A cell entropy inequality similar to (3.6.1) was derived as early as 1971 by Lax [127] for the first-order Lax–Friedrichs finite difference scheme. Since then, many authors have applied and refined these ideas (see e.g. [234], [42], [35], [231], [223] and [120] and references therein).

Proof of Theorem 3.6.1 Note that

$$u_j^{n+1} = \frac{1}{3} \sum_{l=1}^{3} \{u_j - 3\lambda_j [g_{jl} - f(u_j) \cdot \nu_{jl}]\},$$

where $\lambda_j := \Delta t/|T_j|$ and $g_{jl} = g_{jl}(L_j(z_{jl}), L_{jl}(z_{jl}))$. Since U is convex,

$$U(u_j^{n+1}) \leq \frac{1}{3} \sum_{l=1}^{3} U\left(u_j - 3\lambda_j [g_{jl} - f(u_j) \cdot \nu_{jl}]\right).$$

It is thus sufficient to show that for each (j,l)

$$\begin{aligned} E := & U[u_j - 3\lambda_j(g_{jl} - f(u_j) \cdot \nu_{jl})] - U(u_j) \\ & + 3\lambda_j [G_{jl} - F(u_j) \cdot \nu_{jl}] \leq C\|U''\|_{L^\infty(B_M)} h^{2\alpha}. \end{aligned} \tag{3.6.3}$$

Now we fix j and l and define

$$g(u,v) := g_{jl}(u,v), \quad G(u,v) := G_{jl}(u,v), \quad \lambda = \lambda_j,$$

3.6 Proof of convergence for scalar equations

$$u := u_j^n, \quad u_0 := L_j^n(z_{jl}) - u_j^n,$$
$$u_1 := L_{jl}^n(z_{jl}) - u_{jl}^n, \quad w := u_{jl}^n - u_j^n + u_1.$$

Then we have

$$\begin{aligned}
E &= U(u - 3\lambda[g(u+u_0, u+w) - g(u,u)]) \\
&\quad - U(u) + 3\lambda[G(u+u_0, u+w) - G(u,u)] \\
&= \int_0^1 \frac{d}{dt} U(u - 3\lambda[g(u+\tau u_0, u+\tau w) - g(u,u)]) \, d\tau \\
&\quad + 3\lambda \int_0^1 \frac{d}{dt} G(u+\tau u_0, u+\tau w) \, d\tau \\
&= \int_0^1 U'(\ldots)(-3\lambda)(\partial_1 g u_0 + \partial_2 g w) \, d\tau + 3\lambda \int_0^1 (\partial_1 G u_0 + \partial_2 G w) \, d\tau \\
&= \int_0^1 U'(\ldots)(-3\lambda)(\partial_1 g u_0 + \partial_2 g w) \, d\tau \\
&\quad + 3\lambda \int_0^1 [U'(u+\tau u_0) \partial_1 g u_0 + U'(u+\tau w) \partial_2 g w] \, d\tau \\
&= 3\lambda u_0 \int_0^1 \Big\{ U'(u+\tau u_0) - U'(u - 3\lambda[g(u+\tau u_0, u+\tau w) \\
&\quad - g(u,u)]) \Big\} \partial_1 g \, d\tau \\
&\quad - 3\lambda w \int_0^1 \Big\{ U'(u - 3\lambda[g(u+\tau u_0, u+\tau w) - g(u,u)]) \\
&\quad - U'(u+\tau w) \Big\} \partial_2 g \, d\tau \\
&=: E_1 + E_2. \qquad (3.6.4)
\end{aligned}$$

Now let us consider the first term E_1:

$$\begin{aligned}
E_1 &= 3\lambda u_0 \int_0^1 U''(\eta \tau)) \big\{ \tau u_0 + 3\lambda[g(u+\tau u_0, u+\tau w) \\
&\quad - g(u,u)] \big\} \partial_1 g \, d\tau \\
&= 3\lambda u_0 \int_0^1 U''(\eta(\tau)) [\tau u_0 + 3\lambda(\partial_1 g \tau u_0 + \partial_2 g \tau w)] \partial_1 g \, d\tau \\
&= 3\lambda u_0 \int_0^1 U''(\eta(\tau)) [\tau u_0 (1 + 3\lambda \partial_1 g) + 3\lambda \partial_2 g \tau w] \partial_1 g \, d\tau.
\end{aligned}$$

We know that

$$|\lambda| \leq \frac{1}{ch} \quad \text{(see (3.2.3) and (3.2.4))},$$

$$|\partial_1 g|, |\partial_2 g| \leq Ch \quad \text{(see (3.2.8))},$$

$$|3\lambda\partial_1 g| \leq 1 \quad \text{(see (3.2.4) and (3.2.8) for } \Delta t \text{ sufficiently small)},$$

$$|u_0| = |L_j^n(z_{jl}) - u_j^n| \leq Ch^\alpha \quad \text{(see (3.5.14))}, \tag{3.6.5}$$

$$|u_1| = |L_{jl}^n(z_{jl}) - u_{jl}^n| \leq Ch^\alpha \quad \text{(see (3.5.14))}, \tag{3.6.6}$$

$$(L_j^n(z_{jl}) - u_j^n)(u_j^n - u_{jl}^n) \leq Ch^{2\alpha} \quad \text{(see (3.5.15))}.$$

Therefore, using (3.5.15), we obtain

$$E_1 \leq C\|U''\|_{L^\infty(B_M)} h^{2\alpha}.$$

The term E_2 can be treated similarly. □

LEMMA 3.6.2 *Let ν be the Young measure defined by u_h (see (3.5.13) and Theorem 3.3.3). Then for any convex entropy pair (U, F) with $U \in C^2$ we have*

$$\partial_t \langle \nu, U \rangle + \nabla \cdot \langle \nu, F \rangle \leq 0 \tag{3.6.7}$$

in the sense of distributions. Moreover, for all $\varphi \in C_0^\infty(\mathbb{R}^2 \times [0, T[)$ with $\varphi \geq 0$,

$$\int_0^T \int_{\mathbb{R}^2} [\langle \nu_{x,t}, U \rangle \partial_t \varphi(x,t) + \langle \nu_{x,t}, F \rangle \cdot \nabla \varphi(x,t)] \, dx \, dt \tag{3.6.8}$$

$$+ \int_{\mathbb{R}^2} U(u_0(x))\varphi(x,0) \, dx \geq 0.$$

In order to prove this lemma, we need the following lemma.

3.6 Proof of convergence for scalar equations   229

LEMMA 3.6.3 *Let $\varphi \in C_0^\infty(\mathbb{R}^2 \times [0, T[), \varphi_j := \varphi(w_j, t^n), \varphi_{jl} := \varphi(z_{jl}, t^n)$. Then there are constants $C = C(\varphi, T)$ and $h_0 > 0$ such that for all $h \leq h_0$*

$$\Delta t \sum_{n=0}^{N} \sum_{j} \sum_{l=1}^{3} |G_{jl}(L_j(z_{jl}), L_{jl}(z_{jl}))$$

$$- F(u_j) \cdot \nu_{jl}||\varphi_j - \varphi_{jl}| \leq Ch^{\frac{1}{2}\alpha}. \tag{3.6.9}$$

This lemma will be proved after Theorem 3.6.4.

Proof of Lemma 3.6.2 Let B_R be the ball of radius R in $\mathbb{R}^2$, and let χ_R be the indicator function of B_R. Let $\varphi \in C_0^\infty(B_R \times [0, T[), \varphi \geq 0$ and $\varphi_j^n := \varphi(w_j, t^n)$. Multiply (3.6.1) by $|T_j|\varphi_j^n$ and sum over j and $n, n \leq N+1$, where $(N+1)\Delta t = T$:

$$\sum_{n=0}^{N} \sum_{j} |T_j|[U(u_j^{n+1}) - U(u_j^n)]\varphi_j^n + \Delta t \sum_{n=0}^{N} \sum_{j} \sum_{l=1}^{3} G_{jl}\varphi_j^n$$

$$\leq C \sum_{n=0}^{N} \sum_{j} |T_j|\varphi_j^n h^{2\alpha}.$$

We have

$$\sum_{n=0}^{N} \sum_{j} |T_j|[U(u_j^{n+1}) - U(u_j^n)]\varphi_j^n$$

$$= -\Delta t \sum_{n=1}^{N} \sum_{j} |T_j|U(u_j^n) \frac{\varphi_j^n - \varphi_j^{n-1}}{\Delta t} - \sum_{j} |T_j|U(u_j^0)\varphi_j^0$$

$$= -\Delta t \sum_{n=1}^{N} \sum_{j} \int_{T_j} U(u_h(x, t^n))[\partial_t \varphi(x, t^n) + \mathcal{O}(\Delta t)\chi_R(x)] \, dx$$

$$- \sum_{j} \int_{T_j} U(u_h(x, 0))[\varphi(x, 0) + \mathcal{O}(\Delta t)\chi_R(x)] \, dx$$

$$= -\Delta t \sum_{n=1}^{N} \int_{\mathbb{R}^2} U(u_h(x, t^n))\partial_t \varphi(x, t^n) \, dx$$

$$- \int_{\mathbb{R}^2} U(u_h(x, 0))\varphi(x, 0) \, dx + \mathcal{O}(\Delta t)$$

$$= -\int_0^T \int_{\mathbb{R}^2} U(u_h(x,t))\partial_t\varphi(x,t)\,dx\,dt$$
$$-\int_{\mathbb{R}^2} U(u_h(x,0))\varphi(x,0)\,dx + \mathcal{O}(\Delta t). \tag{3.6.10}$$

Since φ is supposed to have compact support and because of (3.3.16), we have (see also Remark 3.3.15)

$$\sum_{n=0}^{N}\sum_{j}\sum_{l=1}^{3} G_{jl}\varphi_{jl}^n = 0,$$

and therefore

$$\Delta t \sum_{n=0}^{N}\sum_{j}\sum_{l=1}^{3} G_{jl}\varphi_j^n = \Delta t \sum_{n=0}^{N}\sum_{j}\sum_{l=1}^{3} G_{jl}(\varphi_j^n - \varphi_{jl}^n)$$
$$= \Delta t \sum_{n=0}^{N}\sum_{j}\sum_{l=1}^{3} [G_{jl} - F(u_j^n)\nu_{jl}](\varphi_j^n - \varphi_{jl}^n)$$
$$- \Delta t \sum_{n=0}^{N}\sum_{j}\sum_{l=1}^{3} F(u_j^n)\nu_{jl}\varphi_{jl}^n.$$

From (3.6.9), the first summand on the right-hand side is $\mathcal{O}(h^{\frac{1}{2}\alpha})$. The second summand is

$$-\Delta t \sum_{n=0}^{N}\sum_{j}\sum_{l=1}^{3} \left(\int_{S_{jl}} F(u_h)n\varphi + \mathcal{O}(h^3)\right)$$
$$= -\Delta t \sum_{n=0}^{N}\sum_{j} \int_{T_j} F(u_h)\cdot\nabla\varphi + TR^2\mathcal{O}(h)$$
$$= -\int_0^T \int_{\mathbb{R}^2} F(u_h)\cdot\nabla\varphi + \mathcal{O}(h) \tag{3.6.11}$$

where the integral $\int_{S_{jl}}$ in (3.6.11) obviously integrates the values of u_h on T_j. Finally, by definition of φ, (3.6.3) and (3.6.4),

$$h^{2\alpha}\sum_{n=0}^{N}\sum_j |T_j|\varphi_j^n \leq Ch^{2\alpha-1}TR^2.$$

Therefore

$$-\int_0^T \int_{\mathbb{R}^2} U(u_h)\partial_t\varphi - \int_0^T \int_{\mathbb{R}^2} F(u_h) \cdot \nabla\varphi - \int_{\mathbb{R}^2} U(u_h(0))\varphi(\cdot,0)$$
$$\leq \mathcal{O}(h^{\frac{1}{2}}\alpha) + \mathcal{O}(h^{2\alpha-1}) + \mathcal{O}(h) = \mathcal{O}(h^\epsilon) \quad \text{for some } \epsilon > 0$$

since $\alpha > \frac{1}{2}$ (see Theorem 3.5.6). Taking the limit as $h \to 0$ yields (3.6.8). For $\varphi \in C_0^\infty(\mathbb{R}^2 \times]0,T[)$, this implies

$$\partial_t \langle \nu, U \rangle + \nabla \cdot \langle \nu, F \rangle \leq 0$$

in the sense of distributions. □

THEOREM 3.6.4 *The Young measure ν in Lemma 3.6.2 is an admissible measure-valued solution of (3.2.1), (3.2.2). Moreover,*

$$\int_0^T \int_{\mathbb{R}^2} [\langle \nu_{x,t}, \mathrm{id}\rangle \partial_t \varphi(x,t) + \langle \nu_{x,t}, f\rangle \cdot \nabla\varphi(x,t)]\,dx\,dt$$
$$+ \int_{\mathbb{R}^2} u_0(x)\varphi(x,0)\,dx = 0. \qquad (3.6.12)$$

Proof We already know from Tartar's theorem (Theorem 3.3.3) that $\operatorname{supp} \nu_{x,t} \subset B_M$ for all $(x,t) \in \mathbb{R}^2 \times [0,T]$. It remains to show (3.3.1) for all Kruzkov entropies $U(\cdot,k) = |\cdot - k|$. Given $k \in \mathbb{R}$ and $\epsilon > 0$, let

$$U_\epsilon(s) := \begin{cases} -\frac{\epsilon}{8}\{[(s-k)/\epsilon]^4 - 6[(s-k)/\epsilon]^2 - 3\} & \text{for } |s-k| < \epsilon, \\ |s-k| & \text{otherwise}, \end{cases}$$

$U_\epsilon \in C^2$, and as $\epsilon \to 0$, $U_\epsilon \to U(\cdot,k)$ uniformly. According to Definition 3.3.4 and Theorem 2.1.16, we choose F_ε. Using the dominated convergence theorem, one can pass to the limit as $\epsilon \to 0$ in (3.6.8) to obtain

$$\int_0^T \int_{\mathbb{R}^2} \langle \nu_{x,t}, U(\mathrm{id},k)\rangle \partial_t\varphi(x,t) + \langle \nu_{x,t}, F(\mathrm{id},k)\rangle \cdot \nabla\varphi(x,t)$$
$$+ \int_{\mathbb{R}^2} \langle \nu_{x,0}, U(\mathrm{id},k)\rangle \varphi(x,0) \geq 0.$$

From here, one immediately obtains (3.3.1) for U and F. Moreover, setting $k = \pm M$, one obtains (3.6.12). □

It remains to prove Lemma 3.6.3.

Proof of Lemma 3.6.3 As before, let $\lambda_j := \Delta t/|T_j|$. In (3.3.57) it is shown for the first order scheme

$$u_j^{n+1} = u_j^n - \sum_{l=1}^{3} \lambda_j g_{jl}(u_j, u_{jl})$$

that for G_{jl} corresponding to the entropy $U(s) = \frac{1}{2}s^2$

$$\frac{1}{2}[(u_j^{n+1})^2 - (u_j^n)^2] + \sum_{l=1}^{3} \lambda_j G_{jl}(u_j, u_{jl})$$
$$+ \sum_{l=1}^{3} \lambda_j^2 [g_{jl}(u_j, u_{jl}) - f(u_j)\nu_{jl}]^2 \leq 0. \qquad (3.6.13)$$

From (3.2.8) and (3.5.14),

$$\sum_{l=1}^{3} \lambda_j [g_{jl}(L_j^n(z_{jl}), L_{jl}^n(z_{jl})) - g_{jl}(u_j, u_{jl})] = \mathcal{O}(h^\alpha).$$

It is now easy to modify the proof of (3.3.57) to show that for the scheme (3.5.6) one has the estimate

$$\frac{1}{2}[(u_j^{n+1})^2 - (u_j^n)^2] + \sum_{l=1}^{3} \lambda_j G_{jl}(u_j, u_{jl})$$
$$+ \sum_{l=1}^{3} (\lambda_j)^2 [g_{jl}(u_j, u_{jl}) - f(u_j)\nu_{jl}]^2 = \mathcal{O}(h^\alpha). \qquad (3.6.14)$$

Multiplying (3.6.14) by $|T_j|$ and summing over j, we obtain

$$\frac{1}{2}\|u_h(t^{n+1})\|_{L^2(\mathbb{R}^2)}^2 - \frac{1}{2}\|u_h(t^n)\|_{L^2(\mathbb{R}^2)}^2$$
$$+ \sum_j \frac{\Delta t^2}{|T_j|} \sum_{l=1}^{3} [g_{jl}(u_j, u_{jl}) - f(u_j) \cdot \nu_{jl}]^2 \leq Ch^\alpha.$$

3.6 Proof of convergence for scalar equations

Using the fact that $(\Delta t)^2/|T_j| \geq (c_1)^2 > 0$ and (3.6.1) with $U(u) := \frac{1}{2}u^2$, multiplying by Δt and summing over $n = 0, \ldots, N$, we obtain

$$\Delta t \sum_{n=0}^{N} \sum_{j} \sum_{l=1}^{3} [g_{jl}(u_j, u_{jl}) - f(u_j) \cdot \nu_{jl}]^2 \leq CTh^{\alpha}. \tag{3.6.15}$$

Now let $U \in C^2$ be any convex entropy with corresponding numerical entropy flux G_{jl} satisfying Assumption 3.5.5. As in the proof of Proposition 3.3.20, we use the compatibility condition (3.3.52) to show that

$$|G_{jl}(u_j, u_{jl}) - F(u_j) \cdot \nu_{jl}| \leq C \|U'\|_{L^\infty(B_M)} |g_{jl}(u_j, u_{jl}) - f(u_j) \cdot \nu_{jl}|.$$

Applying this to (3.6.15), we obtain

$$\Delta t \sum_{n=0}^{N} \sum_{j} \sum_{l=1}^{3} [G_{jl}(u_j, u_{jl}) - F(u_j) \cdot \nu_{jl}]^2 \tag{3.6.16}$$
$$\leq C(\|U'\|_{L^\infty(B_M)})^2 Th^{\alpha}.$$

Since (see (3.6.5) and (3.6.6))

$$[G_{jl}(L_j(z_{jl}), L_{jl}(z_{jl})) - G_{jl}(u_j, u_{jl})] = \mathcal{O}(h^{1+\alpha}),$$

we have

$$\Delta t \sum_{n=0}^{N} \sum_{j} \sum_{l=1}^{3} |[G_{jl}(L_j(z_{jl}), L_{jl}(z_{jl})) - F(u_j) \cdot \nu_{jl}]^2$$
$$\leq 2\Delta t \sum_{n=0}^{N} \sum_{j} \sum_{l=1}^{3} [G_{jl}(u_j, u_{jl}) - F(u_j) \cdot \nu_{jl}]^2 + C\Delta t \sum_{n=0}^{N} \sum_{j} h^{2(1+\alpha)}$$
$$\leq CTh^{\alpha}. \tag{3.6.17}$$

Using (3.6.17) and Hölder's inequality, we derive

$$\Delta t \sum_{n=0}^{N} \sum_{j} \sum_{l=1}^{3} |G_{jl}(L_j(z_{jl}), L_{jl}(z_{jl})) - F(u_j) \cdot \nu_{jl}||\varphi_j - \varphi_{jl}|$$
$$\leq \left(\Delta t \sum_{n=0}^{N} \sum_{j} \sum_{l=1}^{3} |G_{jl} - F(u_j) \cdot \nu_{jl}|^2 \right)^{\frac{1}{2}} \left(\Delta t \sum_{n=0}^{N} \sum_{j} \sum_{l=1}^{3} |\varphi_j - \varphi_{jl}|^2 \right)^{\frac{1}{2}}$$
$$\leq (CTh^{\alpha})^{\frac{1}{2}} (C(\varphi)T)^{\frac{1}{2}} = C(\varphi, T) h^{\frac{1}{2}\alpha} \qquad \square$$

Now we show that the admissible measure-valued solution ν of (3.2.1), (3.2.2) equals the Kruzkov solution of (3.2.1) with initial data (3.2.2). According to the theory outlined in §3.3, it remains to show that (a)–(d) in Theorem 3.3.5 and (3.3.2), (3.3.3) of DiPerna's Theorem 3.3.7 are satisfied. The properties (b) and (c) have been shown already in Theorem 3.6.4 and Lemma 3.6.2.

THEOREM 3.6.5 *For all $t \in [0,T]$,*

$$\int_{\mathbb{R}^2} \langle \nu_{x,t}, |\mathrm{id}| \rangle \, dx \leq \|u_0\|_{L^1(\mathbb{R}^2)},$$

i.e. condition (a) of Theorem 3.3.5 holds.

Proof Since u_h is bounded, we know that

$$u_h \stackrel{*}{\rightharpoonup} \langle \nu, \mathrm{id} \rangle =: u$$

in $L^\infty(\mathbb{R}^2 \times \mathbb{R}_+)$. For almost every t, there is a compactly supported probability measure $\mu_{\cdot,t}$ such that

$$u_h(t) \stackrel{*}{\rightharpoonup} \langle \mu_{\cdot,t}, \mathrm{id} \rangle =: v(\cdot,t).$$

Let $\sigma \in C_0^\infty([0,T])$, $\varphi \in C_0^\infty(\mathbb{R}^2)$ and $\psi(x,t) := \sigma(t)\varphi(x)$. Then

$$\lim_{h \to 0} \int_{\mathbb{R}^2} u_h(x,t)\varphi(x)\, dx = \int_{\mathbb{R}^2} v(x,t)\varphi(x)\, dx,$$

and therefore

$$\lim_{h \to 0} \int_0^T \int_{\mathbb{R}^2} u_h(x,t)\psi(x,t)\, dx\, dt = \int_0^T \int_{\mathbb{R}^2} v(x,t)\psi(x,t)\, dx\, dt.$$

On the other hand,

$$\lim_{h \to 0} \int_0^T \int_{\mathbb{R}^2} u_h(x,t)\psi(x,t)\, dx\, dt = \int_0^T \int_{\mathbb{R}^2} u(x,t)\psi(x,t)\, dx\, dt,$$

3.6 Proof of convergence for scalar equations

and therefore we obtain that $v(x,t) = u(x,t)$ at all Lebesgue points (x,t) of u and v. Therefore we can identify v with u and $\mu_{.,t}$ with $\nu_{.,t}$.

Next, for any convex $U \in C^2$, let (U, F) be an entropy pair and G_{jl} a numerical entropy flux consistent with $F \cdot \nu_{jl}$ and satisfying Assumption 3.5.5. From (3.6.1) we get

$$U(u_j^{n+1}) - U(u_j^n) \leq -\frac{\Delta t}{|T_j|} \sum_{l=1}^{3} G_{jl} + \|U''\|_{L^\infty([-M,M])} \mathcal{O}(h^{2\alpha}).$$

We multiply this by $|T_j|$ and sum over $n = 0, \ldots, N-1$ ($N\Delta t = t$) and j,

$$-\Delta t \sum_{n=0}^{N-1} \sum_{j} \sum_{l=1}^{3} G_{jl} = 0,$$

and obtain

$$\int_{\mathbb{R}^2} U(u_h(x,t)) \, dx \leq \int_{\mathbb{R}^2} U(u_h(x,0)) \, dx$$
$$+ C\|U''\|_{L^\infty([-M,M])} h^{2\alpha-1}. \qquad (3.6.18)$$

Now define U_ε such that $U_\varepsilon(s) \to |s|$ a.e.:

$$U_\varepsilon(s) := \begin{cases} -\dfrac{\varepsilon}{8}\left[\left(\dfrac{s}{\varepsilon}\right)^4 - 6\left(\dfrac{s}{\varepsilon}\right)^2 - 3\right] & \text{for } |s| < \varepsilon, \\ |s| & \text{otherwise}. \end{cases}$$

Let $\varepsilon = \varepsilon(h) := h^{2\beta}$, with $\beta \in]0, \alpha - \frac{1}{2}[$. Note that $\|U''_{\varepsilon(h)}\|_{L^\infty(B_M)} \leq C/\varepsilon = Ch^{-2\beta}$. Therefore (3.6.18) gives

$$\int_{\mathbb{R}^2} U_{\varepsilon(h)}(u_h(x,t)) \, dx \leq \int_{\mathbb{R}^2} U_{\varepsilon(h)}(u_h(x,0)) \, dx$$
$$+ Ch^{2(\alpha-\beta-\frac{1}{2})}. \qquad (3.6.19)$$

Using Tartar's theorem (Theorem 3.3.3) and (3.6.19), we derive

$$\int_{\mathbb{R}^2} \langle \nu_{x,t}, |\mathrm{id}| \rangle \, dx = \lim_{R \to \infty} \int_{B_R} \langle \nu_{x,t}, |\mathrm{id}| \rangle \, dx$$

$$= \lim_{R \to \infty} \lim_{h \to 0} \int_{B_R} |u_h(x,t)| \, dx$$

$$\leq \lim_{R \to \infty} \lim_{h \to 0} \int_{B_R} U_{\epsilon(h)}(u_h(x,t)) \, dx$$

$$\leq \lim_{h \to 0} \int_{\mathbb{R}^2} U_{\epsilon(h)}(u_h(x,t)) \, dx$$

$$\leq \lim_{h \to 0} \left[\int_{\mathbb{R}^2} U_{\epsilon(h)}(u_h(x,0)) \, dx + Ch^{2(\alpha - \beta - \frac{1}{2})} \right]$$

$$= \lim_{h \to 0} \int_{\mathbb{R}^2} U_{\epsilon(h)}(u_h(x,0)) \, dx \, .$$

For every $h > 0$,

$$\int_{\mathbb{R}^2} U_{\epsilon(h)}(u_h(x,0)) \, dx \qquad (3.6.20)$$

$$\leq \int_{\mathrm{supp}(u_h)} c\epsilon \, dx + \int_{\mathbb{R}^2} |u_h(x,0)| \, dx$$

$$\leq Ch^{2\beta} + \int_{\mathbb{R}^2} |u_0(x)| \, dx$$

so

$$\int_{\mathbb{R}^2} \langle \nu_{x,t}, |\mathrm{id}| \rangle \, dx \leq \int_{\mathbb{R}^2} |u_0(x)| \, dx,$$

which is the L^1 bound we were looking for. □

Next we prove that assumption (d) of Theorem 3.3.5 holds. For this purpose, it remains to show that assumptions (3.3.2) and (3.3.3) of Theorem 3.3.7 hold.

LEMMA 3.6.6 *For all $\varphi \in C_0^\infty(\mathbb{R}^2)$*

$$\lim_{t \downarrow 0} \frac{1}{t} \int_0^t \int_{\mathbb{R}^2} \langle \nu_{x,s}, \mathrm{id} \rangle \varphi(x) \, dx \, ds = \int_{\mathbb{R}^2} u_0(x) \varphi(x) \, dx \qquad (3.6.21)$$

i.e. condition (3.3.2) of Theorem 3.3.7 holds.

3.6 Proof of convergence for scalar equations

For the proof see Proposition 3.3.12.

LEMMA 3.6.7 Let $U(u) = \frac{1}{2}u^2$. Then

$$\lim_{t \downarrow 0} \frac{1}{t} \int_0^t \int_{\mathbb{R}^2} \langle \nu_{x,s}, U \rangle \, dx \, ds \leq \int_{\mathbb{R}^2} U(u_0(x)) \, dx \,, \tag{3.6.22}$$

i.e. (3.3.3) of Theorem 3.3.7 holds.

Proof This is a generalization of that of Proposition 3.3.16. As in the proof of Theorem 3.6.5, multiply (3.6.1) by $|T_j|$ and sum over j. Then

$$\sum_j |T_j|[U(u_j^{n+1}) - U(u_j^n)] \leq Ch^{2\alpha} \,. \tag{3.6.23}$$

Let $0 \leq t_1 := n_1 \Delta t \leq t_2 := n_2 \Delta t$. Sum (3.6.23) over $0 \leq n' \leq n$, and let $t := n \Delta t \leq t_2$. Then

$$\sum_j |T_j|[U(u_j^n) - U(u_j^0)] \leq nCh^{2\alpha} \leq t_2 Ch^{2\alpha-1}. \tag{3.6.24}$$

Sum (3.6.24) over $n = n_1, \ldots, n_2 - 1$:

$$\sum_{n=n_1}^{n_2-1} \sum_j |T_j| U(u_j^n) \leq (n_2 - n_1) \sum_j |T_j| U(u_j^0) + (n_2 - n_1) t_2 Ch^{2\alpha-1}.$$

Multiplying this by $\Delta t/(t_2 - t_1)$ gives

$$\frac{1}{t_2 - t_1} \int_{t_1}^{t_2} \int_{\mathbb{R}^2} U(u_h(x,t)) \, dx \, dt \leq \int_{\mathbb{R}^2} U(u_h(x,0)) \, dx + Ct_2 h^{2\alpha-1}$$

Let $h \to 0$. Tartar's theorem (Theorem 3.3.3) implies that

$$\frac{1}{t_2 - t_1} \int_{t_1}^{t_2} \int_{\mathbb{R}^2} \langle \nu_{x,t}, U \rangle \, dx \, dt \leq \int_{\mathbb{R}^2} U(u_0(x)) \, dx.$$

The remaining part of the proof can be found in the proof of Proposition 3.3.16. □

3.6.8 Proof of Theorem 3.5.6 From Theorem 3.6.5, condition (a) of Theorem 3.3.5 holds. From Propositions 3.6.6 and 3.6.7, conditions (3.3.2) and (3.3.3) of Theorem 3.3.7 are satisfied. Therefore we can apply the latter theorem and conclude that condition (d) of Theorem 3.3.5 holds. The properties (b) and (c) of Theorem 3.3.5 follow from Theorem 3.6.4 and Lemma 3.6.2. Using this theorem, we obtain that

$$\nu_{x,t} = \delta_{u(x,t)} \text{ a.e. } (x,t) \in \mathbb{R}^2 \times [0,T],$$

where u is the (unique) Kruzkov solution of (3.1.1), (3.1.2). Corollary 3.3.6 implies that the sequence u_h converges strongly in $L^1_{\text{loc}}(\mathbb{R}^2 \times [0,T])$ to u. This finishes the convergence proof for the higher-order finite volume schemes. □

REMARK 3.6.9 *Convergence of nonconforming finite element approximations to first-order linear hyperbolic equations of the form*

$$b \cdot \nabla u + \alpha u = f$$

is considered in [243].

REMARK 3.6.10 *A method of nonpolynomial reconstructions is described in [222].*

REMARK 3.6.11 *A generalization of these higher-order discontinuous Galerkin methods to systems can also be found in [36] – [38]. A finite element approach can be found in [110].*

3.7 Essentially non-oscillatory schemes

Another method for getting higher-order schemes is that of the ENO (essentially non-oscillatory) schemes, which we shall discuss in this section. For more details we refer to Harten *et al.* [86] and Abgrall [1]. The ENO scheme is composed of a piecewise-polynomial reconstruction from the piecewise-constant values of the last time step. It is based on an adaptive selection of a suitable stencil for each cell such that strong oscillations near discontinuities are avoided. In smooth parts of the solution the scheme is of higher-order. The scheme is called essentially non-oscillatory since small oscillations on the scale of the interpolation error are still possible. More general reconstructions than piecewise-polynomial have been developed and used in [222]. We shall now describe the ENO schemes first for problems in 1-D and later for 2-D.

DEFINITION 3.7.1 (Definition of the algorithm) *We shall describe the ENO scheme of order l_0 first for scalar conservation laws in 1-D. Consider the initial value problem*

$$\partial_t u + \partial_x f(u) = 0 \quad \text{in } \mathbb{R} \times \mathbb{R}^+ ,$$
$$u(x,0) = u_0(x) \quad \text{in } \mathbb{R} .$$

Consider a uniform grid $x_j := j\Delta x$, $t^n := n\Delta t$.

1. *Let*

$$u_j^0 := \frac{1}{\Delta x} \int_{x_{j-1/2}}^{x_{j+1/2}} u_0(x) \, dx .$$

2. *Definition of a reconstruction $R(x)$. Assume that u_j^n, $j \in \mathbb{Z}$, is already defined.*

    (a) *Let*

    $$w_0 := 0 ,$$
    $$w_j := w_{j-1} + h u_j^n , \quad j \in \mathbb{N} ,$$
    $$w_{j-1} := w_j + h u_j^n , \quad -j \in \mathbb{N}_0 .$$

    Formally this means that w_j is obtained by integration of u_j^n.

    (b) *Admissible stencils. Fix j. Define s by*

    $$|[w_s, w_{s+1}, w_{s+2}]|$$
    $$:= \min \{|[w_{j-2}, w_{j-1}, w_j]|, |[w_{j-1}, w_j, w_{j+1}]|\} . \quad (3.7.1)$$

    Remember that $[a] = a$, $[a,b] = b - a$,
    $$[a,b,c] = [b,c] - [a,b] = c - b - (b-a) = a - 2b + c .$$

    Therefore $(1/\Delta x^2)[w_r, w_{r+1}, w_{r+2}]$ is formally a discretization of a second derivative. This means that the stencil defined in (3.7.1) has "minimal curvature". This definition can be generalized. Let s be defined by

$$\|[w_s, w_{s+1}, ..., w_{s+r+1}]\|$$
$$:= \min\{\|[w_{j-r-1}, ..., w_j]\|, ..., \|[w_{j-1}, ..., w_{j+r}]\|\} .$$

(c) Now let $I(x) = I_j(x)$ be the interpolation polynomial of degree $r + 1 = l_0$ for the points

$$(x_s, w_s), ..., (x_{s+r+1}, w_{s+r+1}).$$

(d) Since w is defined formally by integration of u^n, we go back to the level u^n formally if we define

$$R(x) = R_j(x) := \partial_x I(x) .$$

3. Now we need an extrapolation of R in the time direction. First let us give the motivation for this idea. Assume for a moment that the exact solution is smooth. Then

$$u(x,t) = \sum_l \sum_k \frac{(x-x_j)^k}{k!} \frac{(t-t^n)^{l-k}}{(l-k)!} \underbrace{\partial_t^{l-k} \partial_x^k u}_{=: a^{k,l-k}} + ...$$

$$\begin{aligned}
a^{k,0} &= \partial_x^k u ,\\
a^{0,1} &= \partial_t u &&= -f'(u)\partial_x u &&= -f'(a^{0,0})a^{1,0} ,\\
a^{1,1} &= \partial_t \partial_x u &&= -f'(u)\partial_x^2 u - f''(u)(\partial_x u)^2 \\
& &&= -f'(a^{0,0})a^{2,0} - f''(a^{0,0})(a^{1,0})^2 ,\\
&\vdots
\end{aligned} \qquad (3.7.2)$$

Using this motivation, we define

$$v_j(x,t) := \sum_{l=0}^{l_0-1} \sum_{k=0}^{l} a^{k,l-k} \frac{(x-x_j)^k}{k!} \frac{(t-t^n)^{l-k}}{(l-k)!} ,$$

where the $a^{k,l-k}$ are defined recursively if we replace u by R in (3.7.2).

4. Definition of the numerical flux. Let g be a monotone numerical flux and

$$g_{j+\frac{1}{2}} := g(v_j(x_{j+\frac{1}{2}}, t^n), v_{j+1}(x_{j+\frac{1}{2}}, t^n)) .$$

Then the ENO scheme is defined as

$$u_j^{n+1} := u_j^n - \frac{\Delta t}{\Delta x}(g_{j+\frac{1}{2}} - g_{j-\frac{1}{2}}) .$$

3.7 Essentially non-oscillatory schemes

REMARK 3.7.2 *The scheme is in conservation form, and therefore if $u_j^n \to u$ a.e. then the limit u will be a weak solution of the conservation law. The convergence $u_j^n \to u$ for the ENO schemes where u is the entropy solution is an open question.*

REMARK 3.7.3 (Higher-order discretization in time) *Let $\alpha_k, \beta_k \in \mathbb{R}, 0 \leq \beta_k \leq 1$, be given by the following integration formula:*

$$\frac{1}{\tau} \int_{t^n}^{t^{n+1}} h(s) \, ds = \sum_{k=0}^{k_0} \alpha_k h(t^n + \beta_k \tau) + \mathcal{O}(\tau^p),$$

where τ is the time step. Let $g(u, v)$ be defined as before and let

$$g_{j+\frac{1}{2}} = \sum_{k=0}^{k_0} \alpha_k g(v_j(x_{j+\frac{1}{2}}, t^n + \beta_k \tau), v_{j+1}(x_{j+\frac{1}{2}}, t^n + \beta_k \tau)).$$

Then the ENO scheme is defined as

$$u_j^{n+1} := u_j^n - \frac{\Delta t}{\Delta x}(g_{j+\frac{1}{2}} - g_{j-\frac{1}{2}}).$$

In Figures 3.7.1 – 3.7.8 the results of some numerical experiments for the ENO scheme are shown [146]. In Figures 3.7.1 – 3.7.4 we have solved the Burgers equation

$$\partial_t u + \partial_x \left(\frac{u^2}{2}\right) = 0 \quad \text{on } [-1, 1] \times \mathbb{R}^+$$

for the initial values

$$u(x, 0) = \sin(\pi x + \pi) \quad \text{on } [-1, 1]$$

for $\Delta x = \frac{1}{10}$ and $\Delta t = 0.4 \Delta x$. The results are shown for the Lax–Wendroff scheme, ENO schemes of second and third order, and the Lax–Friedrichs, Engquist–Osher and Godunov schemes.

In the second test problem the Riemann problem for the linear equation

$$\partial_t u + \partial_x u = 0 \quad \text{in } [0, 1] \times \mathbb{R}^+$$

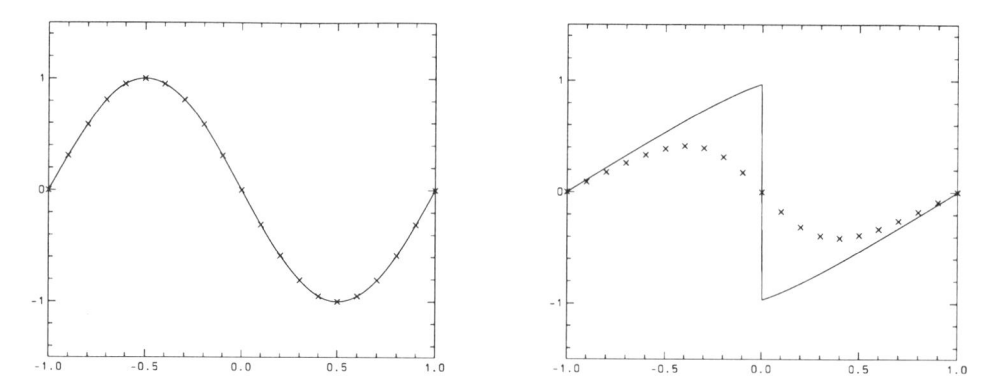

Figure 3.7.1 Initial data and Lax–Fridrichs.

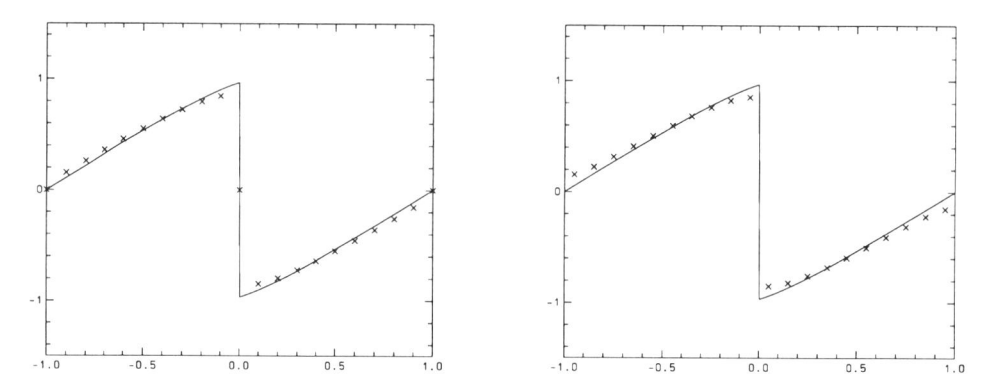

Figure 3.7.2 Engquist–Osher and Godunov.

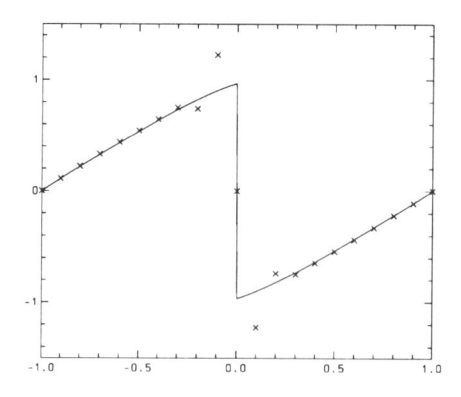

Figure 3.7.3 Lax–Wendroff.

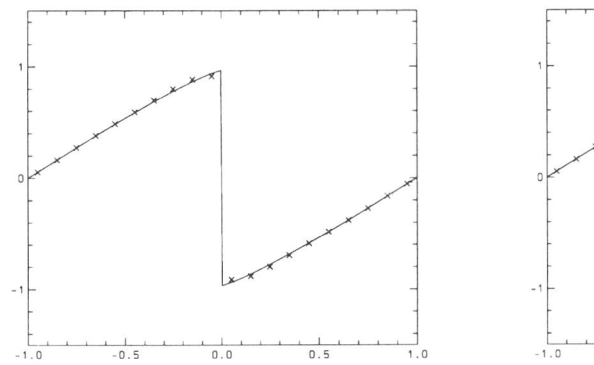

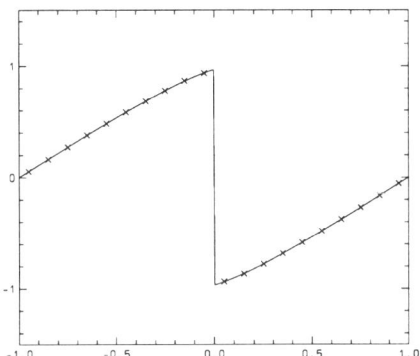

Figure 3.7.4 2nd-order and 3rd-order ENO schemes.

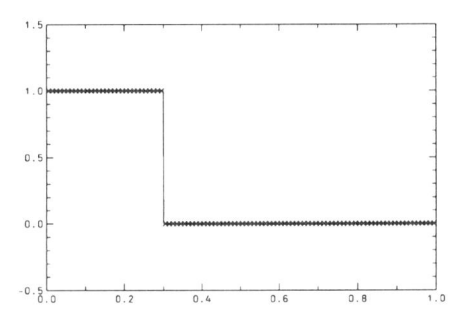

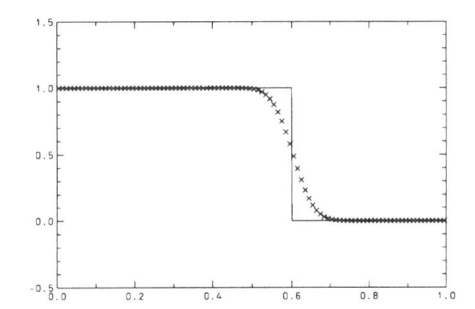

Figure 3.7.5 Initial data and upwind.

with initial values

$$u(x,0) = \begin{cases} 1 & \text{if } x \leq 0.3 \\ 0 & \text{if } x > 0.3 \end{cases}$$

has been treated, for $\Delta x = \frac{1}{100}$ and $\Delta t = 0.4\Delta x$. In Figures 3.7.5 – 3.7.8 we can see the results obtained with the simple upwind method, the Lax–Friedrichs, ENO first, second, and third order, Lax–Wendroff, Beam and Warming, and Godunov schemes.

Now we shall describe the ENO scheme for several space dimensions (see Harten [86] and Abgrall [1]). For simplicity, we describe it only in 2-D.

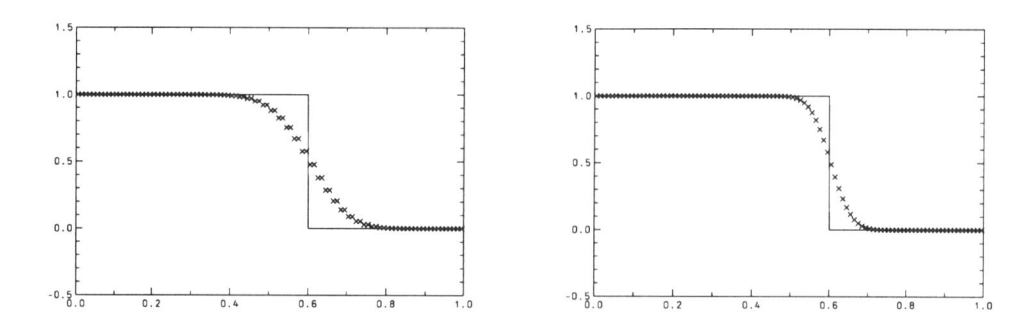

244 Initial value problems for scalar conservation laws in 2-D

Figure 3.7.6 Lax–Friedrichs and Godunov.

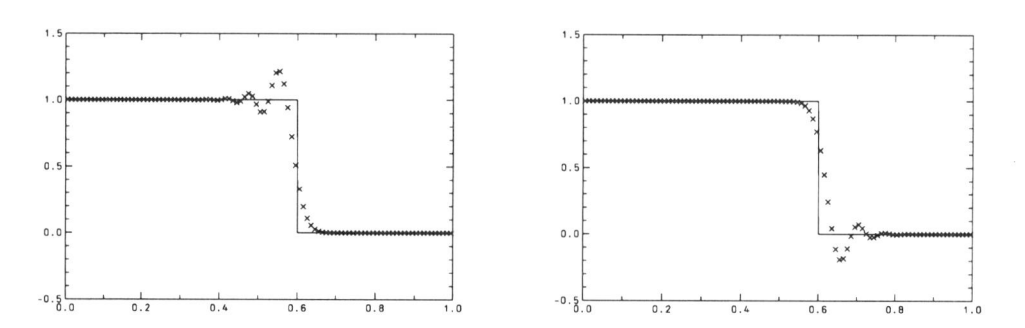

Figure 3.7.7 Lax–Wendroff and Beam–Warming.

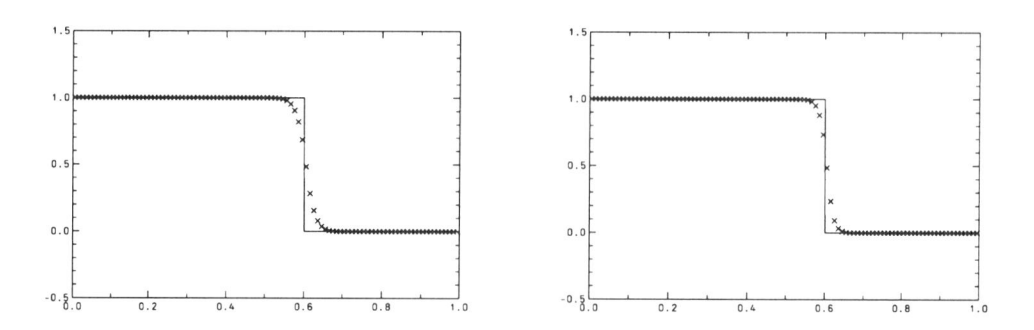

Figure 3.7.8 2nd-order and 3rd-order ENO.

3.7 Essentially non-oscillatory schemes

Consider the problem

$$\partial_t u + \partial_x f(u) + \partial_y g(u) = 0 \quad \text{in } \mathbb{R}^2 \times \mathbb{R}^+ ,$$
$$u(x,y,0) = u_0(x,y) \quad \text{in } \mathbb{R}^2 .$$

Assume that a triangulation $\mathcal{T}_h$ of $\mathbb{R}^2$ is given.

DEFINITION 3.7.4 *The dual cells C_j are defined as follows. Let Q_j be a node of the triangulation. Then in the neighbouring triangles one connects the centres of gravity with the midpoints of the neighbouring edges containing Q_j (see Figure 3.7.9). In this, way one gets the dual cell C_j around Q_j (see also Example 7.1.5).*

DEFINITION 3.7.5 (Problem P) *Let $n \in \mathbb{N}$ be fixed, $u \in L^1(\mathbb{R}^2)$, and $S = \{T_1, \ldots, T_n\}$ be a finite set of triangles $T_1, \ldots, T_n$ or dual cells. Let*

$$P_m = \left\{ p \mid p(x,y) = \sum_{l=0}^{m} \sum_{i+j=l} a_{ij} x^i y^j ; \; a_{ij}, x, y \in \mathbb{R} \right\}$$

Then $p \in P_m$ is called a solution of problem P with respect to S, m and u if

$$\frac{1}{|T_j|} \int_{T_j} u = \frac{1}{|T_j|} \int_{T_j} p$$

for all $T_j \in S$. The set S is called a stencil consisting of $T_1, \ldots, T_n$.

(For the existence of a solution of problem P we refer to [1].)

DEFINITION 3.7.6 (ENO Algorithm in 2-D using quadratic reconstruction)

1. Define u_j^0 by

$$u_j^0 := \frac{1}{|C_j|} \int_{C_j} u_0(x,y) \, dx \, dy \quad \text{for all } j.$$

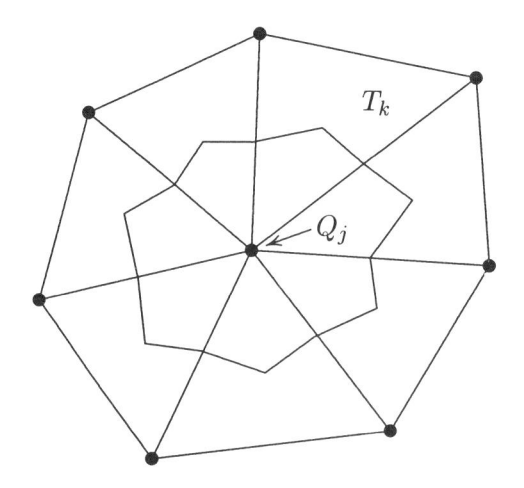

246 Initial value problems for scalar conservation laws in 2-D

Figure 3.7.9

2. Assume that u_j^n is given. We consider u^n as a piecewise-constant function (with jumps along the edges of the dual cells). Fix a grid point (x_0, y_0) and the corresponding dual cell C_j. Consider the triangles $T_1, ..., T_k$, having (x_0, y_0) as a common vertex. Let p_r be the solution of problem P with respect to the three dual cells belonging to the vertices of T_r, u^n and $m = 1$ for $r = 1, ..., k$.

Let

$$p_r(x,y) =: \sum_{l=0}^{1} \sum_{i+j=l} a_{ij}^r (x - x_0)^i (y - y_0)^j$$

and

$$\sigma_r := \sum_{i+j=1} |a_{ij}^r| .$$

Choose that triangle T_q which minimizes σ_r, i.e.

$$\sigma_q := \min_{1 \leq r \leq k} \sigma_r$$

Denote the edges of T_q by S_{q_1}, S_{q_2} and S_{q_3}. There is one neighbouring triangle $\tilde{T}_1$ of T_q that has the edge S_{q_1} in common with T_q (see Figure 3.7.10).

3.7 Essentially non-oscillatory schemes

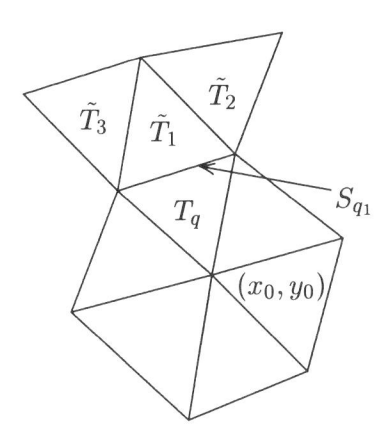

Figure 3.7.10

Then T_q is a neighbouring triangle of $\tilde{T}_1$. The other neighbouring triangles of $\tilde{T}_1$ are denoted by $\tilde{T}_2$ and $\tilde{T}_3$.

Consider the stencil Z_{q_1} consisting of the six dual cells belonging to the vertices of T_q, $\tilde{T}_1, \tilde{T}_2$ and $\tilde{T}_3$ and solve problem P with respect to Z_{q_1} and u^n in P_m with $m = 2$. The solution is denoted by $\tilde{p}_1$ (the index refers to S_{q_1}). Then we repeat these computations for S_{q_2} and S_{q_3} and obtain polynomials $\tilde{p}_1, \tilde{p}_2$ and $\tilde{p}_3$ with

$$\tilde{p}_r(x,y) = \sum_{l=0}^{2} \sum_{i+j=l} b_{ij}^r (x-x_0)^i (y-y_0)^j .$$

Let

$$\tilde{\sigma}_r := \sum_{i+j=2} |b_{ij}^r|$$

and choose the stencil Z_{q_s} that minimizes $\tilde{\sigma}_r$, i.e.

$$\tilde{\sigma}_s := \min_{1 \leq r \leq 3} \tilde{\sigma}_r .$$

Then $\tilde{p}_s$ is called the **reconstruction** concerning the cell C_j.

Define

$$R_j(u^n)(x,y) := \tilde{p}_s(x,y).$$

Let x_{jli} be the quadrature points of S_{jl} (edges of the dual cell) such that

$$\frac{1}{|S_{jl}|}\int_{S_{jl}} u = \sum_{i=1}^{k}\alpha_i u(x_{jli}) + \mathcal{O}(h^\beta),$$

where the α_i are the corresponding weights.

Then the ENO scheme is defined by

$$u_j^{n+1} := u_j^n - \frac{\Delta t}{|C_j|}\sum_{l=1}^{l(j)}\sum_{i=1}^{k}\alpha_i g_{jl}\Big(R_j(u^n)(x_{jli}), R_{jl}(u^n)(x_{jli})\Big),$$

where g_{jl} is a given numerical flux satisfying (3.2.8) – (3.2.10) corresponding to a monotone or an E-scheme, and $l(j)$ denotes the number of edges of C_j (see Figure 3.7.9).

As a test problem for the ENO scheme, the rotating-cone problem has been solved [146]. In Figure 3.7.11 the initial conditions are plotted. The time evolution for a second order ENO scheme on a cartesian grid can be seen in Figures 3.7.12 and 3.7.13. The corresponding results for second-order and third-order ENO schemes on triangular meshes are plotted in Figures 3.7.14 – 3.7.17. In Figures 3.7.18 and 3.7.19 an oblique shock on a triangular mesh has been computed by second- and third-order ENO schemes respectively.

Generally it can be seen that the resolution of the higher-order ENO schemes is much better than for the first-order schemes.

There is still no convergence theory available for ENO schemes. In [194] it was indicated that for the initial value problem for the linear advection equations the convergence rate depends on the initial values and may be less than what is indicated by the local truncation error analysis. There are also cases where a refined grid gives a larger error. For the classical ENO scheme the choice of the stencil depends only on the data of the last time step and not on the conservation law itself. Therefore for suitable chosen initial data the ENO scheme can be forced to choose initially the (unstable) downwind stencil. In [87] it was shown that this leads to stencil switching and to loss of

3.7 Essentially non-oscillatory schemes 249

one order of accuracy. In [216] a modified ENO scheme has been proposed that has the correct order of accuracy.

A very extensive presentation of the definition and properties of ENO schemes can be found in [2].

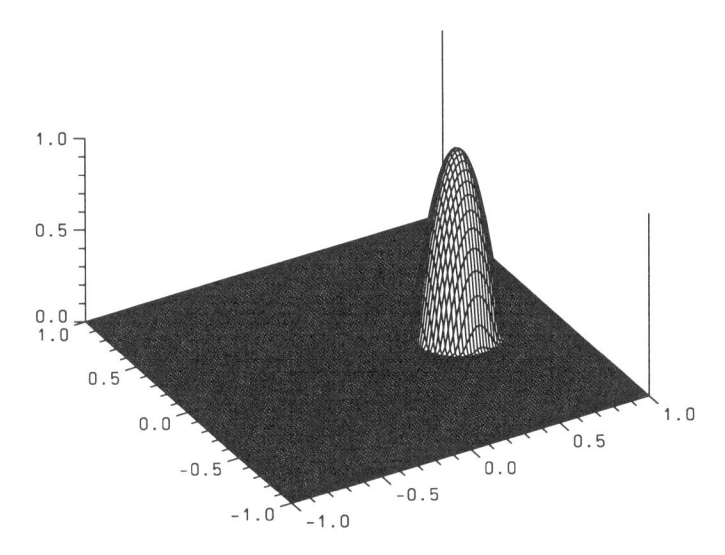

Figure 3.7.11 Initial conditions; $t = 0$, $\Delta h_x = \Delta h_y = 2/114$.

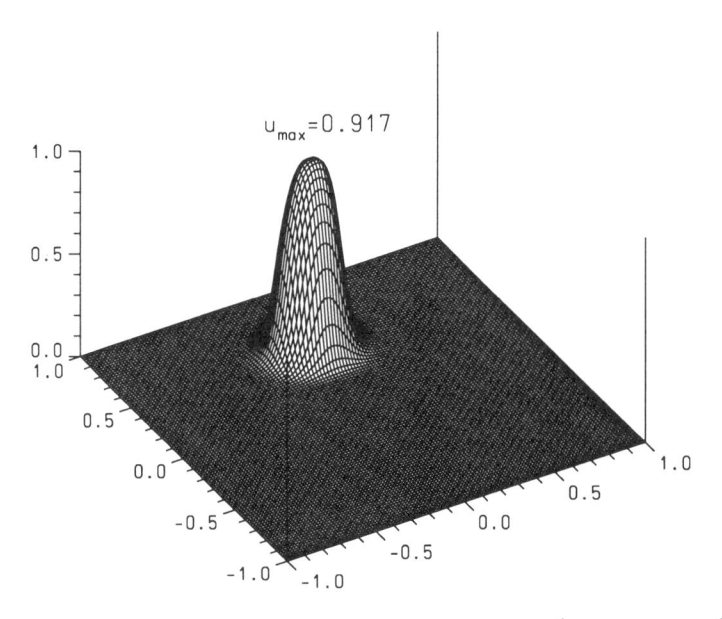

250   Initial value problems for scalar conservation laws in 2-D

Figure 3.7.12 2nd-order ENO solution; $t = \frac{1}{2}\pi$, CFL $= \frac{1}{8}$, $\Delta h_x = \Delta h_y = 2/114$.

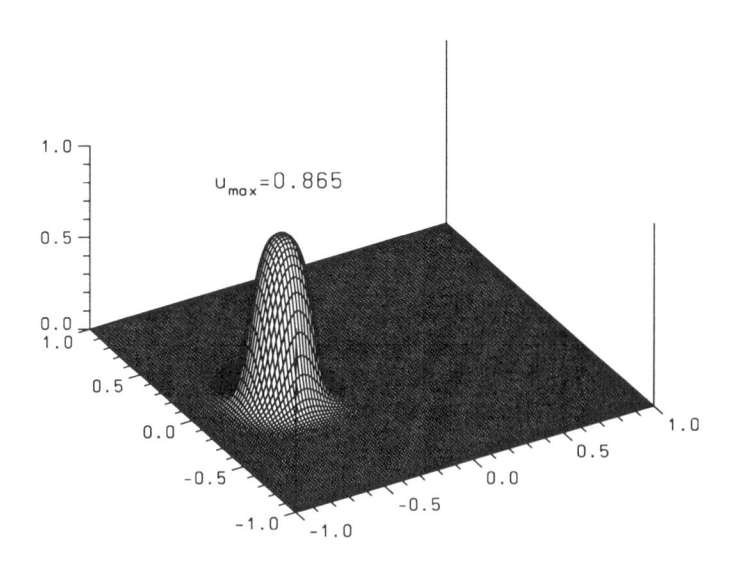

Figure 3.7.13 2nd-order ENO solution; $t = \pi$, CFL $= \frac{1}{8}$, $\Delta h_x = \Delta h_y = 2/114$.

3.7 Essentially non-oscillatory schemes 251

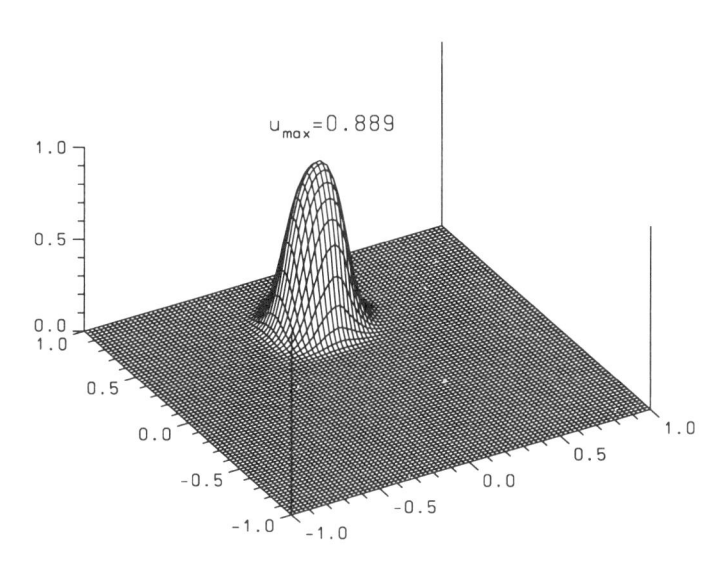

Figure 3.7.14 2nd-order ENO solution; $t = \frac{1}{2}\pi$, CFL $= \frac{1}{8}$, 12800 triangles, 6561 nodes.

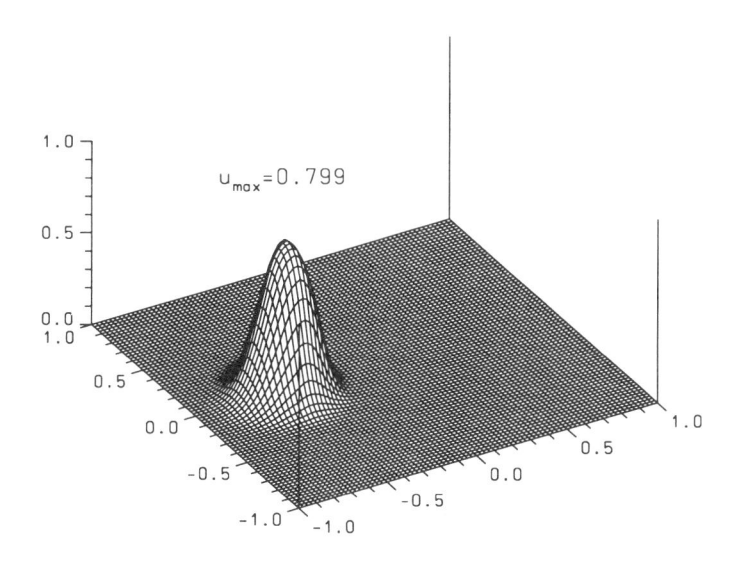

Figure 3.7.15 2nd-order ENO solution; $t = \pi$, CFL $= \frac{1}{8}$, 12800 triangles, 6561 nodes.

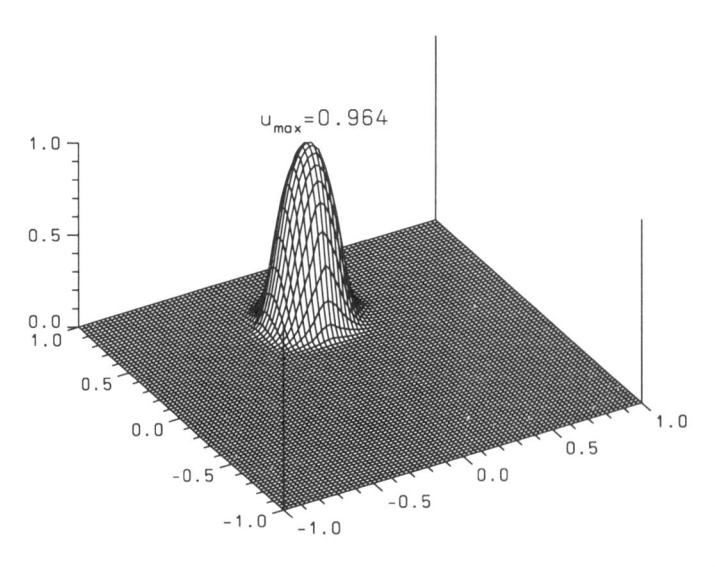

252 Initial value problems for scalar conservation laws in 2-D

Figure 3.7.16 3rd-order ENO solution; $t = \frac{1}{2}\pi$, CFL $= \frac{1}{8}$, **12800 triangles, 6561 nodes.**

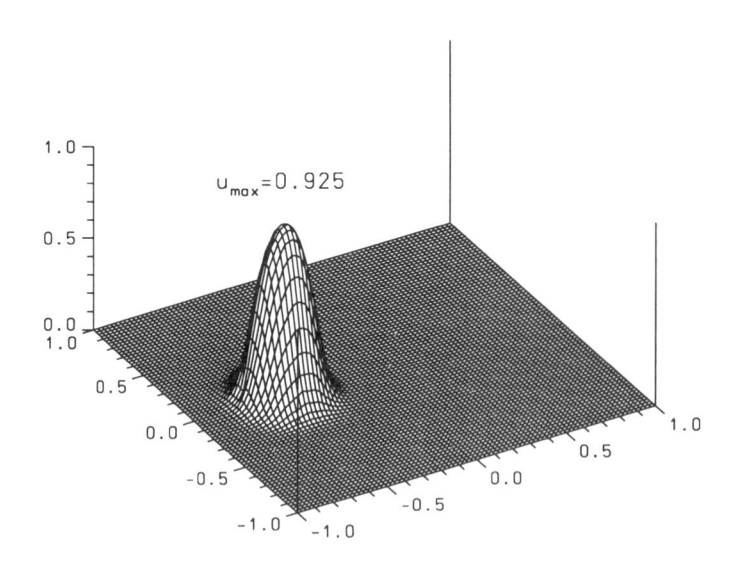

Figure 3.7.17 3rd-order ENO solution; $t = \pi$, CFL $= \frac{1}{8}$, **12800 triangles, 6561 nodes.**

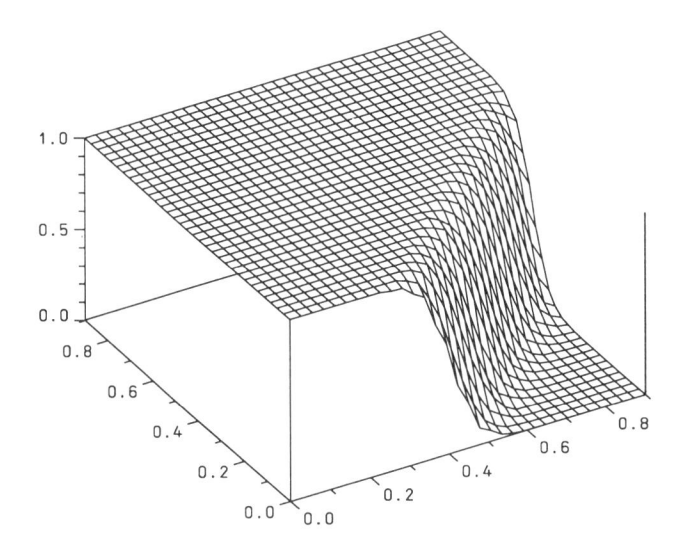

3.7 Essentially non-oscillatory schemes 253

Figure 3.7.18 2nd-order ENO solution; t = 0.2, CFL = 0.4, 1089 nodes, 2048 triangles.

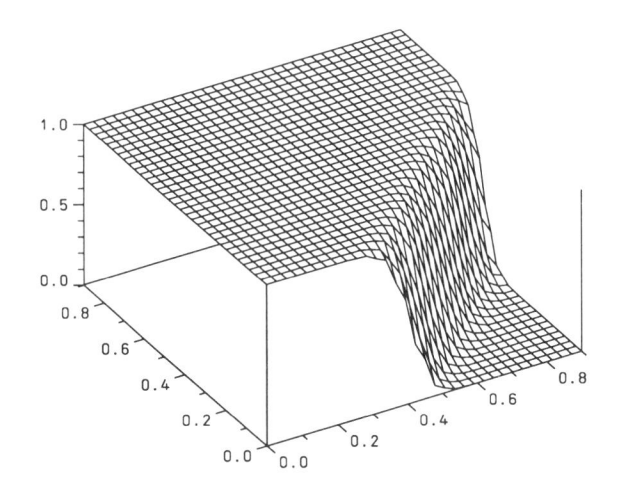

Figure 3.7.19 3rd-order ENO solution; t = 0.2, CFL = 0.4, 1089 nodes, 2048 triangles.

3.8 Fluctuation splitting schemes

Another way to get higher-order schemes for the discretization of conservation laws is via the fluctuation splitting scheme. This is a further example of a scheme where the unknowns are related to the vertices of the grid similar as for the ENO schemes. But the numerical damping is discretized in a completely different way. The new idea is based on a classification of the triangles in two classes: those with one and those with two inflow edges. A very detailed presentation of this method can be found in [48] and [49].

Fluctuation splitting scheme for linear scalar conservation laws.
Let us assume that a regular triangulation $\mathcal{T}$ is given. The nodes of $\mathcal{T}$ are denoted by

$$P_j, \quad j = 1, ..., N.$$

For any $T \in \mathcal{T}$ let

$$P_T := \{j | P_j \in T\}.$$

We consider the linear convection equation

$$\begin{aligned} \partial_t u + a \partial_x u + b \partial_y u &= 0 \quad \text{in } \mathbb{R}^2 \times \mathbb{R}^+, \\ u(\cdot, 0) &= u_0 \quad \text{in } \mathbb{R}^2. \end{aligned} \qquad (3.8.1)$$

Let ν_{Tj} be the scaled outer normal to ∂T with respect to the edge opposite P_j in the triangle T. For any $j = 1, ..., N$ let $\mathcal{T}_j$ be the set of neighbouring triangles of P_j:

$$\mathcal{T}_j := \{T \in \mathcal{T} \mid P_j \in T\}.$$

where C_j denotes the dual cell as defined in Definition (3.7.4) with $|C_j| = \frac{1}{3} \sum_{t \in T_j} |T|$. Let $u_j^0 := (1/|C_j|) \int_{C_j} u_0$ denote the initial values, which are assumed to be linear on each $T \in \mathcal{T}$ and globally continuous. The value of u^0 in the jth-node P_j of $\mathcal{T}$ is denoted by u_j^0. Now we assume that u^n at time level n is already given with values u_j^n in P_j, $j = 1, ..., N$, and that u^n is

3.8 Fluctuation splitting schemes

piecewise-linear and globally continuous. In order to describe the algorithm, we need $\vec{a} := (a,b)^t$ and (see [220])

$$\Phi^T := \frac{1}{2} \sum_{j \in P_T} \vec{a}\, \nu_{Tj}\, u_j^n. \tag{3.8.2}$$

Then the algorithm is given by

$$u_j^{n+1} = u_j^n + \frac{\Delta t}{|C_j|} \sum_{T \in \mathcal{T}_j} \alpha_j^T \Phi^T \tag{3.8.3}$$

where α_j^T are upwind coefficients satisfying $\sum_{j \in P_T} \alpha_j^T = 1$. This condition implies the conservation property

$$\sum_j |C_j| u_j^{n+1} = \sum_j |C_j| u_j^n + \text{ boundary terms}.$$

The explicit definition of the α_j^T is given in [220]. For instance, for triangles with one inflow edge we have $\alpha_{j_1}^T = 0$, $\alpha_{j_2}^T = 1$ and $\alpha_{j_3}^T = 0$ (see Figure 3.8.1).

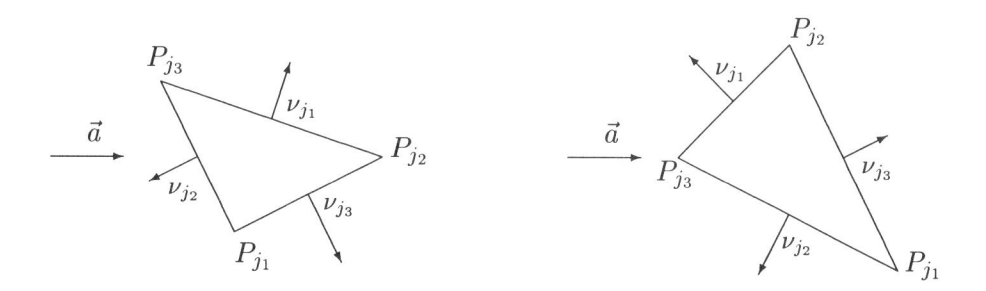

Figure 3.8.1 **Figure 3.8.2**

If T has two inflow edges as in Figure 3.8.2, we have $\alpha_{j_1}^T = k_1(u_{j_1} - u_{j_3})$, $\alpha_{j_2}^T = k_2(u_{j_2} - u_{j_3})$ and $\alpha_{j_3}^T = 0$, where

$$k_l := \frac{\nu_{j_l} \vec{a}}{2\Phi^T} \quad \text{if } \Phi^T \neq 0. \tag{3.8.4}$$

Otherwise, if $\Phi^T = 0$ choose $\alpha_j^T = 0$ in (3.8.3).

Perthame [180] showed that the scheme as defined in (3.8.2)–(3.8.4) converges in the linear case. In order to describe the results of Perthame, we need the following reformulation.

DEFINITION 3.8.1 *Let i, j and k denote the vertices of a triangle T. The triangle T is "one–target of type i" (i.e. $T \in A_1(i)$) if*

$$\vec{a} \cdot \nu_{T_i} < 0, \quad \vec{a} \cdot \nu_{T_j} \geq 0, \quad \vec{a} \cdot \nu_{T_k} \geq 0.$$

Then let

$$\xi(i, T) := \vec{a}, \quad \xi(j, T) = 0, \quad \xi(k, T) = 0.$$

The triangle T is "two–target of type i, j" (i.e. $T \in A_2(i), T \in A_2(j)$) if

$$\vec{a} \cdot \nu_{T_i} < 0, \quad \vec{a} \cdot \nu_{T_j} < 0, \quad \vec{a} \cdot \nu_{T_k} \geq 0.$$

Then define $\xi(i, T), \xi(j, T), \xi(k, T) \in \mathbb{R}^2$ as a solution of the six equations

$$\xi(k, T) = 0, \quad \xi(i, T) + \xi(j, T) = \vec{a},$$

$$\xi(i, T) \cdot \nu_{T_j} = 0, \quad \xi(j, T) \cdot \nu_{T_i} = 0.$$

DEFINITION 3.8.2 (N-scheme) *Let u_j^n be given and let L^n be the globally continuous and piecewise-linear function such that*

$$L^n\big|_T (P_j) = u_j^n$$

Then let

$$u_j^{n+1} := u_j^n - \frac{\Delta t}{|C_j|} \sum_{T \in \mathcal{T}_j} \int_T \xi(j, T) \nabla L^n \, dx \qquad (3.8.5)$$

REMARK 3.8.3 *Let S_{T_l} be the edge of T opposite the vertex l. Then (3.8.5) can be written as*

$$u_j^{n+1} := u_j^n + \frac{\Delta t}{2|C_j|} \sum_{T \in \mathcal{T}_j} \sum_{l=1}^{3} \xi(j, T) \nu_{T_l} u_{T_l}^n.$$

Proof

$$\int_T \xi(j,T)\nabla L^n\, dx = \int_{\partial T} \xi(j,T) n L^n\, d\sigma(x)$$

$$= \sum_{l=1}^{3} \int_{S_{T_l}} \xi(j,T) n_{T_l} L^n d\sigma(x)$$

$$= \sum_{l=1}^{3} \xi(j,T)\nu_{T_l} \frac{1}{2}(u^n_{T_{l-1}} + u^n_{T_{l+1}})$$

$$= -\frac{1}{2}\sum_{l=1}^{3} \xi(j,T)\nu_{T_l} u^n_{T_l}$$

Here we have used the fact that $\nu_{T_1} + \nu_{T_2} + \nu_{T_3} = 0$. $\square$

THEOREM 3.8.4 (Convergence of the N-scheme in the linear case) *Assume that we have a regular triangulation. For given u^n let $u_h(x,t)$ be the piecewise-constant function with value u_j^n on the cell $C_j \times [t^n, t^{n+1}[$, where C_j is the dual cell (see (3.8.3)). Let Δt be sufficiently small and for $h \to 0$*

$$\|u_h(\cdot,0)\|_{L^\infty(\mathbb{R}^2)} \le c\,,\; u_h(\cdot,0) \to u_0 \quad \text{in } L^2(\mathbb{R}^2).$$

Then for all $T > 0$

$$u_h \to u \quad \text{in } L^2_{\text{loc}}(\mathbb{R}^2 \times]0,T[)\;,$$

where u is the exact solution of (3.8.1).

Proof See [180].

REMARK 3.8.5 *Other fluctuation splitting schemes and generalization to nonlinear problems and systems can be found in [48] – [51].*

3.9 Adaptivity and local grid alignment: numerical experiments

In order to optimize the computing time as well as the use of the storage, it is necessary to develop self-adapting algorithms, in particular locally refined grids. For elliptic and parabolic problems there is a well-known theory about error estimators. If the numerical solution u_h is known, they indicate those region where the local error $\|u - u_h\|$ is still large. Using this information, the grid can be refined in these domains successively up to that level if the error $\|u - u_h\|$ is less than a prescribed tolerance. Also, for linear hyperbolic systems theoretical results for error estimators are available (see [224], [152] and [82]). Unfortunately, so far there are no comparable results for nonlinear conservation laws. Therefore in this case one uses more heuristic indicators to find special regions where there are shocks or large gradients in order to refine the grid in these domains and to improve the resolution, in particular for discontinuities like shocks.

Error estimators for elliptic and parabolic equations [63] Let us consider the following elliptic boundary value problem, where $\Omega \subset \mathbb{R}^3$ is a polygonal domain:

$$-\Delta u = f \quad \text{in } \Omega \qquad (3.9.1)$$
$$u = 0 \quad \text{on } \partial\Omega . \qquad (3.9.2)$$

In Ω we use a regular triangulation $\mathcal{T}$ with the grid density $h(x)$ and assume that

$$c_2 h_T \leq h(x) \leq c_3 h_T \qquad \forall T \in \mathcal{T}$$

where $h_T := \operatorname{diam} T$. For the discrete problem let $S_h \subset \overset{\circ}{H}{}^{1,2}(\Omega)$ be a finite-dimensional subspace. Then we have to find $u_h \in S_h$ such that

$$(\nabla u_h, \nabla v) = (f, v) \qquad \forall v \in S_h$$

3.9 Adaptivity and local grid alignment: numerical experiments

where $(\cdot,\cdot)$ denotes the scalar product in $L^2(\Omega)$. Then if the exact solution u of (3.9.1) is in $H^{2,2}(\Omega)$, the following a priori estimates hold [28]:

$$\begin{aligned} \|\nabla(u-u_h)\|_{L^2} &\leq c\|hD^2u\|_{L^2}\,, \\ \|u-u_h\|_{L^2} &\leq c\|h^2D^2u\|_{L^2}\,. \end{aligned} \qquad (3.9.3)$$

These estimates are useless for the numerical control of the grid size, since the right-hand side depends on the exact solution u and therefore cannot be controlled. In the following a posteriori estimates the error $u-u_h$ is bounded from above in terms of u_h [63]:

$$\|\nabla(u-u_h)\|_{L^2} \leq c[\|hf\|_{L^2} + D_1(u_h)]\,, \qquad (3.9.4)$$
$$\|u-u_h\|_{L^2} \leq c[\|h^2f\|_{L^2} + D_2(u_h)]\,, \qquad (3.9.5)$$

where

$$D_m(v) := \left(\sum_{\text{edges}} h^{2m} \,|[\partial_n v]|_{L^2}^2\right)^{\frac{1}{2}}$$

and $[\partial_n v]$ denotes the jump of the normal derivatives across the edges of the triangles and $|\cdot|_{L^2}$ the L^2 norm along the edges. Similar results are available for parabolic equations [63, 106]. Of course, for parabolic equations the error at time t depends on $u_h(s,x)$ for all $0 \leq s \leq t$.

Error estimators for convection-dominated problems [108] Now we consider the following convection-dominated initial value problem:

$$\begin{aligned} \partial_t u + \partial_x f(u) - \varepsilon \partial_x^2 u &= 0 &&\text{in } \mathbb{R} \times \mathbb{R}^+\,, \\ u(x,0) &= u_0(x) &&\text{in } \mathbb{R}\,. \end{aligned} \qquad (3.9.6)$$

For small ε the numerical problems are similar to those of the purely hyperbolic case $\varepsilon = 0$. For the discretization we use the streamline diffusion shock-capturing method [231]. Let u_h denote the discrete solution, $R(u_h)$ the residual and $\hat{u}$ the exact solution of (3.9.6) with

$$\hat{\varepsilon} := \max\left\{\varepsilon, C_1 h \frac{R(u_h)}{|\nabla u_h|}, C_2 h^{\frac{3}{2}}\right\}$$

instead of ε for suitable constants C_1 and C_2. Then

$$\|\hat{u} - u_h\|_{L^2(\Omega\times[0,T])} \leq C \left\| \frac{h^2}{\hat{\varepsilon}} R(u_h) \right\|_{L^2(\Omega\times[0,T])}.$$

Error estimators for conservation laws [164] For conservation laws the only result concerning a posteriori error estimates is due to Tadmor [164]. For finite difference methods in 1-D for general nonlinear conservation laws he has shown that the error $u - u_h$ is small if the residual $\partial_t u_h + \partial_x f(u_h)$, defined in a suitable way, is small. Let us describe this result in more detail. First we have to define the following seminorms.

$$\|\varphi(\cdot)\|_{\text{Lip}} := \operatorname*{ess\,sup}_{x \neq y} \left| \frac{\varphi(x) - \varphi(y)}{x - y} \right|,$$

$$\|\varphi(\cdot,\cdot)\|_{\text{Lip}} := \operatorname*{ess\,sup}_{(x,t) \neq (y,\tau)} \left| \frac{\varphi(x,t) - \varphi(y,\tau)}{|x - y| + |t - \tau|} \right|,$$

$$\|w\|_{\text{Lip}+} := \operatorname*{ess\,sup}_{x \neq y} \left[\frac{w(x) - w(y)}{x - y} \right]^+$$

where $u^+ := \max\{u, 0\}$ and

$$\|w(\cdot)\|_{\text{Lip}'} := \sup_{\varphi \in C_0^\infty} \frac{|(w - \bar{w}, \varphi)|}{\|\varphi(x)\|_{\text{Lip}}}, \quad \text{where } (f, g) := \int f(x) g(x)\, dx,$$

$$\|w(\cdot,\cdot)\|_{\text{Lip}'(x,[0,T])} := \sup_{\varphi \in C_0^\infty(\mathbb{R}\times[0,T])} \frac{|(w - \bar{w}, \varphi)_{x,t}|}{\|\varphi(x,t)\|_{\text{Lip}}},$$

where $(f, g)_{x,t} = \int f(x,t) g(x,t)\, dx\, dt$, and $\bar{w}$ is the mean value of w. Let u_h be a family of conservative approximate solutions of the conservation law

$$\partial_t u + \partial_x f(u) = 0 \quad \text{in } \mathbb{R} \times \mathbb{R}^+,$$
$$u(x, 0) = u_0(x) \quad \text{in } \mathbb{R}$$

defined by a numerical scheme in conservation form such that

$$\|u_0\|_{\text{Lip}+} \leq c, \quad \|u_h(\cdot, t)\|_{\text{Lip}+} \leq C \quad \text{uniformly in } t, h.$$

3.9 Adaptivity and local grid alignment: numerical experiments

Furthermore, for any h there exists an $\varepsilon(h)$ such that $\varepsilon(h) \to 0$ if $h \to 0$ and

$$\|u_h(\cdot,0) - u_0\|_{\mathrm{Lip}'} + \|\partial_t u_h + \partial_x f(u_h)\|_{\mathrm{Lip}'(x,[0,T])} \leq C\,\varepsilon(h)\ . \quad (3.9.7)$$

Then

$$\|u_h(\cdot,t) - u(\cdot,t)\|_{W^{s,p}} = \mathcal{O}\left(\varepsilon^{(1-sp)/2p}\right)$$

for $-1 \leq s \leq 1/p$, $1 \leq p \leq \infty$. In [164] the following improved result has been proved. For all $t \in [0,T]$ we have

$$\|u_h(\cdot,t) - u(\cdot,t)\|_{\mathrm{Lip}'}$$
$$\leq \|u_h(\cdot,0) - u_0\|_{\mathrm{Lip}'} + \|\partial_t u_h + \partial_x f(u_h)\|_{\mathrm{Lip}'(x,[0,T])}\ .$$

For finite volume methods on irregular meshes applied to linear hyperbolic problems we refer to [229].

Extrapolation techniques [22] Assume

$$u_h = u + h\bar{u} + \mathcal{O}(h^2)\ .$$

Then we also have

$$u_{2h} = u + 2h\bar{u} + \mathcal{O}(h^2)\ .$$

Taking the difference of both equations, we get $\bar{u} = (1/h)(u_{2h} - u_h) + \mathcal{O}(h)$, and therefore

$$u_h - u = u_{2h} - u_h + \mathcal{O}(h^2)\ . \quad (3.9.8)$$

Then $u_{2h} - u_h$ can be used as an error estimator for $u_h - u$.

Higher-order differences and $|\nabla_h u_h|$ as mesh indicators Since there are insufficient theoretical results for hyperbolic equations, the $|\nabla_h u_h|$ or higher-order differences are used as mesh indicator in order to find out the regions with shocks and large gradients that should be refined. But notice that $|\nabla_h u_h|$ in general gives no information about the error $\|u_h - u\|$. Of course in the case when we already have the exact solution this mesh indicator is still large. But nevertheless the $|\nabla_h u_h|$ is useful for controlling moving refinement zones, which are necessary for moving shocks. Löhner [150] has proposed mesh indicators for solving the Euler equations based on interpolation theory as (3.9.4) and on limiter functions. In 1-D he defines

$$E_i := \frac{\mid U_{i+1} - 2U_i + U_{i-1} \mid}{\mid U_{i+1} - U_i \mid + \mid U_i - U_{i-1} \mid + C} , \qquad (3.9.9)$$

where

$$C := \Theta \left(\mid U_{i+1} \mid + 2 \mid U_i \mid + \mid U_{i-1} \mid \right) \qquad (3.9.10)$$

for a suitable Θ. The numerator corresponds to the second derivatives in (3.9.4), and the first order differences in the denominator are used for normalization. The constant C is used as a "noise" filter. The generalization of (3.9.9) to 2-D looks like

$$E := \frac{\mid (\nabla U)_i S_{ij} - (\nabla U)_j S_{ij} \mid}{\mid (\nabla U)_i S_{ij} \mid + \mid (\nabla U)_j S_{ij} \mid + C} , \qquad (3.9.11)$$

where

$$C := \Theta \left(\mid U_i \mid + \mid U_j \mid \right) , \quad S_{ij} := x_i - x_j , \qquad (3.9.12)$$

x_j are the nodes of the triangulation, U_j are the values of the discrete solution in x_j and $(\nabla U)_j$ are the difference quotients in x_j with respect to U. Here it is assumed that the data u_j are defined on the cell vertices. Usually high-resolution schemes are based on the limiter functions, which are only active in regions with large gradients. Therefore Löhner [150] has also proposed an error indicator based on these limiter functions. The advantage is that there is no additional computational cost, since in general the limiter will be computed.

All these criteria can be used to mark some triangles $T \in \mathcal{T}$ that should be refined. There are several strategies for local refinement, for instance as shown in Figures 3.9.1 and 3.9.2. In these figures $*$ denotes the triangles that should be refined. In Figure 3.9.1 the marked triangle was refined by bisection. In order to avoid the nonconformal node, the neighbouring triangle was also refined, such that in all two macro triangles are involved in the local refinement process. The situation is very different if the marked triangle is divided into four as shown in Figure 3.9.2. As before, in order to

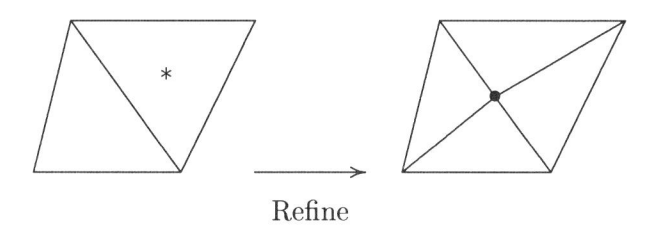

3.9 Adaptivity and local grid alignment: numerical experiments   263

Figure 3.9.1

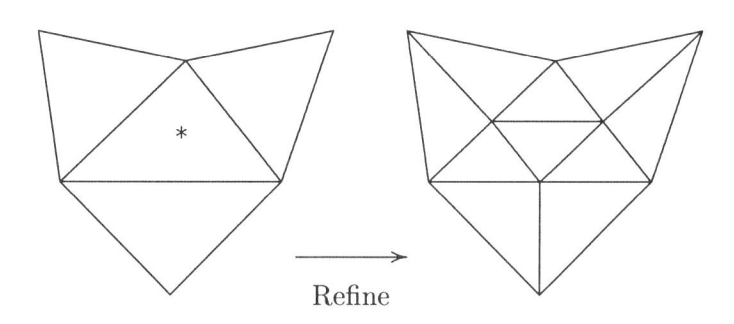

Figure 3.9.2

avoid nonconformal nodes, the neighbouring triangles have to be divided by bisection such that finally in this case four macro triangles are involved in the local refinement process.

For more details we refer to [15], where the first example in Figure 3.9.1 has been investigated in 3-D, also from the theoretical point of view. The first method of refinement has some advantages in particular for coarsening the mesh. Furthermore this process remains more local. In [55] it was shown that for linear elliptic problems under some suitable assumptions local grid refinement will decrease the error $u - u_h$ in each step by a constant factor.

The criterion mentioned in (3.9.8) has been used in [22] on structured, locally refined grids with nonconformal nodes (for more details look at the grids in Figures 3.9.3 – 3.9.7). In Figures 3.9.3 – 3.9.7 we show some results for a rotating cone where the local refinement zone rotates with the solution [210]. The solution has been computed on differently structured grids by an explicit finite volume method. On the finer grid a local time step Δt_{loc} has been used. The values on the boundary of the fine grid were obtained by interpolation of the coarse grid. The number n of time step iteration on the fine grid is given by $n\Delta t_{\text{loc}} = \Delta t_{\text{coarse}}$ where Δt_{coarse} is the time step on the coarse grid. The computation has been performed for one rotation of the cone, and the shape of the cone remains very stable.

For this test problem it was trivial to adapt the grid to the boundary of the domain but in the general case this is much more difficult. A strategy how to do this has been described in [137] and [140].

With the same method as before, a two-dimensional Riemann problem as defined in Figure 3.9.8 has been solved. The results are shown in Figure 3.9.9.

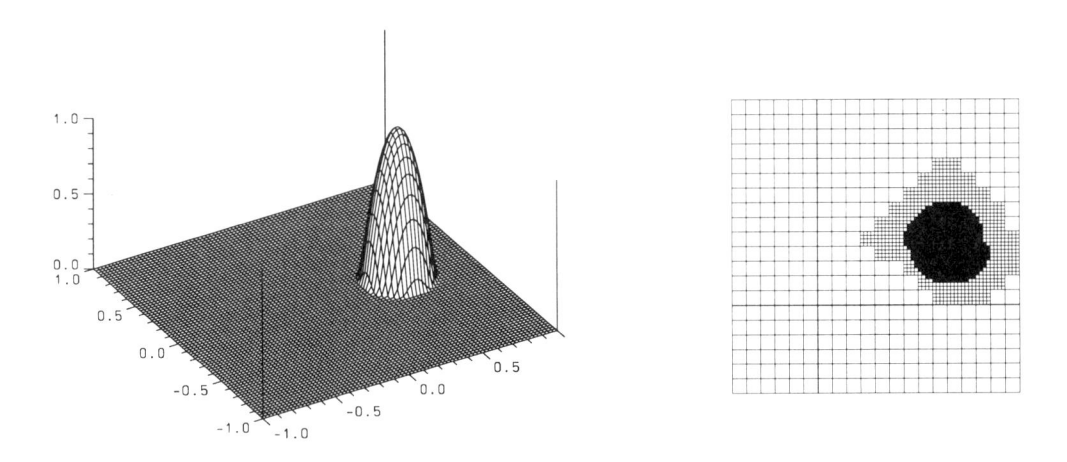

Figure 3.9.3

REMARK 3.9.1 *There are some interesting investigations of estimates like (3.9.4) and (3.9.5) for the nonstationary, incompressible Navier–Stokes equations [105]. The authors have found that the constants in these estimates become very large such that the question arises if there can be any support of these theoretical error estimates for real flow problems.*

REMARK 3.9.2 *Further details of adaptive finite volume and finite element methods can be found in [67] and [158].*

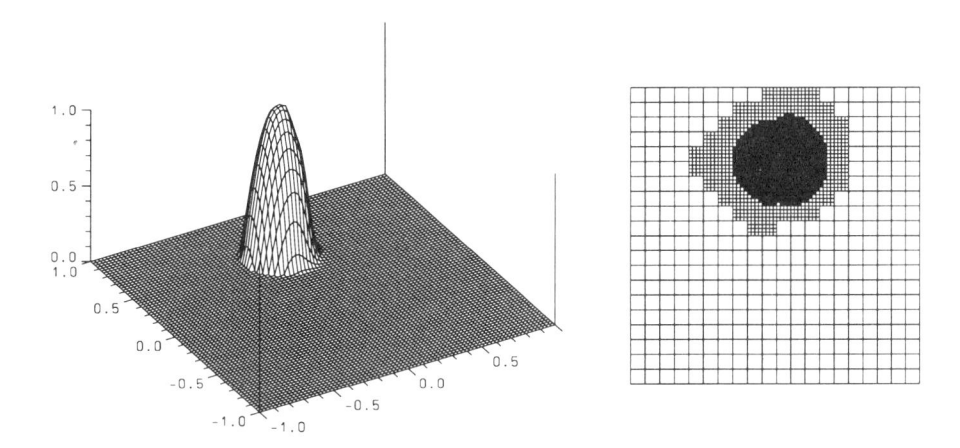

3.9 Adaptivity and local grid alignment: numerical experiments 265

Figure 3.9.4

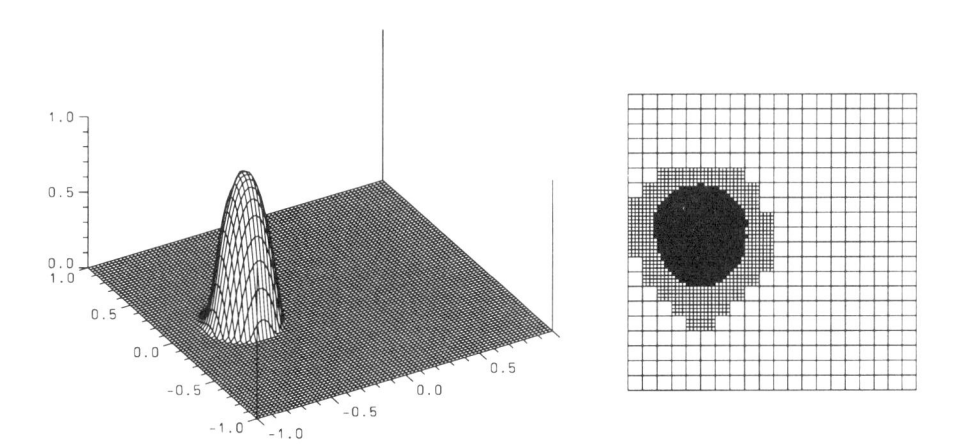

Figure 3.9.5

266 Initial value problems for scalar conservation laws in 2-D

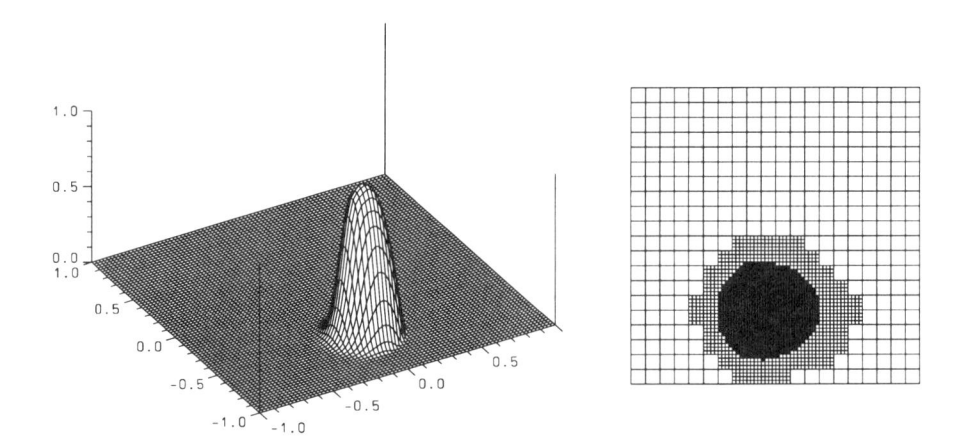

Figure 3.9.6

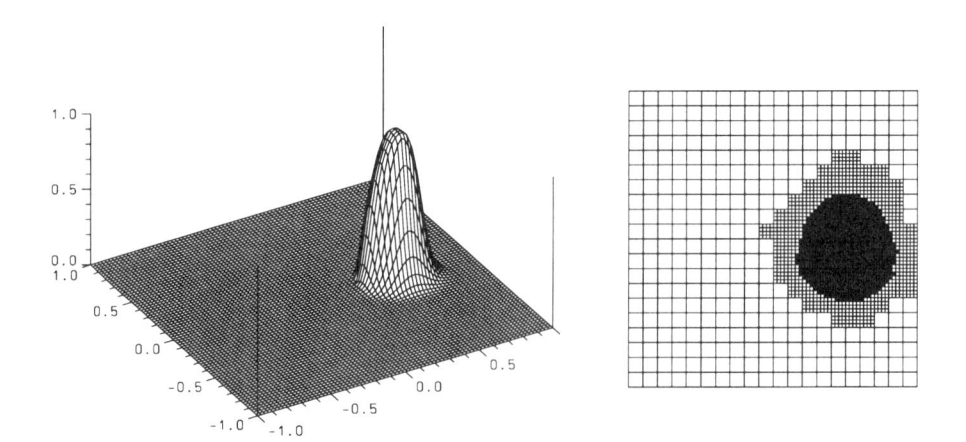

Figure 3.9.7

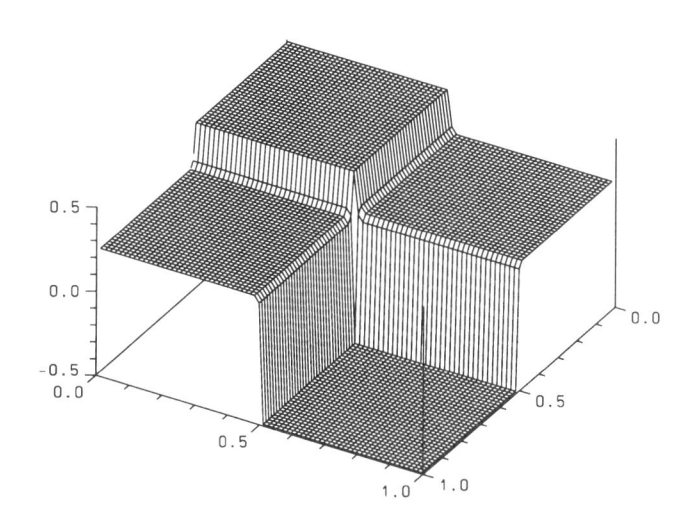

3.9 Adaptivity and local grid alignment: numerical experiments

Figure 3.9.8

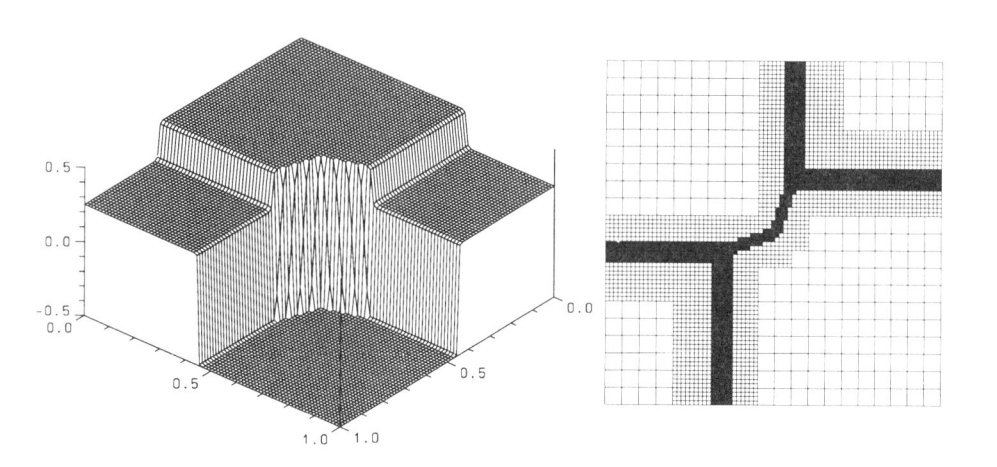

Figure 3.9.9

We shall now consider some numerical experiments concerning grid alignment. It has been found that the resolution of shock fronts on aligned grids is much better than on usually locally refined grids. This will be shown in the following examples.

EXAMPLE 3.9.3 We consider the linear scalar conservation law

$$\partial_t u + \partial_x u + \partial_y u = 0 \quad \text{in} \quad \Omega := [0,1] \times [0,1] \qquad (3.9.13)$$

with initial values

$$u(x,y,0) = 1 \quad \text{if} \quad y = 0, \qquad (3.9.14)$$
$$u(x,y,0) = 0 \quad \text{otherwise}, \qquad (3.9.15)$$

and boundary conditions on the inflow part of the boundary

$$u(x,y,t) = 1 \quad \text{if} \quad y = 0, \qquad (3.9.16)$$
$$u(x,y,t) = 0 \quad \text{otherwise}. \qquad (3.9.17)$$

The algorithm we have used for the discretization is the fluctuation distribution scheme as described in Section 3.8.

The exact stationary solution is given by a straight diagonal line in the unit square, which separates two constant states. The numerical results are shown in Figures 3.9.10 and 3.9.11 respectively on a grid that is aligned and on a grid that is not aligned with the shock. The resolution of the shock on the aligned grid is much better than on the other one. Therefore this example shows that it is necessary to align the triangles in an unstructured grid with the main structures of the solutions.

This effect can be improved even more if we use thin triangles that are aligned with the discontinuities, e.g. shocks. It is necessary to define a mesh indicator or error estimator that is able to do this alignment automatically.

Similar experiments have been done in [24] and [239]. A variational approach for optimal meshes has been studied in [203].

Only a few theoretical results are available. There are some anisotropic interpolation estimates [11, 14], and in [168] the results of [120] were generalized to thin triangles. In [213] error estimates for aligned rectangles for elliptic problems have been shown. In our experiments we have used the

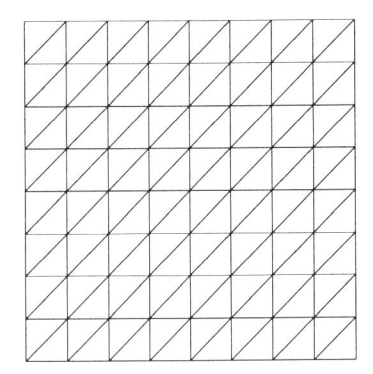

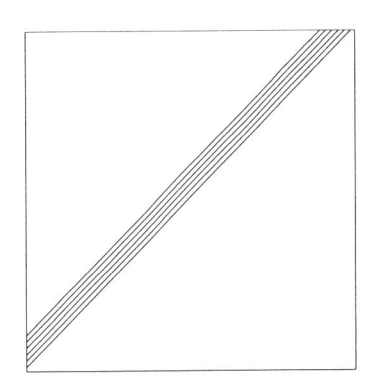

3.9 Adaptivity and local grid alignment: numerical experiments 269

Figure 3.9.10

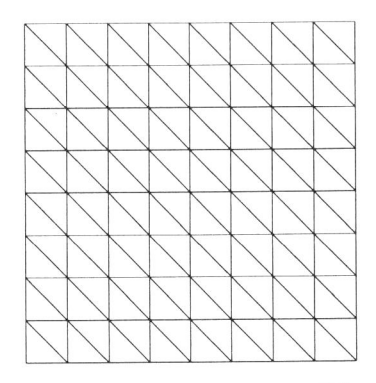

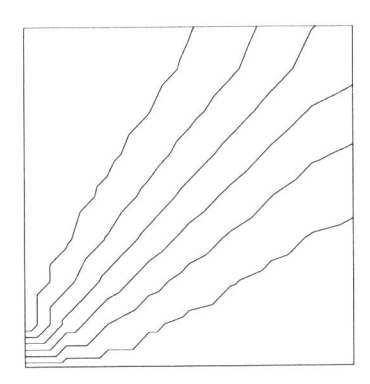

Figure 3.9.11

fluctuation splitting scheme (see [48], [49], [50] and [219], and also §3.8) in combination with the mesh alignment of Kornhuber and Roitzsch [116, 20]. Now we shall explain the numerical scheme and the alignment algorithm in more detail. The algorithm consists of the following steps. First let us describe the alignment of the edges.

(**A1**) Construct a macro triangulation $\mathcal{T}^0$ of Ω.

(**A2**) Construct a globally uniform refined mesh $\mathcal{T}^1$ by refining the macro triangulation $\mathcal{T}^0$.

(**A3**) Compute the discrete solution u_h on $\mathcal{T}^1$ using the algorithm as described in Section 3.8.

(**A4**) For all edges e of all triangles $T \in \mathcal{T}^1$ compute the directional derivative $D(T,e) = \mid m(P_i) - m(P_j) \mid / \|P_i - P_j\|$ of the Mach number m along the edge e. Here e is defined by its endpoints P_j and P_i.

(**A5**) Fix some threshold $\alpha > 0$ and select the set $\mathcal{T}^*$ of all triangles for which we have an edge e such that $\mid D(T,e) \mid \geq \alpha$.

(**A6**) Now consider successively all $T \in \mathcal{T}^*$ and find the edge e_T of T such that

$$\mid D(T,e_T) \mid = \min\Big\{|D(T,e)|\Big| \ e \text{ is an edge of T}\Big\}. \qquad (3.9.18)$$

(**A7**) Consider the local situation as in Figure 3.9.12 and let $T_1, T_2 \in \mathcal{T}^*$. Without restriction we assume $e_{T_1} := \overline{P_1 P_2}$, $e_{T_2} := \overline{Q_1 Q_2}$.
First case: $e_{T_1} \cap e_{T_2} \neq \emptyset$. Without restriction, let $e_{T_1} \cap e_{T_2} = \{P_1\}$. Then move P_2 and Q_2 to $\bar{P}_2$ and $\bar{Q}_2$ respectively such that $\overline{P_1 \bar{P}_2}$ and $\overline{P_1 \bar{Q}_2}$ are nearly parallel to the front of the discontinuity. Then choose another pair of triangles from $\mathcal{T}^*$ (see Figure 3.9.12).
Second case: $e_{T_1} \cap e_{T_2} = \emptyset$. Then choose another pair of triangles from $\mathcal{T}^*$.

(**A8**) Goto (**A2**).

In the examples this type of refinement has been repeated two or three times. Now let us describe the blue refinement and where we apply it.

3.9 Adaptivity and local grid alignment: numerical experiments 271

(**B1**) Use the refinement obtained from (**A1**) and (**A2**) on the whole triangulation. (**B2**) as (**A3**), (**B3**) as (**A4**), (**B4**) as (**A5**), (**B5**) as (**A6**).

(**B6**) Fix a pair T_1, T_2 of triangles.

If $e_{T_1} \cap e_{T_2} \neq \emptyset$ there is no blue refinement; choose another pair of triangles.

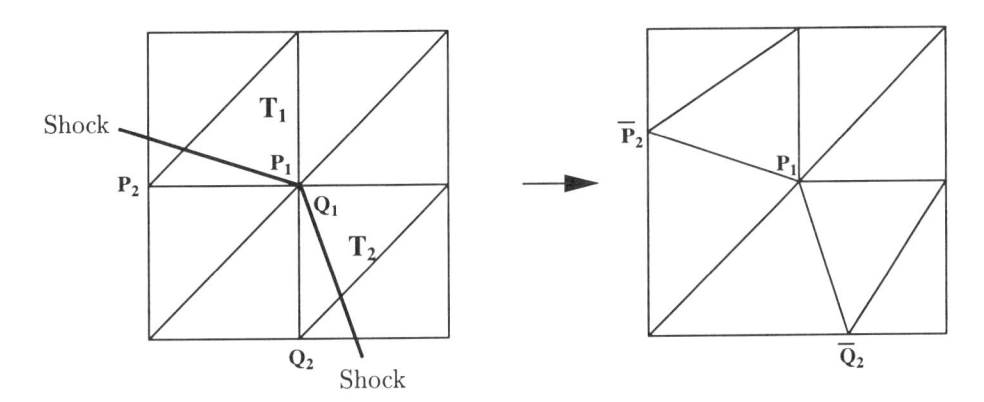

Figure 3.9.12

(**B7**) If
  (a) $e_{T_1} \cap e_{T_2} = \emptyset$ and
  (b) e_{T_1}, e_{T_2} are "nearly" parallel and
  (c) an "angle condition" (see below) is satisfied

then we apply the blue refinement (see Figure 3.9.13); otherwise we choose another pair of triangles.

The angle condition means that (see Figure 3.9.14) if α and β are the largest angles in T_1 and T_2 respectively then γ and δ are the smallest ones in T_1 and T_2 respectively.

This refinement algorithm B is also repeated up to two or three times.

EXAMPLE 3.9.4 The second test problem is to solve

$$\partial_t u + y\partial_x u - x\partial_y u = 0 \quad \text{in} \quad \Omega := [0,1] \times [0,1] \tag{3.9.19}$$

with initial values

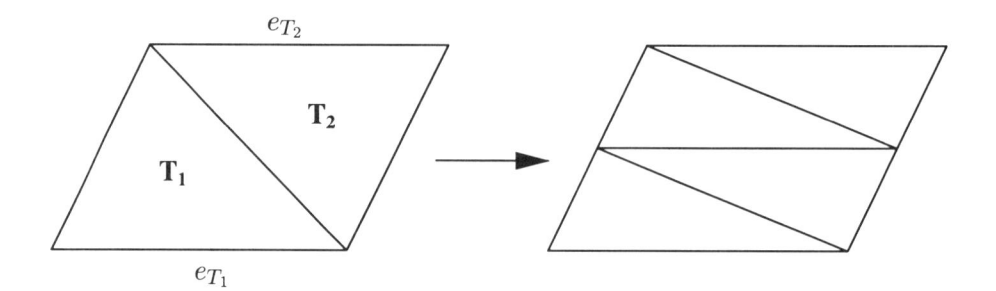

Figure 3.9.13

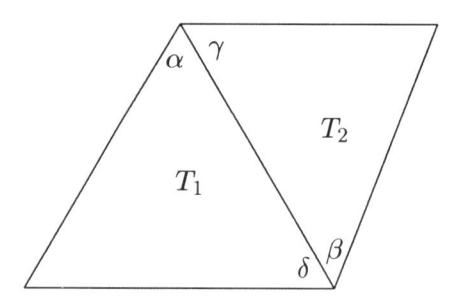

Figure 3.9.14

3.9 Adaptivity and local grid alignment: numerical experiments

$$u(x, y, 0) = 1 \quad \text{if} \quad x = 0, \ y \leq \tfrac{2}{3}, \qquad (3.9.20)$$
$$u(x, y, 0) = 0 \quad \text{otherwise}, \qquad (3.9.21)$$

and the corresponding boundary conditions on the inflow part of the boundary

$$u(x, y, t) = 1 \quad \text{if} \quad x = 0, \ y \leq \tfrac{2}{3}, \qquad (3.9.22)$$
$$u(x, y, t) = 0 \quad \text{otherwise}. \qquad (3.9.23)$$

The exact stationary solution is given by a quarter of a circle, which separates two constant states. The numerical results are shown in Figure 3.9.16 on a locally refined grid and in Figure 3.9.15 on an aligned grid.

REMARK 3.9.5 *For theoretical results concerning anisotropic finite elements we refer to [12] and [13].*

274 Initial value problems for scalar conservation laws in 2-D

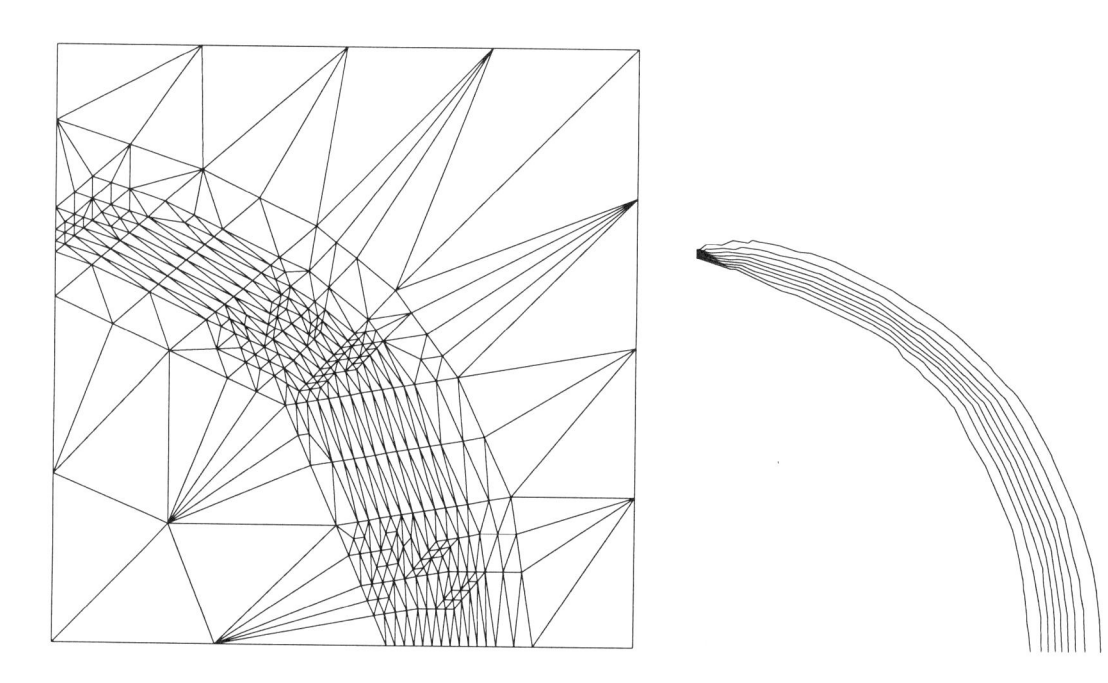

Figure 3.9.15

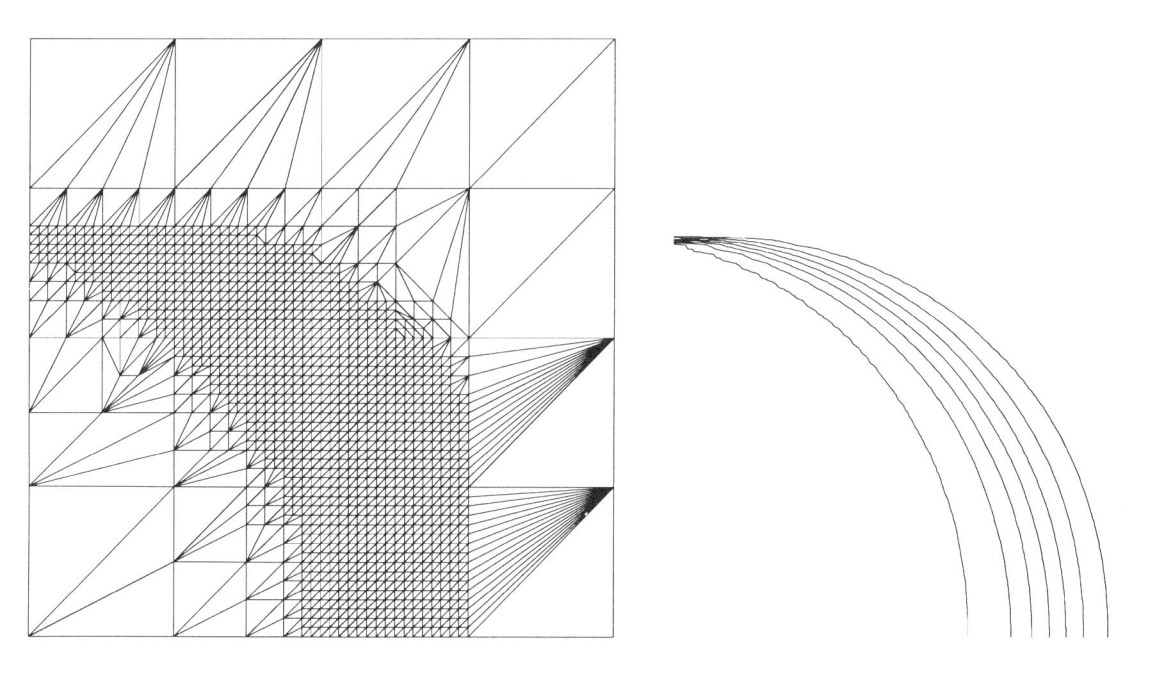

3.9 Adaptivity and local grid alignment: numerical experiments 275

Figure 3.9.16

4 Initial value problems for systems of conservation laws in 1-D

In this chapter we consider the initial value problem for systems of conservation laws in 1-D, in particular for the Euler equations of gas dynamics. First we present some theoretical tools and results that are necessary for obtaining the exact solution of the Riemann problem. This is important because then a special exact solution is available for the validation of the numerical schemes. The most important and most interesting numerical algorithms for solving the Euler equations in 1-D are presented in the remaining part of this chapter.

4.1 Basic results for systems in 1-D

The main theoretical tools that are necessary to solve the general Riemann problem for the one-dimensional Euler equations of gas dynamics will be discussed (see [218]). We consider Riemann invariants, different entropy conditions, k-shocks and k-rarefaction waves and contact discontinuities. Finally we prove the solvability of the general Riemann problem as in [218].

In Chapter 1 we derived the system of Euler equations for an ideal gas in 2-D (see [218]). In conservative variables they are

$$\begin{aligned} \partial_t \rho + \mathrm{div}(\rho \vec{u}) &= 0 \;, \\ \partial_t(\rho u) + \partial_x(\rho u^2 + p) + \partial_y(uv\rho) &= 0 \;, \\ \partial_t(\rho v) + \partial_x(uv\rho) + \partial_y(\rho v^2 + p) &= 0 \;, \\ \partial_t e + \partial_x[u(e+p)] + \partial_y[v(e+p)] &= 0 \;. \end{aligned} \quad (4.1.1)$$

4.1 Basic results for systems in 1-D

Additionally we have to consider the following equation of state:

$$p = (\gamma - 1)\left[e - \frac{\rho}{2}(u^2 + v^2)\right].$$

Here we have used the following notation: p is the pressure, ρ is the density, u and v are the velocity components in the x- and y-directions respectively and e is the total energy. If ε denotes the specific internal energy, we get

$$e = \frac{\rho}{2}\vec{u}^2 + \rho\varepsilon \quad \text{for } \vec{u} = (u, v). \tag{4.1.2}$$

In the following part of this section we shall restrict ourselves to systems in 1-D:

$$\partial_t u(x,t) + \partial_x f(u(x,t)) = 0 \tag{4.1.3}$$

for $t \in [0, T]$, $x \in \mathbb{R}$, $u(x,t) \in \mathbb{R}^m$. In order to be sure that (4.1.3) is a well-posed system, we need some conditions for f. These conditions can be derived even for linear systems

$$\partial_t u + A\partial_x u = 0. \tag{4.1.4}$$

Let us consider special solutions in the form of travelling waves

$$u(x,t) = \xi e^{i(\lambda t + \mu x)} \tag{4.1.5}$$

where $\xi = \text{const} \in \mathbb{R}^m$ and $\lambda, \mu \in \mathbb{C}$. The initial values $u(x,0) = \xi e^{i\mu x}$ are assumed to be bounded. This implies that

$$\text{Im}\,\mu = 0. \tag{4.1.6}$$

Using (4.1.5) in (4.1.4), we obtain

$$(i\lambda + Ai\mu)\xi = 0.$$

Therefore $a + ib := -\lambda/\mu$ with $a, b \in \mathbb{R}$ is an eigenvalue for A. Using a and b instead of λ, we obtain from (4.1.5)

$$u(x,t) = \xi e^{i\mu(x-at)} e^{\mu bt}.$$

Since (4.1.4) should be well posed, u should depend continuously on the initial values $u(x,0)$. This implies that $b = 0$. This can be seen as follows. Assume that $b \neq 0$ and note that $v = 0$ is a solution of (4.1.4). Now we choose $\xi \in \mathbb{R}^m$ such that u as defined in (4.1.5) satisfies

$$\sup_{x>0} |u(x,0) - v(x,0)| < \delta$$

for a given small δ, and then choose $t_0 > 0$ such that

$$\sup_{x>0} |u(x,t_0) - v(x,t_0)| = 1 \ .$$

This can be obtained if $|\xi| = N^{-1}$ for some $N \in \mathbb{N}$, $\mu := N/b \log N$ and λ such that $-\lambda/\mu$ is an eigenvalue of A. Then we consider u for fixed N, μ and λ, and obtain

$$|u(x,0) - v(x,0)| = |u(x,0)| = N^{-1} \ ,$$

$$\left| u\left(x, \frac{1}{N}\right) - v\left(x, \frac{1}{N}\right) \right| = \left| u\left(x, \frac{1}{N}\right) \right| = N^{-1} N = 1 \ .$$

This means that u cannot depend continuously on the initial data. In order to avoid this kind of problem, we shall only consider hyperbolic systems, which are defined as follows.

DEFINITION 4.1.1 (Hyperbolic systems) *The system*

$$\partial_t u + \partial_x f(u) = 0$$

is called (strictly) hyperbolic if $f \in C^1(\mathbb{R}^m, \mathbb{R}^m)$ and $Df(u)$ has only real (distinct real) eigenvalues for all $u \in \mathbb{R}^m$.

EXAMPLE 4.1.2 (Euler equations in 1-D) Consider the system of Euler equations in 1-D for conservative variables

$$\begin{aligned} \partial_t \rho + \partial_x (\rho u) &= 0 \ , \\ \partial_t (\rho u) + \partial_x (\rho u^2 + p) &= 0 \ , \\ \partial_t e + \partial_x (u(e+p)) &= 0 \ , \end{aligned} \qquad (4.1.7)$$

4.1 Basic results for systems in 1-D

and the corresponding equation of state

$$p = (\gamma - 1)\left(e - \frac{\rho}{2}u^2\right) \qquad (\gamma > 1) . \tag{4.1.8}$$

In primitive variables (4.1.7) takes the form

$$\partial_t \rho + \partial_x(\rho u) = 0 , \tag{4.1.9}$$

$$\partial_t u + u \partial_x u + \frac{\partial_x p}{\rho} = 0 , \tag{4.1.10}$$

$$\partial_t p + \rho a^2 \partial_x u + u \partial_x p = 0 , \tag{4.1.11}$$

where $a := \sqrt{\gamma p/\rho}$. Here we have assumed that the functions ρ, u and p are sufficiently smooth and that $p, \rho > 0$. Equation (4.1.10) follows from (4.1.7) on using

$$0 = \rho \partial_t u + u \partial_t \rho + \partial_x(\rho u^2 + p) = \rho \partial_t u + \rho u \partial_x u + \partial_x p .$$

To derive (4.1.11), we differentiate

$$e = \frac{p}{\gamma - 1} + \frac{\rho}{2}u^2$$

with respect to t and

$$u(p + e) = u\left(\frac{p\gamma}{\gamma - 1} + \frac{\rho u^2}{2}\right)$$

with respect to x. We obtain, using (4.1.7) and (4.1.10),

$$\partial_t e = \frac{1}{\gamma - 1}\partial_t p - \frac{1}{2}(u^3 \partial_x \rho + 3\rho u^2 \partial_x u + 2u \partial_x p) ,$$

$$\partial_x[u(p + e)] = \frac{\gamma}{\gamma - 1}(p \partial_x u + u \partial_x p) + \frac{1}{2}\partial_x(\rho u^3) .$$

Now we add the two equations and obtain (4.1.11). Then (4.1.9)–(4.1.11) can be written in the form

$$\partial_t U + A(U)\partial_x U = 0 , \tag{4.1.12}$$

where $U := (\rho, u, p)^t$ and

$$A(U) := \begin{pmatrix} u & \rho & 0 \\ 0 & u & 1/\rho \\ 0 & \rho a^2 & u \end{pmatrix} . \qquad (4.1.13)$$

The eigenvalues of $A(U)$ are

$$u - a , \quad u , \quad u + a ,$$

and they are real. The corresponding eigenvectors are

$$\begin{pmatrix} \rho \\ -a \\ \rho a^2 \end{pmatrix} , \quad \begin{pmatrix} 1 \\ 0 \\ 0 \end{pmatrix} , \quad \begin{pmatrix} \rho \\ a \\ \rho a^2 \end{pmatrix} .$$

This means that the system (4.1.12) is hyperbolic. If $a \neq 0$, the system (4.1.12) is even strictly hyperbolic.

REMARK 4.1.3 *In the following we shall use u as the x-component of the velocity vector in $\mathbb{R}^2$ and U or u as the whole solution vector for systems. But there will be no confusion, since the exact meaning of u will be obvious from the context.*

REMARK 4.1.4 *It can be shown that (4.1.11) is equivalent to*

$$\partial_t S + u \partial_x S = 0 , \qquad (4.1.14)$$

where

$$S := S_0 + c_v \log \frac{p}{\rho^\gamma (\gamma - 1)} \qquad (4.1.15)$$

is the entropy, with constant S_0 and c_v. This can be seen if we use

$$\partial_t S = c_v \left(\frac{\partial_t p}{p} - \gamma \frac{\partial_t \rho}{\rho} \right) \quad \text{and} \quad \partial_x S = c_v \left(\frac{\partial_x p}{p} - \gamma \frac{\partial_x \rho}{\rho} \right)$$

in (4.1.11); that is,

$$\partial_t p + \rho a^2 \partial_x u + u \partial_x p = p \left(\frac{\partial_t p}{p} + \frac{u \partial_x p}{p} - \gamma \frac{\partial_t \rho}{\rho} - \gamma \frac{u \partial_x \rho}{\rho} \right) .$$

In § 2.1 we have seen that the solution of scalar conservation laws can become singular within finite time. The same problem can arise for systems of conservation laws, and therefore we have to consider solutions in the distributional sense.

DEFINITION 4.1.5 (Solutions in the distributional sense) *Let $u_0 \in L^1_{loc}(\mathbb{R}, \mathbb{R}^m) \cap L^\infty(\mathbb{R}, \mathbb{R}^m)$. Then $u \in L^1_{loc}(\mathbb{R} \times \mathbb{R}^+, \mathbb{R}^m)$ is called a solution of*

$$\partial_t u + \partial_x f(u) = 0 \quad \text{on } \mathbb{R} \times \mathbb{R}^+ ,$$
$$u(x,0) = u_0(x) \quad \text{on } \mathbb{R}$$

in the distributional sense if we have

$$\int_\mathbb{R} \int_{\mathbb{R}^+} [u \partial_t \varphi + f(u) \partial_x \varphi] + \int_\mathbb{R} \varphi(\cdot, 0) u_0 = 0 \quad (4.1.16)$$

for all test functions $\varphi \in C_0^\infty(\mathbb{R} \times [0, \infty[)$.

LEMMA 4.1.6 (Rankine–Hugoniot condition) *Let us assume that the half-space $\mathbb{R} \times \mathbb{R}^+$ is separated by a smooth curve $S : t \to (\sigma(t), (t))$ into two parts M_l and M_r. Furthermore, let $u \in L^1_{loc}(\mathbb{R} \times \mathbb{R}^+, \mathbb{R}^m)$ such that $u_l := u|_{M_l} \in C^1(\overline{M_l}, \mathbb{R}^m)$ and $u_r := u|_{M_r} \in C^1(\overline{M_r}, \mathbb{R}^m)$, and u_l and u_r are classical solutions of*

$$\partial_t u + \partial_x f(u) = 0$$

locally in M_l and M_r respectively. Then (4.1.16) holds in the distributional sense if and only if

$$\sigma'(t)[u_l(\sigma(t), t) - u_r(\sigma(t), t)] = f(u_l(\sigma(t), t)) - f(u_r(\sigma(t), t))$$

for all $t > 0$, or in a short form,

$$s(u_l - u_r) = f_l - f_r , \quad (4.1.17)$$

where $s = \sigma'$ is the velocity of S.

282   4 Initial value problems for systems in 1-D

For the proof see Lemma 2.1.4.

Now we shall derive the Rankine–Hugoniot conditions for the Euler equations. We consider the system of Euler equations (4.1.7) using the equation of state:

$$\partial_t \rho + \partial_x(\rho u) = 0 ,$$
$$\partial_t (\rho u) + \partial_x(p + \rho u^2) = 0 , \qquad (4.1.18)$$
$$\partial_t \left(\frac{p}{\gamma-1} + \frac{\rho}{2}u^2 \right) + \partial_x \left[u \left(\frac{p}{\gamma-1} + \frac{\rho}{2}u^2 + p \right) \right] = 0 .$$

LEMMA 4.1.7 (Explicit form of the jump conditions) *Let (ρ, u, p) be a solution of (4.1.18) in the distributional sense and let Γ be a smooth curve in $\mathbb{R} \times \mathbb{R}^+$, $z_0 \in \Gamma$ and V be an open neighbourhood of z_0. Assume that Γ separates V into two parts V_1 and V_2 and that (ρ, u, p) are smooth on $\overline{V_1}$ and $\overline{V_2}$. Let (ρ_i, u_i, p_i), $i = 1, 2$, be the limits of (ρ, u, p) in z_0 with respect to V_1 and V_2. We assume that ρ_1, p_1 and $u_1 - s \neq 0$. Set $\pi := p_2/p_1$, $\rho_1 \neq \rho_2$, $\beta := (\gamma+1)/(\gamma-1)$. Then we have*

$$\frac{\rho_2}{\rho_1} = \frac{1 + \beta\pi}{\pi + \beta} \quad \text{and} \quad \frac{u_2 - u_1}{a_1} = \sqrt{\frac{2}{\gamma(\gamma-1)}} \frac{1-\pi}{\sqrt{1+\pi\beta}} \operatorname{sign}(u_1 - s) , \qquad (4.1.19)$$

where s is according to (4.1.17).

Proof For the jump relations of (4.1.18) as defined in (4.1.17) we have

$$s[\rho] = [\rho u] ,$$
$$s[\rho u] = [p + \rho u^2] , \qquad (4.1.20)$$
$$s\left[\frac{p}{\gamma-1} + \frac{\rho}{2}u^2\right] = \left[u\left(\frac{p}{\gamma-1} + \frac{\rho}{2}u^2 + p\right)\right] ,$$

where here [ ] denotes the jump across Γ, e.g. $[p] = p_2 - p_1$, and s is the velocity of Γ in z_0. Using the notation $v := u - s$ and $m = \rho v$, we get

$$s[\rho] = \rho_2 u_2 - \rho_1 u_1 = \rho_2(v_2 + s) - \rho_1(v_1 + s) = m_2 - m_1 + [\rho]s$$

and therefore

$$[m] = 0 . \qquad (4.1.21)$$

4.1 Basic results for systems in 1-D   283

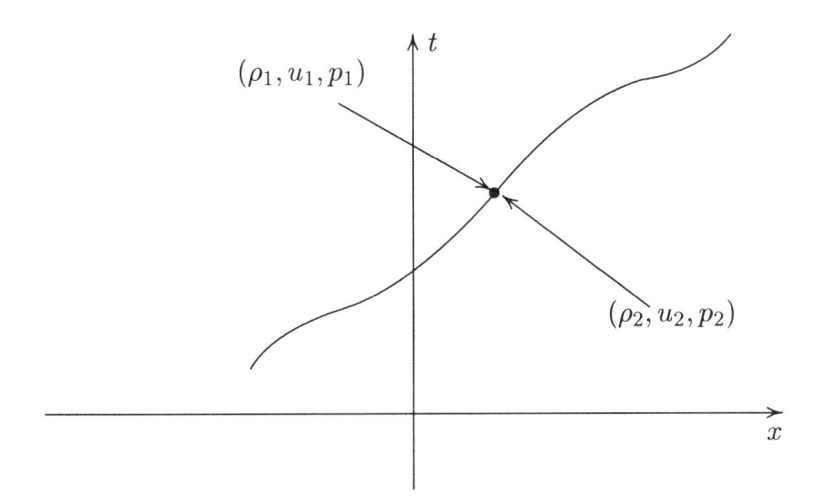

Figure 4.1.1

Similarly we obtain

$$[p + mv] = 0, \qquad (4.1.22)$$

$$m_2 \left[\frac{2}{\gamma - 1} a^2 + v^2 \right] = 0, \qquad (4.1.23)$$

where $a = \sqrt{\gamma p/\rho}$. For $z := \rho_2/\rho_1$ and $\pi = p_2/p_1$ the following equations are obvious:

$$\left(\frac{a_2}{a_1} \right)^2 = \frac{\gamma p_2 \rho_1}{\gamma p_1 \rho_2} = \frac{\pi}{z}, \qquad (4.1.24)$$

$$\frac{1}{z} = \frac{\rho_1}{\rho_2} = \frac{v_2}{v_1} \frac{\rho_1 v_1}{\rho_2 v_2} = \frac{v_2}{v_1} \left(\frac{\rho_1 v_1 - \rho_2 v_2}{\rho_2 v_2} + 1 \right) = \frac{v_2}{v_1}, \qquad (4.1.25)$$

since $\rho_1 v_1 - \rho_2 v_2 = [m] = 0$. Using (4.1.25) in (4.1.23), we derive

$$\frac{2}{\gamma - 1} a_1^2 + v_1^2 = \frac{2}{\gamma - 1} a_2^2 + v_2^2 = \frac{2}{\gamma - 1} \frac{a_2^2}{a_1^2} a_1^2 + \frac{v_2^2}{v_1^2} v_1^2$$

$$= \frac{2}{\gamma - 1} \frac{\pi}{z} a_1^2 + \frac{v_1^2}{z^2},$$

284   4 Initial value problems for systems in 1-D

and therefore
$$\frac{2}{\gamma-1}a_1^2\left(1-\frac{\pi}{z}\right) = v_1^2\left(\frac{1}{z^2}-1\right)$$

and
$$\frac{v_1^2}{a_1^2} = \frac{2z}{\gamma-1}\frac{z-\pi}{1-z^2}. \qquad (4.1.26)$$

Now we use the jump relation (4.1.22) and $p = a^2\rho/\gamma$ in order to get
$$\frac{a_1^2\rho_1}{\gamma} + \rho_1 v_1^2 = \frac{a_2^2\rho_2}{\gamma} + \rho_2 v_2^2 = \frac{a_2^2\rho_2}{\gamma\rho_1}\rho_1 + \frac{\rho_2}{\rho_1}v_2^2\rho_1$$
$$= \left(\frac{a_2^2}{\gamma} + v_2^2\right)z\rho_1,$$

and furthermore (4.1.24) and (4.1.25) imply
$$\frac{a_1^2}{\gamma} + v_1^2 = \left(\frac{a_2^2}{\gamma} + v_2^2\right)z = \left(\frac{a_2^2}{\gamma a_1^2}a_1^2 + \frac{v_2^2}{v_1^2}v_1^2\right)z = \left(\frac{\pi}{z\gamma}a_1^2 + \frac{v_1^2}{z^2}\right)z.$$

Therefore
$$\frac{a_1^2}{\gamma}(1-\pi) = v_1^2\left(\frac{1}{z}-1\right)$$

and
$$\frac{v_1^2}{a_1^2} = \frac{z}{\gamma}\frac{1-\pi}{1-z}. \qquad (4.1.27)$$

From (4.1.26) and (4.1.27) we obtain
$$z = \frac{1+\beta\pi}{\pi+\beta}. \qquad (4.1.28)$$

This implies the first part of (4.1.19). Now we use this expression for z in (4.1.26) and obtain
$$\frac{v_1^2}{a_1^2} = \frac{2}{\gamma-1}(1+\pi\beta)\frac{1+\pi\beta-\pi(\pi+\beta)}{(\pi+\beta)^2-(1+\pi\beta)^2} = \frac{2}{\gamma-1}\frac{1+\pi\beta}{\beta^2-1},$$

and, since $a_1 \geq 0$,

$$\frac{v_1}{a_1} = \sqrt{\frac{2}{\gamma-1}\frac{1+\pi\beta}{\beta^2-1}}\,\text{sign}\,v_1\,. \tag{4.1.29}$$

It is easy to verify the following relations:

$$s = u_1 - v_1\,,\quad u_2 - s = v_2 = \frac{v_2}{v_1}v_1 = \frac{u_1 - s}{z} \quad \text{(see (4.1.25))}$$

$$\frac{1}{z} - 1 = \frac{(\pi-1)(1-\beta)}{1+\beta\pi}\,,\quad 1-\beta = -\frac{2}{\gamma-1}\,.$$

Then we have

$$u_2 - u_1 = u_2 - s + s - u_1 = \frac{u_1 - s}{z} + s - u_1$$

$$= \left(\frac{1}{z}-1\right)(u_1 - s) = \left(\frac{1}{z}-1\right)v_1$$

$$= \frac{(\pi-1)(1-\beta)}{1+\beta\pi}a_1\sqrt{\frac{2}{\gamma-1}\frac{1+\pi\beta}{\beta^2-1}}\,\text{sign}\,v_1\,,$$

and finally

$$u_2 - u_1 = \frac{1-\pi}{\sqrt{1+\pi\beta}}a_1\sqrt{\frac{2}{\gamma(\gamma-1)}}\,\text{sign}\,v_1\,. \tag{4.1.30}$$

This proves the statement of Lemma 4.1.7. □

Now we have to define a k-characteristic. First let us consider it for a linear hyperbolic system

$$\partial_t U + A\partial_x U = 0\,. \tag{4.1.31}$$

If we assume that (4.1.31) is strictly hyperbolic then a regular matrix T exists such that

$$T^{-1}AT = D\,,$$

where D is a diagonal matrix

$$D = \begin{pmatrix} \lambda_1 & & \\ & \ddots & \\ & & \lambda_m \end{pmatrix}$$

and $\lambda_1, \ldots, \lambda_m$ are the eigenvalues of A. Let $w := T^{-1}U$. Then

$$0 = \partial_t U + A\partial_x U = T\partial_t w + AT\partial_x w ,$$

and therefore

$$\partial_t w + D\partial_x w = 0 ,$$

or

$$\partial_t w_i + \lambda_i \partial_x w_i = 0 , \quad i = 1, \ldots, m . \tag{4.1.32}$$

Then the characteristics of (4.1.32) are given by

$$\gamma_i'(t) = \lambda_i , \quad i = 1, \ldots, m . \tag{4.1.33}$$

EXAMPLE 4.1.8 (Diagonal form of the Euler equations) The system of Euler equations in primitive variables (4.1.9)–(4.1.11) can be written in the form

$$\partial_t U + A(U)\partial_x U = 0 , \tag{4.1.34}$$

where $A(U)$ is given in (4.1.13). Now we freeze the coefficient $A(U)$ in (4.1.34) and treat (4.1.34) as a linear system:

$$\partial_t U + A_0 \partial_x U = 0 .$$

Then for

$$T := \begin{pmatrix} \rho & 1 & \rho \\ -a & 0 & a \\ \rho a^2 & 0 & \rho a^2 \end{pmatrix}$$

we have

$$T^{-1}AT = D = \begin{pmatrix} u-a & & \\ & u & \\ & & u+a \end{pmatrix} .$$

Now we use (4.1.33) as a motivation for defining characteristics in the nonlinear case.

DEFINITION 4.1.9 (k-characteristic) *Let u be a smooth solution of the hyperbolic system (4.1.3) and let $\lambda_1(u) \leq \lambda_2(u) \leq \ldots \leq \lambda_m(u)$ be the eigenvalues of $Df(u)$. Assume that there exists a solution $\gamma \in C^1([0,\tau])$ of*

$$\gamma'(t) = \lambda_k(u(\gamma(t), t)), \quad 0 \leq t \leq \tau .$$

Then the curve $\{(\gamma(t), t) \mid 0 \leq t \leq \tau\}$ is called a k-characteristic.

In general, we cannot expect that the solutions are constant along the k-characteristics. But we can show that $w(u)$ is constant along k-characteristics for suitable functions w. Now let us derive necessary conditions for w.

LEMMA 4.1.10 *Let u be a smooth solution of (4.1.3) and let $(\gamma(t), t)$ be a k-characteristic with respect to $\lambda = \lambda_k$,*

$$\gamma'(t) = \lambda(u(\gamma(t), t)) ,$$

and let $w \in C^1(\mathbb{R}^m, \mathbb{R})$ such that

$$Df(u)^t \cdot \nabla w = \lambda \nabla w .$$

Then

$$\frac{d}{dt} w(u(\gamma(t), t)) = 0 .$$

In this case w is called a **Riemann invariant**.

Proof We have

$$\begin{aligned}
\frac{d}{dt}w(u(\gamma(t)),t) &= \sum_{l=1}^{m}[\partial_x u_l \gamma'(t) + \partial_t u_l]\partial_l w \\
&= \sum_{l=1}^{m}[\partial_x u_l \lambda - (Df(u)\partial_x u)_l]\partial_l w \\
&= \sum_{l=1}^{m}[\lambda \partial_x u_l - \sum_j \partial_j f_l(u)\partial_x u_j]\partial_l w \\
&= \lambda \sum_{j=1}^{m}\partial_x u_j \partial_j w - \sum_j \Big[\sum_l \partial_j f_l(u)\partial_l w\Big]\partial_x u_j \\
&= \sum_{j=1}^{m}\Big[\lambda \partial_j w - \sum_l \partial_j f_l(u)\partial_l w\Big]\partial_x u_j \ .
\end{aligned}$$

This term is equal to zero if

$$\lambda \nabla w = Df(u)^t \cdot \nabla w \ ,$$

and this proves the lemma. □

Now let r be a right eigenvector of $Df(u)$ with eigenvalue μ: $Df(u)r = \mu r$. This implies that

$$r^t \lambda \nabla w = r^t Df(u)^t \nabla w = \mu r^t \nabla w$$

Therefore if $\mu \neq \lambda$, we have

$$r^t \cdot \nabla w = 0 \ .$$

If we denote the right eigenvectors and eigenvalues of $Df(u)$ by $r_1, \ldots, r_m$ and $\lambda_1, \ldots, \lambda_m$ respectively, we get

$$r_j^t \cdot \nabla w = 0 \ .$$

for all j with $\lambda_j \neq \lambda$. Now we shall give a definition that is slightly more general.

DEFINITION 4.1.11 (k-Riemann invariant, Riemann invariant) *A function $w \in C^1(\mathbb{R}^m, \mathbb{R})$ is called a k-Riemann invariant if*

$$r_k^t \cdot \nabla w = 0$$

where r_k is the kth right eigenvector of $Df(u)$. w is called a Riemann invariant (with respect to λ) if

$$Df(u)^t \cdot \nabla w = \lambda_k \nabla w \ . \tag{4.1.35}$$

THEOREM 4.1.12 (Linearly independent Riemann invariants) *There are $(m-1)$ k-Riemann invariants $w_1, \ldots, w_{m-1} \in C^1(\mathbb{R}^m, \mathbb{R})$ such that the gradients $\nabla w_1, \ldots, \nabla w_{m-1}$ are linearly independent.*

Proof Let k be fixed, r_k as in Definition 4.1.11, $r_k =: (r_k^1, \ldots, r_k^m)$ and for $w \in C^1(\mathbb{R}^m, \mathbb{R})$

$$R(w) := r_k^t \cdot \nabla w = \sum_{i=1}^m r_k^i \, \partial_i w \ .$$

Choose a regular matrix $A \in \mathbb{R}^{m \times m}$ such that $(Ar_k)_i = \delta_{i1}$. Defining

$$g(y) := w(z) \ ,$$

where $z = A^{-1}y$, we obtain $\partial_{z_i} w(z) = \sum_{l=1}^m \partial_{y_l} g(y) A_{li}$, and therefore

$$\begin{aligned} R(w)(z) &= \sum_{i=1}^m r_k^i \, \partial_{z_i} w(z) = \sum_i \sum_l r_k^i \, \partial_{y_l} g(y) \, A_{li} \\ &= \sum_l \partial_{y_l} g(y) \sum_i r_k^i A_{li} = \sum_l \partial_{y_l} g(y) \delta_{l1} = \partial_{y_1} g(y) \ . \end{aligned}$$

If we define for $j = 1, \ldots, m-1$ the functions $w_j(z) := (Az)_{j+1}$ and $g_j(y) := y_{j+1}$ we have $g_j(y) = y_{j+1} = (Az)_{j+1} = w_j(z)$, and therefore as before

$$R(w_j)(z) = \partial_{y_1} g_j(y) = 0 \ .$$

This means $w_j, j = 1, \ldots, m-1$ are k-Riemann invariants, and since $\partial_{z_i} w_j = \sum_{l=1}^m \delta_{l(j+1)} A_{li} = A_{j+1,i}$, we have

$$\nabla w_j = A_{j+1}$$

for $j = 1, \ldots, m-1$. Because A is regular, $\nabla w_1, \ldots, \nabla w_{m-1}$ are linearly independent. $\square$

EXAMPLE 4.1.13 (Isentropic gas) If the entropy $S = S_0 + c_v \log \frac{p}{\rho^\gamma(\gamma-1)}$ (see (4.1.15)) is constant, the gas is called isentropic. It follows from (4.1.15) that $p = \text{const } \rho^\gamma$. In this case the Euler equations reduce to (see (4.1.9) and (4.1.10))

$$\partial_t \rho + \partial_x(u\rho) = 0 ,$$
$$\partial_t u + u\partial_x u + \frac{\partial_x p}{\rho} = 0 . \tag{4.1.36}$$

This system can also be written in the form

$$\partial_t \begin{pmatrix} \rho \\ u \end{pmatrix} + A \partial_x \begin{pmatrix} \rho \\ u \end{pmatrix} = 0$$

where

$$A = \begin{pmatrix} u & \rho \\ \frac{a^2}{\rho} & u \end{pmatrix} .$$

The eigenvalues and right eigenvectors of A are $u + a$, $u - a$ and

$$r_1 = \begin{pmatrix} 1 \\ \frac{a}{\rho} \end{pmatrix} \quad r_2 = \begin{pmatrix} -1 \\ \frac{a}{\rho} \end{pmatrix}$$

respectively. Then the characteristics $C_{1/2}$ are defined by

$$C_{1/2} : \gamma'(t) = u(\gamma(t), t) \pm a(\gamma(t), t) .$$

In order to get the Riemann invariants, we have to solve (see (4.1.35))

$$A^t \tilde{w} = (u \pm a) \tilde{w} .$$

We obtain

$$\tilde{w}_1 = \begin{pmatrix} \frac{a}{\rho} \\ 1 \end{pmatrix}, \quad \tilde{w}_2 = \begin{pmatrix} -\frac{a}{\rho} \\ 1 \end{pmatrix} .$$

Then the Riemann invariants w_1 and w_2 are given by $\nabla w_i = \tilde{w}_i$, $i = 1, 2$, i.e.

$$w_{1/2}(\rho, u) = u \pm \int_0^\rho \frac{a(s)}{s} ds + c_{1/2} .$$

for some constants c_1 and c_2. Using Lemma 4.1.10, we know that $w_1(\rho, u)$, $w_2(\rho, u)$ are constant along the characteristics $C_{1/2}$.

4.1 Basic results for systems in 1-D

EXAMPLE 4.1.14 (Euler equations) Let us consider the Euler equations (see (4.1.9), (4.1.10) and (4.1.14))

$$\partial_t \rho + \partial_x(\rho u) = 0,$$
$$\partial_t u + u\partial_x u + \frac{\partial_x p}{\rho} = 0, \quad (4.1.37)$$
$$\partial_t S + u\partial_x S = 0.$$

This system can be written in the form

$$\partial_t U + A(U)\partial_x U = 0,$$

where $U = (\rho, u, S)^t$ and

$$A(U) = \begin{pmatrix} u & \rho & 0 \\ \frac{\gamma p}{\rho^2} & u & \frac{p}{\rho c_v} \\ 0 & 0 & u \end{pmatrix}. \quad (4.1.38)$$

This follows from

$$\frac{p}{\rho^2}\gamma\partial_x\rho + \frac{p}{\rho c_v}\partial_x S = \frac{\partial_x p}{\rho}.$$

The eigenvalues and eigenvectors of $A(U)$ are

$$u - a, \quad u, \quad u + a$$

and

$$\begin{pmatrix} \rho \\ -a \\ 0 \end{pmatrix}, \quad \begin{pmatrix} 1 \\ 0 \\ -\frac{c_v \gamma}{p} \end{pmatrix}, \quad \begin{pmatrix} \rho \\ a \\ 0 \end{pmatrix}.$$

Then it is easy to see that

$$\left.\begin{array}{lll} S, u + \frac{2}{\gamma-1}a & \text{are} & 1\text{-} \\ u, p & \text{are} & 2\text{-} \\ S, u - \frac{2}{\gamma-1}a & \text{are} & 3\text{-} \end{array}\right\} \text{Riemann invariants.}$$

Furthermore, we can show that S is also a Riemann invariant, since

$$A(U)^t \nabla S = u \cdot \nabla S.$$

4 Initial value problems for systems in 1-D

DEFINITION 4.1.15 (k-Rarefaction waves) *Let $D \subset \mathbb{R} \times \mathbb{R}^+$ and $U \in C^1(D, \mathbb{R}^m)$ be a solution of the strict hyperbolic system*

$$\partial_t U + \partial_x f(U) = 0 .$$

If all k-Riemann invariants w_j are constant in D, i.e. $(d/dt)w_j(U(\gamma(t), t)) = 0$, $j = 1\ldots, m-1$, and $\gamma'(t) = \lambda_k$, $0 \leq t \leq \tau$, and $(\gamma(t), t) \in D$, $0 \leq t \leq \tau$, the solution U is called a k-rarefaction wave.

For k-rarefaction waves we again have the same property as for scalar equations, that the solution is constant along the k-characteristics. This will be shown in the following theorem.

THEOREM 4.1.16 *Let (4.1.3) be strictly hyperbolic and let u be a k-rarefaction wave in D. Then the k-characteristics in D are straight lines along which u is constant.*

Proof Let l_k be a left eigenvector of $Df(U)$, i.e.

$$l_k Df(U) = \lambda_k l_k ,$$

Let γ be the corresponding k-characteristic and define $n := \binom{\lambda_k}{1}$. Then we see that

$$l_k \frac{d}{dt} U(\gamma(t), t) = l_k(\lambda_k \partial_x U + \partial_t U) = l_k(Df(U)\partial_x U + \partial_t U)$$
$$= 0 . \qquad (4.1.39)$$

By assumption, we know that the k-Riemann invariants $w_1, \ldots, w_{m-1}$ are constant in D. Therefore

$$0 = \frac{d}{dt} w_j(U(\gamma(t), t)) = \nabla w_j(\partial_x U \gamma' + \partial_t U) = \nabla w_j \partial_n U \qquad (4.1.40)$$

for $j = 1, \ldots, m-1$, and (4.1.39) and (4.1.40) imply that

$$\begin{pmatrix} l_k \\ \nabla w_1 \\ \vdots \\ \nabla w_{m-1} \end{pmatrix} \partial_n U = 0 . \qquad (4.1.41)$$

In Theorem 4.1.12 we have shown that $\nabla w_1, \ldots, \nabla w_{m-1}$ are linearly independent, and this is also true for $l_k, \nabla w_1, \ldots, \nabla w_{m-1}$. The last statement follows if we multiply

$$\alpha_0 l_k + \alpha_1 \nabla w_1 + \ldots + \alpha_{m-1} \nabla w_{m-1} = 0$$

by the right eigenvector r_k of $Df(u)$. We obtain $\alpha_0 r_k l_k = 0$. Using

$$\lambda_k l_k r_j = l_k Df r_j = \lambda_j l_k r_j ,$$

we have that $l_k r_j = 0$ if $j \neq k$ and therefore $l_k r_k \neq 0$. This implies that $\alpha_0 = 0$, and thus we have shown that $l_k, \nabla w_1, \ldots, \nabla w_{m-1}$ are linearly independent. But then we get from (4.1.41)

$$\partial_n U = 0 ,$$

and also

$$\gamma'(t) = \lambda_k(U(\gamma(t), t)) = \text{const} .$$
□

EXAMPLE 4.1.17 *Let $w_{1/2}$ and $C_{1/2}$ be as in Example 4.1.13. Let U be a 2-rarefaction wave (i.e. w_2 is constant). Then (u, ρ) are constant along C_2, and the C_2-characteristics are straight lines.*

In (4.1.19) we have derived the jump conditions for piecewise-smooth solutions of the Euler equation. Up to now, the sign in (4.1.19) has not been uniquely defined. But this will be done now using the entropy condition. Furthermore, as in the case of scalar conservation laws, we consider entropy inequalities.

DEFINITION 4.1.18 (Entropy inequality, entropy pair) *Let u be a weak solution of the hyperbolic system*

$$\partial_t u + \partial_x f(u) = 0 \quad \text{in } \mathbb{R} \times \mathbb{R}^+ ,$$

$$u(x, 0) = u_0(x) \quad \text{in } \mathbb{R} .$$
(4.1.42)

Then we say that u satisfies an entropy inequality *if there exists an entropy pair $U, F : \mathbb{R}^m \to \mathbb{R}$ such that U is convex, $DU\, Df = DF$ and*

$$\partial_t U(u) + \partial_x F(u) \leq 0$$
(4.1.43)

in the distributional sense. Then the pair (U, F) is called an entropy pair.

EXAMPLE 4.1.19 (Entropy pair for the p-System) Let us consider the p-system (see (1.0.21))

$$\left. \begin{array}{l} \partial_t v - \partial_x u = 0, \\ \partial_t u + \partial_x p(v) = 0 \end{array} \right\} \quad \text{in } \mathbb{R} \times \mathbb{R}^+ \tag{4.1.44}$$

such that $p' < 0$. Then

$$U(s,t) := \frac{t^2}{2} - \int_0^s p(\tau)\, d\tau, \quad F(s,t) := p(s)t$$

is an entropy pair for the system (4.1.44).

Now we shall investigate the relationship between those solutions of (4.1.42) that are defined as a viscosity limit and those that satisfy an entropy inequality like (4.1.43).

THEOREM 4.1.20 *Let us assume that there are functions $U, F : \mathbb{R}^m \to \mathbb{R}$ such that $DU\, Df = DF$, $D^2 U \cdot A \geq cI$ for a positive-definite matrix $A \in \mathbb{R}^{m \times m}$ and $U \geq 0$. Furthermore, let $u_\varepsilon \in H^{1,2}(\mathbb{R} \times \mathbb{R}^+, \mathbb{R}^m)$ be a solution of*

$$\partial_t u_\varepsilon + \partial_x f(u_\varepsilon) = \varepsilon A \partial_x^2 u_\varepsilon \quad \text{in } \mathbb{R} \times \mathbb{R}^+, \tag{4.1.45}$$

with

$$F(u_\varepsilon(\pm\infty, t)) = 0, \quad DU(u_\varepsilon) A \partial_x u_\varepsilon(\pm\infty, t) = 0.$$

Additionally we assume that U, F, f and u_ε are sufficiently smooth, and

$$u_\varepsilon \to u \ (\varepsilon \to 0) \quad \text{in } L^\infty(\mathbb{R} \times \mathbb{R}^+, \mathbb{R}^m),$$

where u is a solution of

$$\partial_t u + \partial_x f(u) = 0 \quad \text{in } \mathbb{R} \times \mathbb{R}^+$$

in the distributional sense. Then we have

$$\partial_t U(u) + \partial_x F(u) \leq 0 \tag{4.1.46}$$

in the distributional sense.

4.1 Basic results for systems in 1-D

Proof We multiply (4.1.45) from the left by $(DU)(u_\varepsilon)$ and obtain

$$\partial_t U(u_\varepsilon) + \partial_x F(u_\varepsilon) = \varepsilon DU(u_\varepsilon) A \partial_x^2 u_\varepsilon$$
$$= \varepsilon \partial_x [DU(u_\varepsilon) A \partial_x u_\varepsilon]$$
$$- \varepsilon \partial_x u_\varepsilon D^2 U(u_\varepsilon) A \partial_x u_\varepsilon \; . \qquad (4.1.47)$$

Now we integrate (4.1.47) and use the fact that the last term in (4.1.47) is nonnegative:

$$\int_{-\infty}^{\infty} U(u_\varepsilon(x,\tau))\, dx - \int_{-\infty}^{\infty} U(u_\varepsilon(x,0))\, dx = -\varepsilon \int_0^\tau \int_{-\infty}^{\infty} \partial_x u_\varepsilon D^2 U A \partial_x u_\varepsilon$$
$$\leq -c\varepsilon \int_{-\infty}^{\infty} \int_0^\tau (\partial_x u_\varepsilon)^2 \; ,$$

and therefore

$$\int_{-\infty}^{\infty} U(u_\varepsilon(x,\tau))\, dx + \varepsilon c \int_0^\tau \int_{-\infty}^{\infty} (\partial_x u_\varepsilon)^2 \leq \int_{-\infty}^{\infty} U(u_\varepsilon(x,0))\, dx \; . \quad (4.1.48)$$

We multiply (4.1.47) by a nonnegative test function $\varphi \in C_0^\infty(\mathbb{R} \times]0,\infty[)$ and obtain, using $U_\varepsilon := U(u_\varepsilon)$,

$$\int_0^\tau \int_{-\infty}^{\infty} [-U_\varepsilon \partial_t \varphi - F(u_\varepsilon) \partial_x \varphi] \leq -\varepsilon \int_0^\tau \int_{-\infty}^{\infty} DU_\varepsilon A \partial_x u_\varepsilon \partial_x \varphi$$
$$\leq \left(\varepsilon \int_0^\tau \int_{-\infty}^{\infty} |DU_\varepsilon A \partial_x u_\varepsilon|^2 \right)^{\frac{1}{2}} \left(\varepsilon \int_0^\tau \int_{-\infty}^{\infty} |\partial_x \varphi|^2 \right)^{\frac{1}{2}} \; . \qquad (4.1.49)$$

Since u_ε is bounded in L^∞_{loc} and because of (4.1.48), we know that the first term on the right-hand side in (4.1.49) is bounded, and therefore the right-hand side converges to zero. This proves the statement of the theorem. □

REMARK 4.1.21 (Derivation of the Lax entropy condition) *First let us consider the following initial boundary value problem:*

$$\begin{aligned} \partial_t u + a \partial_x u &= 0 && \text{for } x > 0 \text{ and } t > 0 \; , \\ u(x,0) &= u_0(x) && \text{for } x > 0 \; , \\ u(0,t) &= g(t) && \text{for } t > 0 \; , \end{aligned} \qquad (4.1.50)$$

4 Initial value problems for systems in 1-D

where $a \in \mathbb{R}$ is a constant. We know that $u(x,t)$ is constant along the characteristics $\gamma'(t) = a$.

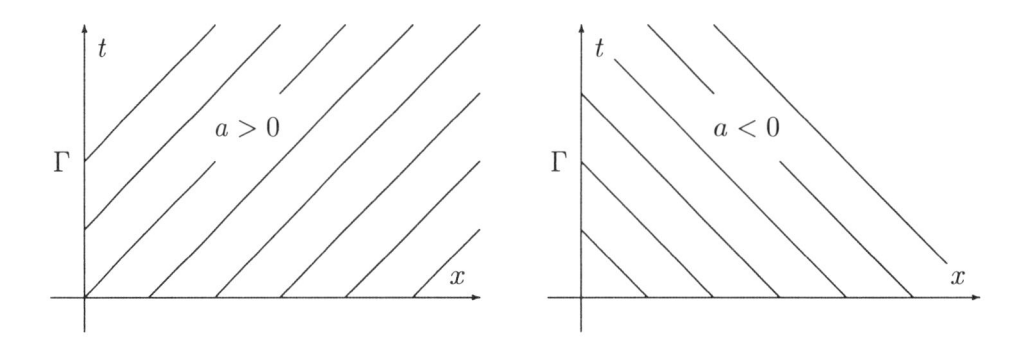

Figure 4.1.2

If $a < 0$ the value of u on the boundary $\Gamma := \{(x,t) \in \mathbb{R}^2 \mid x = 0, \ t > 0\}$ are uniquely defined by the given initial values. If $a > 0$, we are free to impose some boundary values for u on Γ. Now let us consider the linear system

$$\partial_t u + A \partial_x u = 0 \quad \text{for } x > 0 \text{ and } t > 0 ,$$

where $A \in \mathbb{R}^{m \times m}$ has only real eigenvalues $\lambda_1 < \ldots < \lambda_k < 0 < \lambda_{k+1} < \ldots < \lambda_m$ and $u(x,t) \in \mathbb{R}^m$. Let P be defined by

$$P^{-1} A P = \text{diag}(\lambda_1, \ldots, \lambda_m) .$$

Then we obtain for $v := P^{-1} u$ the decoupled system

$$\partial_t v_i + \lambda_i \partial_x v_i = 0 , \quad i = 1, \ldots, m .$$

As in (4.1.50), we can see the following facts. If $i \leq k$, we have $\lambda_i < 0$, and $v_i(0,t)$ is given by the initial values. If $i > k$, we have $\lambda_i > 0$, and $v_i(0,t)$ can be prescribed arbitrary. This means that we can prescribe $m - k$ conditions for v on Γ, and the same is true for $u = Pv$.

4.1 Basic results for systems in 1-D   297

Now we replace the boundary Γ by a moving boundary $S = \{(st, t) \mid t > 0\}$ and assume that for a fixed time t we have

$$\lambda_1 < \ldots < \lambda_k < s < \lambda_{k+1} < \ldots < \lambda_m . \tag{4.1.51}$$

Then we have to prescribe $m - k$ conditions on S in order to define the solution in $\{(x, t) \in \mathbb{R}^2 \mid x - st > 0\}$.

Now let us consider a solution u of

$$\partial_t u + \partial_x f(u) = 0 \quad \text{in } \mathbb{R} \times \mathbb{R}^+ \tag{4.1.52}$$

that is piecewise-smooth in $\mathbb{R} \times \mathbb{R}^+$ and has a jump along $S = \{(x, t) \in \mathbb{R}^2 \mid x = \sigma(t)\}$. Let $(x_0, t_0) \in S$, let u_r and u_l be the limit values of u from the right and from the left in (x_0, t_0), and let $s := \sigma'(t_0)$. We assume that the eigenvalues $\lambda_1(u), \ldots, \lambda_m(u)$ of $Df(u)$ satisfy

$$\lambda_1(u_r) < \ldots < \lambda_k(u_r) < s < \lambda_{k+1}(u_r) < \ldots < \lambda_m(u_r)$$

and

$$\lambda_1(u_l) < \ldots < \lambda_j(u_l) < s < \lambda_{j+1}(u_l) < \ldots < \lambda_m(u_l) .$$

The values u_r and u_l of u on S have to respect k conditions given by the initial values on the right-hand side and $m - j$ conditions given by the initial values on the left-hand side, and they are related by m jump conditions (see (4.1.17))

$$s(u_r - u_l) = f(u_r) - f(u_l) . \tag{4.1.53}$$

Now we have in all $2m + 1$ unknowns

$$u_{l1}, u_{l2}, \ldots, u_{lm}, u_{r1}, \ldots, u_{rm}, s \tag{4.1.54}$$

and $k + m - j + m = 2m + k - j$ conditions. A necessary condition for being able to compute the unknown values in (4.1.54) for arbitrary given initial values is that $2m + 1 = 2m + k - j$, or

$$j = k - 1 . \tag{4.1.55}$$

But this means in particular that

$$\begin{aligned}\lambda_k(u_r) &< s < \lambda_{k+1}(u_r) , \\ \lambda_{k-1}(u_l) &< s < \lambda_k(u_l) .\end{aligned} \tag{4.1.56}$$

DEFINITION 4.1.22 (Lax entropy condition for a k-shock) *The condition (4.1.56) is called the Lax entropy condition or the Lax shock condition. A discontinuity satisfying (4.1.53) and (4.1.56) is called a k-shock.*

COROLLARY 4.1.23 *Let us suppose that the assumption of Theorem 4.1.20 holds and let u be a piecewise-smooth solution of (4.1.3). Then along a discontinuity that moves with velocity s we have*

$$s(U_l - U_r) - [F(u_l) - F(u_r)] \leq 0 . \tag{4.1.57}$$

Proof Replace (2.1.1) by (4.1.44) and repeat the proof of Lemma 2.1.4. □

It can be shown that (4.1.57) is equivalent to the Lax entropy condition (4.1.56).

DEFINITION 4.1.24 (Genuinely nonlinear) *The k-characteristic is called genuinely nonlinear in $D \subset \mathbb{R}^m$ if and only if $\nabla \lambda_k \cdot r_k \neq 0$ in D. We always assume some scaled r_k such that $\nabla \lambda_k \cdot r_k = 1$. The system (4.1.3) is called genuinely nonlinear if each k-characteristic is genuinely nonlinear in some open set $D \subset \mathbb{R}^m$.*

THEOREM 4.1.25 *Let the system (4.1.3) be hyperbolic and genuinely nonlinear. Assume that there exists a strictly convex entropy function U and that u is a solution of (4.1.3) with a weak shock, i.e. the jump is small enough. Then (4.1.57) is equivalent to the Lax entropy condition.*

The proof is similar to that of Theorem 2.1.12, and therefore is omitted. For more details see [218, page 401].

EXAMPLE 4.1.26 Let us consider the p-system that is the flow equation for an isentropic gas in Lagrange coordinates (see (4.1.36) and (1.0.22)):

$$\left. \begin{array}{l} \partial_t v - \partial_x u = 0, \\ \partial_t u + \partial_x p(v) = 0 \end{array} \right\} \quad \text{in } \mathbb{R} \times \mathbb{R}^+ \tag{4.1.58}$$

where $v := 1/\rho$ and $p(v) := k/v^\gamma$. This means that we consider a γ-law gas. For (4.1.58) we shall solve the corresponding Riemann problem for $U := (v, u)$:

$$U(x,0) = U_0(x) = \begin{cases} U_l & \text{for } x < 0, \\ U_r & \text{for } x > 0. \end{cases} \tag{4.1.59}$$

4.1 Basic results for systems in 1-D 299

Now let us consider the following problem. Let U_l be as given in (4.1.59). For which values of U_r can the Riemann problem (4.1.58), (4.1.59) be solved uniquely by an 1-shock $(s < 0)$? Using the notation $F(U) := (-u, p(v))$, we obtain

$$\partial_t U + \partial_x F(U) = 0 ,$$

or

$$\partial_t U + A \partial_x U = 0 , \qquad (4.1.60)$$

where

$$A := \begin{pmatrix} 0 & -1 \\ p'(v) & 0 \end{pmatrix} .$$

The eigenvalues of A are

$$\lambda_1 := -\sqrt{-p'(v)} < 0 < \sqrt{-p'(v)} =: \lambda_2 .$$

The shock conditions (4.1.56) imply for a 1-shock

$$\lambda_1(u_r) < s < \lambda_2(u_r) , \qquad s < \lambda_1(u_l) < 0 , \qquad (4.1.61)$$

and for a 2-shock

$$\lambda_1(u_l) < s < \lambda_2(u_l) , \qquad 0 < \lambda_2(u_r) < s . \qquad (4.1.62)$$

The jump relations (4.1.53) imply

$$s(v_r - v_l) = -(u_r - u_l) , \qquad s(u_r - u_l) = p(v_r) - p(v_l) , \qquad (4.1.63)$$

and therefore

$$u_r - u_l = \pm\sqrt{(v_r - v_l)[p(v_l) - p(v_r)]} . \qquad (4.1.64)$$

From the 1-shock conditions (4.1.61) we obtain

$$\lambda_1(u_r) = -\sqrt{-p'(v_r)} < s < \lambda_1(u_l) = -\sqrt{-p'(v_l)} . \qquad (4.1.65)$$

300   4 Initial value problems for systems in 1-D

Now using $p' < 0$, $p'' > 0$ and $s < 0$, we derive from (4.1.63) and (4.1.65) that $p'(v_r) < p'(v_l)$ and

$$p'(v_r) < p'(v_l) \Rightarrow v_l > v_r \Rightarrow u_r - u_l < 0 .$$

But then the sign in (4.1.64) is uniquely defined, and we get

$$u_r - u_l = -\sqrt{(v_r - v_l)[p(v_l) - p(v_r)]} =: h(v_r) .$$

Then for given U_l we consider the function

$$u(v) := h(v) + u_l .$$

Notice that $h'(v) > 0$ if $v < v_l$. Therefore all points P "below" the given

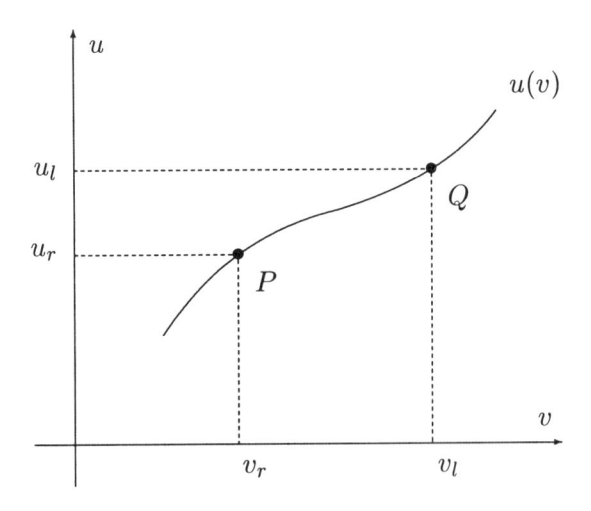

Figure 4.1.3

point $Q = (v_l, u_l)$ on the 1-shock curve gives us admissible values $U_r = (v_r, u_r)$ for which we can solve the Riemann problem (4.1.58), (4.1.59) by a 1-shock (see Figure 4.1.3). Similar considerations lead us to the picture concerning a 2-shock shown in Figure 4.1.4.

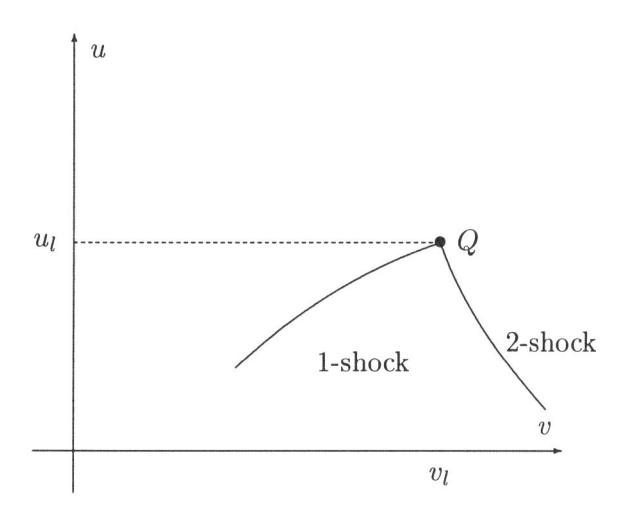

Figure 4.1.4

Now we shall study the situation if $u_r > u_l$. In this case we cannot solve (4.1.58), (4.1.59) by a shock solution but by a rarefaction wave. This will be shown in the following lemma. After the proof of this lemma we shall complete the picture in Figure 4.1.4.

LEMMA 4.1.27 (Rarefaction wave) *Consider the strictly hyperbolic system*

$$\partial_t U + A(U)\partial_x U = 0 \quad \text{in } \mathbb{R} \times \mathbb{R}^+ \tag{4.1.66}$$

with $A \in C^1(\mathbb{R}^m, \mathbb{R}^{m \times m})$, let r_k be a right eigenvector of A with respect the eigenvalue λ_k such that

$$\nabla \lambda_k(U) \cdot r_k(U) = 1 \quad \text{for all } U \in D \subset \mathbb{R}^m,$$

and let $U_l \in D$. Then there exists an $a > 0$ and a function $v \in C^1([\lambda_k(U_l), \lambda_k(U_l) + a]; \mathbb{R}^m)$ such that $v(\lambda_k(U_l)) = U_l$ and such that the Riemann problem for (4.1.66) with respect to the initial values

$$U(x,0) = U_0(x) = \begin{cases} U_l & \text{for } x < 0, \\ v(y) & \text{for } x > 0 \end{cases}$$

for all $y \in [\lambda_k(U_l), \lambda_k(U_l) + a[$ can be solved by a k-rarefaction wave. Furthermore, we have

$$\frac{d}{dy}\lambda_k(v(y)) > 0 \ . \tag{4.1.67}$$

Proof There exists a sufficiently small $a \in \mathbb{R}$ such that the initial value problem

$$v'(y) = r_k(v(y)) \ , \quad \lambda_k(U_l) < y < \lambda_k(U_l) + a \ ,$$
$$v(\lambda_k(U_l)) = U_l$$

has a unique solution. By assumption, we know that in D we have

$$\frac{d}{dy}\lambda_k(v(y)) = \nabla \lambda_k v' = \nabla \lambda_k \cdot r_k = 1 \ ,$$

and since $\lambda_k(v(\lambda_k(U_l))) = \lambda_k(U_l)$,

$$\lambda_k(v(y)) = y \quad \text{for } \lambda_k(U_l) < y < \lambda_k(U_l) + a. \tag{4.1.68}$$

Let w be a k-Riemann invariant in D. Then for the ansatz

$$u(x,t) = \begin{cases} U_l & \text{for } x/t < \lambda_k(U_l) \ , \\ v(x/t) & \text{for } \lambda_k(U_l) < x/t < y \ , \\ v(y) & \text{for } y < x/t \end{cases} \tag{4.1.69}$$

we have

$$\partial_t u + A(u)\partial_x u = 0 \quad \text{in the distributional sense,}$$

and for $\gamma'(t) = \lambda_k(u(\gamma(t), t))$ we have

$$\frac{d}{dt}w(u(\gamma(t),t)) = \nabla w(\partial_x u \gamma' + \partial_t u)$$
$$= \nabla w(\partial_x u \lambda_k + \partial_t u) = 0$$

for any case in (4.1.69). For the second case we need (4.1.68). This means any k-Riemann invariant is constant in D, and therefore u is a k-rarefaction wave. □

Now let us continue with Example 4.1.26. We consider a rarefaction of the form $U(x/t)$. This means that

$$-U'\frac{x}{t^2} + DF(U)\frac{1}{t}U' = 0,$$

and therefore U' is an eigenvector of $DF(U)$ with respect to the eigenvalue $\xi = x/t$ ($x < 0$). This implies (see (4.1.60)) that

$$(A - \lambda_1 Id)U' = \begin{pmatrix} -\lambda_1 & -1 \\ p'(v) & -\lambda_1 \end{pmatrix} \begin{pmatrix} v' \\ u' \end{pmatrix} = 0,$$

and in particular

$$\lambda_1 v' + u' = 0, \quad \text{or} \quad \frac{du}{dv} = -\lambda_1 = \sqrt{-p'(v)}.$$

Integrating the last equality, we obtain

$$u - u_l = \int_{v_l}^{v} \sqrt{-p'(y)}\, dy =: r(v).$$

Since (see Lemma 4.1.27) we have $v(\xi_1) = v_l$, $v(\xi_2) = v$ for $\xi_1 = \lambda_k(U_l)$ and $\xi_1 < \xi_2$. The condition $(d/d\xi)\lambda_k(U(\xi)) > 0$ implies that

$$\lambda_1(v_l) < \lambda_1(v),$$

and, since $p'' > 0$,

$$v > v_l.$$

We obtain similar results for the eigenvalue λ_2. Therefore we obtain the picture shown in Figure 4.1.5.

If we put together the information from Figures 4.1.3 – 4.1.5, we obtain Figure 4.1.6.

304   4 Initial value problems for systems in 1-D

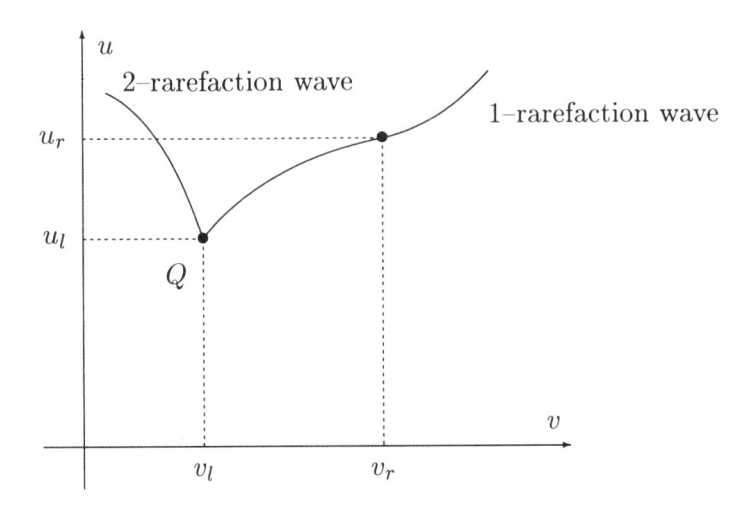

Figure 4.1.5

REMARK 4.1.28 (Change of type) *Consider the special case where*

$$p(v) = \mathcal{K}v - \frac{\gamma+1}{2}v^2,$$

where $\mathcal{K}$ is a constant. Then the p-system (4.1.58) can change its type depending on the sign of $p'(v)$. We have

$$p'(v) = \mathcal{K} - (\gamma+1)v \begin{cases} > 0: & \text{elliptic;} \\ = 0: & \text{degenerate;} \\ < 0: & \text{stricly hyperbolic.} \end{cases}$$

For more details see [218].

Now we should like to answer the following question. Let us assume that two values U_r and U_l are given. Is it possible to connect U_r to U_l by a special combination of shocks and rarefaction waves in the general case for hyperbolic systems?

DEFINITION 4.1.29 (Linearly degenerate) *The k-characteristic is called* linearly degenerate *in D if and only if $\nabla \lambda_k \cdot r_k = 0$ in D.*

4.1 Basic results for systems in 1-D 305

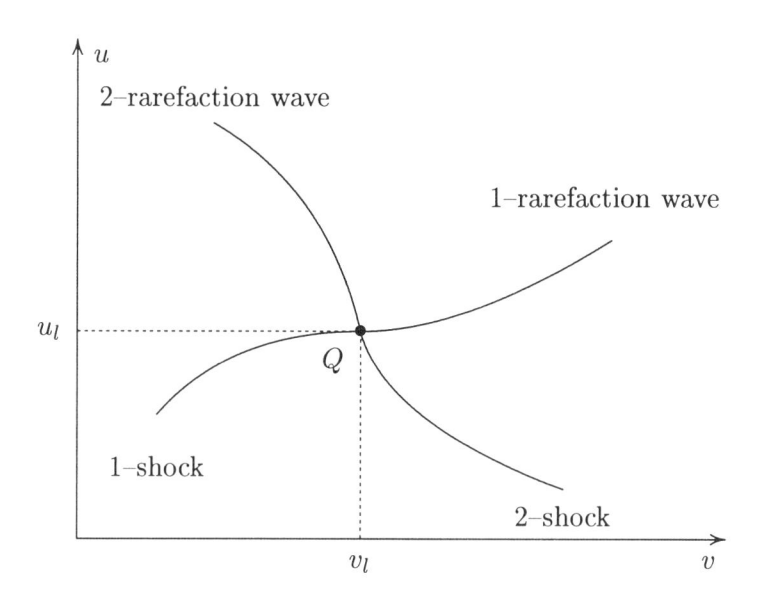

Figure 4.1.6

EXAMPLE 4.1.30 (Example of contact discontinuity, see Definition 4.1.31) Let us assume that the k-characteristic is linearly degenerate. Then, by definition, λ_k is a k-Riemann invariant. Let v be a solution of

$$v'(s) = r_k(v(s)), \qquad v(0) = u_l \ .$$

Then λ_k is constant along v, since

$$\frac{d}{ds}\lambda_k(v(s)) = \nabla\lambda_k v' = \nabla\lambda_k r_k = 0 \ ,$$

and therefore

$$\lambda_k(v(s)) = \lambda_k(v(0)) = \lambda_k(u_l) \ .$$

Define for fixed $s > 0$ and s in the domain of definition of v

$$w(x,t) := \begin{cases} u_l & \text{if } x < t\lambda_k(u_l), \\ v(s) & \text{if } x \geq t\lambda_k(u_l). \end{cases} \qquad (4.1.70)$$

Then w solves
$$\partial_t w + \partial_x f(w) = 0$$
with initial values
$$w(x,0) := \begin{cases} u_l & \text{if } x < 0, \\ v(s) & \text{if } x > 0. \end{cases}$$

To show this, we only have to verify the jump conditions. We have for $\lambda_k = \lambda_k(u_l)$
$$\frac{d}{ds}[f(v(s)) - \lambda_k v(s)] = f'v' - \lambda_k v'(s) = (f' - \lambda_k)v'$$
$$= (f' - \lambda_k)r_k = (f' - \lambda_k)r_k(v(s)) = 0,$$

since r_k is an eigenvector of f'. From this we get
$$f(v(s)) - \lambda_k v(s) = f(u_l) - \lambda_k u_l,$$
and therefore
$$f(v(s)) - f(u_l) = \lambda_k[v(s) - u_l]. \qquad (4.1.71)$$

But the solution w as defined in (4.1.70) has the same type of discontinuity, moving with velocity λ_k. This behaviour (4.1.71), with the velocity of the discontinuity being equal to some eigenvalue, is the motivation for the following definition.

DEFINITION 4.1.31 (Contact discontinuity with respect to λ_k)
Let u be a solution of $\partial_t u + \partial_x f(u) = 0$ in the distributional sense. Let u have a discontinuity along $\Gamma := \{(x,t) \mid x = \gamma(t)\}$. Let $\lambda_k(u^+(\gamma(t),t)) = \gamma'(t)$, where we have used the notation
$$u^+(\gamma(t),t) = \lim_{\varepsilon \to 0} u(\gamma(t) + \varepsilon, t).$$

In this case the discontinuity is called a contact discontinuity *with respect to λ_k. This means that the propagation velocity of the discontinuity and the characteristic speed on one side of the discontinuity are the same. For u^- we define contact discontinuity in a similar way.*

4.1 Basic results for systems in 1-D   307

REMARK 4.1.32 *In the case of a contact discontinuity we obtain*

$$\lambda_k(u_r) \leq s < \lambda_{k+1}(u_r) ,$$
$$\lambda_{k-1}(u_l) < s \leq \lambda_k(u_l) ,$$

where $u_r = v(y)$ for fixed y. At least for one of these inequalities the "$\leq$" part has to be satisfied with the "$=$" sign. This means that the shock conditions are satisfied with "$\leq$" instead of "$<$".

The following theorem will give us sufficient conditions for the initial values such that the corresponding Riemann problem can be solved. We shall not prove this theorem in the general case, but later for the Euler equations.

THEOREM 4.1.33 (Riemann problem) *Let $U_l \in \mathbb{R}^m$ and let D be an open neighbourhood of U_l. The system*

$$\partial_t U + \partial_x f(U) = 0 \qquad (4.1.72)$$

is assumed to be hyperbolic and each characteristic to be either genuinely nonlinear or linearly degenerate in D. Then there exists a neighbourhood $\tilde{D} \subset D$ of U_l such that for all $U_r \in \tilde{D}$ the Riemann problem (4.1.72) with

$$U(x,0) = \begin{cases} U_l(x) & \text{if } x < 0 , \\ U_r(x) & \text{if } x \geq 0 \end{cases} \qquad (4.1.73)$$

is solvable. The solution consists of at most $m+1$ constant states, which can be connected by shocks, rarefaction waves and contact discontinuities. The solution is uniquely defined.

Proof We refer to [218, Theorem 17.18]. □

REMARK 4.1.34 (Nonexistence for general data) *In [115] it was shown that there are strictly hyperbolic and genuinely nonlinear systems of conservation laws in 1-D such that the corresponding Riemann problem for* a pair of constant states U_l and U_r in general has no solutions. In this paper it is shown that

$$\partial_t u + \partial_x (u^2 - v) = 0 ,$$
$$\partial_t v + \partial_x \left(\frac{u^3}{3} - u\right) = 0$$

is strictly hyperbolic and genuinely nonlinear and that the corresponding initial value problem with respect to

$$U(x,0) = \begin{cases} U_l & \text{if } x < 0, \\ U_r & \text{if } x > 0, \end{cases}$$

where $U := \binom{u}{v}$, has no solution if $|U_r - U_l|$ is large.

Now we shall consider a special but most interesting case: the Euler equations of gas dynamics. While there is a smallness condition for the jump $|U_l - U_r|$ in Theorem 4.1.33, we shall now consider solutions of the Riemann problem for the Euler equation, without this restriction on the initial values.

EXAMPLE 4.1.35 (Riemann problem for Euler equations) Let us consider the Euler equations for (ρ, u, S), (4.1.9), (4.1.10) and (4.1.14). For the eigenvalues λ_i, the eigenvectors r_i and the Riemann invariants we obtain the following relationship (see Example 4.1.14); we also consider the products $r_i \nabla \lambda_i$:

	$i=1$	$i=2$	$i=3$
λ_i	$u-a$	u	$u+a$
r_i	$-\frac{2}{(\gamma+1)a}(\rho,-a,0)$	$(\partial_s p, 0, -\partial_\rho p)$	$\frac{2}{(\gamma+1)a}(\rho,a,0)$
$r_i \nabla \lambda_i$	1	0	1
i-Riemann invariant	$\{S, u + \frac{2}{\gamma-1}a\}$	$\{u, p\}$	$\{S, u - \frac{2}{\gamma-1}a\}$

(4.1.74)

Let us prove that $r_1 \nabla \lambda_1 = 1$ and $u + [2/(\gamma-1)]a$ is a 1-Riemann invariant. The other properties can be shown similarly. For p we have (see (4.1.15))

$$p = (\gamma-1)\rho^\gamma \exp\left(\frac{S-S_0}{c_v}\right),$$

and therefore

$$a = \sqrt{\frac{\gamma p}{\rho}} = \sqrt{\gamma(\gamma-1)} \exp\left(\frac{S-S_0}{2c_v}\right) \rho^{\frac{1}{2}(\gamma-1)}. \qquad (4.1.75)$$

For $\partial_\rho(u-a)$ and $\partial_u(u-a)$ we obtain

$$\partial_\rho(u-a) = -\frac{\gamma-1}{2}\frac{a}{\rho}, \quad \partial_u(u-a) = 1,$$

and therefore

$$\nabla \lambda_1 r_1 = 1.$$

Furthermore,

$$\partial_\rho \left(u + \frac{2}{\gamma-1}a\right) = \frac{2}{\gamma-1}\frac{\gamma-1}{2}\frac{a}{\rho} = \frac{a}{\rho}, \quad \partial_u \left(u + \frac{2}{\gamma-1}a\right) = 1,$$

and therefore
$$\nabla\left(u + \frac{2}{\gamma - 1}a\right) r_1 = 0 .$$

In Lemma 4.1.7 we have already derived the jump conditions for the Euler equations (see (4.1.19)). Now we shall use the shock conditions in order to get the sign of the expression $u_2 - u_1$ in (4.1.19). For the 1-shock we obtain (notice that $u_2 = u_r$, $u_1 = u_l$)

$$\lambda_1(u_2) < s < \lambda_2(u_2), \quad s < \lambda_1(u_1), \tag{4.1.76}$$
$$u_2 - a_2 < s < u_2, \quad s < u_1 - a_1 . \tag{4.1.77}$$

This implies that

$$a_1 < u_1 - s = v_1 \quad \text{and} \quad \frac{v_1}{a_1} > 1 , \tag{4.1.78}$$

where v is defined as in the proof of Lemma 4.1.7. On the other hand, we have (see (4.1.29))

$$\frac{v_1}{a_1} = \text{sign}(v_1)\sqrt{\frac{2}{\gamma - 1}\frac{1 + \pi\beta}{\beta^2 - 1}} ,$$

and therefore $\text{sign}(v_1) = 1$. Using (4.1.30), we obtain

$$\frac{u_2 - u_1}{a_1} = \sqrt{\frac{2}{\gamma(\gamma - 1)}}\frac{1 - \pi}{\sqrt{1 + \pi\beta}} . \tag{4.1.79}$$

For a 3-shock we can use similar arguments to obtain

$$\frac{u_2 - u_1}{a_1} = \sqrt{\frac{2}{\gamma(\gamma - 1)}}\frac{\pi - 1}{\sqrt{1 + \pi\beta}} . \tag{4.1.80}$$

Therefore we can show the following lemma.

4.1 Basic results for systems in 1-D

LEMMA 4.1.36 *Using the assumption of Lemma 4.1.7, we obtain for a 1-shock*

$$\frac{\rho_2}{\rho_1} = \frac{1+\beta\pi}{\pi+\beta}, \quad \frac{u_2 - u_1}{a_1} = \sqrt{\frac{2}{\gamma(\gamma-1)}} \frac{1-\pi}{\sqrt{1+\pi\beta}},$$

where $\pi := p_2/p_1 > 1$ and $\beta := (\gamma+1)/(\gamma-1)$ as in Lemma 4.1.7.

For a 3-shock we have

$$\pi = \frac{p_2}{p_1} < 1, \quad \frac{\rho_2}{\rho_1} = \frac{1+\pi\beta}{\pi+\beta}, \quad \frac{u_2 - u_1}{a_1} = \sqrt{\frac{2}{\gamma(\gamma-1)}} \frac{\pi-1}{\sqrt{1+\pi\beta}}.$$

These types of shocks are called compression shocks.

Proof It remains to compute the sign of $\pi - 1$ for both types of shocks. For the 1-shock we obtain from (4.1.78) and (4.1.77)

$$v_1 > a_1 \quad \text{and} \quad v_2 = u_2 - s < a_2. \tag{4.1.81}$$

Furthermore, we have $a_1 < a_2$ since

$$\frac{2}{\gamma-1}a_1^2 + a_1^2 < \frac{2}{\gamma-1}a_1^2 + v_1^2 = \frac{2}{\gamma-1}a_2^2 + v_2^2 < \frac{2}{\gamma-1}a_2^2 + a_2^2. \tag{4.1.82}$$

Here we have used (4.1.81) and the jump relation (4.1.23). In particular, (4.1.82) and $a_1 < a_2$ imply that $v_1^2 > v_2^2$ and $|v_1| > |v_2|$. Since $m = \rho v$ is constant across the discontinuity, we get $\rho_1 < \rho_2$, and therefore

$$\pi = \frac{p_2}{p_1} = \frac{\rho_2}{\rho_1}\left(\frac{a_2}{a_1}\right)^2 > 1.$$

The result for the 3-shock can be shown in a similar way. □

Now we should like to consider 1- and 3-rarefaction waves (see Figure 4.1.7).

Since S is a 1-Riemann invariant, we know that by definition (see Definition 4.1.15) S is constant in a 1-rarefaction wave, i.e.

$$S_1 = S_2. \tag{4.1.83}$$

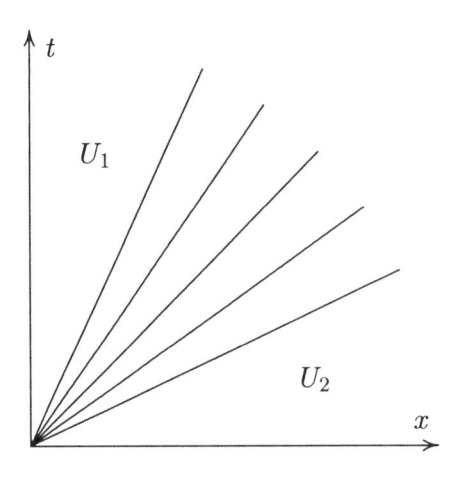

Figure 4.1.7

For the same reason we obtain for the other 1-Riemann invariant $u + \frac{2}{\gamma-1}a$:

$$u_1 + \frac{2}{\gamma-1}a_1 = u_2 + \frac{2}{\gamma-1}a_2 \ . \tag{4.1.84}$$

From (4.1.83) and since $p = (\gamma-1)e^{(S-S_0)/c_v}\rho^\gamma$, we get

$$\frac{p_2}{p_1} = \left(\frac{\rho_2}{\rho_1}\right)^\gamma, \tag{4.1.85}$$

and from (4.1.84)

$$\frac{u_2 - u_1}{a_1} = \frac{2}{\gamma-1}\left(1 - \frac{a_2}{a_1}\right) \ . \tag{4.1.86}$$

For a 1-rarefaction wave (4.1.67) implies that the eigenvalue is monotonically increasing, $\lambda_1(U_2) \geq \lambda_1(U_1)$; this means that

$$u_2 - u_1 \geq a_2 - a_1 \ . \tag{4.1.87}$$

Now (4.1.86) and (4.1.87) imply

$$\frac{a_2}{a_1} - 1 = \frac{a_2 - a_1}{a_1} \leq \frac{u_2 - u_1}{a_1} = \frac{2}{\gamma - 1}\left(1 - \frac{a_2}{a_1}\right),$$

and therefore

$$0 \leq \left(\frac{2}{\gamma - 1} + 1\right)\left(1 - \frac{a_2}{a_1}\right).$$

But this means that $a_2/a_1 \leq 1$, and, using (4.1.85),

$$\frac{a_2}{a_1} = \left(\frac{p_2}{p_1}\right)^{\frac{1}{2} - \frac{1}{2\gamma}} \leq 1.$$

Thus we obtain

$$\frac{p_2}{p_1} \leq 1.$$

Now let us put together these results concerning the 1-rarefaction waves in the following lemma. The statements concerning the 3-rarefaction wave can be obtained in a similar way.

LEMMA 4.1.37 *Let (ρ, u, p) be a solution of (4.1.37) in the distributional sense. In particular, let (ρ, u, p) be a 1-rarefaction wave connecting the states (ρ_1, u_1, p_1) and (ρ_2, u_2, p_2). Then we have*

$$\pi := \frac{p_2}{p_1} \leq 1, \quad \frac{\rho_2}{\rho_1} = \pi^{1/\gamma}, \quad \frac{u_2 - u_1}{a_1} = \frac{2}{\gamma - 1}(1 - \pi^\tau)$$

where $\tau := \frac{1}{2} - \frac{1}{2\gamma}$. For a 3-rarefaction wave we obtain

$$\pi := \frac{p_2}{p_1} \geq 1, \quad \frac{\rho_2}{\rho_1} = \pi^{1/\gamma}, \quad \frac{u_2 - u_1}{a_1} = \frac{2}{\gamma - 1}(\pi^\tau - 1).$$

Now we shall discuss the situation where there is a contact discontinuity. This case can be treated very easily if the k-characteristics is linearly degenerate. This is true for the Euler equations for $k = 2$ (see Example 4.1.30). Let U_1 be given and let v be a solution of

$$v'(y) = r_2(v(y)), \qquad 0 < y < y_0, \qquad v(0) = U_1,$$

where r_2 is the second eigenvector of A (see (4.1.74)). Then p and u are constant along $\{v(y) \mid y \in [0, y_0[\}$, since we have for $p(U)$, where $U = (\rho, u, S)$,

$$\frac{d}{dy} p(v(y)) = \nabla p \, v'(y) = \nabla p \cdot r_2(v(y)) = 0,$$

and for u itself

$$\frac{d}{dy} u(v(y)) = \nabla u \, v'(y) = \nabla u \cdot r_2(v(y)) = 0.$$

Here we have considered $p = (\gamma - 1)e^{(S-S_0)/c_v} \rho^\gamma$ as a function of $U = (\rho, u, S)$. This follows because u and p are 2-Riemann invariants. This means that the admissible states $v(y)$ that can be connected to U_1 by a contact discontinuity are given by ($v = (\rho, u, S)$)

$$v'(y) = \begin{pmatrix} \partial_S p(v(y)) \\ 0 \\ -\partial_\rho p(v(y)) \end{pmatrix}, \qquad v(0) = U_1.$$

Since

$$p = ke^{S/c_v} \rho^\gamma, \qquad \partial_S p_1 = \frac{p_1}{c_v}, \qquad \partial_\rho p_1 = \frac{\gamma p_1}{\rho}, \qquad (4.1.88)$$

we have to solve

$$\begin{aligned} v_1'(y) &= \rho'(y) = \frac{p_1}{c_v}, & \rho(0) &= \rho_1, \\ v_2'(y) &= u'(y) = 0, & u(0) &= u_1, \\ v_3'(y) &= S'(y) = -\frac{\gamma p_1}{\rho}, & S(0) &= S_1. \end{aligned}$$

The solutions are given by

$$\rho(y) = \frac{p_1}{c_v} y + \rho_1 \ , u(y) = u_1 \ , S(y) = c_v \ln \frac{p_1}{k\rho(y)^\gamma}. \quad (4.1.89)$$

The jump conditions (4.1.20) are satisfied with $u = s = \lambda_2$. This means that u and p are constant and ρ can have jumps. In order to see this for p, we use $S(y)$ as given in (4.1.89) in the expression for p in (4.1.88). Then we obtain

$$p = k\rho^\gamma \exp\left(\ln \frac{p_1}{k\rho^\gamma}\right) = p_1 \ .$$

Therefore we have proved the following result.

LEMMA 4.1.38 *Let (ρ, u, p) be a solution of (4.1.37) in the distributional sense. In particular, let (ρ, u, p) be a contact discontinuity connecting the states (ρ_1, u_1, p_1) and (ρ_2, u_2, p_2). Thus we have*

$$\frac{p_2}{p_1} = 1 \ , \quad u_2 - u_1 = 0 \ ,$$

and no condition for ρ_1 and ρ_2.

Now we put together the results of Lemmas 4.1.36 – 4.1.38 in the following theorem.

THEOREM 4.1.39 (Elementary wave solutions) *Let (ρ, u, p) be a solution of (4.1.37) in the distributional sense. In particular, let (ρ, u, p) be an elementary wave (i.e. k-shock, k-rarefaction wave or a contact discontinuity) connecting the states (ρ_1, u_1, p_1) and (ρ_2, u_2, p_2). Then we obtain for $\beta = (\gamma+1)/(\gamma-1)$ and $\tau = \frac{1}{2} - 1/2\gamma$:*

(a) In the case of a 1-shock $(x < 0)$ or a 1-rarefaction wave $(x \geq 0)$

$$\frac{p_2}{p_1} =: e^{-x} =: g_1(x)$$

$$\frac{\rho_2}{\rho_1} = \begin{cases} e^{-x/\gamma} & (x \geq 0) \\ \dfrac{\beta + e^x}{1 + \beta e^x} & (x < 0) \end{cases} =: f_1(x) \ ,$$

$$\frac{u_2 - u_1}{a_1} = \begin{cases} \dfrac{2}{\gamma-1}(1 - e^{-\tau x}) & (x \geq 0) \\ \sqrt{\dfrac{2}{\gamma(\gamma-1)}} \dfrac{1 - e^{-x}}{(1 + \beta e^{-x})^{\frac{1}{2}}} & (x < 0) \end{cases} =: h_1(x) \ .$$

316   4 Initial value problems for systems in 1-D

(b) In the case of a contact discontinuity,
$$\frac{p_2}{p_1} = 1, \quad u_2 - u_1 = 0,$$
and no condition for ρ_1 and ρ_2.

(c) In the case of a 3-shock ($x < 0$) or a 3-rarefaction wave ($x \geq 0$)
$$\frac{p_2}{p_1} =: e^x =: g_3(x),$$
$$\frac{\rho_2}{\rho_1} = \frac{1}{f_1(x)} =: f_3(x),$$
$$\frac{u_2 - u_1}{a_1} = \begin{cases} \dfrac{2}{\gamma-1}(e^{\tau x} - 1) & (x \geq 0) \\ \sqrt{\dfrac{2}{\gamma(\gamma-1)}} \dfrac{e^x - 1}{\sqrt{1+\beta e^x}} & (x < 0) \end{cases} =: h_3(x).$$

Now we should like to investigate the general Riemann problem. Up to now we have considered special states U_r and U_l that could be connected by one type of elementary wave. But now we shall consider arbitrary states U_l and U_r, and we ask if the corresponding Riemann problem can be solved.

THEOREM 4.1.40 (Existence for the Riemann problem in case of the Euler equations) *Consider the system (4.1.9), (4.1.10), (4.1.14) together with the following equations of state:*
$$p = R\rho T, \quad e = c_v T, \quad p = k e^{s/c_v} \rho^\gamma$$
(ideal gas). Then for all $U_l = (\rho_l, u_l, p_l)$ and $U_r = (\rho_r, u_r, p_r)$ there exists a solution of the Riemann problem (4.1.9), (4.1.10), (4.1.14) with respect to the initial values
$$U(x,0) = \begin{cases} U_l & (x \leq 0), \\ U_r & (x > 0) \end{cases} \tag{4.1.90}$$
in the class consisting of a 1-shock or 1-rarefaction wave, or a contact discontinuity followed by a 3-shock or 3-rarefaction wave if and only if
$$u_r - u_l < \frac{2}{\gamma - 1}(a_l + a_r). \tag{4.1.91}$$

4.1 Basic results for systems in 1-D   317

Proof We shall use the following notation: $v = (v_1, v_2, v_3) = (\rho, p, u)$. We define the map $T_x^i : \mathbb{R}^3 \to \mathbb{R}^3$, $i = 1, 2, 3$, by

$$T_x^1(v) := \left(f_1(x)v_1, e^{-x}v_2, v_3 + \left(\frac{\gamma v_2}{v_1} \right)^{\frac{1}{2}} h_1(x) \right),$$

$$T_x^2(v) := (e^x v_1, v_2, v_3),$$

$$T_x^3(v) := \left(f_3(x)v_1, e^x v_2, v_3 + \left(\frac{\gamma v_2}{v_1} \right)^{\frac{1}{2}} h_3(x) \right),$$

where h_1, h_3, f_1 and f_3 are defined as in Theorem 4.1.39. Then solving the Riemann problem (4.1.9), (4.1.10), (4.1.14), with (4.1.90) is equivalent to finding a point $x = (x_1, x_2, x_3) \in \mathbb{R}^3$ such that

$$U_r = T_{x_3}^3 T_{x_2}^2 T_{x_1}^1 U_l ,$$

or, written explicitly,

$$\begin{pmatrix} \rho_r \\ p_r \\ u_r \end{pmatrix} = \begin{pmatrix} f_3(x_3) e^{x_2} f_1(x_1) \rho_l \\ e^{x_3 - x_1} p_l \\ u_l + a_l \left[h_1(x_1) + \sqrt{\frac{e^{-(x_1+x_2)}}{f_1(x_1)}} h_3(x_3) \right] \end{pmatrix}. \quad (4.1.92)$$

Using $A := \rho_r/\rho_l$, $B = p_r/p_l$ and $C = (u_r - u_l)/a_l$, we obtain from (4.1.92)

$$x_3 - x_1 = \log B , \quad (4.1.93)$$
$$f_1(x_1) e^{x_2} f_3(x_3) = A , \quad (4.1.94)$$
$$h_1(x_1) + \sqrt{\frac{e^{-(x_1+x_2)}}{f_1(x_1)}} h_3(x_3) = C . \quad (4.1.95)$$

The last two relations imply that

$$C = h_1(x_1) + \sqrt{\frac{e^{-x_1}}{A f_1(x_3)}} h_3(x_3) . \quad (4.1.96)$$

Now it is easy to verify the following lemma.

LEMMA 4.1.41

$$h_1' > 0, \quad h_1(\mathbb{R}) = \left]-\infty, \frac{2}{\gamma-1}\right[, \quad (4.1.97)$$

$$h_3(x) = \sqrt{f_1(x)} e^{\frac{1}{2}x} h_1(x). \quad (4.1.98)$$

Using (4.1.98) in (4.1.96), we can derive

$$C = h_1(x_1) + \frac{1}{\sqrt{A}} e^{\frac{1}{2}(x_3-x_1)} h_1(x_3),$$

and (4.1.93) implies that

$$C = h_1(x_1) + \sqrt{\frac{B}{A}} h_1(x_1 + \log B) =: F(x_1). \quad (4.1.99)$$

Because of (4.1.97), we have $F'(x_1) > 0$, and therefore

$$F(\mathbb{R}) = \left]-\infty, \left(1 + \sqrt{\frac{B}{A}}\right) \frac{2}{\gamma-1}\right[.$$

This means that a sufficient condition for the solvability of (4.1.99) (C is given) is

$$C < \left(1 + \sqrt{\frac{B}{A}}\right) \frac{2}{\gamma-1}$$

This is just the condition (4.1.91). □

4.2 Chorin's method for solving the Riemann problem

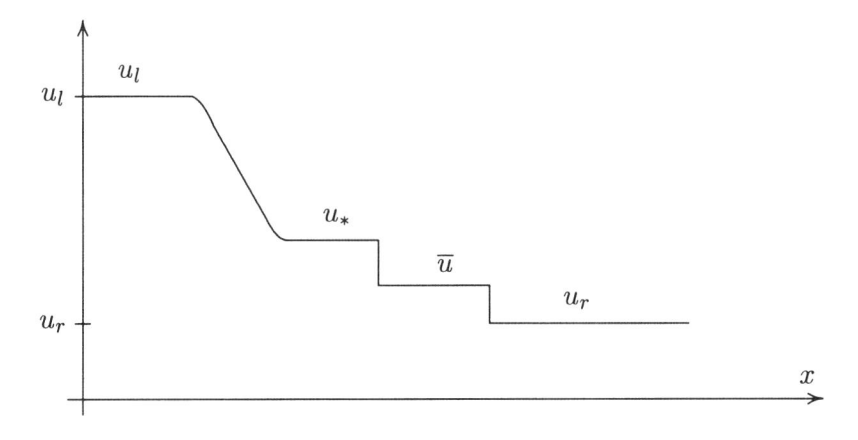

Figure 4.2.1

4.2 Chorin's method for solving the Riemann problem for systems

We should like to compute the exact solution of the Riemann problem for the Euler equations (4.1.37), (4.1.38) with

$$U(x,0) = \begin{cases} U_l & (x < 0) \\ U_r & (x \geq 0) \end{cases}$$

such that U_l is connected to U_r by an 1-wave, a contact discontinuity and a 3-wave (see Figure 4.2.1). This means that we have to compute an approximation of the intermediate values U_* and $\overline{U}$. Since U_* should be connected to $\overline{U}$ by a contact discontinuity, we have $p_* = \overline{p}$ and $u_* = \overline{u}$.

Now we shall describe an iteration scheme to get p_* and u_*. This is known as Chorin's method [26]. Using Theorem 4.1.39, we get for a 1-wave ($u_l = u_1, u_2 = u_*$)

$$u_* - u_l = h_1(x)a_* ,$$
$$p_* - p_l = g_1(x) \quad \text{(see Theorem 4.1.39)} ,$$

and for a 3-wave

$$u_r - u_* = h_3(x)a_3 ,$$
$$p_r - p_* = g_3(x) \quad \text{(see Theorem 4.1.39)} .$$

Now let
$$M_l := \frac{p_l - p_*}{u_l - u_*}, \quad M_r := \frac{p_r - p_*}{u_r - u_*}. \tag{4.2.1}$$

It can be shown that M_l and M_r do not depend on u_*.

LEMMA 4.2.1

$$M_l = -\sqrt{\rho_l p_l}\, \Phi\left(\frac{p_*}{p_l}\right), \quad M_r = \sqrt{\rho_r p_r}\, \Phi\left(\frac{p_*}{p_r}\right), \tag{4.2.2}$$

where

$$\Phi(w) = \begin{cases} \left(\frac{\gamma+1}{2}w + \frac{\gamma-1}{2}\right)^{\frac{1}{2}} & (w > 1), \\ \frac{\gamma-1}{2\sqrt{\gamma}} \frac{1-w}{1-w^{(\gamma-1)/(2\gamma)}} & (w < 1), \\ \gamma^{\frac{1}{2}} & (w = 1). \end{cases}$$

Proof See 4.2.2.

Now we continue to describe Chorin's method. From (4.2.1) we obtain

$$p_* = p_r - (u_r - u_*) M_r, \quad u_* = u_l - \frac{p_l - p_*}{M_l}, \tag{4.2.3}$$

and therefore

$$p_* = p_r - \left(u_r - u_l + \frac{p_l - p_*}{M_l}\right) M_r,$$

$$p_*\left(\frac{M_l - M_r}{M_l}\right) = p_r + M_r(u_l - u_r) - p_l \frac{M_r}{M_l},$$

$$p_* = \frac{M_l p_r + M_l M_r(u_l - u_r) - p_l M_r}{M_l - M_r}$$

$$= \frac{M_l(p_*) p_r + M_l(p_*) M_r(p_*)(u_l - u_r) - p_l M_r(p_*)}{M_l(p_*) - M_r(p_*)}. \tag{4.2.4}$$

Equation (4.2.4) has to be solved iteratively in the following way.

(1) Choose initial values $p_*^0 := \frac{1}{2}(p_r + p_l)$ and compute M_l^0 and M_r^0 according to (4.2.2). Now we assume that p_*^q, M_l^q and M_r^q are already given.

(2) Define (see 4.2.4)
$$\tilde{p} := \frac{M_l^q p_r + M_l^q M_r^q(p_*)(u_l - u_r) - p_l M_r^q}{M_l^q - M_r^q}.$$

(3) Define
$$p_*^{q+1} := \max\{\varepsilon_1, \tilde{p}\} \quad (\varepsilon_1 \approx 10^{-6})$$
in order to avoid the pressure becoming too small,
$$M_r^{q+1} := \sqrt{p_r \rho_r} \, \Phi\left(\frac{p_*^{q+1}}{p_r}\right)$$
$$M_l^{q+1} := \sqrt{p_l \rho_l} \, \Phi\left(\frac{p_*^{q+1}}{p_l}\right).$$

(4) IF $\max\{|M_r^{q+1} - M_r^q|, |M_l^{q+1} - M_l^q|\} \leq \varepsilon_2$ STOP
ELSE $q = q + 1$ and GOTO (2).

Then we have an approximating value p^*, using (4.2.3) for u^*. Now we have the situation shown in Figure 4.2.2.

322   4 Initial value problems for systems in 1-D

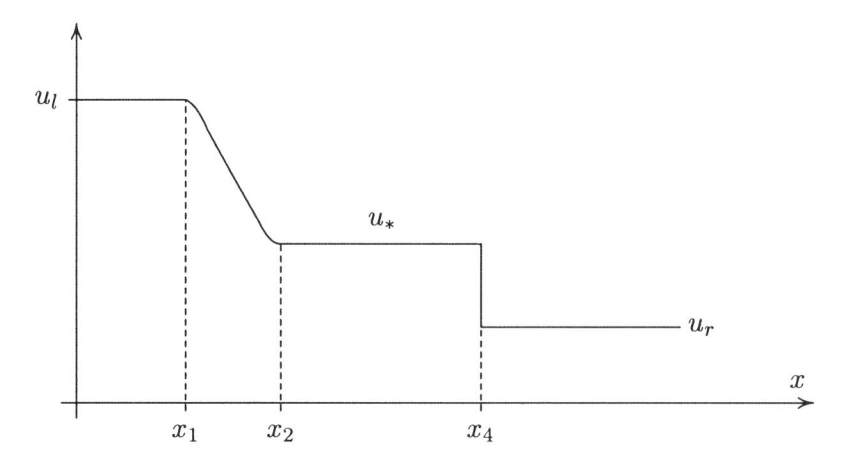

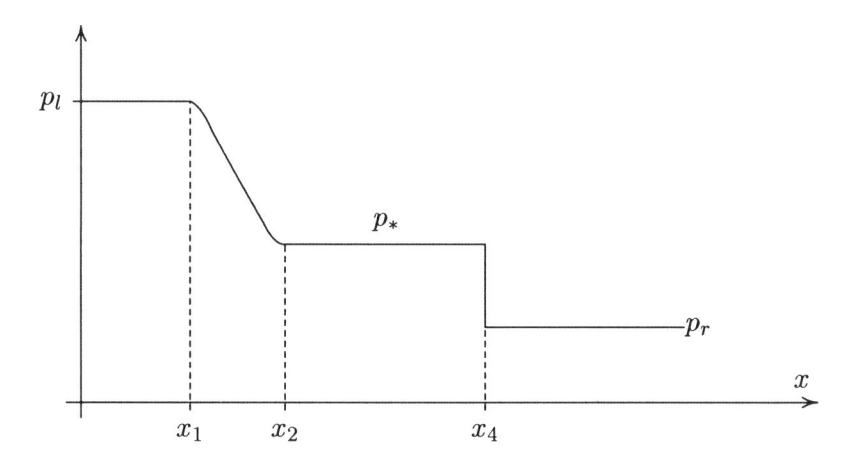

Figure 4.2.2

We still have to compute ρ_{*l} and ρ_{*r}, the values of the density in the contact discontinuity, and the velocities of the jumps x_2 and x_4 and the region $[x_1, x_2]$ of the rarefaction wave. In order to get ρ_{*r}, we use (4.1.28) and obtain

$$\frac{\rho_r}{\rho_{*r}} = \frac{1 + \beta\pi}{\pi + \beta}, \quad \text{where } \pi = \frac{p_r}{p_*}, \quad \beta = \frac{\gamma + 1}{\gamma - 1}.$$

Then we obtain s_4 from (see (4.1.21))

$$\rho_{*r}(u_* - s_4) = \rho_r(u_r - s_4),$$

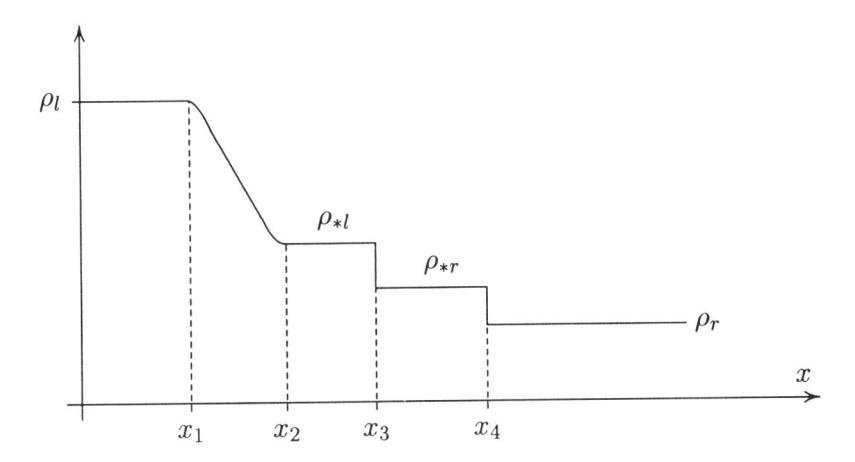

Figure 4.2.3

and therefore $x_4 = s_4 t$. By definition, we have $x_3 = \lambda_2(u_*)t = u_* t$. The rarefaction wave starts at $x_1 = \lambda_1(u_l)t$ (see (4.1.69)). In order to compute ρ_{*l}, we use the fact that

$$u + \frac{2}{\gamma - 1} a$$

is a 1-Riemann invariant (see 4.1.74). Therefore

$$u_l + \frac{2}{\gamma - 1} a_l = u_{*l} + \frac{2}{\gamma - 1} a_{*l}, \qquad u_{*l} = u_*,$$

$$a_{*l} = \left(u_l - u_{*l} + \frac{2}{\gamma - 1} a_l \right) \frac{\gamma - 1}{2},$$

$$\rho_{*l} := \frac{\gamma p_*}{a_{*l}^2}.$$

From (4.1.69) we get a y_0 such that

$$(y_0) = \begin{pmatrix} \rho_{*l} \\ u_* \\ p_* \end{pmatrix}.$$

Then the rarefaction wave runs up to

$$x_2 = \lambda_1(v(y_0)) = u_* - a_{*l}.$$

The convergence of this iteration is not studied in [26], and there will be some difficulties for strong rarefaction waves. Therefore a modification that is also prescribed in [26] should sometimes be used. In Figures 4.4.2 – 4.4.5 we have plotted the numerical results for the shock tube problem. The solid line is the exact solution that is computed as described in this section (see also [146]).

4.2.2 Proof of Lemma 4.2.1 First let us consider a 1-shock. Then we have $\pi = p_*/p_l > 1$,

$$u_l - u_* = a_l \frac{\pi - 1}{\sqrt{1 + \pi\beta}} \sqrt{\frac{2}{\gamma(\gamma - 1)}}$$

and $p_l - p_* = p_l(1 - \pi)$. We obtain for M_l

$$M_l := \frac{p_l - p_*}{u_l - u_*} = \frac{p_l(1 - \pi)}{a_l(\pi - 1)} \sqrt{1 + \pi\beta} \sqrt{\frac{\gamma(\gamma - 1)}{2}}$$

$$= -\frac{p_l}{\sqrt{\frac{\gamma p_l}{\rho_l}}} \sqrt{1 + \pi\beta} \sqrt{\frac{\gamma(\gamma - 1)}{2}} \qquad (\beta := \frac{\gamma + 1}{\gamma - 1})$$

$$= -\sqrt{p_l \rho_l} \sqrt{\left(1 + \pi \frac{\gamma + 1}{\gamma - 1}\right) \frac{\gamma - 1}{2}}$$

$$= -\sqrt{p_l \rho_l} \sqrt{\frac{\gamma - 1}{2} + \frac{p_*}{p_l} \frac{\gamma + 1}{2}}.$$

For a 3-rarefaction wave M_r can be obtained as follows for $\pi = p_r/p_*$:

$$M_r = \frac{p_r - p_*}{u_r - u_*}$$

$$= -\frac{p_r(1 - 1/\pi)}{a_*(1 - \pi^\tau)} \frac{\gamma - 1}{2}$$

$$= -\frac{p_r(1 - 1/\pi)}{a_r(1 - \pi^\tau)} \sqrt{\pi} \left(\frac{\rho_*}{\rho_r}\right)^{1/2} \frac{\gamma - 1}{2}$$

$$= -\sqrt{p_r \rho_r} \frac{1 - 1/\pi}{(1 - \pi^\tau)} \sqrt{\pi} \frac{\gamma - 1}{2\sqrt{\gamma}} \pi^{-1/2\gamma}$$

$$= -\sqrt{p_r \rho_r} \frac{\gamma - 1}{2\sqrt{\gamma}} \frac{\pi - 1}{\pi^{1-\tau} - \pi}$$

$$= -\sqrt{p_r \rho_r} \frac{\gamma-1}{2\sqrt{\gamma}} \frac{1-1/\pi}{1/\pi^\tau - 1}$$

$$= \sqrt{p_r \rho_r} \frac{\gamma-1}{2\sqrt{\gamma}} \frac{1/\pi - 1}{1/\pi^\tau - 1} \, .$$

The case $w = 1$ follows by using the l'Hôpital's lemma. □

4.3 The basis of Glimm's existence proof for systems in 1-D

The first proof of the existence of a solution of the Euler equations in 1-D globally in time was given by Glimm using a random choice method. The most interesting point of this result is the fact that the idea for the theoretical existence proof came from the numerical algorithm. This algorithm works very well and gives sharp shocks, resolved only on one cell of the grid. Unfortunately the theoretical basis of the proof as well as the nice numerical properties cannot be generalized to 2-D or n-D. In this section we describe the basic ideas of the Glimm scheme and how it is used for the existence proof (see [218] and [78]).

Let us repeat the basic ideas of the Godunov scheme (see Example 2.3.17). We assume that we have a uniform mesh $x_i = ih$, $i \in \mathbb{Z}$, on $\mathbb{R}$ and $t^n := n\Delta t$, $n \in \mathbb{N}_0$, on $\mathbb{R}^+$. For given initial values $u_0 \in L^\infty(\mathbb{R})$ we define piecewise-constant discrete initial values u_i^0 by

$$u_i^0 := \frac{1}{h} \int_{x_{i-1/2}}^{x_{i+1/2}} u_0(x)\, dx \, . \tag{4.3.1}$$

Now we assume that the discrete function u_i^n for the time level n is already defined. Then in the second step we solve the Riemann problems on the cells $Z_i :=]x_i, x_{i+1}[\times]t^n, t^{n+1}[$ with respect to the piecewise-constant initial values $\{u_i^n, u_{i+1}^n\}$ for $i \in \mathbb{Z}$. We denote by $v_{i+\frac{1}{2}}(x,t)$ the corresponding solution of

$$\partial_t u + \partial_x f(u) = 0 \quad \text{in } Z_i \tag{4.3.2}$$

with initial data

$$u(x, t^n) = \begin{cases} u_i^n & (x < x_{i+\frac{1}{2}}), \\ u_{i+1}^n & (x > x_{i+\frac{1}{2}}). \end{cases} \quad (4.3.3)$$

The solutions $v_{i+\frac{1}{2}}(x, t)$ of the local Riemann problems are now used to define the function v for all $x \in \mathbb{R}$ by

$$v(x, \Delta t) := \begin{cases} v_{i-\frac{1}{2}}(x, t^{n+1}) & \text{if } x_{i-\frac{1}{2}} < x < x_i, \\ v_{i+\frac{1}{2}}(x, t^{n+1}) & \text{if } x_i \le x < x_{i+\frac{1}{2}}. \end{cases} \quad (4.3.4)$$

Then in the third step we define

$$u_i^{n+1} := \frac{1}{h} \int_{x_{i-1/2}}^{x_{i+1/2}} v(x, \Delta t) \, dx. \quad (4.3.5)$$

For scalar equations or for special systems (see Examples 2.3.17 and 4.4.3) we can solve the Riemann problem explicitly, and then the Godunov scheme is defined by (4.3.1)–(4.3.5).

EXAMPLE 4.3.1 (Glimm scheme) Let $(\alpha_n)_n$ be a sequence of random numbers in $]-\frac{1}{2}, \frac{1}{2}[$. Then for given initial values $u_0 \in L^\infty(\mathbb{R})$ we define piecewise-constant values u_i^0 by

$$u_i^0 := u_0((i + \alpha_0)h). \quad (4.3.6)$$

Assume that u_i^n is defined already for the time level n. Then in the second step we solve the Riemann problems on the cells $Z_i :=]x_i, x_{i+1}[\times]t^n, t^{n+1}[$ with respect to the piecewise-constant initial values $\{u_i^n, u_{i+1}^n\}$ for $i \in \mathbb{Z}$ as in the Godunov scheme (see (4.3.3)). We define $v_{i+\frac{1}{2}}$ and v as in (4.3.3) and (4.3.4). Then in the third step we set

$$u_i^{n+1} := v((i + \alpha_{n+1})h, \Delta t) \quad \text{on }]x_{i-\frac{1}{2}}, x_{i+\frac{1}{2}}[. \quad (4.3.7)$$

Let us consider the following example.

4.3 The basis of Glimm's existence proof for systems in 1-D

EXAMPLE 4.3.2 Let

$$\alpha_0 = \tfrac{1}{4}, \ \alpha_1 = -\tfrac{1}{3}, \ \alpha_2 = 0, \ \alpha_3 = -\tfrac{1}{4}, \ \alpha_4 = \tfrac{3}{8}, \ \alpha_5 = -\tfrac{1}{5}, \ \alpha_6 = \tfrac{1}{8}, \ \ldots$$

be a sequence of random numbers, and let

$$f(u) := \tfrac{1}{2}u^2, \ u_L := 4, \ u_R := 2 \ \text{and} \ s = 3, \ s\Delta t = \tfrac{1}{4}h \ .$$

Then the exact solution of

$$\partial_t u + \partial_x f(u) = 0 \quad \text{in } \mathbb{R} \times \mathbb{R}^+ \ ,$$

$$u(x,0) = \begin{cases} u_L & \text{if } x < 0 \ , \\ u_R & \text{if } x \geq 0 \ , \end{cases}$$

is given by

$$u(x,t) = \begin{cases} u_L & (x < st) \ , \\ u_R & (x \geq st) \ , \end{cases}$$

where $s(u_L - u_R) = f(u_L) - f(u_R)$. Now let us consider the Glimm scheme. We obtain

Step 1 (Definition of the initial values with $\alpha_0 := \tfrac{1}{4}$)

$$u_i^0 = u((i + \tfrac{1}{4})h, 0) = \begin{cases} u_R & (\ i \geq 0) \ , \\ u_L & (\ i < 0) \ , \end{cases} \tag{4.3.8}$$

and let u_0 be piecewise-constant on the cells $]x_{i-\frac{1}{2}}, x_{i+\frac{1}{2}}[$.

328   4 Initial value problems for systems in 1-D

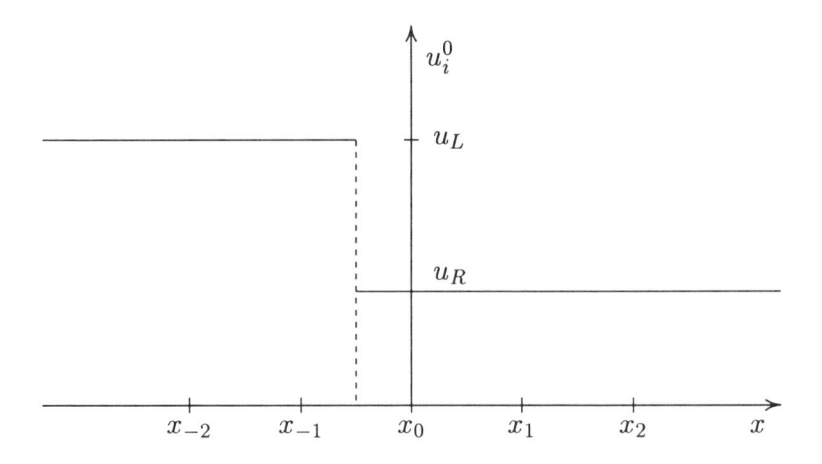

Step 2 (Solution of the local Riemann problem) For $0 < t \leq \Delta t$

$$v_{-\frac{1}{2}}(x,t) := \begin{cases} u_L & \text{if } x < st + x_{-\frac{1}{2}}, \\ u_R & \text{if } x \geq st + x_{-\frac{1}{2}}, \end{cases} \qquad (4.3.9)$$

is the solution of the local Riemann problem in $x_{-\frac{1}{2}}$.

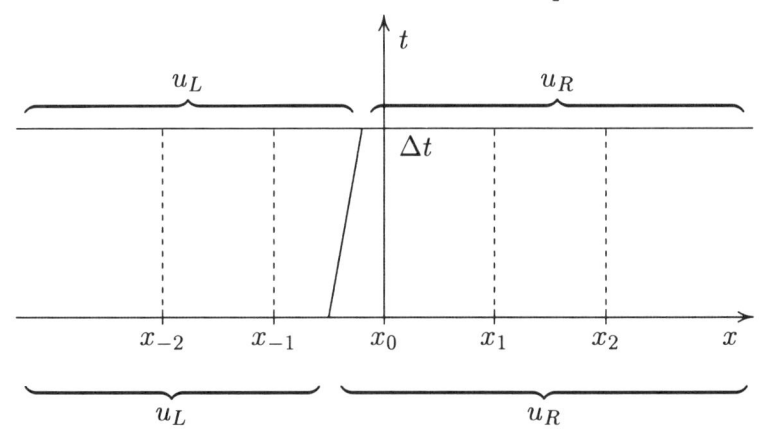

Step 3 (Definition of the global function v)

$$v(x, \Delta x) := \begin{cases} u_L & \text{if } x \leq x_{-1}, \\ v_{-\frac{1}{2}}(x, \Delta t) & \text{if } x_{-1} < x < x_0, \\ u_R & \text{if } x_0 \leq x. \end{cases} \qquad (4.3.10)$$

4.3 The basis of Glimm's existence proof for systems in 1-D   329

Step 4 (Definition of u^1)

$$u_i^1 = v((i+\alpha_1)h, \Delta t) = \begin{cases} u_L & (i \leq 0), \\ u_R & (i \geq 1). \end{cases} \quad (4.3.11)$$

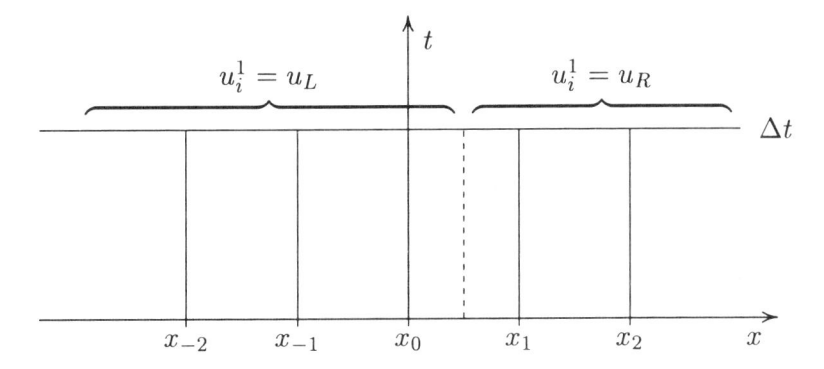

Putting several steps together, we obtain

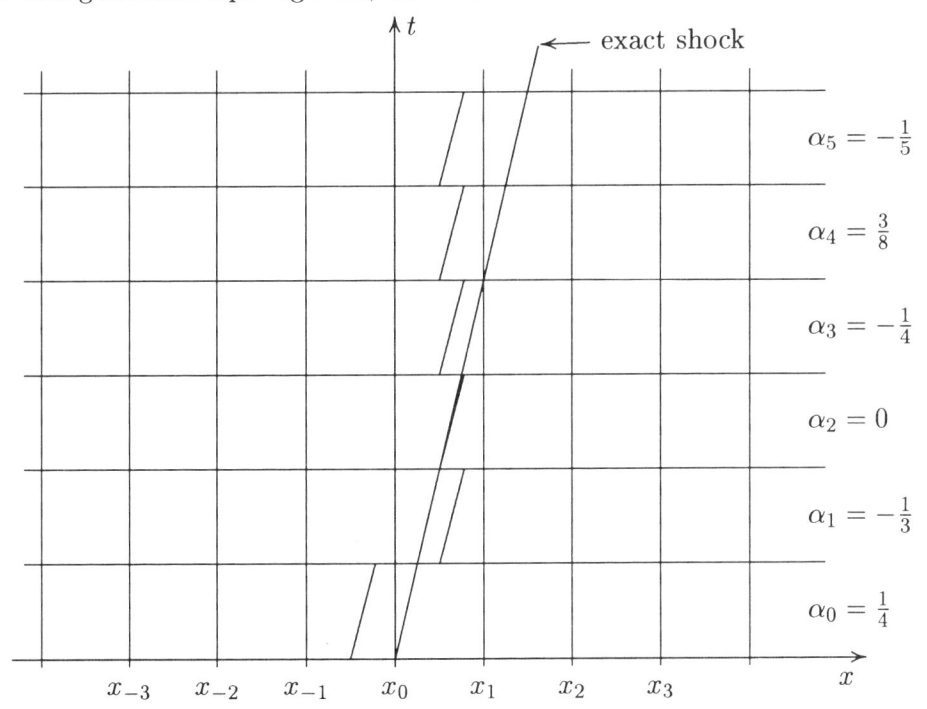

$\alpha_5 = -\frac{1}{5}$

$\alpha_4 = \frac{3}{8}$

$\alpha_3 = -\frac{1}{4}$

$\alpha_2 = 0$

$\alpha_1 = -\frac{1}{3}$

$\alpha_0 = \frac{1}{4}$

330   4 Initial value problems for systems in 1-D

Now let us assume that the random numbers α_n are uniformly distributed over the interval $]-\frac{1}{2},\frac{1}{2}[$. Consider the single steps (4.3.8)–(4.3.11) of the Glimm scheme. From (4.3.11) we obtain for $i=0$

$$u_0^1 = v(\alpha_1 h, \Delta t) \ .$$

Case 1 $\alpha_1 h < (s\Delta t + x_{-\frac{1}{2}})$ (see (4.3.9)). Obviously we have $s\Delta t + x_{-\frac{1}{2}} = -\left(\frac{1}{2}h - s\Delta t\right)$. The probability p for this case is given by

$$p = \frac{\mu\left(]-\frac{1}{2},-\frac{1}{2}+s\frac{\Delta t}{h}[\right)}{\mu\left(]-\frac{1}{2},\frac{1}{2}[\right)} = \frac{s\Delta t}{h} \ ,$$

where μ denotes the Lebesque measure on $\mathbb{R}$. This means that

$$u_0^1 = u_L \quad \text{with probability} \quad s\frac{\Delta t}{h} \ .$$

In this case the shock will start in $(x_{0+\frac{h}{2}}, \Delta t)$. This means that the shock moves one step, i.e. h, to the right.

Case 2 $\alpha_1 h \geq -(\frac{1}{2}h - s\Delta t)$, and then

$$u_0^1 = u_R \quad \text{with probability} \quad \frac{h - s\Delta t}{h} = 1 - s\frac{\Delta t}{h} \ .$$

In this case the shock will start in $(x_{0-\frac{h}{2}}, \Delta t)$. This means that the shock remains in the same position. In general, we obtain for the shock position

$$x_{i+\frac{h}{2}} \quad \text{with probability} \quad s\frac{\Delta t}{h} \ ,$$

$$x_{i-\frac{h}{2}} \quad \text{with probability} \quad 1 - s\frac{\Delta t}{h} \ .$$

Then the expected value for the shock position is given by

$$\left(x_i + \frac{h}{2}\right)\frac{s\Delta t}{h} + \left(x_i - \frac{h}{2}\right)\left(1 - \frac{s\Delta t}{h}\right) = x_i - \frac{h}{2} + s\Delta t \ .$$

4.3 The basis of Glimm's existence proof for systems in 1-D

Now we consider a sequence of n time steps. By definition, we have

$$u_i^n := v(ih + h\alpha_n, n\Delta t),$$

where v is the function consisting of the solution of the local Riemann problems.

Define

$$J_n := \#\left\{\alpha_j \Big| 1 \leq j \leq n, \alpha_j < \frac{s\Delta t}{h} - \frac{1}{2}\right\}.$$

J_n denotes the number of timesteps for which the shock moves one step to the right. Let us assume that for $t = 0$ the discrete shock starts in $-\frac{1}{2}h$. The "weak law of large numbers" implies that

$$J_n = n\frac{s\Delta t}{h} + R_n \quad \text{with } \frac{R_n}{n} \to 0 \text{ as } n \to \infty.$$

Let us assume that $\lambda = \Delta t/\Delta x$ is fixed. Therefore after n time steps we get for the position $x(t^n)$ of the shock

$$x(t^n) - x(0) = J_n h = st^n + R_n h$$
$$= st^n + R_n \Delta t \frac{1}{\lambda}$$
$$= st^n + R_n \frac{t^n}{n} \frac{1}{\lambda} = st^n + \mathcal{O}\left(\frac{R_n}{n}\right) t^n.$$

Now if $t = t^n$ is fixed and $n \to \infty$, we have

$$x(t^n) = x(0) + st.$$

But this is also the correct shock position for the exact solution.

The Glimm scheme can also be generalized to systems. In the same way as for a scalar equation, the mean value step (4.4.17) (Godunov scheme for systems) is replaced as in (4.3.7).

THEOREM 4.3.3 (Boundedness of the total variation) *Consider the initial value problem $(u \in \mathbb{R}^m)$.*

$$\partial_t u + \partial_x f(u) = 0 \quad \text{in } \mathbb{R} \times \mathbb{R}^+ \tag{4.3.12}$$
$$u(\cdot, 0) = u_0 \quad \text{in } \mathbb{R}. \tag{4.3.13}$$

Assume that the system (4.3.12) is hyperbolic and genuinely nonlinear. Let u_h denote the sequence of approximating functions given by the Glimm scheme. Furthermore, we suppose that there exist $\delta \in \mathbb{R}^+$ and $c \in \mathbb{R}^m$ such that

$$\|u_h(\cdot, 0) - c\|_{L^\infty(\mathbb{R})} \leq \delta,$$
$$\|u_h(\cdot, 0)\|_{BV(\mathbb{R})} \leq \delta$$

and that $\Delta t / \Delta x$ is a fixed constant. Then the Glimm scheme is well defined and there exists a constant K such that for all $t \in \mathbb{R}^+$

$$\|u_h(\cdot, t) - c\|_{L^\infty(\mathbb{R})} \leq K\|u_h(\cdot, 0) - c\|_{L^\infty(\mathbb{R})},$$
$$\|u_h(\cdot, t)\|_{BV(\mathbb{R})} \leq K\|u_h(\cdot, 0)\|_{BV(\mathbb{R})},$$

$$\int_{\mathbb{R}} |u_h(x, t^1) - u_h(x, t^2)| \, dx \leq K[|t^2 - t^1| + 4\Delta t]\|u_h(\cdot, 0)\|_{BV}.$$

We refer to [78] for the proof and also for the proof of the following theorem, which guarantees the convergence of the Glimm scheme.

THEOREM 4.3.4 (Convergence of the Glimm scheme) *Let μ be the multidimensional Lebesgue measure on*

$$A := \prod_{k,n} [(k - \tfrac{1}{2})h, \ (k + \tfrac{1}{2})h] \times n\Delta t.$$

We suppose that the assumptions of Theorem 4.3.3 hold. Then there exists a set $N \subset A$ such that $\mu(N) = 0$ and a sequence $h \to 0$ such that for all $a \in A \setminus N$ the corresponding sequence u_h defined by the Glimm scheme converges to a weak solution of the initial value problem (4.3.12), (4.3.13).

This result can also be generalized to the system of Euler equations.

REMARK 4.3.5 *Liu [142] generalizes the convergence proof of Glimm to strictly hyperbolic systems, which are allowed to be linearly degenerate. With these methods, we have, especially for the Euler equations, convergence of the Glimm scheme.*

REMARK 4.3.6 (Approximate Riemann solver for the Glimm scheme) *In [88] it was shown that the exact Riemann solver in the Glimm scheme can be replaced by an approximate Riemann solver such that the scheme is still convergent.*

4.4 Numerical schemes for hyperbolic systems in 1-D

In this section we shall consider different numerical schemes for solving the initial value problem ($u \in \mathbb{R}^m$) for the following system:

$$\partial_t u + \partial_x f(u) = 0 \quad \text{in } \mathbb{R} \times \mathbb{R}^+ , \qquad (4.4.1)$$

$$u(x, \cdot) = u_0(x) \quad \text{in } \mathbb{R} . \qquad (4.4.2)$$

The numerical schemes are formally of the same form as for scalar equations and they also have the discrete conservation property. The special form of these schemes imply that if they define a convergent sequence of discrete solution then the sequence will converge to a solution of the conservation law (4.4.1), (4.4.2). We shall introduce schemes for the Riemann problem for a linear equation, and several solvers for nonlinear problems.

Let u_i^0 be defined as

$$u_i^0 := \frac{1}{\Delta x} \int_{x_{i-1/2}}^{x_{i+1/2}} u_0(x) \, dx . \qquad (4.4.3)$$

For systems we use the same notation as for scalar equations. On a grid $x_j = j\Delta x$ and $t^n := n\Delta t$ we consider u_j^n as an approximation of $u(x_j, t^n)$, where u is the exact solution of (4.4.1), (4.4.2). A numerical scheme is said to be in conservation form if it can be written as

$$u_j^{n+1} = u_j^n - \frac{\Delta t}{\Delta x}[g(u_j^n, u_{j+1}^n) - g(u_{j-1}^n, u_j^n)] . \qquad (4.4.4)$$

The numerical flux is consistent with the conservation law (4.4.1) if

$$g(u, u) = f(u) .$$

Furthermore, we shall assume that $g, G \in C^{0,1}(\mathbb{R}^m, \mathbb{R}^m)$. The numerical entropy flux is consistent with an entropy condition if

$$U_j^{n+1} \leq U_j^n - \frac{\Delta t}{\Delta x}[G(u_j^n, u_{j+1}^n) - G(u_{j-1}^n, u_j^n)] ,$$

where U is convex. Furthermore, we have $G(U, U) = F(U)$ and (U, F) is an entropy pair. Here we have used the notation

$$U_j^n := U(u_j^n) .$$

The justification of these schemes in conservation form is given by the following Lax–Wendroff theorem.

THEOREM 4.4.1 (Lax–Wendroff theorem [128]) *Assume that $(k_r)_r$ and $(h_r)_r$ converge to zero for $r \in \mathbb{N}$ as $r \to \infty$ such that $k_r/h_r = $ constant and let $(u_r)_r$ be a sequence of discrete solutions as defined in (4.4.3) and (4.4.4) with respect to $\Delta x = h_r$, $\Delta t = k_r$ and the initial values u_i^0. Assume that there exists a constant K such that*

$$\sup_r \sup_{\mathbb{R} \times \mathbb{R}^+} |u_r(x, t)| \leq K$$

and $u_r \longrightarrow u$ almost everywhere in $\mathbb{R} \times \mathbb{R}^+$. Then u is a solution of

$$\partial_t u + \partial_x f(u) = 0 \quad \text{in } \mathbb{R} \times \mathbb{R}^+, \quad u(\cdot, 0) = u_0 \quad \text{in } \mathbb{R} \quad (4.4.5)$$

in the distributional sense.

Proof This is the same as for Theorem 2.3.1.

4.4 Numerical schemes for hyperbolic systems in 1-D

EXAMPLE 4.4.2 (Simple upwind scheme for linear systems) First let us repeat this method for scalar equations. For

$$\partial_t u + a \partial_x u = 0 \quad \text{in } \mathbb{R} \times \mathbb{R}^+,$$

with $a =$ constant, the simple upwind scheme is defined as $\left(\lambda = \Delta t / \Delta x\right)$

$$v_j^{n+1} = v_j^n - \lambda a \begin{cases} v_{j+1}^n - v_j^n & \text{if } a < 0, \\ v_j^n - v_{j-1}^n & \text{if } a > 0. \end{cases}$$

This simple upwind scheme can also be written in conservation form:

$$\begin{aligned} v_j^{n+1} &= v_j^n - \lambda[a^+(v_j^n - v_{j-1}^n) + a^-(v_{j+1}^n - v_j^n)] \\ &= v_j^n - \lambda[a^+ v_j^n + a^- v_{j+1}^n - (a^+ v_{j-1}^n + a^- v_j^n)], \end{aligned} \quad (4.4.6)$$

or, in viscosity form,

$$v_j^{n+1} = v_j^n - \frac{\lambda}{2} a(v_{j+1}^n - v_{j-1}^n) + \frac{\lambda}{2}|a|(v_{j+1}^n - 2v_j^n + v_{j-1}^n). \quad (4.4.7)$$

Now let us use these ideas for the discretization of linear hyperbolic systems in 1-D. Consider the following linear and strictly hyperbolic system:

$$\partial_t u + A \partial_x u = 0, \quad A \in \mathbb{R}^{m \times m} \quad \text{constant.} \quad (4.4.8)$$

Since (4.4.8) is assumed to be hyperbolic, A can be diagonalized, i.e. there exists a matrix T such that

$$T^{-1} A T = \Lambda,$$

where $\Lambda = \text{diag}(\lambda_1, \ldots, \lambda_m)$ is a diagonal matrix with entries the eigenvalues λ_i of A, $i = 1, \ldots, m$. Then we obtain for $w := T^{-1} u$

$$\partial_t w + \Lambda \partial_x w = 0. \quad (4.4.9)$$

This system is decoupled. The components of w are called characteristic variables. Now we apply (4.4.6) to any of the decoupled equations in (4.4.9) and obtain for $v := w_k$ and $\lambda = \lambda_k$

$$v_j^{n+1} = v_j^n - \frac{\Delta t}{\Delta x}[\lambda^+ v_j^n + \lambda^- v_{j+1}^n - (\lambda^+ v_{j-1}^n + \lambda^- v_j^n)]$$

or, in matrix form,

$$w_j^{n+1} = w_j^n - \frac{\Delta t}{\Delta x}[\Lambda^+ w_j^n + \Lambda^- w_{j+1}^n - (\Lambda^+ w_{j-1}^n + \Lambda^- w_j^n)]. \quad (4.4.10)$$

In order to get (4.4.10) again in terms of u, we define $u_j^n = T w_j^n$. Then (4.4.10) implies

$$u_j^{n+1} = u_j^n - \frac{\Delta t}{\Delta x}\left[A^+ u_j^n + A^- u_{j+1}^n - (A^+ u_{j-1}^n + A^- u_j^n)\right],$$

where

$$A^\pm = T\Lambda^\pm T^{-1},$$

or

$$u_j^{n+1} = u_j^n - \frac{\Delta t}{\Delta x}\left[g(u_j^n, u_{j+1}^n) - g(u_{j-1}^n, u_j^n)\right],$$

where

$$g(u,v) = A^+ u + A^- v. \quad (4.4.11)$$

This solves the linear system (4.4.8).

Now we should like to study the Godunov scheme for hyperbolic systems. As we have already seen in Example 2.3.17 for scalar equations, the basic idea of the Godunov scheme in 1-D is to use the exact solutions of local Riemann problems. Therefore we shall first study the exact solution of the Riemann problem for linear systems.

EXAMPLE 4.4.3 (Riemann problem for a linear system) Consider the system (4.4.8) with respect to the initial conditions

$$u(0,x) = \begin{cases} u_L & \text{if } x < 0, \\ u_R & \text{if } x \geq 0. \end{cases} \quad (4.4.12)$$

4.4 Numerical schemes for hyperbolic systems in 1-D

Denote the right eigenvectors of A by $r_1, ..., r_m$, the eigenvalues by $\lambda_1 < ... < \lambda_m$ and define α_j, $j = 1, \ldots, m$, by

$$u_R - u_L =: \sum_{j=1}^{m} \alpha_j r_j \ . \tag{4.4.13}$$

Let $u_k := u_L + \sum_{j=1}^{k} \alpha_j r_j$. Then the solution of the Riemann problem (4.4.8), (4.4.12) is given by

$$u(x,t) = \begin{cases} u_L \text{ if } x/t < \lambda_1 \ , \\ u_k \text{ if } \lambda_k \leq x/t < \lambda_{k+1} \text{for} \quad k = 1, \ldots, m-1 \ , \\ u_R \text{ if } \lambda_m \leq x/t \ , \end{cases} \tag{4.4.14}$$

Since u as defined above is piecewise-constant, we have to verify only the jump conditions. For this, consider the states u_k and u_{k+1}. We get

$$[u] = u_k - u_{k+1} = \sum_{j=1}^{k} \alpha_j r_j - \sum_{j=1}^{k+1} \alpha_j r_j = -\alpha_{k+1} r_{k+1}$$

$$[Au] = Au_k - Au_{k+1} = A[u] = -\lambda_{k+1} \alpha_{k+1} r_{k+1} \ .$$

Therefore the jump condition is satisfied, and therefore u is the unique solution of (4.4.8) and (4.4.12) (since the problem is linear).

Now let us repeat the Godunov scheme for the following hyperbolic system:

$$\partial_t u + \partial_x f(u) = 0 \quad \text{in } \mathbb{R} \times \mathbb{R}^+ \ ,$$
$$u(x,0) = u_0(x) \quad \text{in } \mathbb{R} \ .$$

The discrete initial values are defined as

$$u_i^0 := \int_{x_{i-1/2}}^{x_{i+1/2}} u_0 \, dx \ . \tag{4.4.15}$$

Now let us assume that u^n is already known and that u^n is piecewise-constant on $[x_{i-\frac{1}{2}}, x_{i+\frac{1}{2}}]$ for $i \in \mathbb{N}$. Then we solve the local Riemann problem

$$\partial_t u + \partial_x f(u) = 0 \quad \text{on } [x_i, x_{i+1}] \times [t^n, t^{n+1}],$$
$$u(x, 0) = \begin{cases} u_i^n & \text{for } x < x_{i+\frac{1}{2}}, \\ u_{i+1}^n & \text{for } x \geq x_{i+\frac{1}{2}}. \end{cases} \qquad (4.4.16)$$

We denote the solution by $w_i^n(x, t)$. Of course w_i^n depends on u_i^n and u_{i+1}^n. Using the solutions w_i^n of the local Riemann problems, we define a global function v^n by

$$v^n(x, t) := w_i^n(x, t) \quad \text{if } t^n \leq t \leq t^{n+1} \text{ and } x_i \leq x \leq x_{i+1}.$$

Then we define the new value u_i^{n+1} by

$$u_i^{n+1} := \frac{1}{\Delta x} \int_{x_{i-1/2}}^{x_{i+1/2}} v^n(x, t^{n+1}) \, dx. \qquad (4.4.17)$$

In order to avoid two neighbouring Riemann problems interacting, we assume that

$$\max |\lambda_i| \leq \frac{\Delta x}{2\Delta t},$$

where λ_i denote the eigenvalues of $f'(u)$.

REMARK 4.4.4 The values u_i^{n+1} as defined in (4.4.17) can be written as

$$u_i^{n+1} = \frac{1}{\Delta x} \int_{x_{i-1/2}}^{x_{i+1/2}} v^n(x, t^{n+1}) \, dx$$
$$= \frac{1}{\Delta x} \int_{x_{i-1/2}}^{x_i} w_{i-1}^n(x, t^{n+1}) \, dx$$
$$+ \frac{1}{\Delta x} \int_{x_i}^{x_{i+1/2}} w_i^n(x, t^{n+1}) \, dx. \qquad (4.4.18)$$

4.4 Numerical schemes for hyperbolic systems in 1-D

Since v^n is locally the exact solution, we have (see(2.2.5))

$$\int_{x_i}^{x_{i+1}} v^n(x, t^{n+1})\, dx = \int_{x_i}^{x_{i+1}} v^n(x, t^n)\, dx - \int_{t^n}^{t^{n+1}} f(v^n(x_{i+1}, s))\, ds$$

$$+ \int_{t^n}^{t^{n+1}} f(v^n(x_i, s))\, ds \; . \tag{4.4.19}$$

LEMMA 4.4.5 *The Godunov scheme is in conservation form, consistent and consistent with the entropy condition.*

Proof The conservation form can be proved as in §2.2 for scalar equations, and the consistency is obvious. Let us prove the consistency with the entropy condition. The exact solution $v^n(x, t)$ of the local Riemann problem satisfies the entropy condition, for all entropy pairs (U, F) we have

$$\int_{x_{i-1/2}}^{x_{i+1/2}} U(v^n(x, t^{n+1}))\, dx - \int_{x_{i-1/2}}^{x_{i+1/2}} U(v^n(x, t^n))\, dx$$

$$+ \int_{t^n}^{t^{n+1}} F(v^n(x_{i+1/2}, t))\, dt - \int_{t^n}^{t^{n+1}} F(v^n(x_{i-1/2}, t))\, dt \leq 0 \; , \tag{4.4.20}$$

and therefore, since $v^n(\cdot, t^n)$ is constant on $[x_{i-\frac{1}{2}}, x_{i+\frac{1}{2}}]$,

$$\int_{x_{i-1/2}}^{x_{i+1/2}} U(v^n(x, t^{n+1}))\, dx \leq \Delta x U(u_i^n)$$

$$- \Delta t [G(u_i^n, u_{i+1}^n) - G(u_{i-1}^n, u_i^n)] \; ,$$

where

$$u_i^n := 1\Delta x \int_{x_{i-1/2}}^{x_{i+1/2}} v^n(x, t^n)\, dx \; ,$$

$$G(u_i^n, u_{i+1}^n) := \frac{1}{\Delta t} \int_{t^n}^{t^{n+1}} F(v^n(x_{i+\frac{1}{2}}, t))\, dt.$$

Therefore we have $G(u,u) = F(u)$ and, since U is convex,

$$U(u_i^{n+1}) = U\left(\frac{1}{\Delta x} \int_{x_{i-1/2}}^{x_{i+1/2}} v^n(x, t^{n+1})\, dx\right)$$

$$\leq \frac{1}{\Delta x} \int_{x_{i-1/2}}^{x_{i+1/2}} U(v^n(x, t^{n+1}))\, dx$$

$$\leq U(u_i^n) - \lambda\Big[G(u_i^n, u_{i+1}^n) - G(u_{i-1}^n, u_i^n)\Big]. \qquad (4.4.21)$$

Therefore the Godunov scheme is consistent with the entropy condition. $\square$

Now we shall give some examples for Godunov-like schemes where the exact Riemann solver is replaced by an approximation Riemann solver.

EXAMPLE 4.4.6 (Linearization of scalar equations) The exact solution of the scalar initial value problem

$$\partial_t u + \partial_x f(u) = 0 \quad \text{in } \mathbb{R} \times \mathbb{R}^+,$$
$$u(x, 0) = u_0(x) = \begin{cases} u_L & \text{for } x < 0, \\ u_R & \text{for } x \geq 0 \end{cases} \qquad (4.4.22)$$

for $u_L > u_R$ is given by

$$u(x, t) := \begin{cases} u_L & \text{for } x/t < s, \\ u_R & \text{for } x/t \geq s, \end{cases}$$

where

$$s := s(u_L, u_R) = \frac{f(u_L) - f(u_R)}{u_L - u_R}.$$

This means that the optimal linearization or (4.4.22) is given by

$$\partial_t u + s\partial_x u = 0.$$

In particular, we have

$$s(u_L, u_R)(u_L - u_R) = f(u_L) - f(u_R),$$

4.4 Numerical schemes for hyperbolic systems in 1-D

$$s(u_L, u_R) \to f'(u_L) \quad \text{as} \quad u_R \to u_L .$$

This way of linearizing nonlinear equations is the basis of the following definition of the Roe scheme.

DEFINITION 4.4.7 (Riemann solver of Roe [187]) *We replace the Riemann problem*

$$\partial_t u + \partial_x f(u) = 0 \quad \text{in } \mathbb{R} \times \mathbb{R}^+ , \tag{4.4.23}$$

$$u(x,0) = \begin{cases} u_L & \text{for } x < 0 , \\ u_R & \text{for } x \geq 0 \end{cases}$$

by the following linear Riemann problem:

$$\partial_t w + A_{LR} \partial_x w = 0 \quad \text{in } \mathbb{R} \times \mathbb{R}^+ ,$$

$$w(x,0) = \begin{cases} u_L & \text{for } x < 0 , \\ u_R & \text{for } x \geq 0 \end{cases} \tag{4.4.24}$$

where $A_{LR} \in \mathbb{R}^{m \times m}$, which should depend on u_L and u_R ($A_{LR} = A(u_L, u_R)$), has to satisfy the following conditions

$$f(v) - f(w) = A(v,w)\,(v-w) , \tag{4.4.25}$$
$$A(v,w) \to f'(v) \text{ in the operator norm as } w \to v , \tag{4.4.26}$$
$$A(v,w) \text{ has only real eigenvalues} \tag{4.4.27}$$
$$A(v,w) \text{ has a complete system of eigenvectors .}$$

The last condition implies the solvability of (4.4.24). Then the Roe scheme is defined like the Godunov scheme (4.4.15)–(4.4.17), but the exact local Riemann solution (4.4.16) is replaced by the exact solution of (4.4.24), which is given by (4.4.14).

LEMMA 4.4.8 (Conservation form of the Roe scheme) *Let us assume that there exists a matrix $A(v,w)$ satisfying (4.4.25)–(4.4.27). Then the Roe scheme as defined in Definition 4.4.7 can be written in conservation form*

$$w_j^{n+1} = w_j^n - \frac{\Delta t}{\Delta x}[g(w_j^n, w_{j+1}^n) - g(w_{j-1}^n, w_j^n)] ,$$

342   4 Initial value problems for systems in 1-D

where the numerical flux $g : \mathbb{R}^{m \times m} \to \mathbb{R}^m$ is defined as

$$g(v, w) := \frac{1}{2}[f(v) + f(w)] - \frac{1}{2} \sum_{i=1}^{m} |\lambda_i| \alpha_i r_i . \tag{4.4.28}$$

The λ_i and the r_i are the eigenvalues and eigenvectors of $A(v, w)$ respectively, and the coefficients α_i are defined by

$$w - v = \sum_{i=1}^{m} \alpha_i r_i. \tag{4.4.29}$$

Proof Let $u(x, t, u_L, u_R)$ denote the exact solution of the Riemann problem (4.4.24). According to the Definition 4.4.7 of the Roe scheme, we have

$$w_j^{n+1} = w_j^n - \frac{\Delta t}{\Delta x} \Big\{ f_j + \frac{\Delta x}{2\Delta t} w_j^n$$

$$- \frac{1}{\Delta t} \int_{x_j}^{x_{j+1/2}} u(x - x_{j+\frac{1}{2}}, \Delta t, w_j^n, w_{j+1}^n) dx$$

$$- \Big[f_j - \frac{\Delta x}{2\Delta t} w_j^n + \frac{1}{\Delta t} \int_{x_{j-1/2}}^{x_j} u(x - x_{j-\frac{1}{2}}, \Delta t, w_{j-1}^n, w_j^n) dx \Big] \Big\} ,$$

$$\tag{4.4.30}$$

where $f_j := f(w_j^n)$. Let $\tilde{f}(u) := A_{LR} u$ and $u(x - x_{j+\frac{1}{2}}, t) := u(x - x_{j+\frac{1}{2}}, t, w_j^n, w_{j+1}^n)$. The integral form of the conservation law (see (2.2.5)) implies

$$\int_{x_j}^{x_{j+1/2}} u(x - x_{j+\frac{1}{2}}, \Delta t) \, dx - \int_{x_j}^{x_{j+1/2}} u(x - x_{j+\frac{1}{2}}, 0) \, dx$$

$$+ \int_0^{\Delta t} \tilde{f}(u(0, s)) \, ds - \int_0^{\Delta t} \tilde{f}(u(-\frac{1}{2}\Delta x, s)) \, dx = 0 .$$

4.4 Numerical schemes for hyperbolic systems in 1-D

Therefore we obtain

$$\int_{-\Delta x/2}^{0} u(x, \Delta t)\, dx - \frac{1}{2}\Delta x w_j^n$$
$$+ \int_{0}^{\Delta t} \tilde{f}(u(0, s, w_j^n, w_{j+1}^n))\, ds - \Delta t \tilde{f}(w_j^n) = 0 . \quad (4.4.31)$$

Similarly we get in $]x_{j+\frac{1}{2}}, x_{j+1}[$

$$\int_{0}^{\Delta x/2} u(x, \Delta t, w_j^n, w_{j+1}^n)\, dx - \frac{1}{2}\Delta x w_{j+1}^n + \Delta t \tilde{f}(w_{j+1}^n)$$
$$- \int_{0}^{\Delta t} \tilde{f}(u(0, s, w_j^n, w_{j+1}^n))\, ds = 0 . \quad (4.4.32)$$

Then (4.4.31) and (4.4.32) imply that

$$\int_{0}^{\Delta x/2} u(x, \Delta t, w_j^n, w_{j+1}^n)\, dx - \frac{1}{2}\Delta x w_{j+1}^n + \Delta t \tilde{f}(w_{j+1}^n)$$
$$= - \int_{-\Delta x/2}^{0} u(x, \Delta t, w_j^n, w_{j+1}^n)\, dx + \frac{1}{2}\Delta x w_j^n + \Delta t \tilde{f}(w_j^n) ,$$

and therefore, using $\tilde{f}(w_{j+1}) - \tilde{f}(w_j^n) = f(w_{j+1}^n) - f(w_j^n)$ (see (4.4.25)),

$$\Delta t f_j + \frac{1}{2}\Delta x w_j^n - \int_{x_j}^{x_{j+1/2}} u(x - x_{j+\frac{1}{2}}, \Delta t, w_j^n, w_{j+1}^n)\, dx$$
$$= \Delta t f_{j+1} - \frac{1}{2}\Delta x w_{j+1}^n + \int_{x_{j+1/2}}^{x_{j+1}} u(x - x_{j+\frac{1}{2}}, \Delta t, w_j^n, w_{j+1}^n)\, dx \quad (4.4.33)$$
$$=: \Delta t g(w_j^n, w_{j+1}^n) .$$

4 Initial value problems for systems in 1-D

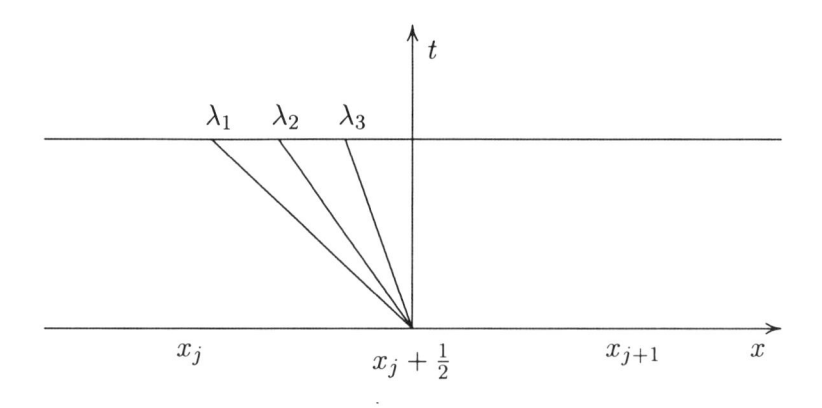

Figure 4.4.1

Then the scheme (4.4.30) can be written in conservation form, i.e.

$$w_j^{n+1} = w_j^n - \frac{\Delta t}{\Delta x}[g(w_j^n, w_{j+1}^n) - g(w_{j-1}^n, w_j^n)]. \quad (4.4.34)$$

Recall that (see (4.4.14))

$$u(x, \Delta t) = \begin{cases} u_0 & \text{if } x < \lambda_1 \Delta t, \\ u_0 + \sum_{j=1}^{k} \alpha_j r_j & \text{if } \Delta t \lambda_k \leq x < \lambda_{k+1} \Delta t, \\ & k = 1, \ldots, m-1, \\ u_0 + \sum_{j=1}^{m} \alpha_j r_j & \text{if } \Delta t \lambda_m \leq x. \end{cases}$$

Using this we get

$$\int_{x_j}^{x_{j+1/2}} u(x - x_{j+\frac{1}{2}}, \Delta t, w_j^n, w_{j+1}^n) \, dx$$

$$= \int_{-\Delta x/2}^{0} u(x, \Delta t, w_j^n, w_{j+1}^n) \, dx$$

4.4 Numerical schemes for hyperbolic systems in 1-D   345

$$= \int_{-\Delta x/2}^{\lambda_1^- \Delta t} u(x, \Delta t)\, dx + \sum_{k=1}^{m-1} \int_{\lambda_k^- \Delta t}^{\lambda_{k+1}^- \Delta t} u(x, \Delta t)\, dx + \int_{\lambda_n^- \Delta t}^{0} u(x, \Delta t)\, dx$$

$$= u_0 \left[\lambda_1^- \Delta t - \left(-\frac{1}{2}\Delta x\right)\right] + \sum_{k=1}^{m-1} \Delta t (\lambda_{k+1}^- - \lambda_k^-)\left(u_0 + \sum_{j=1}^{k} \alpha_j r_j\right)$$

$$- \lambda_m^- \Delta t \left(u_0 + \sum_{j=1}^{m} \alpha_j r_j\right)$$

$$= \frac{1}{2}\Delta x u_0 + \sum_{k=2}^{m} \Delta t \lambda_k^- \sum_{j=1}^{k-1} \alpha_j r_j - \sum_{k=1}^{m-1} \Delta t \lambda_k^- \sum_{j=1}^{k} \alpha_j r_j -$$

$$\lambda_m^- \Delta t \sum_{j=1}^{m} \alpha_j r_j$$

$$= \frac{1}{2}\Delta x u_0 - \Delta t \sum_{j=1}^{m} \lambda_j^- \alpha_j r_j\ .$$

Then we can show that the numerical flux $g(u,v)$ as defined in (4.4.33) can be written as

$$g(w_j, w_{j+1}) = f_j + \frac{\Delta x}{2\Delta t} w_j - \frac{\Delta x}{2\Delta t} w_j + \sum_{i=1}^{m} \alpha_i \lambda_i^- r_i$$

$$= \frac{1}{2}(f_j + f_{j+1}) - \frac{1}{2}(f_{j+1} - f_j) + \sum_{i=1}^{m} \alpha_i \lambda_i^- r_i$$

$$= \frac{1}{2}(f_j + f_{j+1}) - \frac{1}{2} A(w_j, w_{j+1})(w_{j+1} - w_j)$$
$$+ \sum_{i=1}^{m} \alpha_i \lambda_i^- r_i$$

$$= \frac{1}{2}(f_j + f_{j+1}) - \sum_{i=1}^{m} \left(\frac{\lambda_i}{2} - \lambda_i^-\right) \alpha_i r_i$$

$$= \frac{1}{2}(f_j + f_{j+1}) - \frac{1}{2} \sum_{i=1}^{m} |\lambda_i| \alpha_i r_i. \qquad (4.4.35)$$

□

REMARK 4.4.9 *In general the discrete functions defined by the Roe scheme do not approximate the entropy solution.*

4 Initial value problems for systems in 1-D

Proof We consider the special case of a scalar equation

$$\partial_t u + \partial_x f(u) = 0 \quad \text{in } \mathbb{R} \times \mathbb{R}^+ ,$$

$$u(x,0) = \begin{cases} u_L & \text{in } x < 0 , \\ u_R & \text{in } x \geq 0 . \end{cases}$$

Then the corresponding linear approximation is

$$\partial_t u + \lambda \partial_x u = 0 .$$

Then the numerical flux is given by (4.4.35). Let as assume that the dimension of the system is equal to 1. As in (4.4.14), we get for the exact solution of (4.4.24) the following intermediate steps

$$u_0 = u_L , \quad u_1 = u_R ,$$

with λ given by $\lambda(u_L - u_R) = f(u_L) - f(u_R)$. Let us assume $f(u) = u^2$ and λ given by $\lambda(u_L - u_R) = f(u_L) - f(u_R)$, $u_L = -1$, $u_R = 1$, and therefore $\lambda = 0$. Then we get for the numerical flux

$$g(u_L, u_R) = \tfrac{1}{2}[f(u_L) + f(u_R)] - \tfrac{1}{2}|\lambda|(u_L - u_R)$$
$$= \tfrac{1}{2}[f(u_L) + f(u_R)].$$

But this means that central differences may occur, and we have seen that in this case in general the entropy condition is not satisfied (see Example 2.1.20a).

EXAMPLE 4.4.10 Consider the Euler equations of gas dynamics (4.4.1) in conservation form

$$\partial_t U + \partial_x F(U) = 0.$$

We define

$$A(U_L, U_R) := DF(\bar{U}) ,$$

where

$$\bar{U} = \begin{pmatrix} \bar{\rho} \\ \bar{\rho u} \\ \bar{e} \end{pmatrix}$$

and

$$\bar{\rho} = \sqrt{\rho_R \rho_L},$$
$$\bar{u} = \frac{\sqrt{\rho_L} u_L + \sqrt{\rho_R} u_R}{\sqrt{\rho_L} + \sqrt{\rho_R}}, \qquad (4.4.36)$$
$$\bar{H} := \frac{\sqrt{\rho_L} H_L + \sqrt{\rho_R} H_R}{\sqrt{\rho_L} + \sqrt{\rho_R}},$$
$$H := \frac{e + p}{\rho}.$$

Then $A(U_L, U_R)$ satisfies the conditions (4.4.25) – (4.4.27) (see [193]). The values $\bar{\rho}, \bar{u}$ and $\bar{H}$ are called the Roe mean values.

EXAMPLE 4.4.11 (Application of the Roe scheme) We consider the Euler equation in the form (4.1.9)–(4.1.11). Then for $U = (\rho, u, p)^t$ this system can be written as

$$\partial_t U + A \partial_x U = 0, \qquad (4.4.37)$$

where (see (4.1.13))

$$A = A(U) = \begin{pmatrix} u & \rho & 0 \\ 0 & u & 1/\rho \\ 0 & \rho a^2 & u \end{pmatrix}.$$

The eigenvalues and eigenvectors of A are $u - a$, u, $u + a$ and

$$\begin{pmatrix} \rho \\ -a \\ \rho a^2 \end{pmatrix}, \quad \begin{pmatrix} 1 \\ 0 \\ 0 \end{pmatrix}, \quad \begin{pmatrix} \rho \\ a \\ \rho a^2 \end{pmatrix} \qquad (4.4.38)$$

respectively. Then the Roe scheme for (4.1.9)–(4.1.11) is given by

$$w_j^{n+1} = w_j^n - \frac{\Delta t}{\Delta x}[g(w_j^n, w_{j+1}^n) - g(w_{j-1}^n, w_j^n)]$$

where

$$g(v, w) = \tfrac{1}{2}[F(v) + F(w)] - \tfrac{1}{2}\sum_{i=1}^{3} |\lambda_i| \alpha_i r_i. \qquad (4.4.39)$$

348   4 Initial value problems for systems in 1-D

The function $\tilde{F}$ is given by the following definition. Define the Roe mean value $\bar{U}$ of v and w, and let

$$\tilde{F}(u) := A(\bar{U})u . \qquad (4.4.40)$$

λ_i, α_i and r_i in (4.4.39) are defined as follows. The eigenvalues λ_i of $A(\bar{U})$ are

$$\lambda_1 = \bar{u} - \bar{a}, \quad \lambda_2 = \bar{u}, \quad \lambda_3 = \bar{u} + \bar{a}.$$

The eigenvectors r_i of $A(\bar{U})$ are

$$r_1 = \begin{pmatrix} \bar{\rho} \\ -\bar{a} \\ \bar{\rho}\bar{a}^2 \end{pmatrix}, \quad r_2 = \begin{pmatrix} 1 \\ 0 \\ 0 \end{pmatrix}, \quad r_3 = \begin{pmatrix} \bar{\rho} \\ \bar{a} \\ \bar{\rho}\bar{a}^2 \end{pmatrix},$$

and the α_i are defined by

$$w - v = \sum_{i=1}^{3} \alpha_i r_i.$$

REMARK 4.4.12 *Let us consider the Riemann problem for the Euler equations of gas dynamics in conservation variables $U = (\rho, \rho u, e)$:*

$$\begin{aligned} \partial_t U + \partial_x f(U) &= 0 \quad \text{in } \mathbb{R} \times \mathbb{R}^+ , \\ U(x, 0) &= \begin{cases} U_l & \text{if } x < 0 , \\ U_r & \text{if } x \geq 0 , \end{cases} \end{aligned} \qquad (4.4.41)$$

with $U_l = (\rho_l, (\rho u)_l, e_l)$ and $U_r = (\rho_r, (\rho u)_r, e_r)$. For smooth solutions the partial differential equation in (4.4.41) can be written as

$$\partial_t U + A(U)\partial_x U = 0 \quad \text{in } \mathbb{R} \times \mathbb{R}^+ ,$$

where

$$A(U) = \begin{pmatrix} 0 & 1 & 0 \\ -\frac{1}{2}(3-\gamma)u^2 & (3-\gamma)u & \gamma - 1 \\ (\gamma-1)u^3 - \frac{\gamma u e}{\rho} & \frac{\gamma e}{\rho} - \frac{3}{2}(\gamma-1)u^2 & \gamma u \end{pmatrix}$$

4.4 Numerical schemes for hyperbolic systems in 1-D

(see [97, Vol. II, page 144]). The eigenvectors r_1, r_2, r_3 and eigenvalues $\lambda_1, \lambda_2, \lambda_3$ of A are

$$r_1 = (1, u-a, H-ua)^t,$$
$$r_2 = (1, u, \tfrac{1}{2}u^2)^t,$$
$$r_3 = (1, u+a, H+ua)^t,$$
$$\lambda_1 = u-a, \quad \lambda_2 = u, \quad \lambda_3 = u+a,$$

where $H = \frac{\gamma}{\gamma-1}\frac{p}{\rho} + \frac{u^2}{2}$. As for the Roe scheme, we consider the following linear problem for $W = (\rho, \rho u, e)$:

$$\partial_t W + A(\hat{W})\partial_x W = 0 \quad \text{in } \mathbb{R} \times \mathbb{R}^+,$$
$$W(x,0) = \begin{cases} U_l & \text{if } x < 0, \\ U_r & \text{if } x \geq 0, \end{cases} \quad (4.4.42)$$

where $\hat{W}$ is a suitable constant state. The basic idea of the Roe scheme consists in choosing $\hat{W}$ such that W as a solution of (4.4.42) is a good approximation of U, the solution of (4.4.41). Now we shall prove the following lemma.

LEMMA 4.4.13 (Existence for the Riemann problem) *Consider the Riemann problem (4.4.41) for initial values $U_l := (\tilde{\rho}, -\tilde{u}\tilde{\rho}, \tilde{e})$ and $U_r := (\tilde{\rho}, \tilde{u}\tilde{\rho}, \tilde{e})$ such that $\tilde{\rho}, \tilde{u} > 0$ and*

$$(\gamma-1)\frac{\tilde{e}}{\tilde{\rho}\tilde{u}} < \tilde{u} < \frac{4\gamma}{3\gamma-1}\frac{\tilde{e}}{\tilde{\rho}\tilde{u}} \quad \text{and } \gamma > 1. \quad (4.4.43)$$

Then we have $\tilde{p} > 0$, and the Riemann problem has a solution.

Proof First we shall show that the condition (4.1.89) is satisfied, i.e.

$$\tilde{u} < \frac{2}{\gamma-1}\tilde{a}.$$

We obtain from (4.4.43)

$$\tilde{u}^2 < \frac{4\gamma}{\gamma-1}\left(\frac{\tilde{e}}{\tilde{\rho}} - \frac{\tilde{u}^2}{2}\right) = \frac{4}{(\gamma-1)^2}\frac{\gamma\tilde{p}}{\tilde{\rho}} = \frac{4}{(\gamma-1)^2}\tilde{a}^2.$$

Furthermore, we have $\tilde{p} > 0$, since

$$\tilde{p} = (\gamma - 1)\left(\tilde{e} - \frac{\tilde{\rho}}{2}\tilde{u}^2\right) = \frac{\gamma - 1}{2}\tilde{\rho}\tilde{u}\left(\frac{2\tilde{e}}{\tilde{\rho}\tilde{u}} - \tilde{u}\right) > 0,$$

because $2 > 4\gamma/(3\gamma - 1)$. □

Now we should like to show that for the Roe scheme with mean values as in (4.4.36) it may happen that the density ρ and the pressure p may become negative.

LEMMA 4.4.14 (Nonpositive density for the Roe scheme) *Let U_l and U_r be as in Lemma 4.4.13. Then for any $\hat{W} = (\hat{\rho}, \hat{\rho}\hat{u}, \hat{e})$ with $\hat{u} = 0$ the solution W of (4.4.42) has a solution such that $\rho < 0$ or $p < 0$. In particular, this may happen for the Roe mean values (4.4.36) in the case of the initial data as described in Example 4.4.13.*

Proof (See [59], where the argument works only with $\hat{u} = 0$.) We compute the exact solution of the Riemann problem (4.4.42) with U_l and U_r as defined in Lemma 4.4.13. Using (4.4.13), we compute the coefficients α_1, α_2 and α_3 such that

$$U_r - U_l = \sum_{i=1}^{3} \alpha_i r_i.$$

It turns out that

$$\begin{pmatrix} 0 \\ 2(\tilde{u}\tilde{\rho}) \\ 0 \end{pmatrix} = \alpha_1 \begin{pmatrix} 1 \\ \hat{u} - \hat{a} \\ \hat{H} - \hat{u}\hat{a} \end{pmatrix} + \alpha_2 \begin{pmatrix} 1 \\ \hat{u} \\ \frac{1}{2}\hat{u}^2 \end{pmatrix} + \alpha_3 \begin{pmatrix} 1 \\ \hat{u} + \hat{a} \\ \hat{H} + \hat{u}\hat{a} \end{pmatrix}$$

$$= \alpha_1 \begin{pmatrix} 1 \\ -\hat{a} \\ \hat{H} \end{pmatrix} + \alpha_2 \begin{pmatrix} 1 \\ 0 \\ 0 \end{pmatrix} + \alpha_3 \begin{pmatrix} 1 \\ \hat{a} \\ \hat{H} \end{pmatrix},$$

since $\hat{u} = 0$. Therefore we obtain

$$\alpha_1 = -\frac{\tilde{u}\tilde{\rho}}{\hat{a}}, \quad \alpha_2 = 0, \quad \alpha_3 = \frac{\tilde{u}\tilde{\rho}}{\hat{a}}.$$

4.4 Numerical schemes for hyperbolic systems in 1-D

As in Example 4.4.3 the exact solution is given by

$$W(x,t) = \begin{cases} U_l & \text{if } x/t < \hat{u} - \hat{a}, \\ U_1 & \text{if } \hat{u} - \hat{a} \le x/t < \hat{u}, \\ U_2 & \text{if } \hat{u} \le x/t < \hat{u} + \hat{a}, \\ U_r & \text{if } \hat{u} + \hat{a} \le x/t, \end{cases}$$

where $U_k := U_l + \sum_{i=1}^{k} \alpha_i r_i$ for $k = 1, 2$. Explicitly, we obtain

$$U_1 = \begin{pmatrix} \rho_1 \\ u_1 \rho_1 \\ e_1 \end{pmatrix} = U_l + \alpha_1 r_1 = \begin{pmatrix} \tilde{\rho} \\ -\tilde{u}\tilde{\rho} \\ \tilde{e} \end{pmatrix} - \frac{\tilde{\rho}\tilde{u}}{\hat{a}} \begin{pmatrix} 1 \\ -\hat{a} \\ \hat{H} \end{pmatrix}$$

$$= \begin{pmatrix} \tilde{\rho} - \frac{\tilde{\rho}\tilde{u}}{\hat{a}} \\ 0 \\ \tilde{e} - \frac{\tilde{\rho}\tilde{u}}{\hat{a}}\hat{H} \end{pmatrix}$$

and $U_2 = U_1$. We should like to show that $\rho_1 < 0$ or $p_1 < 0$. Therefore assume that

$$\rho_1 \ge 0 \quad \text{and} \quad p_1 \ge 0. \tag{4.4.44}$$

Since

$$\rho_1 = \tilde{\rho} - \frac{\tilde{\rho}\tilde{u}}{\hat{a}}$$

and

$$p_1 = (\gamma - 1)e_1$$
$$= (\gamma - 1)\left(\tilde{e} - \frac{\tilde{\rho}\tilde{u}}{\hat{a}}\hat{H}\right) = (\gamma - 1)\left(\tilde{e} - \frac{\tilde{\rho}\tilde{u}}{\hat{a}}\frac{\hat{a}^2}{\gamma - 1}\right),$$

we obtain from (4.4.44)

$$\hat{a} \ge \tilde{u} \quad \text{and} \quad \hat{a} \le (\gamma - 1)\frac{\tilde{e}}{\tilde{\rho}\tilde{u}},$$

and then obviously

$$\tilde{u} \le (\gamma - 1)\frac{\tilde{e}}{\tilde{\rho}\tilde{u}}$$

but this contradicts (4.4.43). □

Because of this lemma, the Roe scheme has to be modified in order to avoid this phenomena. In [59] the reader will find some ideas for this.

EXAMPLE 4.4.15 (Riemann solver of Harten, Lax and van Leer)
Again we consider the Riemann problem for (4.4.37) with respect to the values (U_L, U_R) and let a_m and a_M denote the smallest and largest eigenvalues of $Df(\bar{U})$,

$$a_m := \bar{u} - \bar{a} \quad \text{and} \quad a_M := \bar{u} + \bar{a} \,,$$

where $\bar{u}$ and $\bar{a}$ are defined in (4.4.36) as the Roe mean values. Then the approximate Riemann solver of Harten, Lax and van Leer is given by

$$U\left(x, t, u_L, u_R\right) = \begin{cases} U_L & (x/t < a_m) \,, \\ \tilde{U} & (a_m < x/t < a_M) \,, \\ U_R & (a_M < x/t) \,, \end{cases}$$

where $\tilde{U}$ is defined such that

$$\tilde{U} = \frac{a_M U_R - a_m U_L}{a_M - a_m} - \frac{f(U_R) - f(U_L)}{a_M - a_m}.$$

The numerical flux can be obtained in a similar way as in Lemma 4.4.8:

$$g(U_L, U_R) = \frac{a_M^- - a_m^-}{a_M - a_m} f(U_R) + \frac{a_M^+ - a_m^+}{a_M - a_m} f(U_L)$$
$$- \frac{1}{2} \frac{a_M |a_m| - a_m |a_M|}{a_M - a_m} (U_R - U_L).$$

It is very simple to implement this scheme, and the results are quite good. As for the Roe scheme, there may be some problems with the entropy condition.

Now we shall describe some further schemes for solving the Euler equations in 1-D. First we need some preparatory lemmas.

LEMMA 4.4.16 *For the Euler equations*

$$\partial_t U + \partial_x F(U) = 0 \tag{4.4.45}$$

we have

$$\alpha F(U) = F(\alpha U).$$

4.4 Numerical schemes for hyperbolic systems in 1-D

Proof This is left as an exercise.

COROLLARY 4.4.17 *For the Euler equations (4.4.45) we have*
$$F(U) = DF(U)U.$$

Proof Using Lemma 4.4.16, we obtain
$$F(U) = F(U)\frac{1+h-1}{h} = \frac{F(U)(1+h) - F(U)}{h}$$
$$= \frac{F(U(1+h)) - F(U)}{h} = DF(U)U. \qquad \square$$

Now we can define the flux vector splitting scheme.

DEFINITION 4.4.18 (Flux vector splitting scheme) *Let us describe the flux vector splitting scheme for solving (4.4.45). We assume a uniform grid $x_i := i\Delta x$ in the x-direction. Let*
$$U_j^0 := \frac{1}{\Delta x} \int_{x_{j-1/2}}^{x_{j+1/2}} U_0(x)\,dx,$$

where U_0 are the initial values for (4.4.45). Assume that U_i^n is already defined as a piecewise-constant function on $[x_{i-\frac{1}{2}}, x_{i+\frac{1}{2}}]$.

Since the system (4.4.45) is hyperbolic, there exists a matrix $T = T(U)$ such that
$$T^{-1}DF(U)T = \operatorname{diag} = \begin{pmatrix} \lambda_1 & & \\ & \lambda_2 & \\ & & \lambda_3 \end{pmatrix},$$

where $\lambda_i = \lambda_i(U)$ are the eigenvalues of $DF(U)$. Define
$$A^\pm := DF(U)^\pm := T \begin{pmatrix} \lambda_1^\pm & & \\ & \lambda_2^\pm & \\ & & \lambda_3^\pm \end{pmatrix} T^{-1}.$$

Then the flux vector splitting scheme is defined as

$$U_i^{n+1} = U_i^n - \frac{\Delta t}{\Delta x}[g(U_i^n, U_{i+1}^n) - g(U_{i-1}^n, U_i^n)], \qquad (4.4.46)$$

where

$$g(u, v) := A^+(w_1)u + A^-(w_2)v.$$

with $w_1 = u$ and $w_2 = v$ for the scheme of Steger and Warming [226], $w_1 = w_2 = \frac{1}{2}(u+v)$ for the scheme of Vijayasundaram [237]. For the flux vector splitting of van Leer we have

$$g(u,v) = \frac{1}{2}\left[A(u) + \left|A\left(\frac{u+v}{2}\right)\right|\right]u + \frac{1}{2}\left[A(v) - \left|A\left(\frac{u+v}{2}\right)\right|\right]v.$$

If the problem is linear, this corresponds to (4.4.11).

This scheme can be motivated as follows. We assume that U is smooth and integrate (4.4.45) with respect to space and time:

$$0 = \int_{t^n}^{t^{n+1}} \int_{x_i}^{x_{i+1}} \partial_t U + \partial_x F(U) = \int_{x_i}^{x_{i+1}} [U(x, t^{n+1}) - U(x, t^n)]$$

$$+ \int_{t^n}^{t^{n+1}} \left[F(U(x_{i+1}, t)) - F(U(x_i, t))\right]$$

$$= \int_{x_i}^{x_{i+1}} [U(x, t^{n+1}) - U(x, t^n)]$$

$$+ \int_{t^n}^{t^{n+1}} [A_{i+1} U(x_{i+1}, t) - A_i U(x_i, t)],$$

where $A_{i+1} := F'(U(x_{i+1}, t))$ and where we have used Corollary 4.4.17. So

$$0 = \int_{x_i}^{x_{i+1}} \ldots + \int_{t^n}^{t^{n+1}} (A_{i+1}^+ + A_{i+1}^-)U_{i+1}^n - (A_i^+ + A_i^-)U_i^n$$

$$= \Delta x(U_i^{n+1} - U_i^n) + \Delta t[A_{i+1}^+ U_{i+\frac{1}{2}}^n + A_{i+1}^- U_{i+\frac{3}{2}}^n$$

$$- (A_i^+ U_{i-\frac{1}{2}}^n + A_i^- U_{i+\frac{1}{2}}^n)] + \mathcal{O}(\Delta t).$$

Using this as motivation for a numerical scheme we get

$$U_i^{n+1} = U_i^n - \frac{\Delta t}{\Delta x}[A_{i+1}^+ U_{i+\frac{1}{2}}^n + A_{i+1}^- U_{i+\frac{3}{2}}^n - (A_i^+ U_{i-\frac{1}{2}}^n + A_i^- U_{i+\frac{1}{2}}^n)] \,.$$

Or, using

$$g(u,v) := A^+(u)u + A^-(v)v \,,$$

we obtain (4.4.46). A similar method was developed by Vijayasundaram (see [237]):

$$g(u,v) = A^+(w)u + A^-(w)v$$

where $w = \frac{1}{2}(u+v)$.

REMARK 4.4.19 *For numerical examples in 1-D we refer to [221]. Test problems in 2-D and 3-D on unstructured grids are given in §5.2. Numerical experiments for the shock tube problem with ENO schemes as described in §3.7 generalized to systems are shown in Figures 4.4.2 - 4.4.5.*

REMARK 4.4.20 *In [212] convergence of finite difference schemes for the p-system and the Navier–Stokes equations in 1-D in the case of smooth solutions are considered.*

4.5 The Osher–Solomon scheme

In this section we describe the Osher–Solomon scheme. First let us recall the Engquist–Osher scheme for the scalar equation ($f(0) = 0$)

$$\partial_t u + \partial_x f(u) = 0 \quad \text{in } \mathbb{R} \times \mathbb{R}^+ \,,$$
$$u(x,0) = u_0(x) \quad \text{in } \mathbb{R}.$$

Then the Engquist–Osher scheme is given by

$$u_j^{n+1} = u_j^n - \frac{\Delta t}{\Delta x}[f^+(u_j^n) - f^+(u_{j-1}^n) + f^-(u_{j+1}^n) - f^-(u_j^n)] \,, (4.5.1)$$

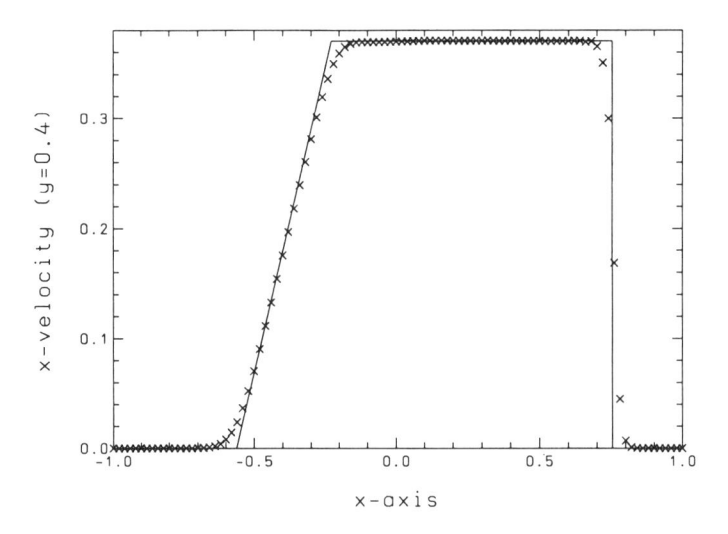

Figure 4.4.2 Shock Tube Problem: 2nd-order ENO solution, t = 0.75, CFL = $\frac{1}{2}$, 10 000 triangles, 5151 nodes.

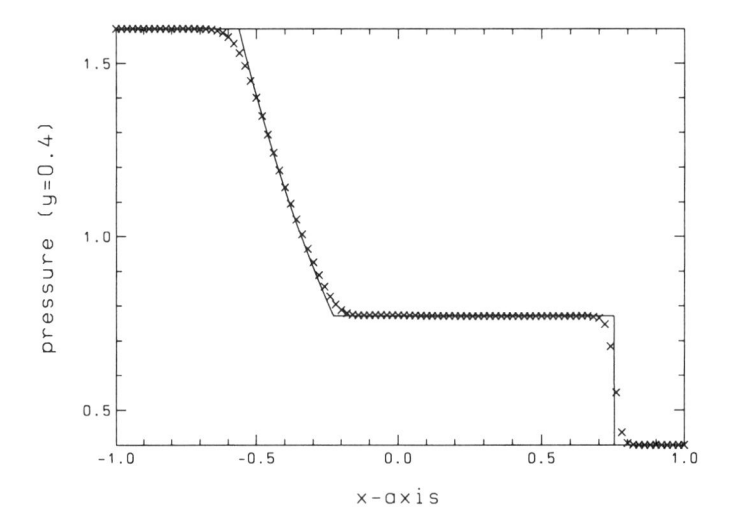

Figure 4.4.3 Shock Tube Problem: 2nd-order ENO solution, t = 0.75, CFL = $\frac{1}{2}$, 10 000 triangles, 5151 nodes.

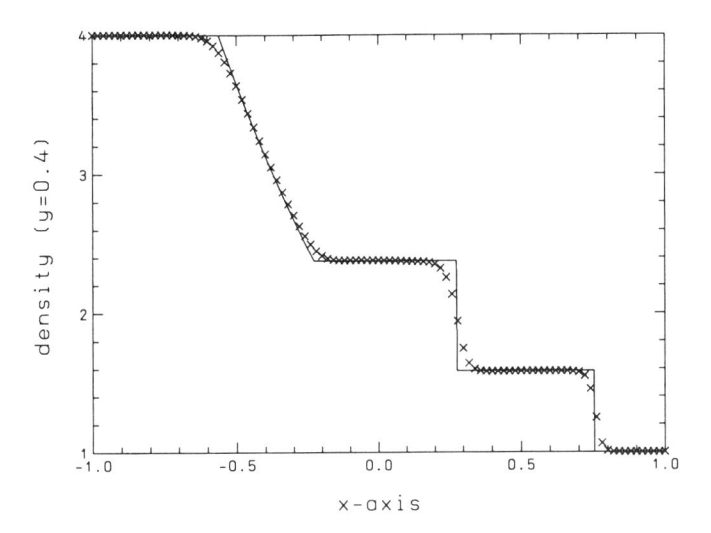

Figure 4.4.4 Shock Tube Problem: 2nd-order ENO solution, $t = 0.75$, CFL $= \frac{1}{2}$, 10 000 triangles, 5151 nodes.

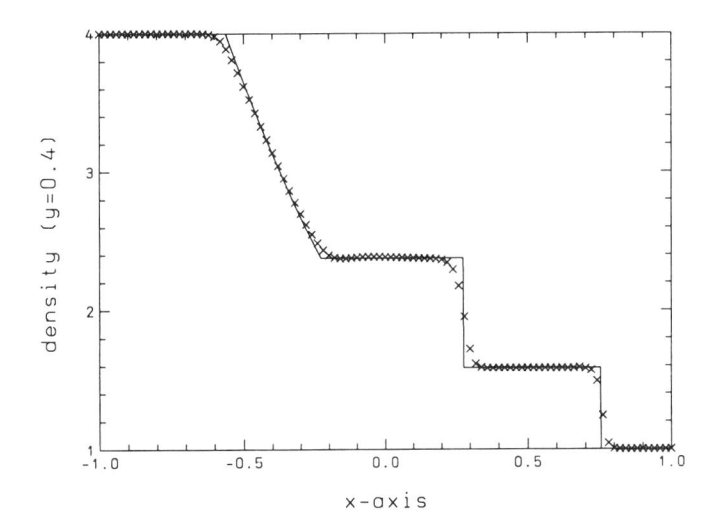

Figure 4.4.5 3rd-order ENO solution, $t = 0.75$, CFL $= \frac{1}{2}$, 10 000 triangles, 5151 nodes.

where $f^{\pm}(s) := \int_0^s (f'(\tau))^{\pm} d\tau$. Using the characteristic function χ of the set $\{\tau \big| f'(\tau) > 0\}$, we obtain

$$f^+(s) = \int_0^s \chi(\tau) f'(\tau) d\tau , \qquad f^-(s) = \int_0^s [1 - \chi(\tau)] f'(\tau) \, d\tau. \quad (4.5.2)$$

Then (4.5.1) can be written as

$$u_j^{n+1} = u_j^n - \frac{\Delta t}{\Delta x} \left[\int_{u_{j-1}^n}^{u_j^n} \chi(\tau) f'(\tau) \, d\tau + \int_{u_j^n}^{u_{j+1}^n} [1 - \chi(\tau)] f'(\tau) \, d\tau \right]. (4.5.3)$$

Now this special form of the Engquist–Osher scheme will be generalized to hyperbolic systems of the form

$$\partial_t v + \partial_x f(v) = 0 \quad \text{in } \mathbb{R} \times \mathbb{R}^+ ,$$
$$v(x, 0) = v_0(x) \quad \text{in } \mathbb{R} ,$$

where $v : \mathbb{R} \times \mathbb{R}^+ \to \mathbb{R}^m$ and $f : \mathbb{R}^m \to \mathbb{R}^m$. The original ideas have been published in [175]. Let $\lambda_1 < \ldots < \lambda_m$ be the right eigenvalues and $r_1, \ldots, r_m$ the right eigenvectors of $Df(v)$, and let $T(v) \in \mathbb{R}^{m \times m}$ be such that

$$T^{-1} Df T = \text{diag}(\lambda_1, \ldots, \lambda_m) =: \Lambda .$$

In order to use a similar notation as in (4.5.3), we define

$$\chi(v) := T(v) \text{diag}\{\tfrac{1}{2} + \tfrac{1}{2} \text{sign } \lambda_k(v)\} T^{-1}(v).$$

Then we obtain for $\Lambda^* := \text{diag}\{\lambda_j^+\}$ and $\text{diag} = \text{diag}\{\tfrac{1}{2} + \tfrac{1}{2} \text{sign } \lambda_k(v)\}$.

$$\chi(v) Df(v) = T \, \text{diag} \, T^{-1} \, T \Lambda T^{-1}$$
$$= T \Lambda^* T^{-1} =: Df(v)^+ .$$

Similarly, we get

$$[1 - \chi(v)] Df(v) =: [Df(v)]^- .$$

Let $I(0,s) := [0, s^+] \cup [s^-, 0]$ and let $w : \mathbb{R}^m \to \mathbb{R}^m$ be defined in the following way. For given v_{j-1}, v_j and $s_1, \ldots, s_m$ let $w^{m+1} := v_{j-1}$ and for $k = m, \ldots, 1$ (see Figure 4.5.1)

$$(w^k)'(s) = r_k(w^k) \quad \text{for } s \in I(0, s_k) ,$$

$$w^k(0) = \begin{cases} w^{(k+1)}(s_{k+1}) & \text{if } k < m , \\ w^{m+1} & \text{if } k = m . \end{cases} \quad (4.5.4)$$

We assume that $r_k(w)$ is Lipschitz-continuous in w. Then $w^1(s_1)$ depends of course on the choice of $s_1, \ldots, s_m$ and we define

$$w(s_1, \ldots, s_m) := w^1(s_1). \quad (4.5.5)$$

Now we should like to choose $s_1, \ldots, s_m$ such that

$$w(s_1, \ldots, s_m) = v_j. \quad (4.5.6)$$

Notice that

$$w(0, \ldots, 0) = w^1(0) = w^2(s_2)\big|_{s_2=0}$$
$$= w^3(s_3)\big|_{s_3=0} = \ldots = w^m(s_m)\big|_{s_m=0}$$
$$= w^{m+1}(s_{m+1}) = v_{j-1} . \quad (4.5.7)$$

The Jacobian of w in 0 is equal to

$$Dw = (r_1(v_{j-1}), \ldots, r_m(v_{j-1})) , \quad (4.5.8)$$

and therefore $\det Dw \neq 0$. The equation in (4.5.8) can be seen as follows.

$$\partial_1 w\big|_{\vec{s}=0} = \partial_1 w^1(s_1)\big|_{\vec{s}=0} = r_1(w^1(s_1))\big|_{\vec{s}=0}$$
$$= r_1(w^1(0)) = r_1(v_{j-1}).$$

The last equality follows from (4.5.7).
Since

$$w^1(s_1) = w^1(s_2) + \int_0^{s_1} r_2(w^1(s)) \, ds ,$$

we have

$$\partial_2 w^1(s_1) = \partial_2 w^2(s_2) = r_2(w^2(s_2)),$$
$$\partial_2 w\big|_{\bar{s}=0} = r_2(w^2(0)) = r_2(v_{j-1}).$$

Continuing this calculation, we finally obtain (4.5.8). Therefore (4.5.8) implies that for sufficiently small $|v_j - v_{j-1}|$ there exists a $(s_1, \ldots, s_m)$ such that

$$w(s_1, \ldots, s_m) = v_j. \tag{4.5.9}$$

Therefore we assume that $|v_j - v_{j-1}|$ is small enough. Since the $(s_1, \ldots, s_m)$ satisfying (4.5.8) depend on j (see (4.5.3)), we denote them by $(s_1^j, \ldots, s_m^j)$. Let Γ_k be the curve $\{(w^k(s), s) | s \in I(0, s_k)\}$. Then we have

$$\int_{\Gamma_k} \chi(\tau) Df(\tau) \, d\tau := \int_0^{s_k} \chi(w^k(s)) Df(w^k(s))(w^k)'(s) \, ds$$

$$= \int_0^{s_k} \chi(w^k(s)) \lambda_k(w^k(s)) r_k(w^k(s)) \, ds$$

$$= \int_0^{s_k} \lambda_k(w^k(s))^+ r_k(w^k(s)) \, ds$$

since we assume that $r_1, \ldots, r_m$ are orthonormal. Using this property, we define the Osher–Solomon scheme [175] as

$$v_j^{n+1} = v_j^n - \frac{\Delta t}{\Delta x} \sum_{k=1}^m \left[\int_0^{s_k^j} \lambda_k(w^k(s))^+ r_k(w^k(s)) \, ds \right. \tag{4.5.10}$$
$$\left. + \int_0^{s_k^{j+1}} \lambda_k(w^k(s))^- r_k(w^k(s)) \, ds \right].$$

The time step has to be restricted such that

$$\frac{\Delta t}{\Delta x} \sup_{k, \bar{s}} |\lambda_k| < 1.$$

4.5 The Osher–Solomon scheme

LEMMA 4.5.1 (Conservation form) *The Osher–Solomon scheme is in conservation form.*

Proof

$$v_j^{n+1} = v_j^n - \frac{\Delta t}{\Delta x} \sum_{k=1}^{m} \Bigg[\int_0^{s_k^j} \lambda_k(w^k(s))^+ r_k(w^k(s))\, ds$$

$$+ \int_0^{s_k^{j+1}} \lambda_k(w^k(s))^- r_k(w^k(s))\, ds \Bigg]$$

$$= v_j^n - \frac{\Delta t}{\Delta x} \sum_{k=1}^{m} \Bigg\{ \int_{\Gamma_k^j} \chi(w) Df(w) + \int_{\Gamma_k^{j+1}} [1-\chi(w)] Df(w) \Bigg\}$$

and furthermore we can show that ($\Gamma^j := \Gamma_k^j$, k fixed)

$$\int_{\Gamma^j} \chi(w) Df(w) + \int_{\Gamma^{j+1}} [1-\chi(w)] Df(w)$$

$$= \int_{\Gamma^{j+1}} Df(w) - \int_{\Gamma^{j+1}} \chi(w) Df(w) + \int_{\Gamma^j} \chi(w) Df(w)$$

$$= -\Delta_+ \int_{\Gamma^j} \chi(w) Df(w) + \int_{\Gamma^{j+1}} Df(w)$$

$$= \frac{1}{2}\Bigg\{ \Delta_+ f(w_j) - \Delta_+ \int_{\Gamma^j} \chi(w) Df(w) + \Delta_- f(w_j)$$

$$+ \Delta_+ \int_{\Gamma^j} (1-\chi(w)) Df(w) \Bigg\}$$

where

$$\Delta_+ f(w_j) := \int_{\Gamma^{j+1}} Df(w), \quad \Delta_- f(w_j) := \int_{\Gamma^j} Df(w)$$

and

$$\Delta_+ \int_{\Gamma^j} = \int_{\Gamma^{j+1}} - \int_{\Gamma^j}.$$

We continue:

$$= \frac{1}{2}(\Delta_+ + \Delta_-)f(w_j) - \frac{1}{2}\Delta_+ \int_{\Gamma^j} \{\chi(w)Df(w) - [1-\chi(w)]Df(w)\}$$

$$= \frac{1}{2}(\Delta_+ + \Delta_-)f(w_j) - \frac{1}{2}\Delta_+ \int_{\Gamma^j} |Df(w)|$$

$$= \frac{1}{2}[\Delta_+ f(w_j) + \Delta_- f(w_j)] - \frac{1}{2}\left(\int_{\Gamma^{j+1}} |Df| - \int_{\Gamma^j} |Df|\right).$$

Therefore

$$v_j^{n+1} = v_j^n - \frac{\Delta t}{\Delta x}\sum_{k=1}^{m}\int_{\Gamma_k^j} \chi(w)Df(w) + \int_{\Gamma_k^{j+1}} [1-\chi(w)]Df(w)$$

$$= v_j^n - \frac{\Delta t}{\Delta x}\left[f(v_{j+1}) - f(v_{j-1}) - \sum_{k=1}^{m}\left(\int_{\Gamma_k^{j+1}} |Df| - \int_{\Gamma_k^j} |Df|\right)\right]\frac{1}{2}$$

$$= v_j^n - \frac{\Delta t}{\Delta x}[g(v_j, v_{j+1}) - g(v_{j-1}, v_j)],$$

where

$$g(v_j, v_{j+1}) = \frac{1}{2}[f(v_{j+1}) + f(v_j)] - \frac{1}{2}\sum_{k=1}^{m}\int_{\Gamma_k^{j+1}} |Df|.$$

This completes the proof of the lemma. □

REMARK 4.5.2 *For the numerical flux of the Osher–Solomon scheme we obtain*

$$g(v_j, v_{j+1}) = f(v_j) + \int_{\Gamma^{j+1}} [Df(w)]^-.$$

4.5 The Osher–Solomon scheme

Proof By simple calculation we obtain

$$g(v_j, v_{j+1})$$
$$= \frac{1}{2}[f(v_{j+1}) + f(v_j)] - \frac{1}{2}\sum_k \int_{\Gamma_k^{j+1}} |Df|$$

$$= \frac{1}{2}\left\{f(v_{j+1}) + f(v_j) - \sum_k \int_{\Gamma_k^{j+1}} \left[Df(w) - 2Df(w)^-\right]\right\}$$

$$= \frac{1}{2}\left\{f(v_{j+1}) + f(v_j) - \left[f(v_{j+1}) - f(v_j) - 2\sum_k \int_{\Gamma_k^{j+1}} Df(w)^-\right]\right\}$$

$$= \frac{1}{2}\left[2f(v_j) + 2\int_{\Gamma^{j+1}} Df(w)^-\right]$$

$$= f(v_j) + \int_{\Gamma^{j+1}} Df(w)^- .$$

□

The form (4.5.10) is not very useful for practical computations, since the w^k and $s_1, \ldots, s_m$ are not known explicitly. Therefore we assume that each k-field is either genuinely nonlinear, i.e.

$$\frac{d}{ds}\lambda_k(w(s)) = \nabla\lambda_k(w(s))r_k(w(s)) = 1 , \qquad (4.5.11)$$

or linearly degenerate, i.e.

$$\frac{d}{ds}\lambda_k(w(s)) = \nabla\lambda_k(w(s))r_k(w(s)) = 0. \qquad (4.5.12)$$

In the second case, if $\lambda_k(w(s)) = \text{const} > 0$, we have

$$\int_{\Gamma_k} \chi(\tau)Df(\tau) \, d\tau = \int_0^{s_k^j} \lambda_k(w^k(s))^+ r_k(w^k(s)) \, ds$$

$$= \int_0^{s_k^j} Df(w_k(s))(w^k)'(s) \, ds$$

$$= f(w^k(s_k^j)) - f(w^k(0))$$

$$= f(w^{k-1}(0)) - f(w^k(0)) ,$$

and is zero otherwise. Altogether, we have

$$\int_0^{s_k^j} \lambda_k(w^k(s))^+ r_k(w^k(s)) \, ds$$
$$= \begin{cases} f(w^{k-1}(0)) - f(w^k(0)) & \text{if } \lambda_k > 0, \\ 0 & \text{if } \lambda_k \leq 0. \end{cases} \qquad (4.5.13)$$

The computation of the w^k will be explained in Example 4.5.3 for a special situation. Now let us consider the genuinely nonlinear case (see (4.5.11)). Then $\lambda_k(w^k(\cdot))$ is strictly monotonically increasing. If $\lambda_k(w^k(\cdot))$ does not change sign then we can proceed as in the linearly degenerate case. Otherwise there exists at most one point t_k such that

$$\lambda_k(w^k(t_k)) = 0. \qquad (4.5.14)$$

Let $\bar{w}^k := w^k(t_k)$. Then we obtain

$$\int_0^{s_k^j} \lambda_k(w^k(s))^+ r_k(w^k(s)) \, ds = f(a) - f(b),$$

where a and b are defined as follows:

$$a = \begin{cases} w^{k-1}(0) = w^k(s_k) & \text{if } \lambda_k(w^{k-1}(0)) > 0, \\ \bar{w}^k & \text{if } \lambda_k(w^{k-1}(0)) \leq 0, \end{cases}$$
$$b = \begin{cases} w^k(0) & \text{if } \lambda_k(w^k(0)) > 0, \\ \bar{w}^k & \text{if } \lambda_k(w^k(0)) \leq 0. \end{cases} \qquad (4.5.15)$$

In order to derive this, we have to consider the following cases.

(1a) $\lambda_k(w^k(0)) \leq \lambda_k(w^k(s_k)) \leq 0$,

(1b) $\lambda_k(w^k(0)) \leq 0 \leq \lambda_k(w^k(s_k))$,

(1c) $0 \leq \lambda_k(w^k(0)) \leq \lambda_k(w^k(s_k))$.

Of course for (1a) we have

$$\int_0^{s_k^j} \lambda_k(w^k(s))^+ r_k(w^k(s)) \, ds = 0 \, ,$$

which corresponds to (4.5.15).

For (1b) we obtain

$$\int_0^{s_k^j} \lambda_k(w^k(s))^+ r_k(w^k(s)) \, ds = \int_{t_k}^{s_k^j} \lambda_k(w^k(s)) r_k(w^k(s)) \, ds$$

$$= \int_{t_k}^{s_k^j} Df(w^k(s))(w^k)'(s) \, ds = f(w^k(s_k^j)) - f(\bar{w}^k) \, .$$

Case (1c) and the result for

$$\int_0^{s_k^{j+1}} \lambda_k(w^k(s))^- r_k(w^k(s)) \, ds$$

follow similarly.

EXAMPLE 4.5.3 (Application of the Osher–Solomon scheme to the Euler equations of gas dynamics.) Now we shall describe the scheme more explicitly for the Euler equations of gas dynamics. The curves Γ_k^j are given by (see (4.5.4))

$$\begin{aligned}(w^k)' &= r_k(w^k) \quad \text{for } s \in I(0, s_k^j) \, , \\ w^k(0) &= w^{k+1}(s_{k+1}^j)\end{aligned} \tag{4.5.16}$$

for $k = 1, 2, 3$. Then $\Gamma^j = \bigcup_{k=1}^3 \Gamma_k^j$ connects v_{j-1} and v_j in the following way (see Figure 4.5.1 and notice that $v_{j-1} = w^4 = w^3(0)$):

$$v_{j-1} = w^3(0) \underset{\Gamma_3^j}{\to} w^3(s_3) = w^2(0) \underset{\Gamma_2^j}{\to} w^2(s_2) = w^1(0) \underset{\Gamma_1^j}{\to}$$

$$w^1(s_1) = v_j \, , \tag{4.5.17}$$

4 Initial value problems for systems in 1-D

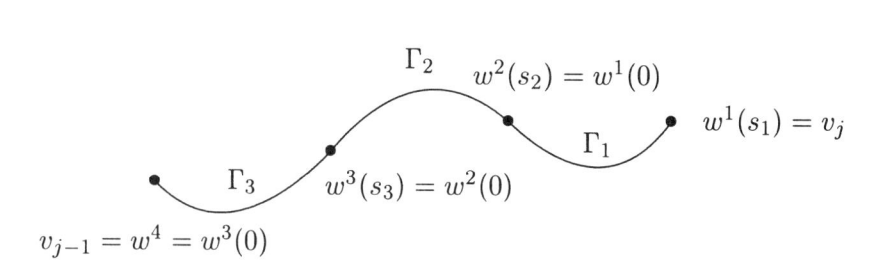

$$\Gamma_2 \quad w^2(s_2) = w^1(0)$$
$$w^1(s_1) = v_j$$
$$\Gamma_1$$
$$\Gamma_3 \quad w^3(s_3) = w^2(0)$$
$$v_{j-1} = w^4 = w^3(0)$$

Figure 4.5.1

Now we have to compute s_1, s_2 and s_3. In (4.5.6) the existence of s_1, s_2, s_3 has been shown. Now we shall use the k-Riemann invariants to get them explicitly. The k-Riemann invariants are (see Example 4.1.36)

$$\begin{array}{ll} \psi_2^1 = \frac{p}{\rho^\gamma}, & \psi_3^1 = u + \frac{2}{\gamma-1}a, \\ \psi_1^2 = u, & \psi_3^2 = p, \\ \psi_1^3 = \frac{p}{\rho^\gamma}, & \psi_2^3 = u - \frac{2}{\gamma-1}a. \end{array} \quad (4.5.18)$$

By definition of the k-Riemann invariants, we know that ψ_ν^k are constant along $w^k(s)$ as defined in (4.5.16):

$$\begin{aligned} \frac{d}{ds}\psi_\nu^k(w^k(s)) &= \nabla \psi_\nu^k(w^k(s)) \cdot (w^k)'(s) \\ &= \nabla \psi_\nu^k(w^k(s)) \cdot r_k(w^k(s)) = 0. \end{aligned} \quad (4.5.19)$$

Notice that w^k is a parametrization of Γ_k^j. Let us use the notation (see (4.5.17))

$$v_{j-\frac{1}{3}} := w^2(s_2) = w^1(0), \quad v_{j-\frac{2}{3}} = w^3(s_3) = w^2(0).$$

Then (4.5.19) implies

$$\begin{array}{ll} \psi_2^1(w^1(0)) = \psi_2^1(w^1(s_1)), & \text{i.e. } \psi_2^1(v_{j-\frac{1}{3}}) = \psi_2^1(v_j), \\ \psi_3^1(w^1(0)) = \psi_3^1(w^1(s_1)), & \text{i.e. } \psi_3^1(v_{j-\frac{1}{3}}) = \psi_3^1(v_j), \\ \psi_1^2(w^2(0)) = \psi_1^2(w^2(s_2)), & \text{i.e. } \psi_1^2(v_{j-\frac{2}{3}}) = \psi_1^2(v_{j-\frac{1}{3}}), \\ \psi_3^2(w^2(0)) = \psi_3^2(w^2(s_2)), & \text{i.e. } \psi_3^2(v_{j-\frac{2}{3}}) = \psi_3^2(v_{j-\frac{1}{3}}), \\ \psi_1^3(w^3(0)) = \psi_1^3(w^3(s_3)), & \text{i.e. } \psi_1^3(v_{j-1}) = \psi_1^3(v_{j-\frac{2}{3}}), \\ \psi_2^3(w^3(0)) = \psi_2^3(w^3(s_3)), & \text{i.e. } \psi_2^3(v_{j-1}) = \psi_2^3(v_{j-\frac{2}{3}}). \end{array} \quad (4.5.20)$$

4.5 The Osher–Solomon scheme

Using the definition (4.5.18) in (4.5.20), we obtain (see Figure 4.5.2)

$$\frac{p_{j-\frac{1}{3}}}{\rho_{j-\frac{1}{3}}^{\gamma}} = \frac{p_j}{\rho_j^{\gamma}}, \quad u_{j-\frac{1}{3}} + \frac{2}{\gamma-1}a_{j-\frac{1}{3}} = u_j + \frac{2}{\gamma-1}a_j, \quad (4.5.21)$$

$$u_{j-\frac{1}{3}} = u_{j-\frac{2}{3}}, \quad p_{j-\frac{1}{3}} = p_{j-\frac{2}{3}}, \quad (4.5.22)$$

$$\frac{p_{j-\frac{2}{3}}}{\rho_{j-\frac{2}{3}}^{\gamma}} = \frac{p_{j-1}}{\rho_{j-1}^{\gamma}}, \quad u_{j-\frac{2}{3}} - \frac{2}{\gamma-1}a_{j-\frac{2}{3}} = u_{j-1} - \frac{2}{\gamma-1}a_{j-1}. \quad (4.5.23)$$

This is a system of six equations with six unknowns $u_{j-\frac{2}{3}}, u_{j-\frac{1}{3}}, p_{j-\frac{2}{3}}, p_{j-\frac{1}{3}},$

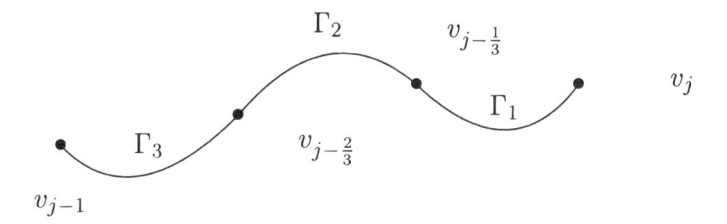

Figure 4.5.2

$\rho_{j-\frac{2}{3}}$ and $\rho_{j-\frac{1}{3}}$. It can be solved, and we obtain

$$\rho_{j-\frac{1}{3}} = \left\{ \frac{\frac{1}{2}(\gamma-1)(u_j - u_{j-1}) + a_j + a_{j-1}}{a_j\left[1 + (p_{j-1}/p_j)^{1/2\gamma}(\rho_{j-1}/\rho_j)^{-1/2}\right]} \right\}^{2/(\gamma-1)} \rho_j,$$

$$\rho_{j-\frac{2}{3}} = \left\{ \frac{\frac{1}{2}(\gamma-1)(u_j - u_{j-1}) + a_j + a_{j-1}}{a_{j-1}\left[1 + (p_j/p_{j-1})^{1/2\gamma}(\rho_j/\rho_{j-1})^{-1/2}\right]} \right\}^{2/(\gamma-1)} \rho_{j-1},$$

$$p_{j-\frac{2}{3}} = p_{j-\frac{1}{3}} = p_{j-1}\left(\frac{\rho_{j-1}}{\rho_{j-\frac{2}{3}}}\right)^{-\gamma},$$

$$u_{j-\frac{2}{3}} = u_{j-\frac{1}{3}} = u_{j-1} - \frac{2}{\gamma-1}\left[a_{j-1} - \left(\frac{\gamma p_{j-\frac{2}{3}}}{\rho_{j-\frac{2}{3}}}\right)^{\frac{1}{2}}\right].$$

4 Initial value problems for systems in 1-D

In the linearly degenerate case we use (4.5.12) with $k = 2$ and we obtain from (4.5.13)

$$\int_{\Gamma_2^j} [Df(w)]^+ = \int_0^{s_2^j} \lambda_2(w^2(s))^+ r_2(w^2(s))\, ds$$

$$= \begin{cases} f(w_{j-\frac{1}{3}}) - f(w_{j-\frac{2}{3}}) & \text{if } \bar{u} > 0, \\ 0 & \text{if } \bar{u} \leq 0, \end{cases} \quad (4.5.24)$$

(because of (4.5.12), λ_2 is constant along Γ_2^j). Since f can be written as

$$f(w) = \begin{pmatrix} \rho u \\ \rho u^2 + p \\ u\left(\frac{\gamma}{\gamma-1}p + \rho\frac{u^2}{2}\right) \end{pmatrix}$$

and, since $u_{j-\frac{2}{3}} = u_{j-\frac{1}{3}}$, $p_{j-\frac{2}{3}} = p_{j-\frac{1}{3}}$, we obtain for (4.5.24)

$$\int_{\Gamma_2^j} [Df(w)]^+ = (\rho_{j-\frac{1}{3}} - \rho_{j-\frac{2}{3}}) \max\{u_{j-\frac{2}{3}}, 0\} \begin{pmatrix} 1 \\ u_{j-\frac{2}{3}} \\ \frac{1}{2}u_{j-\frac{2}{3}}^2 \end{pmatrix},$$

and similarly

$$\int_{\Gamma_2^{j+1}} [Df(w)]^- = (\rho_{j+\frac{2}{3}} - \rho_{j+\frac{1}{3}}) \min\{u_{j+\frac{1}{3}}, 0\} \begin{pmatrix} 1 \\ u_{j+\frac{1}{3}} \\ \frac{1}{2}u_{j+\frac{1}{3}}^2 \end{pmatrix}.$$

Now we have to consider the genuinely nonlinear fields. Let $\hat{w}$ and $\tilde{w}$ be defined by (see (4.5.14))

$$\lambda_1(\hat{w}) = 0 \quad \text{and} \quad \lambda_3(\tilde{w}) = 0.$$

Similarly as in (4.5.20), the properties of the Riemann invariants imply also in this case that

$$\psi_2^1(\hat{w}) = \psi_2^1(v_j), \quad \psi_3^1(\hat{w}) = \psi_3^1(v_j) \quad (4.5.25)$$

and
$$\psi_1^3(\hat{w}) = \psi_1^3(v_{j-1}), \quad \psi_2^3(\hat{w}) = \psi_2^3(v_{j-1}). \tag{4.5.26}$$

Then using $\lambda_1(\hat{w}) = \hat{u} - \hat{a}$ and (4.5.18), we obtain

$$\hat{u} = \hat{a} \qquad \hat{p} = p_j(\frac{\hat{\rho}}{\rho_j})^\gamma$$
$$\hat{u} + \frac{2}{\gamma - 1}\hat{a} = u_j + \frac{2}{\gamma - 1}a_j$$

and therefore

$$\hat{a} = \frac{\gamma - 1}{\gamma + 1}\left(u_j + \frac{2}{\gamma - 1}a_j\right)$$

$$\hat{a} = \left(\frac{\gamma\hat{p}}{\hat{\rho}}\right)^{\frac{1}{2}} = \left(\frac{\gamma p_j \hat{\rho}^\gamma}{\hat{\rho}\rho_j^\gamma}\right)^{\frac{1}{2}} = \left(\frac{\gamma p_j}{\rho_j}\frac{\hat{\rho}^{\gamma-1}}{\rho_j^{\gamma-1}}\right)^{\frac{1}{2}}$$

$$= a_j \cdot \frac{\hat{\rho}^{(\gamma-1)/2}}{\rho_j^{(\gamma-1)/2}}$$

and consequently

$$\hat{\rho}^{(\gamma-1)/2} = \frac{\gamma - 1}{\gamma + 1}\left(u_j + \frac{2}{\gamma - 1}a_j\right)\frac{\rho_j^{(\gamma-1)/2}}{a_j}$$

$$\hat{u} = u_j + \frac{2}{\gamma - 1}(a_j - \hat{a}).$$

Similarly

$$\tilde{\rho}^{(\gamma-1)/2} = -\frac{\gamma - 1}{\gamma + 1}\frac{u_{j-1} - \frac{2}{\gamma-1}a_{j-1}}{a_{j-1}}\rho_{j-1}^{(\gamma-1)/2},$$

$$\tilde{p} = p_{j-1}\left(\frac{\tilde{\rho}}{\rho_{j-1}}\right)^\gamma,$$

$$\tilde{u} = u_{j-1} - \frac{2}{\gamma - 1}(a_{j-1} - \tilde{a}).$$

Therefore we obtain from (4.5.15)

$$\int_{\Gamma_1^j} (Df(w))^+ = f(a) - f(b) ,\qquad (4.5.27)$$

where

$$\begin{aligned}
a &= v_j & \text{if } \lambda_1(v_j) > 0, &\quad \text{i.e. if } \rho_j < \hat{\rho}, \\
a &= \hat{w} & \text{if } \lambda_1(v_j) \leq 0, &\quad \text{i.e. if } \rho_j \geq \hat{\rho}, \\
b &= v_{j-\frac{1}{3}} & \text{if } \lambda_1(v_{j-\frac{1}{3}}) > 0, &\quad \text{i.e. if } \rho_{j-\frac{1}{3}} < \hat{\rho}, \\
b &= \hat{w} & \text{if } \lambda_1(v_{j-\frac{1}{3}}) \leq 0, &\quad \text{i.e. if } \rho_{j-\frac{1}{3}} \geq \hat{\rho}.
\end{aligned} \qquad (4.5.28)$$

The first line in (4.5.28) can be seen as follows. Since $a = \sqrt{\gamma p/\rho}$ and $1 < \gamma \leq 2$, the function λ_1 is a decreasing function of ρ and $\lambda_1(\hat{\rho}) = 0$. Therefore $\lambda_1(v_j) > 0$ if and only if $\rho_j < \hat{\rho}$ and $\lambda_1(v_{j-\frac{1}{3}}) > 0$ if and only if $\rho_{j-\frac{1}{3}} < \hat{\rho}$. Now the expressions (4.5.27) and (4.5.28) can be written as

$$\int_{\Gamma_1^j} [Df(w)]^+ = f(a) - f(b) ,$$

where

$$a = \begin{cases} v_j & \text{if } \rho_j < \hat{\rho}, \\ \hat{w} & \text{if } \rho_j \geq \hat{\rho}, \end{cases}$$

$$b = \begin{cases} v_{j-\frac{1}{3}} & \text{if } \rho_{j-\frac{1}{3}} < \hat{\rho}, \\ \hat{w} & \text{if } \rho_{j-\frac{1}{3}} \geq \hat{\rho}. \end{cases}$$

Similarly, we obtain

$$\int_{\Gamma_1^j} [Df(w)]^- dw = f(a) - f(b) ,\qquad (4.5.29)$$

where

$$a = \begin{cases} v_j & \text{if } \rho_j \geq \hat{\rho}, \\ \hat{w} & \text{if } \rho_j < \hat{\rho}, \end{cases}$$

$$b = \begin{cases} v_{j-\frac{1}{3}} & \text{if } \rho_{j-\frac{1}{3}} \geq \hat{\rho}, \\ \hat{w} & \text{if } \rho_{j-\frac{1}{3}} < \hat{\rho}. \end{cases}$$

Since λ_3 is an increasing function of ρ on Γ_3^j, we have

$$\int_{\Gamma_3^j} [Df(w)]^+ dw = f(a) - f(b) , \qquad (4.5.30)$$

where

$$a = \begin{cases} v_{j-\frac{2}{3}} & \text{if } \rho_{j-\frac{2}{3}} > \tilde{\rho} , \\ \tilde{w} & \text{if } \rho_{j-\frac{2}{3}} \leq \tilde{\rho} , \end{cases}$$

$$b = \begin{cases} v_{j-1} & \text{if } \rho_{j-1} > \tilde{\rho} , \\ \tilde{w} & \text{if } \rho_{j-1} \leq \tilde{\rho} . \end{cases}$$

Furthermore,

$$\int_{\Gamma_3^j} [Df(w)]^- dw = f(a) - f(b) , \qquad (4.5.31)$$

where

$$a = \begin{cases} v_{j-\frac{2}{3}} & \text{if } \rho_{j-\frac{2}{3}} \leq \tilde{\rho} , \\ \tilde{w} & \text{if } \rho_{j-\frac{2}{3}} > \tilde{\rho} , \end{cases}$$

$$b = \begin{cases} v_{j-1} & \text{if } \rho_{j-1} \leq \tilde{\rho} , \\ \tilde{w} & \text{if } \rho_{j-1} > \tilde{\rho} . \end{cases}$$

For numerical experiments with this schemes we refer to Example 5.3.7.

5 Initial value problems for systems of conservation laws in 2-D

5.1 Finite volume methods for systems in several space dimensions

Let us use the same notation as in §3.2. In this section we shall describe the algorithm to solve systems of conservation laws in several space dimensions. We restrict ourselves to problems in 2-D. Problems in 3-D can be treated similarly. Let us write the systems considered in the form

$$\partial_t u + \partial_x f_1(u) + \partial_y f_2(u) = 0 \quad \text{in } \mathbb{R}^2 \times \mathbb{R}^+ \,, \ u \in \mathbb{R}^m \,. \tag{5.1.1}$$

Unfortunately there is no general existence result for the solution of (5.1.1) globally in time. Of course, this implies that there are no convergence results for numerical schemes. Therefore we can only prescribe some algorithms for solving (5.1.1), which have been successful in numerical test problems. As for scalar equations, we can use structured or unstructured grids. In the first case we use the schemes, described in §4.4 and 4.5 in each space dimension in order to get an approximation of (5.1.1). The details of these algorithms are similar to those described in §3.1. Therefore we concentrate on the second case, i.e. on unstructured grids. At least we can prove that if the finite volume scheme as in (5.1.7) defines a convergent sequence, the limit is a weak solution of the conservation law (5.1.1). At the end of this section we discuss the details for the numerical schemes and present some numerical testproblems, including local grid refinement and coarsening. In particular, we are interested in solving the Euler equations of gas dynamics.

$$\partial_t \begin{pmatrix} \rho \\ \rho u \\ \rho v \\ e \end{pmatrix} + \partial_x \begin{pmatrix} \rho u \\ \rho u^2 + p \\ \rho u v \\ u(e+p) \end{pmatrix} + \partial_y \begin{pmatrix} \rho v \\ \rho u v \\ \rho v^2 + p \\ v(e+p) \end{pmatrix} = 0 \text{ in } \mathbb{R}^2 \times \mathbb{R}^+ \quad (5.1.2)$$

with suitable initial and boundary conditions and the additional equation of state

$$p = (\gamma - 1)\left[e - \frac{\rho}{2}(u^2 + v^2)\right].$$

Below we only consider strictly hyperbolic systems, which are defined as follows.

DEFINITION 5.1.1 (Strictly hyperbolic systems) *The system (5.1.1) is called (strictly) hyperbolic if all eigenvalues of $\alpha f_1'(u) + \beta f_2'(u)$ are real (and distinct) for all $\alpha, \beta \in \mathbb{R}, u \in \mathbb{R}^m$.*

ASSUMPTION 5.1.2 For all $l = 1, \ldots, k$ we assume that for any bounded domain Ω

$$\sum_{T_j \cap \Omega \neq \emptyset} \big||T_j| - |T_{jl}|\big| = o(1) \quad (5.1.3)$$

if $h \to 0$ and that there are constants $c_1, c_2 \geq 0$ such that

$$0 < c_1 \leq \frac{\Delta t}{h} \leq c_2 \quad (5.1.4)$$

as $\Delta t, h \to 0$. Moreover, we assume that there exists a constant $c_3 > 0$ such that

$$\sup_j \frac{h^2}{|T_j|} \leq c_3. \quad (5.1.5)$$

A similar condition as (5.1.3) is used in [25].

374 Initial value problems for systems of conservation laws in 2-D

DEFINITION 5.1.3 *For given initial values $u_0 \in L^\infty(\mathbb{R}^2, \mathbb{R}^m)$ let u_j^n be defined by the following numerical scheme:*

$$u_j^0 := \frac{1}{|T_j|} \int_{T_j} u_0, \tag{5.1.6}$$

$$u_j^{n+1} := u_j^n - \frac{\Delta t}{|T_j|} \sum_{l=1}^{k} g_{jl}(u_j^n, u_{jl}^n), \tag{5.1.7}$$

where for g_{jl}, $l = 1, \ldots, k$ we assume that for any $R > 0$ and for all $u, v, u', v' \in B_R(0)$ we have (see also (3.2.8)-(3.2.10)).

$$| g_{jl}(u,v) - g_{jl}(u',v') | \leq c(R) \, h \, (| u - u' | + | v - v' |), \tag{5.1.8}$$

$$g_{j,\alpha(j,l)}(u,v) = -g_{\alpha(j,l),j}(v,u), \tag{5.1.9}$$

$$g_{j,\alpha(j,l)}(u,u) = \nu_{j,\alpha(j,l)} f(u), \tag{5.1.10}$$

where $f(u) := \begin{pmatrix} f_1(u) \\ f_2(u) \end{pmatrix}$, $\nu \cdot f = \nu^x f_1 + \nu^y f_2$ and we have used the notation $g_{j,\alpha(j,l)} := g_{jl}$.

EXAMPLE 5.1.4 (Flux vector splitting [226, 237, 134]) Assume that the nonlinearities in (5.1.1) satisfy

$$f_1(u) = f_1'(u) \, u \quad \text{and} \quad f_2(u) = f_2'(u) \, u. \tag{5.1.11}$$

This is for instance valid for the Euler equations of gas dynamics (see [226]). Let

$$C_{jl}(w) := n_{jl} f'(w). \tag{5.1.12}$$

Since the system (5.1.1) is hyperbolic, there exists a nonsingular matrix $Q(= Q_{jl}(w))$ such that

$$Q^{-1} C_{jl}(w) Q = D = \text{diag}(\lambda_1, \ldots, \lambda_m)$$

where $D = D(w)$ is a diagonal matrix consisting of the eigenvalues $\lambda_i = \lambda_i(w)$ of $C_{jl}(w)$. Now let

$$D^+ := \operatorname{diag}(\lambda_1^+, \ldots, \lambda_m^+) \quad \text{and} \quad C_{jl}^+(w) := QD^+Q^{-1}.$$

The matrix $C_{jl}^-(w)$ is defined similarly. Then the flux vector splitting can be written in the form (5.1.7) with

$$g_{jl}(u,v) = |S_{jl}| \left[C_{jl}^+(u)u + C_{jl}^-(v)v \right] \tag{5.1.13}$$

for the scheme of Steger and Warming [226]. Using the more general form

$$g_{jl}(u,v) = |S_{jl}|[C_1(u,v)u + C_2(u,v)v],$$

we get for

$$C_1(u,v) := C_{jl}^+\left(\frac{u+v}{2}\right), \quad C_2(u,v) := C_{jl}^-\left(\frac{u+v}{2}\right) \tag{5.1.14}$$

the scheme of Vijayasundaram [237], and for

$$\begin{array}{l} C_1(u,v) := C_{jl}(u) + |C_{jl}(\frac{u+v}{2})|, \\ C_2(u,v) := C_{jl}(v) - |C_{jl}(\frac{u+v}{2})| \end{array} \tag{5.1.15}$$

the scheme of van Leer [134].

The schemes defined by (5.1.13)–(5.1.15) satisfy the conditions (5.1.8)–(5.1.10) respectively if (5.1.11) is satisfied.

The proof can be given in a similar way as in Example 3.2.7.

The following generalization of the classical Lax–Wendroff theorem will show that if there is a L^2-convergent sequence defined by a scheme of the form (5.1.6), (5.1.7) then the limit will be a solution of the system (5.1.1) in the distributional sense, i.e. the following vector identity holds:

$$\int_{\mathbb{R}^2 \times \mathbb{R}^+} u \partial_t \varphi + \int_{\mathbb{R}^2 \times \mathbb{R}^+} [f_1(u)\partial_x \varphi + f_2(u)\partial_y \varphi]$$

$$+ \int_{\mathbb{R}^2} u_0 \varphi(\cdot, 0) = 0 \tag{5.1.16}$$

for all test functions $\varphi \in C_0^\infty(\mathbb{R}^2 \times [0, \infty[)$.

THEOREM 5.1.5 *Let us assume that there is a mesh satisfying 3.2.1 – 3.2.4 and 5.1.3 and initial values $u_0 \in L^\infty(\mathbb{R}^2, \mathbb{R}^m)$. Then let u_j^n be defined by a numerical scheme of the form (5.1.7) and u_h be defined as in Notation 3.2.2. Furthermore, we assume*

$$\sup_{h} \sup_{x,y,t} | u_h(x,y,t) | \leq \text{const}, \tag{5.1.17}$$

$$\| u - u_h \|_{L^2(\mathbb{R}^2 \times \mathbb{R}^+, \mathbb{R}^m)} \to 0 . \tag{5.1.18}$$

Then u is a solution of (5.1.1) in the distributional sense.

A sufficient condition for (5.1.17) in the case of scalar equations is given in (3.3.36). The convergence to the entropy solution for all scalar equations has been considered in § 3.2 and 3.4.

To prove Theorem 5.1.5, we have to multiply (5.1.7) by $\varphi(w_j, t^n)$, where φ is a smooth test function, and sum over n and j. First we shall consider the term in (5.1.7) corresponding to the time derivative, and later the sum in (5.1.7) corresponding to the spatial derivatives. For the following propositions and lemmas we always assume the same conditions as in Theorem 5.1.5.

PROPOSITION 5.1.6 *Let $\varphi \in C_0^\infty(\mathbb{R}^2 \times [0, \infty[)$ and $\psi \in C^{0,1}(\mathbb{R})$. Then we have*

$$\sum_j \sum_{n=0}^N | T_j | [\psi(u_j^{n+1}) - \psi(u_j^n)] \varphi(w_j, t^n)$$

$$= -\int_{\mathbb{R}^2 \times \mathbb{R}^+} \psi(u_h) \partial_t \varphi - \int_{\mathbb{R}^2} \psi(u_0) \varphi(\cdot, 0) + \mathcal{O}(h). \tag{5.1.19}$$

Here $N \in \mathbb{N}$ is chosen such that $\varphi(\cdot, N\Delta t) = 0$, and the sum over j is taken such that $\varphi(w_j, t^n) \neq 0$.

Proof Using summation by parts we obtain

$$\sum_{j}\sum_{n=0}^{N} |T_j| \left[\psi(u_j^{n+1}) - \psi(u_j^n)\right] \varphi(w_j, t^n)$$

$$= \sum_{j}\sum_{n=0}^{N} |T_j| \, \psi(u_j^{n+1}) \left[\varphi(w_j, t^n) - \varphi(w_j, t^{n+1})\right]$$

$$- \sum_{j} |T_j| \, \psi(u_j^0)\varphi(w_j, 0) \, . \tag{5.1.20}$$

In the first sum of (5.1.20) we use

$$\varphi(w_j, t^n) - \varphi(w_j, t^{n+1}) = -\int_{t^{n+1}}^{t^{n+2}} \partial_t\varphi(w_j, t) \, dt + \mathcal{O}(\Delta t^2)$$

to obtain

$$\sum_{j}\sum_{n=0}^{N} |T_j| \, \psi(u_j^{n+1}) \left[\varphi(w_j, t^n) - \varphi(w_j, t^{n+1})\right]$$

$$= -\sum_{j}\sum_{n=0}^{N} \int_{T_j} \int_{t^{n+1}}^{t^{n+2}} \psi(u_j^{n+1})\partial_t\varphi(w_j, t) \, dt \, dx + \mathcal{O}(\Delta t)$$

$$= -\sum_{j}\sum_{n=0}^{N} \int_{T_j} \int_{t^n}^{t^{n+1}} \psi(u_h)\partial_t\varphi + \mathcal{O}(h)$$

$$= -\int_{\mathbb{R}^2 \times \mathbb{R}^+} \psi(u_h)\partial_t\varphi + \mathcal{O}(h) \, . \tag{5.1.21}$$

Then the second sum in (5.1.20) is equal to

$$\sum_{j} \int_{T_j} \psi(u_j^0)\varphi(w_j, 0) = \sum_{j} \int_{T_j} \psi(u_h(\cdot, 0))\varphi(\cdot, 0) + \mathcal{O}(h) \, .$$

This and (5.1.21) are just the right-hand side of (5.1.19), which gives us the desired result. □

Now let us consider the sum in (5.1.7) corresponding to the spatial derivatives.

PROPOSITION 5.1.7 Let u_j^n be defined by (5.1.6) and (5.1.7) and let u_h be as in Notation 3.2.2. Then for all test functions $\varphi \in C_0^\infty(\mathbb{R}^2 \times [0, \infty[)$ we have

$$\Delta t \sum_j \sum_{n=0}^N \sum_{l=1}^k g_{jl}(u_j^n, u_{\alpha(j,l)}^n)\varphi_j^n = -\int_{\mathbb{R}^2 \times \mathbb{R}^+} f(u_h)\nabla\varphi + o(1),$$

where $\varphi_j^n := \varphi(w_j, t^n)$.

This proposition will now be proved in several steps. In the following we suppress the time index n.

LEMMA 5.1.8 For all n we have

$$\sum_j \sum_{l=1}^k g_{jl}(u_j, u_{\alpha(j,l)})\varphi_j^n = \sum_j \sum_{l=1}^k g_{jl}(u_j, u_{\alpha(j,l)}) \left[\varphi_j^n - \varphi^n(z_{jl})\right],$$

where z_{jl} is the midpoint of S_{jl}.

Proof Let $j \in \mathbb{N}, l \in \{1, \ldots, k\}$ and $i := \alpha(j,l)$. Then there is a uniquely determined m such that $j = \alpha(i,m)$ (see Remark 3.2.3). In particular, by definition, $z_{jl} = z_{im}$. Using this property and the notation $\varphi(z_{jl}) := \varphi(z_{jl}, t^n)$, it turns out that

$$\sum_j \sum_{l=1}^k g_{jl}(u_j, u_{\alpha(j,l)})\varphi(z_{jl})$$

$$= -\sum_j \sum_{l=1}^k g_{\alpha(j,l),j}(u_{\alpha(j,l)}, u_j)\varphi(z_{jl})$$

$$= -\sum_i \sum_{m=1}^k g_{im}(u_i, u_{\alpha(i,m)})\varphi(z_{im})$$

$$= -\sum_j \sum_{l=1}^k g_{jl}(u_j, u_{\alpha(j,l)})\varphi(z_{jl}). \tag{5.1.22}$$

This proves the lemma. □

Now let us repeat (see 3.2.5) the basic idea of the finite volume methods. For smooth functions φ we have

$$\nabla\varphi(w_j) \sim \frac{1}{|T_j|}\int_{T_j}\nabla\varphi = \frac{1}{|T_j|}\int_{\partial T_j} n\varphi \qquad (5.1.23)$$

$$= \frac{1}{|T_j|}\sum_{l=1}^{k}\int_{S_{jl}} n\varphi \sim \frac{1}{|T_j|}\sum_{l=1}^{k}\nu_{j,\alpha(j,l)}\varphi(z_{jl}).$$

In the following lemma we shall make precise the $\sim$ symbol in (5.1.23).

LEMMA 5.1.9
$$|T_j|\nabla\varphi(w_j) \doteq \sum_{l=1}^{k}\nu_{j,\alpha(j,l)}\varphi(z_{jl}) + \mathcal{O}(h^3).$$

Proof By Taylor series expansions and (5.1.5), we obtain

$$\int_{T_j}\partial_x\varphi(z)\,dz = |T_j|\,\partial_x\varphi(w_j) + \mathcal{O}(h^4).$$

On the other hand using the midpoint rule,

$$\int_{T_j}\partial_x\varphi(z)\,dz = \int_{\partial T_j} n_x\varphi(x) = \sum_{l=1}^{k}\int_{S_{jl}} n_x\varphi(x) = \sum_{l=1}^{k}\nu_x\varphi(z_{jl}) + \mathcal{O}(h^3).$$

A similar formula can be derived for $\int_{T_j}\partial_y\varphi(z)\,dz$. Putting both together we get Lemma 5.1.9. $\square$

As a corollary we obtain the following.

COROLLARY 5.1.10

$$\int_{\mathbb{R}^2} f(u_h(\cdot,t^n))\nabla\varphi^n = \sum_j\sum_{l=1}^{k}\nu_{j,\alpha(j,l)}f(u_j)\varphi^n(z_{jl}) + \mathcal{O}(h).$$

Proof Using Lemma 5.1.9 and the assumption (5.1.17), we get

$$\sum_j \sum_{l=1}^k \nu_{j,\alpha(j,l)} f(u_j) \varphi^n(z_{jl}) = \sum_j f(u_j) \sum_{l=1}^k \nu_{j,\alpha(j,l)} \varphi^n(z_{jl})$$

$$= \sum_j f(u_j) \left[|T_j| \nabla \varphi^n(w_j) + \mathcal{O}(h^3) \right]$$

$$= \sum_j \int_{T_j} [f(u_j) \nabla \varphi^n(w_j)] + \mathcal{O}(h) = \int_{\mathbb{R}^2} f(u_h(\cdot, t^n)) \nabla \varphi^n + \mathcal{O}(h) .$$

$\square$

Now we show how to control the convergence of (5.1.7) as h tends to zero.

LEMMA 5.1.11 *Suppose that u_j^n and u_h are defined as in Theorem 5.1.5, satisfying (5.1.17) and (5.1.18). Then*

$$\Delta t \left| \sum_n \sum_j \sum_{l=1}^k [g_{jl}(u_j, u_{jl}) - \nu_{jl} f(u_j)] [\varphi_j - \varphi(z_{jl})] \right| \to 0$$

as $h\Delta t \to 0$.

Proof Using the smoothness of φ, (5.1.8) and (5.1.10), we get the following estimate:

$$\left| \sum_j \sum_{l=1}^k [g_{jl}(u_j, u_{jl}) - \nu_{jl} f(u_j)] [\varphi_j - \varphi(z_{jl})] \right|$$

$$\leq ch \sum_j \sum_{l=1}^k | g_{jl}(u_j, u_{jl}) - \nu_{jl} f(u_j) |$$

$$= ch \sum_j \sum_{l=1}^k | g_{jl}(u_j, u_{jl}) - g_{jl}(u_j, u_j) |$$

$$\leq ch^2 \sum_j \sum_{l=1}^k | u_{jl} - u_j | . \qquad (5.1.24)$$

For $x \in T_j$ and l fixed define

$$v_h^l(x,t) := u_h(w_{jl}, t) = u_{jl}^n \qquad (t^n \leq t < t^{n+1}) .$$

Then, using (5.1.5), we obtain

$$h^2 \Delta t \sum_{n,j} \sum_{l=1}^{k} | u_{jl} - u_j |$$

$$= h^2 \Delta t \sum_{n,j} \sum_{l} \frac{1}{|T_j|} \int_{T_j} | v_h^l(z, t^n) - u_h(z, t^n) | \, dz \qquad (5.1.25)$$

$$\leq c \sum_{l=1}^{k} \int_{\Omega} | v_h^l(z, t) - u_h(z, t) | \, dz \, dt, \qquad (5.1.26)$$

where $\Omega := (\mathbb{R}^2 \times \mathbb{R}^+) \cap \text{supp}(\varphi)$. If we can show that the last integral in (5.1.26) tends to zero, the lemma is proved. Since Ω is bounded, it is sufficient to show that

$$\int_{\Omega} | v_h^l(z, t) - u_h(z, t) |^2 \, dz \, dt \to 0 \quad \text{as} \quad h \to 0 \qquad (5.1.27)$$

for $l = 1, \ldots, k$. The functions v_h^l and u_h are vector-valued. But it follows from

$$\int_{\Omega} | v_h^l(z, t) - u_h(z, t) |^2 \, dz \, dt = \int_{\Omega} \sum_{i=1}^{n} | v_{h,i}^l(z, t) - u_{h,i}(z, t) |^2 \, dz \, dt$$

that it is sufficient to prove (5.1.27) componentwise. In the following we drop the indices l and i, writing v_h instead of $v_{h,i}^l$ and similarly for $u_{h,i}$. Since u_h is uniformly bounded in $L^\infty(\mathbb{R}^2 \times \mathbb{R}^+)$ (see (5.1.17)), so is v_h, and there exists a function $v \in L^\infty(\mathbb{R}^2 \times \mathbb{R}^+)$ such that

$$v_h \to v \quad \text{weak-} * .$$

We prove that $v = u$, where u is the L^2 limit of u_h (see (5.1.18)). Using (5.1.3), we obtain for a fixed $\psi \in C_0^\infty(\mathbb{R}^2 \times \mathbb{R}^+)$

$$\int_{\mathbb{R}^2 \times \mathbb{R}^+} v_h \psi = \int_{\mathbb{R}^+} \sum_j \int_{T_j} u_{jl} \psi(z) \, dz$$

$$= \int_{\mathbb{R}^+} \sum_j |T_j| \, u_{jl} \psi(w_{jl}) + \mathcal{O}(h)$$

$$= \int_{\mathbb{R}^+} \sum_j |T_{jl}| u_{jl}\psi(w_{jl})$$

$$+ \int_{\mathbb{R}^+} \sum_j (|T_j| - |T_{jl}|) u_{jl}\psi(w_{jl}) \quad + \mathcal{O}(h)$$

$$= \int_{\mathbb{R}^+} \sum_j |T_j| u_j \psi(w_j) + o(1)$$

$$= \int_{\mathbb{R}^+} \sum_j \int_{T_j} u_j \psi(w_j) \, dz + o(1)$$

$$= \int_{\mathbb{R}^+} \sum_j \int_{T_j} u_j \psi(z) \, dz + o(1)$$

$$= \int_{\mathbb{R}^2 \times \mathbb{R}^+} u_h \psi + o(1) \quad \to \quad \int_{\mathbb{R}^2 \times \mathbb{R}^+} u\psi \; .$$

This implies that $u = v$. Similarly, we can prove that

$$v_h^2 \to u^2 \quad \text{weak-} * \; .$$

Therefore

$$\int_{\mathbb{R}^2 \times \mathbb{R}^+} |v_h - u|^2 \psi = \int v_h^2 \psi - 2 \int v_h u \psi + \int u^2 \psi$$

$$\to \int u^2 \psi - 2 \int u^2 \psi + \int u^2 \psi = 0 \; .$$

This proves (5.1.27) (since ψ can be chosen such that $\psi \equiv 1$ on Ω) and completes the proof of Lemma 5.1.11. □

Now we can prove Proposition 5.1.7.

Proof of proposition 5.1.7. We apply Lemma 5.1.8 and obtain

$$\sum_j \sum_l g_{jl}\varphi_j = \sum_j \sum_l g_{jl} [\varphi_j - \varphi(z_{jl})]$$

$$= \sum_j \sum_l [g_{jl} - \nu_{jl} f(u_j)] [\varphi_j - \varphi(z_{jl})]$$

$$- \sum_j \sum_l \nu_{jl} f(u_j) \varphi(z_{jl}) \; . \tag{5.1.28}$$

Here we have used the fact that $\sum_l \nu_{jl} = 0$. Now we multiply by Δt and sum over n. By Lemma 5.1.11, the first part in (5.1.28) converges to 0. Corollary 5.1.10 implies that the second term in (5.1.28) is equal to

$$\int_{\mathbb{R}^2} f(u_h(\cdot, t^n)) \nabla \varphi^n + \mathcal{O}(h).$$

Then we obtain the desired result in Proposition 5.1.7. □

Proof of Theorem 5.1.5 We multiply (5.1.7) by $\varphi_j^n = \varphi(w_j, t^n)$ and $|T_j|$ and sum with respect to j and n. Since g_{jl} and f satisfy (5.1.8)–(5.1.10) we can apply Proposition 5.1.7, which implies that

$$\Delta t \sum_j \sum_{n=0}^N \sum_{l=1}^k g_{jl}(u_j, u_{\alpha(j,l)}) \varphi_j^n = -\int_{\mathbb{R}^2 \times \mathbb{R}^+} f(u_h) \nabla \varphi + o(1). \quad (5.1.29)$$

For the discretization of the time derivative in (5.1.7) we obtain from Proposition 5.1.6 with $\psi = \mathrm{id}$

$$\sum_j \sum_{n=0}^N |T_j| (u_j^{n+1} - u_j^n) \varphi(w_j, t^n) \quad (5.1.30)$$

$$= -\int_{\mathbb{R}^2 \times \mathbb{R}^+} u_h \partial_t \varphi - \int_{\mathbb{R}^2} u_0 \varphi(\cdot, 0) + \mathcal{O}(h).$$

Now as $h \to 0$ we get from (5.1.29) and (5.1.30)

$$\int_{\mathbb{R}^2 \times \mathbb{R}^+} u \partial_t \varphi + \int_{\mathbb{R}^2 \times \mathbb{R}^+} f(u) \nabla \varphi + \int_{\mathbb{R}^2} u_0 \varphi(\cdot, 0) = 0.$$

This proves Theorem 5.1.5. □

REMARK 5.1.12 *The classification of the Riemann problem for two-dimensional gas dynamics is established in [213], where 16 different configurations are obtained for the isentropic gas. Numerical solutions of two-dimensional Riemann problems are given in [214].*

5.2 The discretization matrices for systems in two and three space dimensions

The Euler equations in 2-D are

$$\partial_t u + \partial_x f_1(u) + \partial_y f_2(u) = 0 , \qquad (5.2.1)$$

where

$$u = (\rho, \rho u_1, \rho u_2, e)^t$$

and

$$f_i(u) = \begin{pmatrix} \rho u_i \\ \rho u_i u_1 + \delta_{i1} p \\ \rho u_i u_2 + \delta_{i2} p \\ u_i(e + p) \end{pmatrix} ,$$

$$p = (\gamma - 1)\left[e - \frac{\rho}{2}(u_1^2 + u_2^2)\right] .$$

For smooth solutions (5.2.1) can be written as

$$\partial_t u + A_1(u)\partial_x u + A_2(u)\partial_y u = 0 ,$$

where

$$A_1(u) = \begin{pmatrix} 0 & 1 & 0 & 0 \\ [\frac{\gamma-3}{2}u_1^2 + \frac{\gamma-1}{2}u_2^2] & (3-\gamma)u_1 & -(\gamma-1)u_2 & \gamma-1 \\ -u_1 u_2 & u_2 & u_1 & 0 \\ [-\gamma u_1 E + (\gamma-1)u_1|u|^2] & [\gamma E - \frac{\gamma-1}{2}(u_2^2 + 3u_1^2)] & -(\gamma-1)u_1 u_2 & \gamma u_1 \end{pmatrix} ,$$

The discretization matrices for systems in two and three space dimensions

$$A_2(u) = \begin{pmatrix} 0 & 0 & 1 & 0 \\ -u_1 u_2 & u_2 & u_1 & 0 \\ [\frac{\gamma-3}{2}u_2^2 + \frac{\gamma-1}{2}u_1^2] & -(\gamma-1)u_1 & (3-\gamma)u_2 & \gamma-1 \\ [-\gamma u_2 E + (\gamma-1)u_2|u|^2] & -(\gamma-1)u_1 u_2 & \gamma E - \frac{\gamma-1}{2}(u_1^2 + 3u_2^2) & \gamma u_2 \end{pmatrix},$$

where $|u|^2 = u_1^2 + u_2^2$ and $E = e/\rho$. The numerical flux g_{jl} for the Euler equations in 2-D can be obtained by using (see (5.1.12))

$$C_{jl}(u) := k_1 A_1(u) + k_2 A_2(u) ,$$

where $k_1 := n_{jlx}$, $k_2 := n_{jly}$. We have to find a matrix Q such that

$$Q^{-1} C_{jl} Q = D = \text{diag}(d_1, \ldots, d_4) .$$

This can be done much more easily if we consider primitive rather than conservative variables. The equations in primitive variables $w = (\rho, u_1, u_2, p)^t$ are

$$\partial_t w + B_1(w) \partial_x w + B_2(w) \partial_y w = 0 ,$$

where

$$B_1(w) := \begin{pmatrix} u_1 & \rho & 0 & 0 \\ 0 & u_1 & 0 & 1/\rho \\ 0 & 0 & u_1 & 0 \\ 0 & \rho a^2 & 0 & u_1 \end{pmatrix}, \quad B_2(w) := \begin{pmatrix} u_2 & 0 & \rho & 0 \\ 0 & u_2 & 0 & 0 \\ 0 & 0 & u_2 & 1/\rho \\ 0 & 0 & \rho a^2 & u_2 \end{pmatrix},$$

with $a = \sqrt{\gamma p/\rho}$. Then there exists a regular matrix M such that

$$B_j(w) = M^{-1} A_j(u) M , \quad j = 1, 2 , \qquad (5.2.2)$$

where

$$M := \begin{pmatrix} 1 & 0 & 0 & 0 \\ u_1 & \rho & 0 & 0 \\ u_2 & 0 & \rho & 0 \\ \frac{1}{2}|u|^2 & \rho u_1 & \rho u_2 & \frac{1}{\gamma-1} \end{pmatrix},$$

$$M^{-1} = \begin{pmatrix} 1 & 0 & 0 & 0 \\ -\frac{u_1}{\rho} & \frac{1}{\rho} & 0 & 0 \\ -\frac{u_2}{\rho} & 0 & \frac{1}{\rho} & 0 \\ \frac{\gamma-1}{2}|u|^2 & -(\gamma-1)u_1 & -(\gamma-1)u_2 & \gamma-1 \end{pmatrix}.$$

Now we have to consider the eigenvalues and eigenvectors of

$$P := k_1 B_1 + k_2 B_2 . \tag{5.2.3}$$

where $k_1^2 + k_2^2 < 1$. There exists a regular matrix T such that

$$T^{-1} P T = D = \text{diag}(\lambda_1, \lambda_2, \lambda_3, \lambda_4) \tag{5.2.4}$$

where λ_i, $i = 1, \ldots, 4$, are the eigenvalues of P and

$$T = \begin{pmatrix} 1 & 0 & \frac{\rho}{2a} & \frac{\rho}{2a} \\ 0 & k_2 & \frac{k_1}{2} & -\frac{k_1}{2} \\ 0 & -k_1 & \frac{k_2}{2} & -\frac{k_2}{2} \\ 0 & 0 & \frac{\rho a}{2} & \frac{\rho a}{2} \end{pmatrix},$$

$$T^{-1} = \begin{pmatrix} 1 & 0 & 0 & -\frac{1}{a^2} \\ 0 & k_2 & -k_1 & 0 \\ 0 & k_1 & k_2 & \frac{1}{\rho a} \\ 0 & -k_1 & -k_2 & \frac{1}{\rho a} \end{pmatrix}.$$

The discretization matrices for systems in two and three space dimensions

Then we obtain using (5.2.2)–(5.2.4)

$$C_{jl}(u) = \sum_{i=1}^{2} k_i A_i(u) = \sum_{i=1}^{2} k_i M B_i(u) M^{-1}$$
$$= MPM^{-1} = MTD(MT)^{-1} . \qquad (5.2.5)$$

Using the notation $n := \binom{k_1}{k_2}$, the matrices MT and $(MT)^{-1}$ are given by

$$MT = \begin{pmatrix} 1 & 0 & \frac{\rho}{2a} & \frac{\rho}{2a} \\ u_1 & \rho k_2 & \frac{\rho}{2a}(u_1 + ak_1) & \frac{\rho}{2a}(u_1 - ak_1) \\ u_2 & -\rho k_1 & \frac{\rho}{2a}(u_2 + ak_2) & \frac{\rho}{2a}(u_2 - ak_2) \\ \frac{|u|^2}{2} & \rho(u_1 k_2 - u_2 k_1) & \frac{\rho}{2a}(H + au \cdot n) & \frac{\rho}{2a}(H - au \cdot n) \end{pmatrix}$$

$$(MT)^{-1} =$$

$$\begin{pmatrix} 1 - \frac{\gamma-1}{2}M^2 & (\gamma-1)\frac{u_1}{a^2} & (\gamma-1)\frac{u_2}{a^2} & -\frac{\gamma-1}{a^2} \\ \frac{1}{\rho}(u_2 k_1 - u_1 k_2) & \frac{k_2}{\rho} & -\frac{k_1}{\rho} & 0 \\ \frac{a}{\rho}[\frac{(\gamma-1)}{2}M^2 - \frac{u \cdot n}{a}] & \frac{1}{\rho}[k_1 - (\gamma-1)\frac{u_1}{a}] & \frac{1}{\rho}[k_2 - (\gamma-1)\frac{u_2}{a}] & \frac{\gamma-1}{\rho a} \\ \frac{a}{\rho}[\frac{(\gamma-1)}{2}M^2 + \frac{u \cdot n}{a}] & -\frac{1}{\rho}[k_1 + (\gamma-1)\frac{u_1}{a}] & -\frac{1}{\rho}[k_2 + (\gamma-1)\frac{u_2}{a}] & \frac{\gamma-1}{\rho a} \end{pmatrix}$$

where $H := (e + p)/\rho$, $M := \sqrt{|u|^2/a^2}$. The eigenvalues of C_{jl} are

$$\lambda_1 = \lambda_2 = u \cdot n, \quad \lambda_3 = u \cdot n + a, \quad \lambda_4 = u \cdot n - a .$$

Then $C_{jl}^{\pm}$, which defines the numerical flux using (5.1.12), is given by (see (5.2.5))

$$C_{jl}^{\pm} := MTD^{\pm}(MT)^{-1} ,$$

where
$$D^{\pm} := \begin{pmatrix} \lambda_1^{\pm} & & & 0 \\ & \lambda_2^{\pm} & & \\ & & \lambda_3^{\pm} & \\ 0 & & & \lambda_4^{\pm} \end{pmatrix}.$$

The corresponding matrices in 3-D are given below. Similarly to before, let
$$C_{jl}(w) := k_1 f_1'(w) + k_2 f_2'(w) + k_3 f_3'(w) ,$$
where
$$n = \begin{pmatrix} k_1 \\ k_2 \\ k_3 \end{pmatrix}, \kappa := \sqrt{\sum_{j=1}^{3} k_j^2}, \quad a := \sqrt{\frac{\gamma p}{\rho}} .$$

The eigenvalues are
$$\lambda_1 := \sum_{j=1}^{3} k_j u_j , \qquad \lambda_2 := \lambda_1 , \qquad \lambda_3 := \lambda_1 ,$$

$$\lambda_4 := \lambda_1 + a\kappa , \qquad \lambda_5 := \lambda_1 - a\kappa ,$$

$$D^{\pm} := \begin{pmatrix} \lambda_1^{\pm} & & & & 0 \\ & \lambda_2^{\pm} & & & \\ & & \lambda_3^{\pm} & & \\ & & & \lambda_4^{\pm} & \\ 0 & & & & \lambda_5^{\pm} \end{pmatrix},$$

$$M := \begin{pmatrix} 1 & 0 & 0 & 0 & 0 \\ u_1 & \rho & 0 & 0 & 0 \\ u_2 & 0 & \rho & 0 & 0 \\ u_3 & 0 & 0 & \rho & 0 \\ q^2/2 & \rho u_1 & \rho u_2 & \rho u_3 & 1/(\gamma - 1) \end{pmatrix}$$

The discretization matrices for systems in two and three space dimensions

where $q^2 := u_1^2 + u_2^2 + u_3^2$,

$$M^{-1} = \begin{pmatrix} 1 & 0 & 0 & 0 & 0 \\ -\frac{u_1}{\rho} & \frac{1}{\rho} & 0 & 0 & 0 \\ -\frac{u_2}{\rho} & 0 & \frac{1}{\rho} & 0 & 0 \\ -\frac{u_3}{\rho} & 0 & 0 & \frac{1}{\rho} & 0 \\ -\frac{\gamma-1}{2}q^2 & -(\gamma-1)u_1 & -(\gamma-1)u_2 & -(\gamma-1)u_3 & \gamma-1 \end{pmatrix}.$$

Let $r_j := k_j/\kappa$; then

$$T := \begin{pmatrix} r_1 & r_2 & r_3 & \frac{\rho}{a\sqrt{2}} & \frac{\rho}{a\sqrt{2}} \\ 0 & -r_3 & r_2 & \frac{r_1}{\sqrt{2}} & -\frac{r_1}{\sqrt{2}} \\ r_3 & 0 & -r_1 & \frac{r_2}{\sqrt{2}} & -\frac{r_2}{\sqrt{2}} \\ -r_2 & r_1 & 0 & \frac{r_3}{\sqrt{2}} & -\frac{r_3}{\sqrt{2}} \\ 0 & 0 & 0 & \frac{\rho a}{\sqrt{2}} & \frac{\rho a}{\sqrt{2}} \end{pmatrix},$$

$$T^{-1} := \begin{pmatrix} r_1 & 0 & r_3 & -r_2 & -\frac{r_1}{a^2} \\ r_2 & -r_3 & 0 & r_1 & -\frac{r_1}{a^2} \\ r_3 & r_2 & -r_1 & 0 & -\frac{r_1}{a^2} \\ 0 & \frac{r_1}{\sqrt{2}} & \frac{r_2}{\sqrt{2}} & \frac{r_3}{\sqrt{2}} & \frac{1}{\rho a\sqrt{2}} \\ 0 & -\frac{r_1}{\sqrt{2}} & -\frac{r_2}{\sqrt{2}} & -\frac{r_3}{\sqrt{2}} & \frac{1}{\rho a\sqrt{2}} \end{pmatrix},$$

and finally

$$C_{jl}^{\pm}(w) := (MT)D^{\pm}(MT)^{-1} , \qquad (5.2.6)$$

where the matrices MT and $(MT)^{-1}$ are given on the following pages. Then the numerical fluxes for the schemes mentioned in (5.1.13)–(5.1.15) can be computed explicitly using (5.2.6).

REMARK 5.2.1 *For the numerical algorithms one should multiply several lines in the matrices MT and $(MT)^{-1}$ gy ρ and a respectively.*

For further results concerning implicit algorithms and the corresponding matrices we refer to [185].

$$MT = \begin{pmatrix} r_1 & r_2 & r_3 & \frac{\rho}{a\sqrt{2}} & \frac{\rho}{a\sqrt{2}} \\ u_1 r_1 & u_1 r_2 - \rho r_3 & u_1 r_3 + \rho r_2 & \frac{u_1 \rho}{a\sqrt{2}} + \frac{\rho r_1}{\sqrt{2}} & \frac{u_1 \rho}{a\sqrt{2}} - \frac{\rho r_1}{\sqrt{2}} \\ u_2 r_1 + \rho r_3 & u_2 r_2 & u_2 r_3 - \rho r_1 & \frac{u_2 \rho}{a\sqrt{2}} + \frac{\rho r_2}{\sqrt{2}} & \frac{u_2 \rho}{a\sqrt{2}} - \frac{\rho r_2}{\sqrt{2}} \\ u_3 r_1 - \rho r_2 & u_3 r_2 + \rho r_1 & u_3 r_3 & \frac{u_3 \rho}{a\sqrt{2}} + \frac{\rho r_3}{\sqrt{2}} & \frac{u_3 \rho}{a\sqrt{2}} - \frac{\rho r_3}{\sqrt{2}} \\ \frac{r_1}{2} q^2 + \rho u_2 r_3 & \frac{r_2}{2} q^2 - \rho u_1 r_3 & \frac{r_3}{2} q^2 + \rho u_1 r_2 & \frac{1}{\sqrt{2}}\left(\frac{q^2 \rho}{2a} + \rho u_1 r_1 + \frac{\rho a}{\gamma - 1}\right) & \frac{1}{\sqrt{2}}\left(\frac{q^2 \rho}{2a} - \rho u_1 r_1 + \frac{\rho a}{\gamma - 1}\right) \\ -\rho u_3 r_2 & +\rho u_3 r_1 & -\rho u_2 r_1 & +\rho u_2 r_2 + \rho u_3 r_3 & -\rho u_2 r_2 - \rho u_3 r_3 \end{pmatrix}$$

$$(MT)^{-1} = T^{-1}M^{-1} =$$

$$\begin{pmatrix}
r_1 - \dfrac{r_3 u_2}{\rho} + \dfrac{r_2 u_3}{\rho} - \dfrac{r_1(\gamma-1)q^2}{2a^2} & \dfrac{r_1(\gamma-1)u_1}{a^2} & \dfrac{r_3}{\rho} + \dfrac{r_1(\gamma-1)u_2}{a^2} & -\dfrac{r_2}{\rho} + \dfrac{r_1(\gamma-1)u_3}{a^2} & -\dfrac{r_1(\gamma-1)}{a^2} \\[2ex]
r_2 + \dfrac{r_3 u_1}{\rho} - \dfrac{r_1 u_3}{\rho} - \dfrac{r_2(\gamma-1)q^2}{2a^2} & -\dfrac{r_3}{\rho} + \dfrac{(\gamma-1)u_1 r_2}{a^2} & \dfrac{r_2(\gamma-1)u_2}{a^2} & \dfrac{r_1}{\rho} + \dfrac{r_2(\gamma-1)u_3}{a^2} & -\dfrac{r_2(\gamma-1)}{a^2} \\[2ex]
r_3 - \dfrac{r_2 u_1}{\rho} + \dfrac{r_1 u_2}{\rho} - \dfrac{r_3(\gamma-1)q^2}{2a^2} & \dfrac{r_2}{\rho} + \dfrac{(\gamma-1)u_1 r_3}{a^2} & -\dfrac{r_1}{\rho} + \dfrac{r_3(\gamma-1)u_2}{a^2} & \dfrac{r_3(\gamma-1)u_3}{a^2} & -\dfrac{r_3(\gamma-1)}{a^2} \\[2ex]
-\dfrac{u_1 r_1}{\rho\sqrt{2}} - \dfrac{u_2 r_2}{\rho\sqrt{2}} + \dfrac{(\gamma-1)q^2}{2\sqrt{2}\rho a} - \dfrac{u_3 r_3}{\rho\sqrt{2}} & \dfrac{r_1}{\rho\sqrt{2}} - \dfrac{(\gamma-1)u_1}{\rho a\sqrt{2}} & \dfrac{r_2}{\rho\sqrt{2}} - \dfrac{(\gamma-1)u_2}{\rho a\sqrt{2}} & \dfrac{r_3}{\rho\sqrt{2}} - \dfrac{(\gamma-1)u_3}{\rho a\sqrt{2}} & \dfrac{\gamma-1}{\rho a\sqrt{2}} \\[2ex]
\dfrac{u_1 r_1}{\rho\sqrt{2}} + \dfrac{u_2 r_2}{\rho\sqrt{2}} + \dfrac{(\gamma-1)q^2}{2\sqrt{2}\rho a} + \dfrac{u_3 r_3}{\rho\sqrt{2}} & -\dfrac{r_1}{\rho\sqrt{2}} - \dfrac{(\gamma-1)u_1}{\rho a\sqrt{2}} & -\dfrac{r_2}{\rho\sqrt{2}} - \dfrac{(\gamma-1)u_2}{\rho a\sqrt{2}} & -\dfrac{r_3}{\rho\sqrt{2}} - \dfrac{(\gamma-1)u_3}{\rho a\sqrt{2}} & \dfrac{\gamma-1}{\rho a\sqrt{2}}
\end{pmatrix},$$

REMARK 5.2.2 (Implementation of higher order schemes for the Euler equations of gas dynamics) *In order to get better results, in particular for the resolution of the shocks, we apply the higher-order methods as described in §3.5 to systems. The reconstruction as defined in Examples 3.5.7 and 3.5.8 has to be applied to each component of the unknown vector. In the case of the system of Euler equations in [250], the reconstruction has been applied to the conservative variables. After the limiting process the pressure is calculated with the limited quantities. If the pressure becomes negative on one of the neighbouring triangles of T_j by these procedures, the constant α in Example 3.5.18 is set equal to zero on T_j. For the density there are no problems. If all densities from the last time step are greater than zero, than the corresponding values after the reconstruction process (Example 3.5.7 and 3.5.8) will also be greater than zero.*

REMARK 5.2.3 (Some hints for the implementation [250, 191]) *If the numerical flux is computed as described in (5.1.13)–(5.1.15) and (5.2.6), too many floating point operations are necessary. Since there are some zeros on the diagonal of $D^\pm(w) = D_{jl}^\pm(w)$, the number of operations can be reduced for some special cases. The eigenvalues of $D_{jl}(w)$ are of the form*

$$\lambda_1 = \lambda_2 = u \cdot n_{jl}, \quad \lambda_3 = u \cdot n_{jl} + a, \quad \lambda_4 = u \cdot n_{jl} - a$$

where $a = \sqrt{\gamma p/\rho}$ denotes the sound velocity. Let us consider the following cases.

(a) All eigenvalues $\lambda_1, \ldots, \lambda_4 > 0$: in this case we have, using (5.2.6), (5.2.5) and (5.1.11), (5.1.12),

$$C_{jl}^+(w) \cdot w = C_{jl}(w) \cdot w = n_{jl} f'(w) \cdot w$$
$$= n_{jl} f(w).$$

(b) $\lambda_3 > 0$, $\lambda_1, \lambda_2, \lambda_4 < 0$: In this case D_{jl}^+ has only one nonzero component:

$$C_{jl}^+(w)w = \lambda_3 \begin{pmatrix} \frac{1}{2a} \\ \frac{u_1 + an_{jl}^x}{2a} \\ \frac{u_2 + an_{jl}^y}{2a} \\ \left[\frac{e+p}{\rho} + a(u \cdot n_{jl})\right]\frac{1}{2a} \end{pmatrix} \begin{pmatrix} (\gamma-1)\frac{|u|}{2a} - u \cdot n_{jl} \\ n_{jl}^x - (\gamma-1)\frac{u_1}{a} \\ n_{jl}^y - (\gamma-1)\frac{u_2}{a} \\ \frac{\gamma-1}{a} \end{pmatrix}^t w,$$

where $v = \begin{pmatrix} u_1 \\ u_2 \end{pmatrix}$.

c) $\lambda_4 < 0$, λ_1 , λ_2 , $\lambda_3 > 0$: In this case we have

$$C_{jl}^+(w)w = C_{jl}(w)w - C_{jl}^-(w)w = n_{jl}f(w) - C_{jl}^-(w)w$$

where $C_{jl}^-(w)w$ can be computed similarly as before

$$C_{jl}^-(w)w = \lambda_4 \begin{pmatrix} \frac{1}{2a} \\ \frac{u_1 - an_{jl}^x}{2a} \\ \frac{u_2 - an_{jl}^y}{2a} \\ \left(\frac{e+p}{\rho} - a(u \cdot n_{jl})\right)\frac{1}{2a} \end{pmatrix} \begin{pmatrix} (\gamma-1)\frac{|u|}{2a} + u \cdot n_{jl} \\ -n_{jl}^x - (\gamma-1)\frac{u_1}{a} \\ -n_{jl}^y - (\gamma-1)\frac{u_2}{a} \\ \frac{\gamma-1}{a} \end{pmatrix}^t w$$

where $u = \begin{pmatrix} u_1 \\ u_2 \end{pmatrix}$. Using these limits for the implementation the computing time can be reduced considerably.

REMARK 5.2.4 *The same simplification can be done for general numerical fluxes in particular for the Vijayasundaram and the van Leer scheme (see [191]).*

REMARK 5.2.5 *Another interesting algorithm for solving the Euler equations in multi-dimensions has been considered in [102, 103]. Here the state vector is decomposed into three multidimensional waves.*

5.3 Numerical examples for the Euler equations of gas dynamics

In this section we shall present several examples of the numerical calculation of approximating functions for the Euler equations of gas dynamics in many dimensions on structured and unstructured grids. We shall use the schemes as described before. For finite element methods we refer to [110], for multigrid methods to [157], and for other methods to [183], [102] and [103].

EXAMPLE 5.3.1 (Numerical tests for the system of the Euler equations in 2-D [250]) The following example refers to the shock tube problem in 2-D, i.e. the jump of the initial data only occurs in the x-direction. Therefore the solution should not depend on the y-direction. This problem has been chosen as a test problem in 2-D, since the exact solution

5.3 Numerical examples for the Euler equations of gas dynamics

can be computed very accurately using the method described in § 4.2. It was solved numerically on an unstructured grid in 2-D using a finite volume method based on ansatz C as described in Examples 3.5.7 and 3.5.8. The Riemann problem for the Euler equations in 2-D,

$$\partial_t u + \partial_x f_1(u) + \partial_y f_2(u) = 0 \text{ in } [-1,1] \times [0,1] \times [0,T] \qquad (5.3.1)$$

with

$$u = \begin{pmatrix} \rho \\ \rho u_1 \\ \rho u_2 \\ e \end{pmatrix}, \quad f_i(u) = \begin{pmatrix} \rho u_i \\ \rho u_i u_1 + \delta_{i1} p \\ \rho u_i u_2 + \delta_{i2} p \\ (e+p) u_i \end{pmatrix} \qquad (5.3.2)$$

and

$$p = (\gamma - 1)\left[e - \frac{\rho}{2}(u_1^2 + u_2^2)\right],$$

with initial conditions

$$\rho(x,y,0) = \begin{cases} 4 & \text{if } x < 0, \\ 1 & \text{if } x > 0, \end{cases}$$

$$p(x,y,0) = \begin{cases} 1.6 & \text{if } x < 0, \\ 0.4 & \text{if } x > 0, \end{cases} \qquad (5.3.3)$$

$$u_1(x,y,0) = u_2(x,y,0) = 0,$$

has been solved. The solution of the shock tube problem in 1-D can be calculated to any order of accuracy with a Newton solver (see [26] and Example 4.1.40). Then it is obvious to get the reference solution of (5.3.1)–(5.3.3) in 2-D. For solving (5.3.1)–(5.3.3) the schemes (3.5.5) and (3.5.6) are used, with the three numerical fluxes of Steger and Warming, Vijayasundaram and van Leer respectively. These fluxes are given in (5.1.13)–(5.1.15) for the 3-D case. To avoid boundary effects on the order of convergence, we use reflection boundary conditions at the upper and lower boundaries.

In the following tables the numerical errors in the density for the different schemes on different grids C (Figure 5.3.1) and D (Figure 5.3.2) are presented. In the first two tables are results obtained with the numerical flux of Steger and Warming. For the notation see Example 3.5.10.

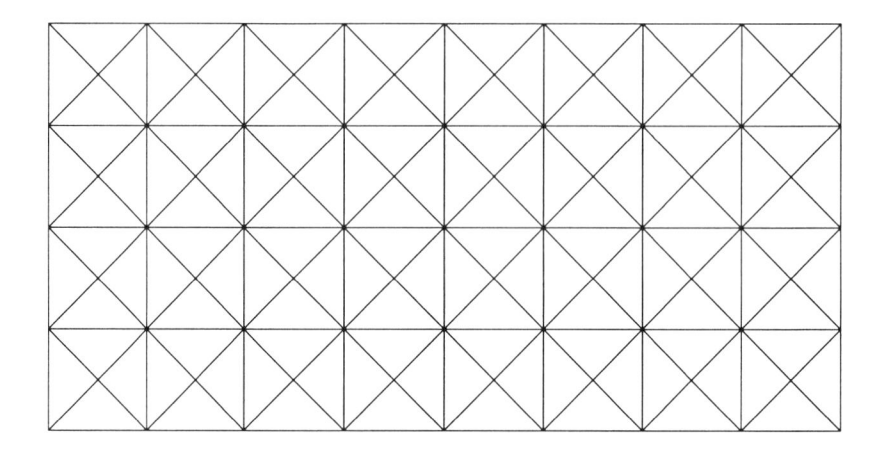

Figure 5.3.1 Grid C.

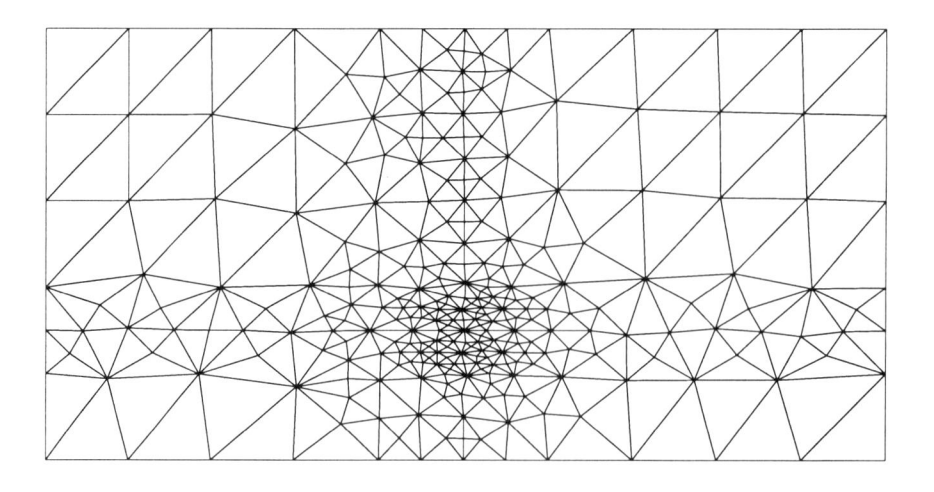

Figure 5.3.2 Grid D.

5.3 Numerical examples for the Euler equations of gas dynamics

Results on grid C (CFL $= 0.3, T = 0.7$)

h	$\|\rho - \rho_h\|_{L^1}$	EOC	$\|\rho - \rho_h^C\|_{L^1}$	EOC	$\|\rho - \rho_h^S\|_{L^1}$	EOC
0.25	0.3910		0.2790		0.2757	
0.13	0.2975	0.39	0.1345	1.05	0.1407	0.97
0.06	0.2132	0.48	0.0677	0.99	0.0774	0.86
0.03	0.1438	0.57	0.0346	0.97	0.0446	0.79
0.02	0.0938	0.62	0.0172	1.01	0.0327	0.45

Results on grid D (CFL $= 0.3, T = 0.7$)

h	$\|\rho - \rho_h\|_{L^1}$	EOC	$\|\rho - \rho_h^C\|_{L^1}$	EOC	$\|\rho - \rho_h^S\|_{L^1}$	EOC
0.15	0.3254		0.1934		0.2704	
0.08	0.2460	0.40	0.0995	0.96	0.1429	0.92
0.04	0.1713	0.52	0.0502	0.99	0.0752	0.93
0.02	0.1138	0.59	0.0267	0.91	0.0394	0.93

We again get the best results with ansatz C. The improvement of the L^1 error in the velocity component is of the same quality. In Figure 5.3.3 we see the graph on a line ($y = 0.5$) of the exact solution (sharp discontinuities) and the approximated piecewise-constant solutions calculated with the first- and second-order schemes on the finest level of grids C and D.

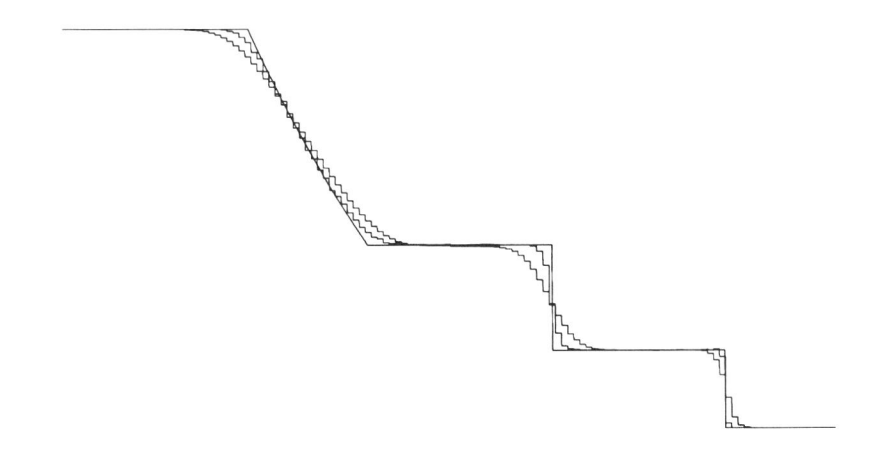

Figure 5.3.3

In the next table we see the results using the numerical flux of van Leer and Vijayasundaram for the same calculations on grid C.

Results on grid C with the numerical flux of Vijayasundaram
(CFL $= 0.3, T = 0.7$)

h	$\|\rho - \rho_h\|_{L^1}$	EOC	$\|\rho - \rho_h^C\|_{L^1}$	EOC	$\|\rho - \rho_h^S\|_{L^1}$	EOC
0.25	0.3676		0.2387		0.2405	
0.13	0.2675	0.46	0.1231	0.96	0.1367	0.81
0.06	0.1874	0.51	0.0628	0.97	0.0833	0.71
0.03	0.1251	0.58	0.0354	0.83	0.0498	0.74
0.02	0.0803	0.64	0.0211	0.75	0.0339	0.55

Results on grid C with the numerical flux of van Leer
(CFL $= 0.3, T = 0.7$)

h	$\|\rho - \rho_h\|_{L^1}$	EOC	$\|\rho - \rho_h^C\|_{L^1}$	EOC	$\|\rho - \rho_h^S\|_{L^1}$	EOC
0.25	0.3250		0.1984		0.2088	
0.13	0.2302	0.50	0.1058	0.90	0.1092	0.93
0.06	0.1585	0.54	0.0534	0.99	0.0637	0.78
0.03	0.1051	0.59	0.0286	0.90	0.0400	0.67
0.02	0.0669	0.65	0.0150	0.93	0.0273	0.55

The best results for the first-order scheme are obtained with the numerical flux of van Leer.

5.3 Numerical examples for the Euler equations of gas dynamics

EXAMPLE 5.3.2 (Supersonic flow around a cylinder, Figures 5.3.4 and 5.3.5 [210]) In this example the inviscid compressible flow around a circle in 2-D is simulated. For symmetry reasons, we consider only a halfcircle. The underlying grid is a locally refined structured grid with hanging nodes. The values in these points are obtained by interpolation. In the locally refined region the explicit scheme advances with a local time step Δt_{loc} and is synchronized with the timestep Δt in the unrefined steps region if $\Delta t_{\text{loc}} l = \Delta t$. The grid indicator as defined in § (3.9.8) has been used for the local refinement. The used scheme is the dimensional splitting scheme as described in § 3.1. For the details of the boundary conditions we refer to § 6.1, in particular to Examples 6.1.1 and 6.1.2.

Initial conditions: $\rho = 1.4$, $u\rho = 7$, $v\rho = 0$, $e = 20$.
Boundary conditions:
  left boundary: inflow;
  right boundary: outflow;
  upper boundary: outflow;
  lower boundary: symmetry condition.
Scheme: dimensional splitting, Steger and Warming.
Grid indicator: see (3.9.8).

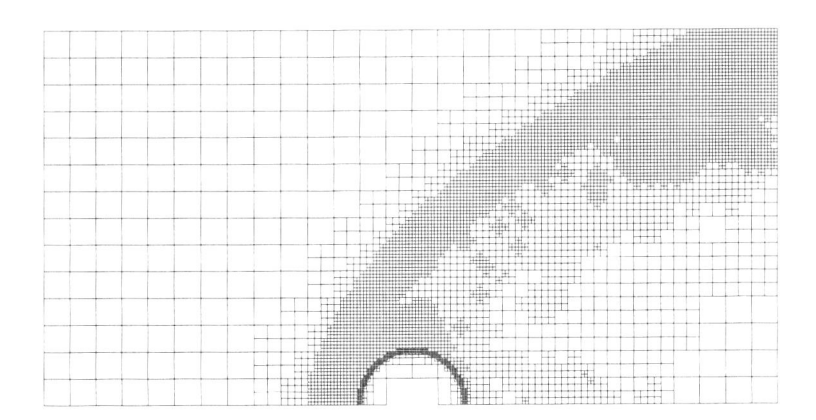

Figure 5.3.4

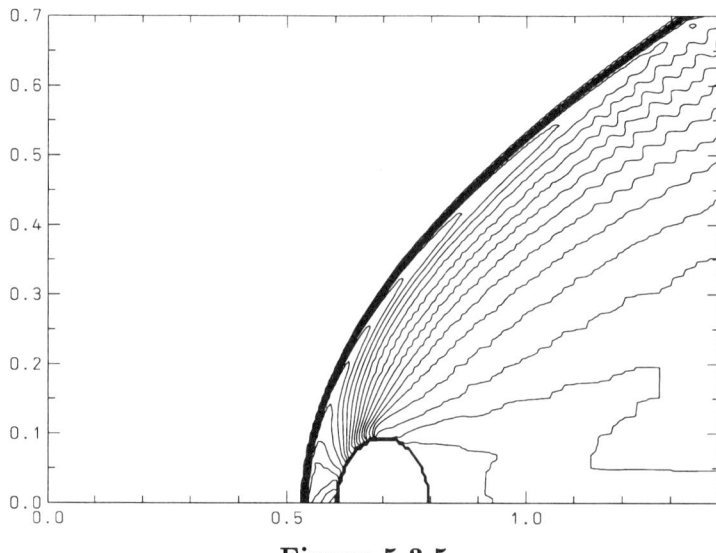

Figure 5.3.5

5.3 Numerical examples for the Euler equations of gas dynamics

EXAMPLE 5.3.3 (Shock reflection, Figures 5.3.6 and 5.3.7 [210])
For the shock problem we have used the same numerical scheme as in Example 5.3.2. The data are chosen such that we have a stationary solution consisting of three states separated by shocks.

Initial conditions: $\rho = 1.4$, $u = 2.9$, $v = 0.0$, $p = 1.0$.
Boundary conditions:
   left boundary: $\rho = 1.4$, $u = 2.9$, $v = 0.0$, $p = 1.0$;
   right boundary: outflow;
   upper boundary: $\rho = 2.4739$, $u = 2.5876$, $v = -0.5438$, $p = 2.2685$;
   lower boundary: reflection.
Scheme: dimensional splitting, Steger and Warming.
Grid indicator: see (3.9.8).

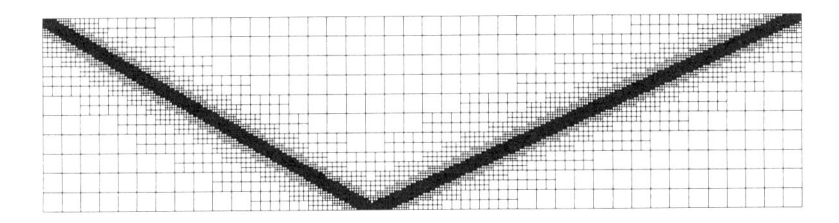

Figure 5.3.6

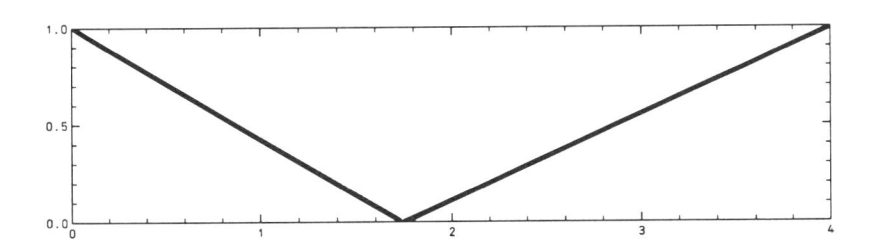

Figure 5.3.7

EXAMPLE 5.3.4 (Two passing "supersonic" trains in a tunnel, Figures 5.3.9 – 5.3.13 [210]) In Examples 5.3.3 and 5.3.2 we have treated problems with stationary shock fronts. Here we have studied problems with moving fronts. In this case we have had to develop a grid indicator that is able to control moving refinement zones. As a test problem we have treated the problem of two passing supersonic (unrealistic) "trains" in a tunnel. The left train moves with velocity $(3.0, 0.0)$ and the right one with $(-3.0, 0.0)$. Both "trains" have bow shocks in front of them, which are resolved on a moving locally refined grid zone.

Initial conditions: $\rho = 1.4$, $u = v = 0.0$, $p = 1.0$.
Boundary conditions:
  left boundary: outflow;
  right boundary: outflow;
  upper boundary: reflection;
  lower boundary: reflection;
  inner boundary: reflection.
Scheme: dimensional splitting, Steger and Warming.
Grid indicator: see (3.9.8).

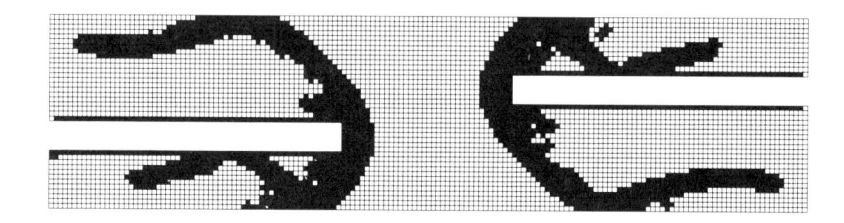

Figure 5.3.8

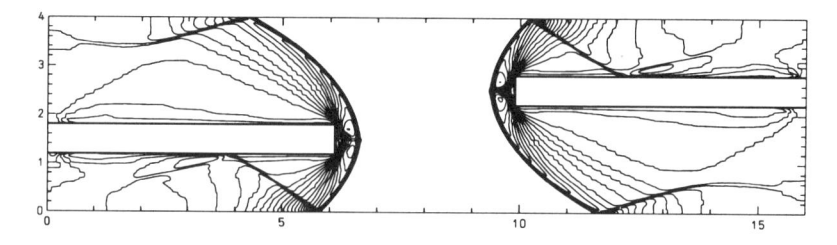

Figure 5.3.9

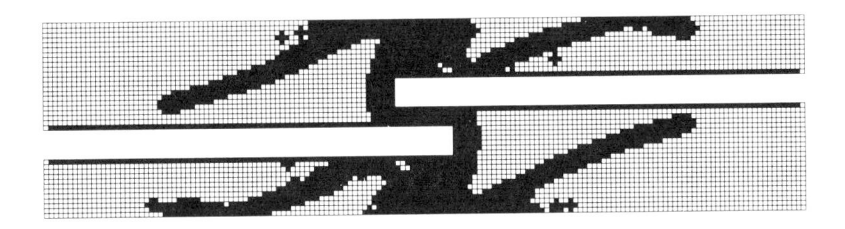

5.3 Numerical examples for the Euler equations of gas dynamics 403

Figure 5.3.10

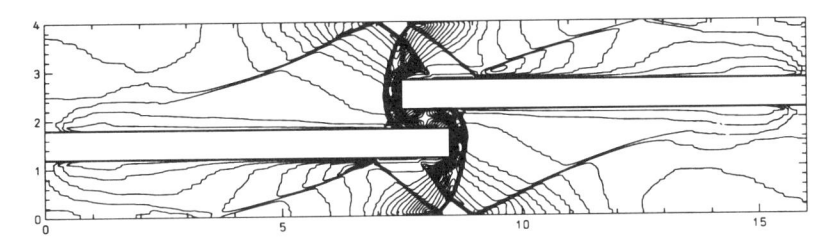

Figure 5.3.11

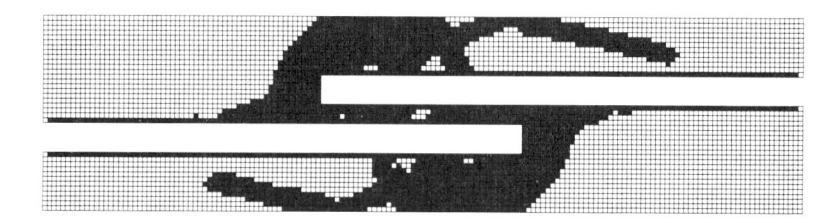

404   Initial value problems for systems of conservation laws in 2-D

Figure 5.3.12

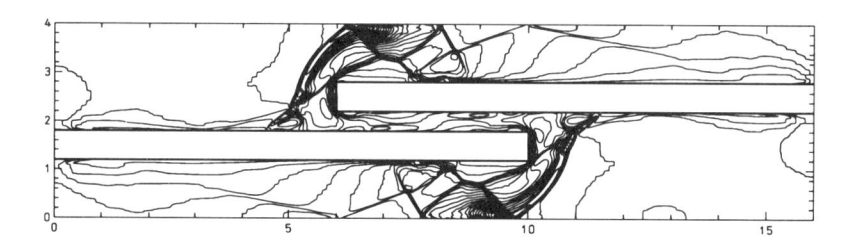

Figure 5.3.13

5.3 Numerical examples for the Euler equations of gas dynamics

In Figures 5.3.14 – 5.3.18 we show some applications of the Osher–Solomon scheme [19]. In Figure 5.3.14 results for the foward-facing step (see [39]) and in Figure 5.3.15 those for the shock reflection problem are presented. The supersonic flow around a double ellipsoid is treated on an unstructured grid in Figures 5.3.17 and 5.3.18.

EXAMPLE 5.3.5 (Forward-facing step, Figure 5.3.14 [19]) We consider the famous "forward-facing step" problem, as described in [39]. Here we have used the same data as in [39]. The underlying is now a locally refined unstructured triangulation. As a grid indicator (3.9.8) has been used.

Initial conditions: $\rho = 1.0$, $u = 2.0$, $v = 0.0$, $p = 7.0$.
Boundary conditions:
  left boundary: inflow;
  right boundary: outflow;
  upper boundary: reflection;
  lower boundary: reflection.
Scheme: van Leer finite volume on unstructured grid.
Grid indicator: $|\nabla \rho|$, where ρ is the density.

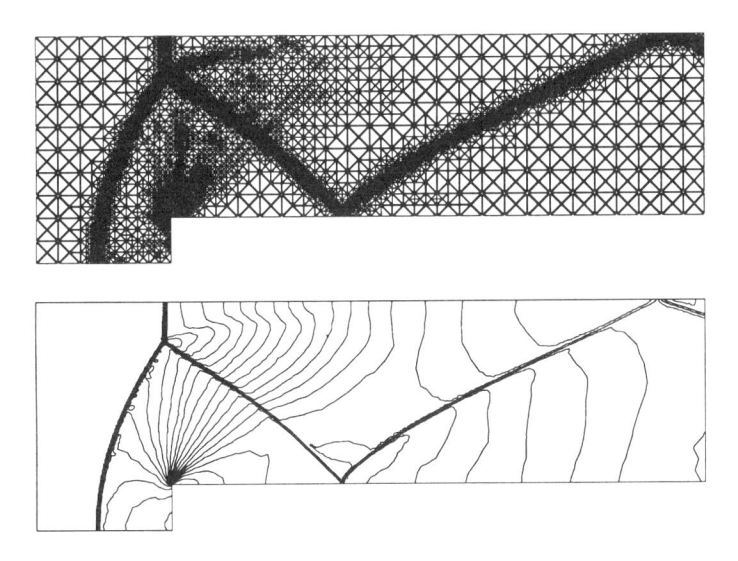

Figure 5.3.14

EXAMPLE 5.3.6 (Flow over a circular arc bump, Figures 5.3.15 and 5.3.16 [19]) This test problem consists of flow over a ramp that is part of a circle. We have used the same numerical scheme as in Example 5.3.5.

Initial conditions: $\rho = 1.4$, $u = 1.4$, $v = 0.0$, $p = 1.0$.
Boundary conditions:
  left boundary: inflow;
  right boundary: outflow;
  upper boundary: $n \cdot \vec{u} = 0$;
  lower boundary: $n \cdot \vec{u} = 0$.
Scheme: Osher–Solomon.
Grid indicator: $|\nabla \rho|$, where ρ is the density.

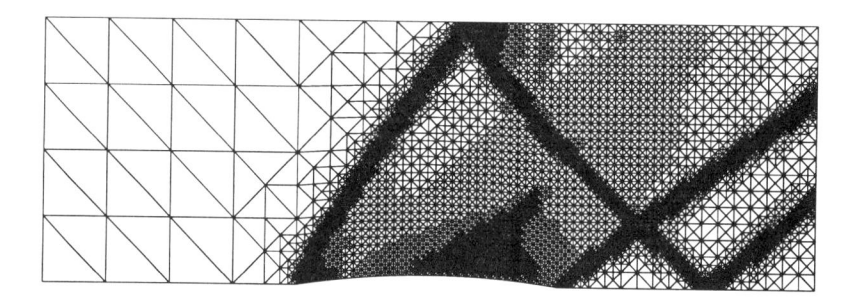

Figure 5.3.15

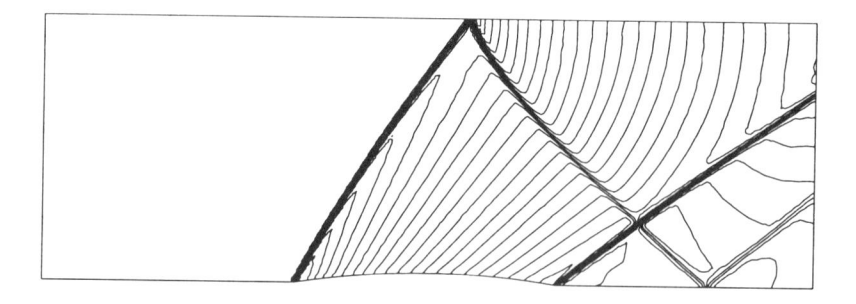

Figure 5.3.16

5.3 Numerical examples for the Euler equations of gas dynamics 407

EXAMPLE 5.3.7 (Flow around a double ellipsoid, Figures 5.3.17 and 5.3.18 [19]) The geometry for this problem consists of two intersecting ellipsoids. Also in this case we have used the Osher–Solomon finite volume scheme on an unstructured grid consisting of triangles.

Initial conditions: $\rho = 1.4$, $u = 3.0$, $v = 0.0$, $p = 1.0$.
Boundary conditions:
  left boundary: inflow;
  right boundary: outflow;
  upper boundary: outflow;
  lower boundary: outflow;
  inner boundary: $n \cdot \vec{u} = 0$.
Scheme: Osher–Solomon.
Grid indicator: $|\nabla \rho|$, where ρ is the density.

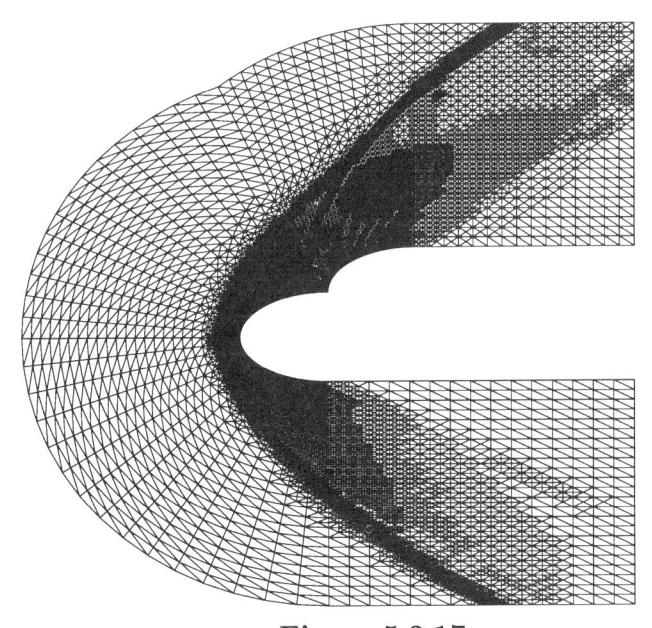

Figure 5.3.17

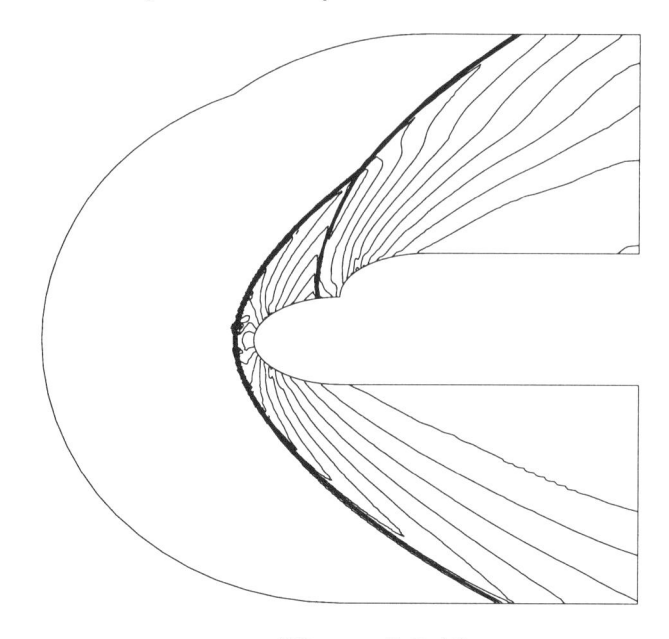

Figure 5.3.18

EXAMPLE 5.3.8 (Flow around two cylinders, Figure 5.3.19 [74])
In this test problem a supersonic flow with two cylindrical obstacles is calculated. The underlying grid is a locally refined unstructured triangulation.

Initial conditions: $\rho = 0$, $\rho u = 5.0$, $\rho v = 0.0$, $e = 15.0$.
Boundary conditions:
   left boundary: inflow;
   right boundary: outflow;
   upper boundary: outflow;
   lower boundary: outflow;
   inner boundary: $n \cdot \vec{u} = 0$.
Scheme: Steger and Warming finite volume on unstructured grid.
Grid indicator: $|\nabla \rho|$, where ρ is the density.

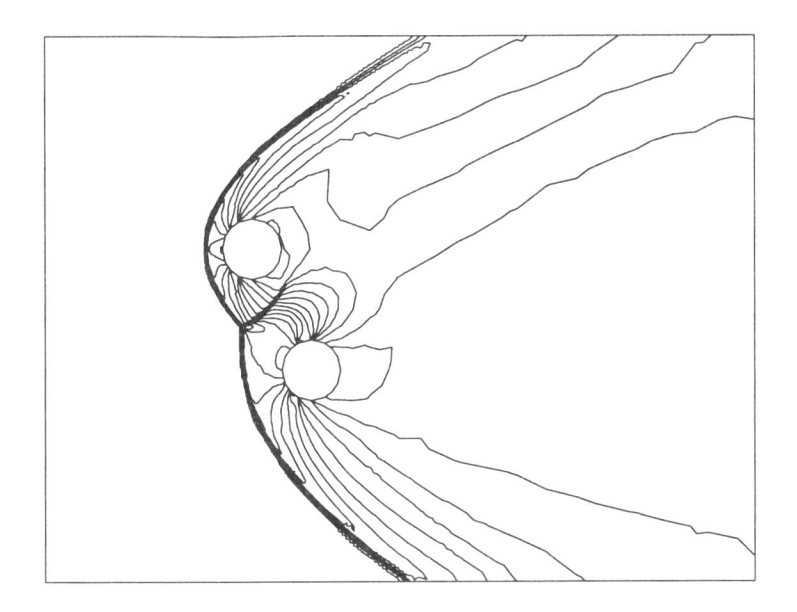

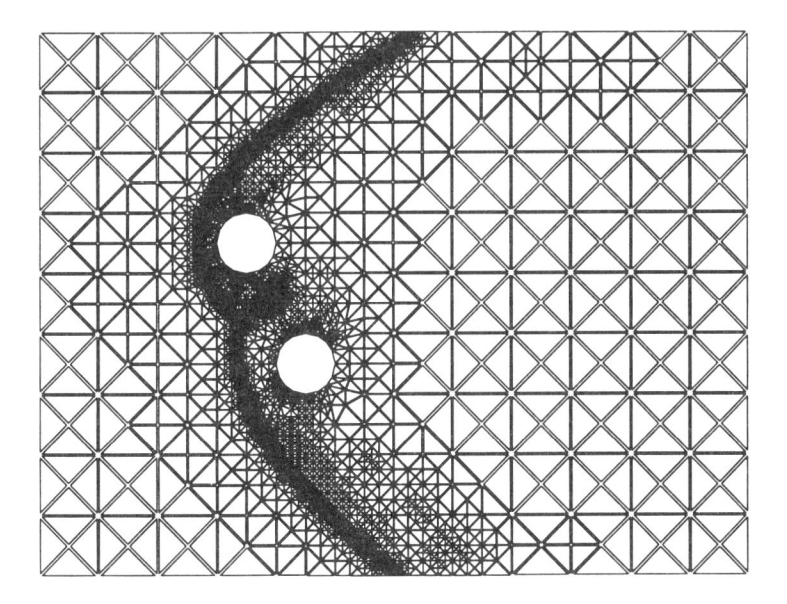

5.3 Numerical examples for the Euler equations of gas dynamics 409

Figure 5.3.19

EXAMPLE 5.3.9 (Transient phase for the forward-facing step in 2-D, Figures 5.3.20 – 5.3.23 [74]) The same method as described in Example 5.3.9 has been used to get a time-accurate solution for the transient phase between the constant initial values and the state plotted in Figures 5.3.24 and 5.3.25. In this case the mesh indicator has to control a moving refinement zone around the moving shock in 2-D. In particular, the grid had to be refined and to be coarsened after the shock has passed through.

Initial conditions: $\rho = 1.0$, $u = 2.0$, $v = 0.0$, $p = 7.0$.
Boundary conditions:
  left boundary: inflow;
  right boundary: outflow;
  upper boundary: reflection;
  lower boundary: reflection.
Scheme: van Leer finite volume on unstructured grid.
Grid indicator: $|\nabla \rho|$, where ρ is the density.

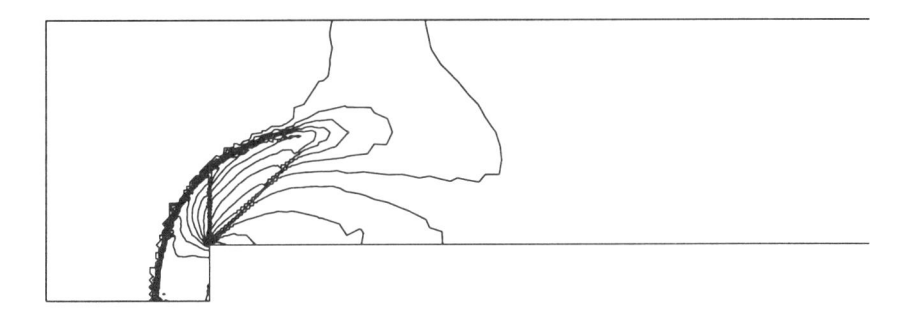

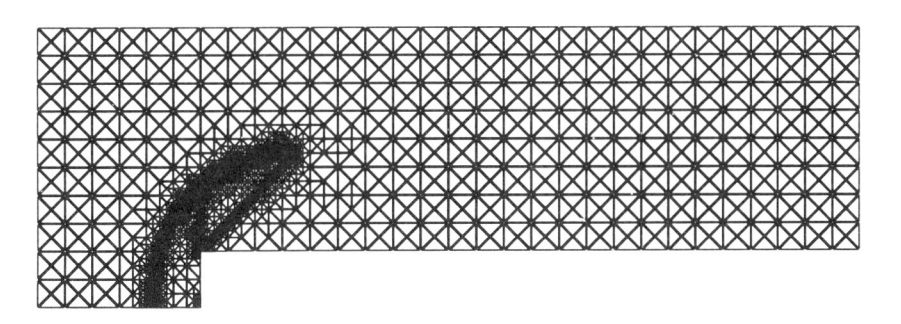

5.3 Numerical examples for the Euler equations of gas dynamics 411

Figure 5.3.20

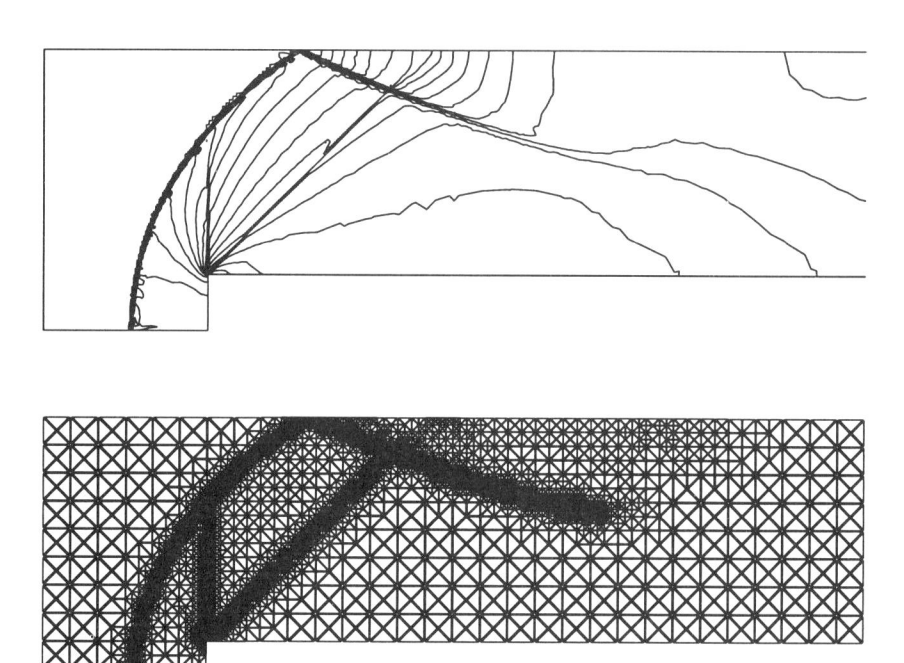

Figure 5.3.21

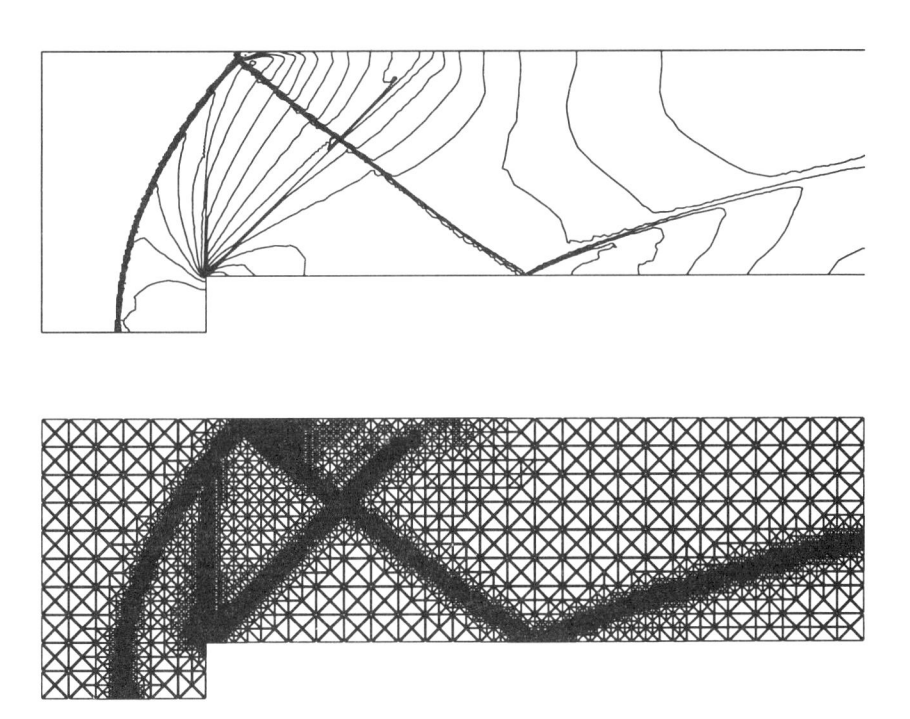

5.3 Numerical examples for the Euler equations of gas dynamics 413

Figure 5.3.22

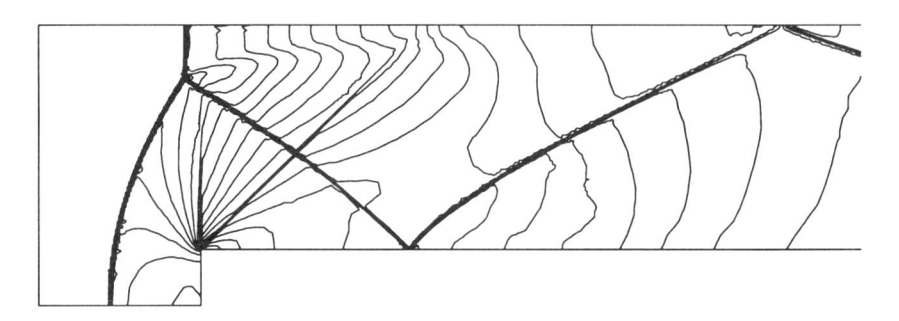

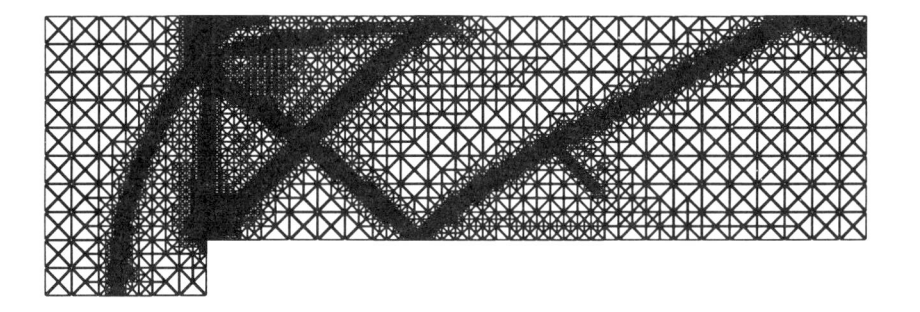

414 Initial value problems for systems of conservation laws in 2-D

Figure 5.3.23

5.3 Numerical examples for the Euler equations of gas dynamics

EXAMPLE 5.3.10 (Forward-facing step for first and higher order in 3-D, Figures 5.3.24 and 5.3.25 [250]) In Figures 5.3.24 and 5.3.25 the results for the forward-facing step problem computed with a first- and a higher-order finite volume scheme respectively are shown. The scheme used is the Steger and Warming method, and for higher order with additional reconstruction and limiting process as described in §3.5. The isolines of the density are plotted on a vertical clipping plane. The mesh is locally refined and adapted to the main structures of the discrete solution. The following expression was used as a mesh indicator (see [225]). The triangle T_j has to be refined (respectively) if

$$\eta_j(u_h^n) := |S_j^n| \sum_{l=1}^{4} |(r_j^n)_l|_{L^2(T_j)}$$

exceeds (respectively remains below) some precribed parameter, where S_j^n is the largest face of T_j and

$$r_j^n := \frac{u_j^{n+1} - u_j^n}{\Delta t^n} + \sum_{l=1}^{3} f_l'(u_j^n) \partial_l L_j^n(x_j) \ .$$

For the implementation

$$\partial_k L_j^n(x_j) = \frac{1}{|T_j|} \sum_{l=1}^{4} \int_{S_{jl}} L_j^n(x) n_{jl}^k$$

$$= \frac{1}{|T_j|} \sum_{l=1}^{4} |S_{jl}| L_j^n(x) n_{jl}^k$$

has been used, where L_j^n is the admissible piecewise-linear reconstruction as defined in (3.5.4).

Initial conditions: $\rho = 1.0$, $u = 2.0$, $v = 0.0$, $p = 7.0$.
Boundary conditions:
  left boundary: inflow;
  right boundary: outflow;
  upper boundary: reflection;
  lower boundary: reflection;

416 Initial value problems for systems of conservation laws in 2-D

front boundary: constant extension;
rear boundary: constant extension.

Scheme: Steger and Warming finite volume on unstructured grid.

Grid indicator: $|\nabla \rho|$, where ρ is the density.

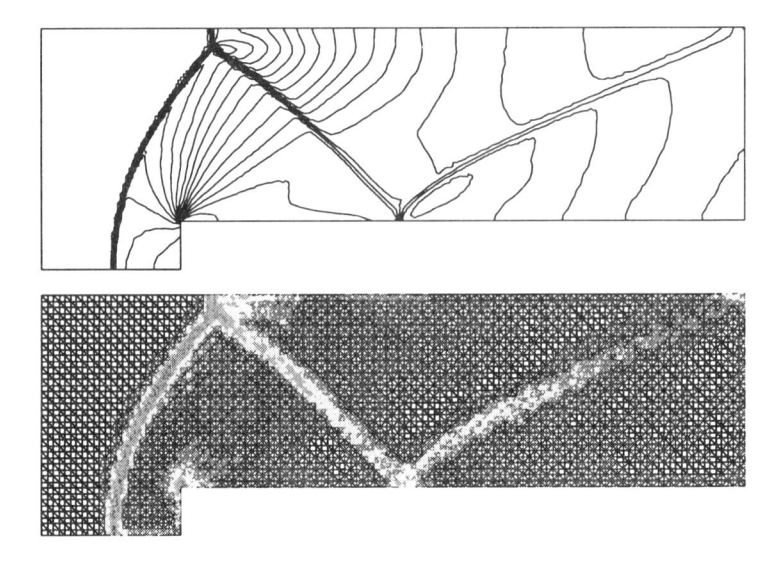

Figure 5.3.24

The states plotted in Figures 5.3.24 and 5.3.25 correspond to the time $T = \frac{5}{3}$. For the computation corresponding to Figure 5.3.24, 6033 time steps were necessary and for Figure 5.3.25, 10 500 timesteps. The CPU time for Figure 5.3.24 was 3.68 times faster than that for Figure 5.3.25.

REMARK 5.3.11 *For more details concerning the numerical solution of the forward-facing step problem we refer to [104].*

5.3 Numerical examples for the Euler equations of gas dynamics 417

Figure 5.3.25

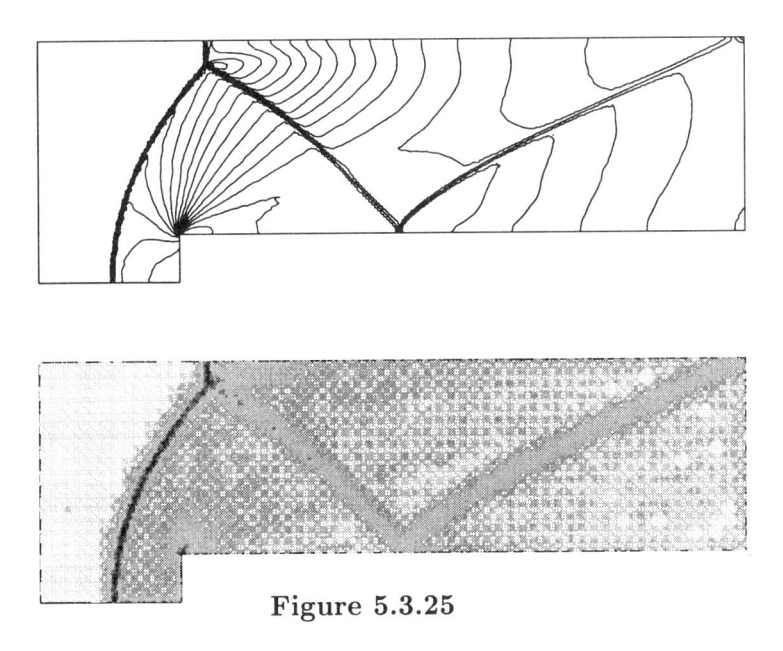

EXAMPLE 5.3.12 (Flow through an engine with a moving piston, Figure 5.3.26 [250, 249]) In this example the flow through an engine with and without a moving piston has been simulated. Chemical reactions, viscous effects and two-phase properties have not been considered. The mathematical model consists of the Euler equations in 3-D. Local refinement as described in Example 5.3.9 has been used. Numerical experiments indicate that an upwind method as described in this book has to be used. This seems to be obvious, since the situation in the inflow channel, which is partially closed by the piston, is similar to the forward-facing step problem. Also, in the phase where the inflow channel is nearly closed large velocities will appear. In Figure 5.3.26 the geometry is plotted, and in Figures 5.3.27 – 5.3.30 a time-dependent solution, obtained in the case with a moving piston, is plotted for four different times during one period. The projection of the velocity field on a plane near the symmetry plane is plotted. In Figure 5.3.31 particle traces indicate the global flow behavior.

For visualization of the results in this example GRAPE [204, 252, 182, 205, 206] has been used.

418 Initial value problems for systemsof conservation laws in 2-D

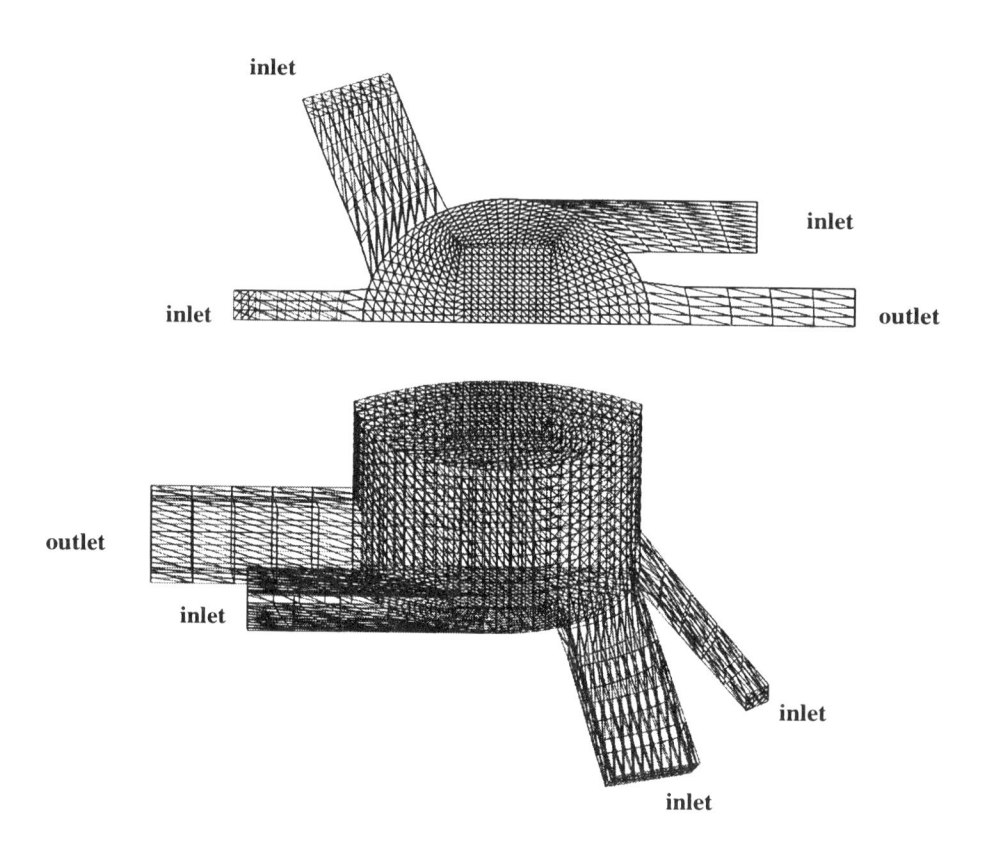

Figure 5.3.26

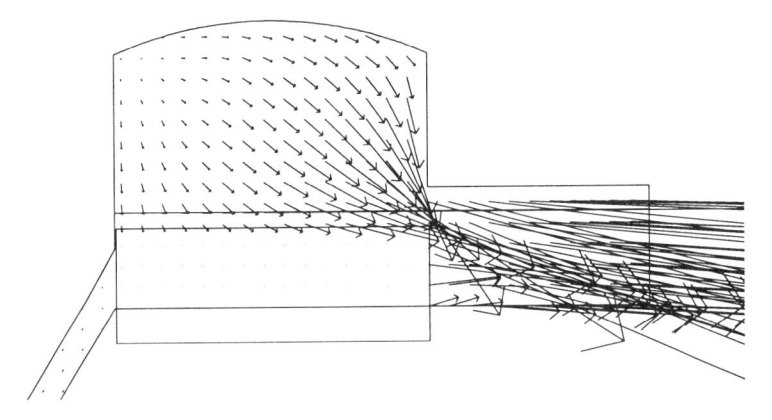

5.3 Numerical examples for the Euler equations of gas dynamics 419

Figure 5.3.27

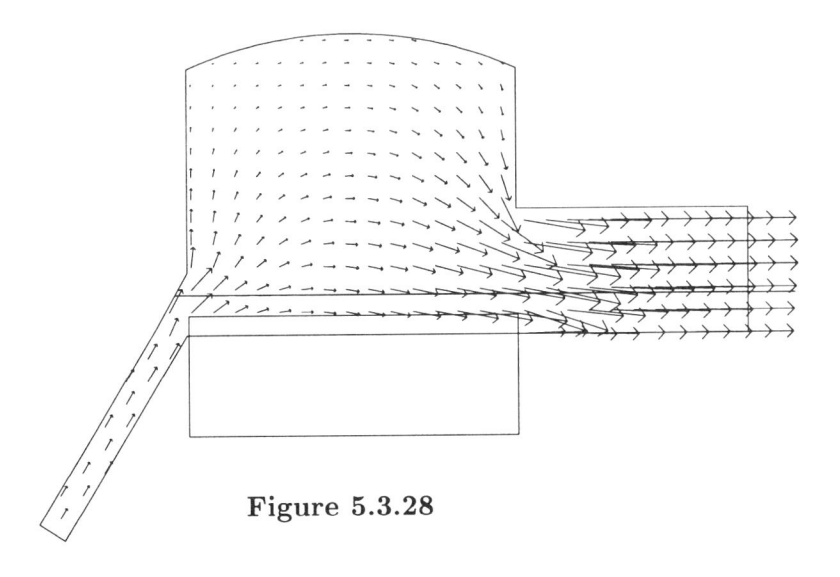

Figure 5.3.28

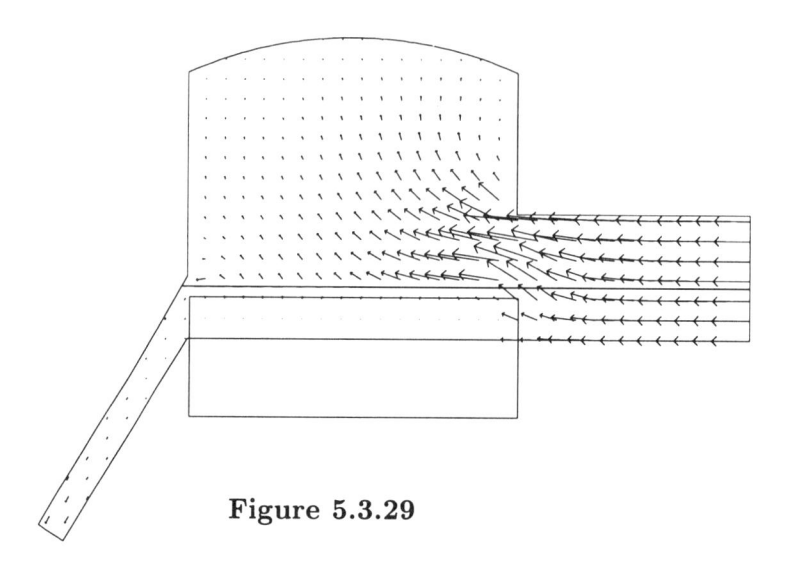

Figure 5.3.29

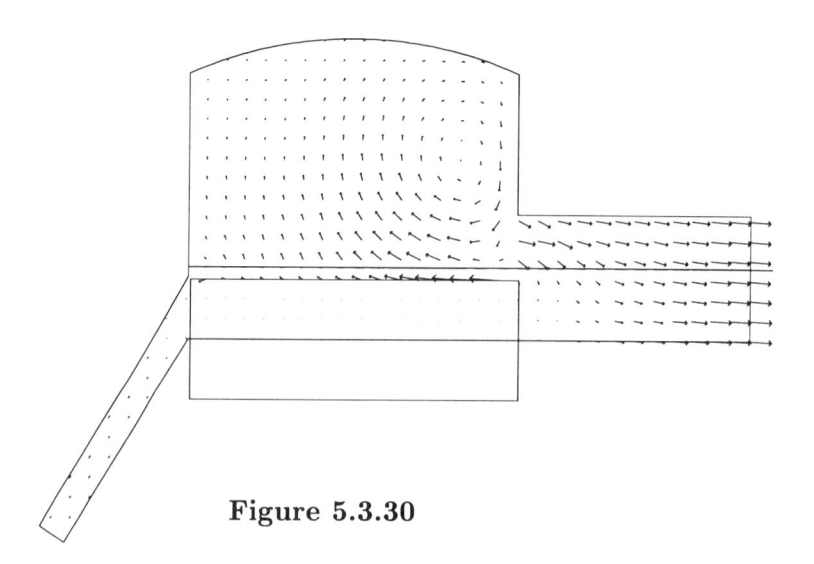

Figure 5.3.30

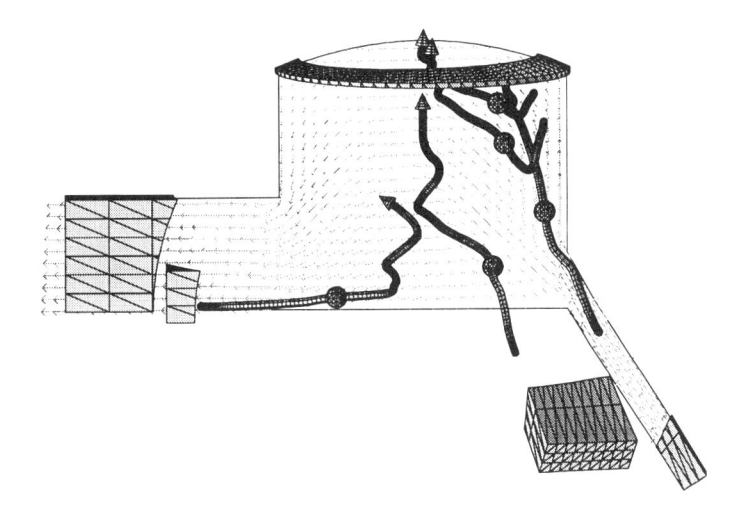

5.3 Numerical examples for the Euler equations of gas dynamics 421

Figure 5.3.31

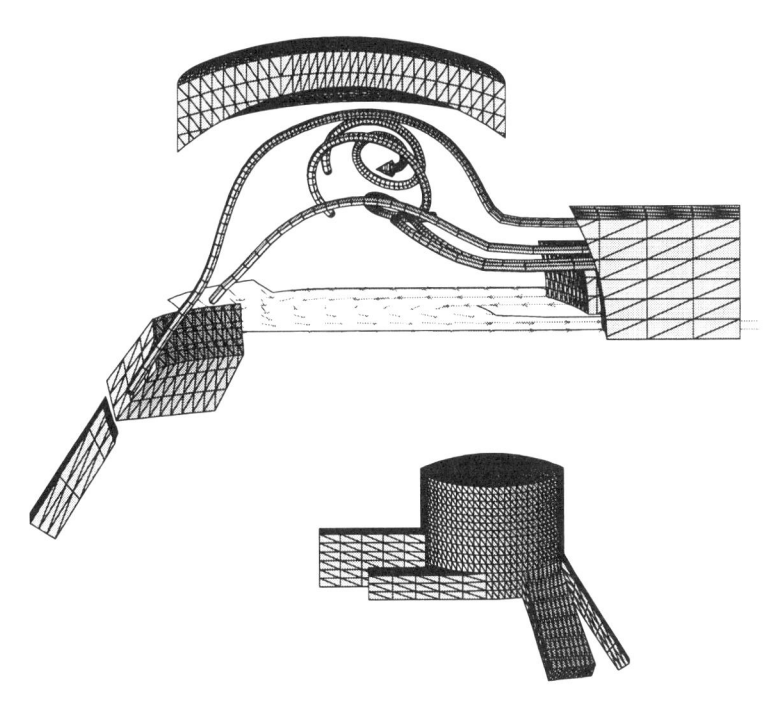

Figure 5.3.32

6 Initial boundary value problems for conservation laws

Up to now we always have considered the initial value problem for conservation laws. For the numerical treatment of initial value problems it is always necessary to work on a finite domain and to impose boundary conditions on the boundary of this finite domain. But in general it is not possible to prescribe arbitrary boundary values on the whole part of the boundary. Inflow and outflow parts of the boundary have to be considered differently. For linear problems, these parts can be discovered very easily by analysing the characteristics. The situation is much more difficult for nonlinear problems. In this case the direction of the characteristics depends on the solution and therefore also the in- and outflow parts of the boundary. The linear case will be considered in §6.1, and the nonlinear one in §6.2.

6.1 Boundary conditions for linear systems

For linear systems we discuss the different types of boundary conditions which are admissible on the basis of the analysis of the characteristics. The details depend on the sign of the eigenvalues of the Jacobians evaluated on the boundary. This concerns the in- and outflow parts of the boundary. On a solid wall no-flux boundary conditions have to be imposed. For problems in outer domains, absorbing boundary conditions are necessary. In the remaining part of this section we discuss and motivate the uniform Kreiss condition, a sufficient condition for the well-posedness of an initial boundary value problem for a linear system.

First let us repeat the situation in the linear scalar case in 1-D:

$$\partial_t u + a \partial_x u = 0 \ .$$

6.1 Boundary conditions for linear systems

If $a > 0$ boundary conditions are allowed in $x = 0$. In general, if $a < 0$ then no solution exists for prescribed boundary values in $x = 0$.

In this case a convergent numerical scheme is given by

$$u_j^{n+1} := u_j^n - a\frac{\Delta t}{\Delta x}(u_j^n - u_{j-1}^n), \qquad j = 1, ..., M.$$

The scheme needs u_0^n but not u_M^n.

Now let us consider linear systems:

$$\partial_t u = -A\partial_x u + Cu \text{ in } \mathbb{R}^+ \times \mathbb{R}^+ . \quad u(x,t) \in \mathbb{R}^n . \tag{6.1.1}$$

We assume $A, C \in C^1(\mathbb{R}^+ \times \mathbb{R}, \mathbb{R}^{n\times n})$. A is supposed to be hyperbolic. Then, as usual, there exists a matrix Q such that

$$Q^{-1}AQ = \text{diag}\{\lambda_1, ..., \lambda_n\} =: D .$$

For $w := Q^{-1}u$ we obtain

$$\partial_t w = -D\partial_x w + (Q^{-1}C + Q_t^{-1} + Q^{-1}AQ_x Q^{-1})Qw .$$

Let us assume that $Q^{-1}C + Q_t^{-1} + Q^{-1}AQ_x Q^{-1} = 0$. Then (6.1.1) has the form

$$\partial_t w_j + \lambda_j \partial_x w_j = 0 , \quad j = 1, \ldots, n .$$

Boundary values can be imposed in $x = 0$ if $\lambda_j > 0$. Otherwise no solution will exist in general for the jth equation.

EXAMPLE 6.1.1 (Euler equations in 1-D)

$$U = (\rho, u, p)^t ,$$

$$\partial_t U + A(U)\partial_x U = 0 \quad \text{in } \mathbb{R}^+ \times \mathbb{R}^+ ,$$

$$A(U) = \begin{pmatrix} u & \rho & 0 \\ 0 & u & 1/\rho \\ 0 & \rho a^2 & u \end{pmatrix} .$$

6 Initial boundary value problems for conservation laws

We assume that U is smooth near the boundary $x = 0$. Let

$$Q = \begin{pmatrix} 1, & \frac{\rho}{2a} & -\frac{\rho}{2a} \\ 0 & \frac{1}{2} & \frac{1}{2} \\ 0 & \frac{1}{2}\rho a & -\frac{1}{2}\rho a \end{pmatrix}.$$

Then

$$Q^{-1} = \begin{pmatrix} 1 & 0 & -\frac{1}{a^2} \\ 0 & 1 & \frac{1}{\rho a} \\ 0 & 1 & -\frac{1}{\rho a} \end{pmatrix}$$

and

$$Q^{-1}AQ = \begin{pmatrix} u & & \\ & u+a & \\ & & u-a \end{pmatrix} = D.$$

If we assume that Q is constant then $w := Q^{-1}U$ satisfies

$$Q(\partial_t w + D\partial_x w) = Q(\partial_t Q^{-1}U + DQ^{-1}\partial_x U)$$
$$= \partial_t U + A\partial_x U = 0,$$

or

$$\partial_t w_i + \lambda_i \partial_x w_i = 0.$$

We obtain for w

$$w = Q^{-1}U = \begin{pmatrix} 1 & 0 & -\frac{1}{a^2} \\ 0 & 1 & \frac{1}{\rho a} \\ 0 & 1 & -\frac{1}{\rho a} \end{pmatrix} \begin{pmatrix} \rho \\ u \\ p \end{pmatrix} = \begin{pmatrix} \rho - \frac{p}{a^2} \\ u + \frac{p}{\rho a} \\ u - \frac{p}{\rho a} \end{pmatrix} \qquad (6.1.2)$$

and, furthermore, with

$$\lambda_1 = u,$$
$$\lambda_2 = u + a,$$
$$\lambda_3 = u - a,$$

we have the following cases.

1. *Subsonic inflow:* $0 \leq u < a$

$$\lambda_1 > 0, \quad \lambda_2 > 0, \quad \lambda_3 < 0.$$

This means that we cannot prescribe boundary values on $x = 0$ for w_3. For w_1 and w_2 it is possible. From (6.1.2) we obtain

$$w_3 = u - \frac{p}{\rho a} = u - \sqrt{\frac{p}{\rho \gamma}}. \tag{6.1.3}$$

Therefore two of the functions ρ, u, p can be prescribed and the third can be computed from (6.1.3).

2. *Supersonic inflow:* $a < u$

$$\lambda_1 > 0, \quad \lambda_2 > 0, \quad \lambda_3 > 0,$$

and w_1, w_2, w_3 can be prescribed. This means that we can choose (ρ, u, p) arbitrarily.

The values that cannot be prescribed on the boundary but that are necessary for numerical schemes are obtained by extrapolation from the values in the interior.

EXAMPLE 6.1.2 (No-flux boundary condition in $\mathbb{R}^2$) Now we consider systems in two space dimensions:

$$\partial_t U + A \partial_x U + P \partial_y U = 0.$$

If we consider the Euler equations of gas dynamics in a domain Ω with a solid wall, a boundary condition that is convenient from the physical point of view is

$$n \cdot \begin{pmatrix} u \\ v \end{pmatrix} = 0 \quad \text{on } \partial \Omega. \tag{6.1.4}$$

426 6 Initial boundary value problems for conservation laws

For small data and locally in time Zajaczkowski and Kantiem have shown the existence of smooth solutions for the Euler equation with respect to the boundary condition (6.1.4) and suitable initial values (see [255]). Now we should like to point out how we have to discretize this boundary condition. Let us consider a triangle near the boundary (see Figure 6.1.1) and let us

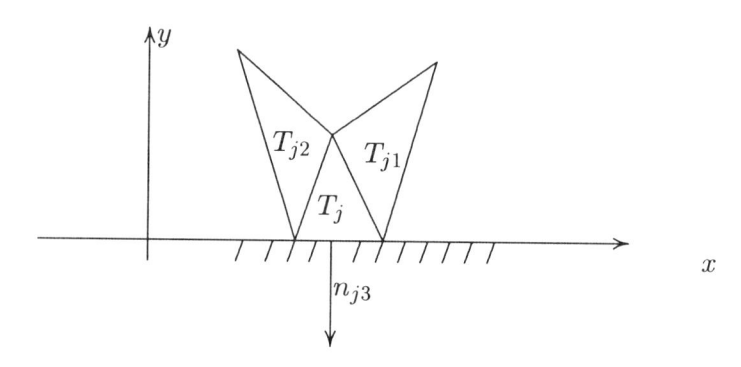

Figure 6.1.1

consider the Euler equations in conservative variables:

$$\partial_t U + \partial_x f_1(U) + \partial_y f_2(U) = 0 .$$

If we integrate this equation (we assume that the solution is sufficiently smooth) over the triangle T_j, we obtain for $f = \binom{f_1}{f_2}$

$$\partial_t \int_{T_j} U + \int_{T_j} \mathrm{div}\, f = 0 ,$$

$$\partial_t \int_{T_j} U + \sum_l \int_{S_{jl}} n_{jl} f(U) = 0.$$

Let us consider

$$n_{jl} f(u) = n_{jl}^x \begin{pmatrix} \rho u \\ \rho u^2 + p \\ \rho u v \\ u(e+p) \end{pmatrix} + n_{jl}^y \begin{pmatrix} \rho v \\ \rho u v \\ \rho v^2 + p \\ v(e+p) \end{pmatrix}.$$

The edges $j1$ and $j2$ are treated as usual. For $j3$ we have

$$n_{j3}^x = 0, \quad n_{j3}^y = -1, \quad v = 0.$$

This means that

$$n_{j3} f(u) = n_{j3}^y \begin{pmatrix} 0 \\ 0 \\ p \\ 0 \end{pmatrix}, \qquad (6.1.5)$$

and therefore p is necessary for computing $n_{j3}f(u)$. For a finite volume scheme working on a triangle T_j, as indicated in Figure 6.1.1, the fluxes g_{jl} in (5.1.7) are evaluated as usual for $l = 1$ and $l = 2$, and for $l = 3$ the expression g_{j3} is replaced by (6.1.5). Usually for this pressure one takes p_j or another value obtained by extrapolation from the interior values.

REMARK 6.1.3 (Absorbing boundary conditions) *For initial value problems in an outer domain such as the flow around a cylinder, a sphere or an airfoil there is an infinite computational domain G. For practical computations one has to consider only a bounded domain Ω and one has to impose boundary conditions on the outer boundary Γ such that the solution of the initial boundary value problem on Ω is a good approximation to the exact solution of the initial value problem on the unbounded domain G.*

The main idea is to choose boundary conditions on Γ such that a perturbation coming from the interior of Ω will not be reflected at Γ but will be absorbed. In general, the condition

$$\partial_n u = 0 \quad on \ \Gamma$$

is a good approximation to the "global" absorbing boundary condition that was derived in [61] and [122].

Now we shall study initial boundary value problems for linear hyperbolic systems in 2-D. Since the existence theory is extremely technical, we shall only present some necessary and sufficient conditions for well-posedness. For more details we refer the reader to [96] and [95].

6 Initial boundary value problems for conservation laws

Let $A, B_j, C \in C^1(\mathbb{R}^+ \times \mathbb{R}^m \times \mathbb{R}^+, \mathbb{R}^{n \times n})$ be symmetric, let $F \in L^2(\Omega_T)$, where $\Omega_T := \mathbb{R}^+ \times \mathbb{R}^m \times]0, T[$ and assume that

$$\sigma A + \sum_{j+1}^{m} \omega_j B_j$$

has real distinct eigenvalues for all real $\sigma, \omega_1, \ldots, \omega_m$, where $|\sigma| + |\omega| \neq 0$. In order to fix notation, these functions are assumed to depend on $(x, y, t) \in \mathbb{R}^+ \times \mathbb{R}^m \times \mathbb{R}^+$. Then consider the initial boundary value problem for the following linear hyperbolic system:

$$\partial_t u = A \partial_x u + \sum_{j=1}^{m} B_j \partial_j u + Cu + F \quad \text{in } \mathbb{R}^+ \times \mathbb{R}^m \times \mathbb{R}^+,$$
$$u(x, y, 0) = u_0(x, y),$$
$$Eu(0, y, t) = u_1(y, t),\qquad\qquad (6.1.6)$$

where $E \in \mathbb{R}^{l \times n}$ and l is the number of negative eigenvalues of A. This means that we are studying the initial boundary value problem in the set shown in Figure 6.1.2 with prescribed initial values u_0 for $t = 0$ and boundary values $Eu = u_1$ on the boundary $x = 0$.

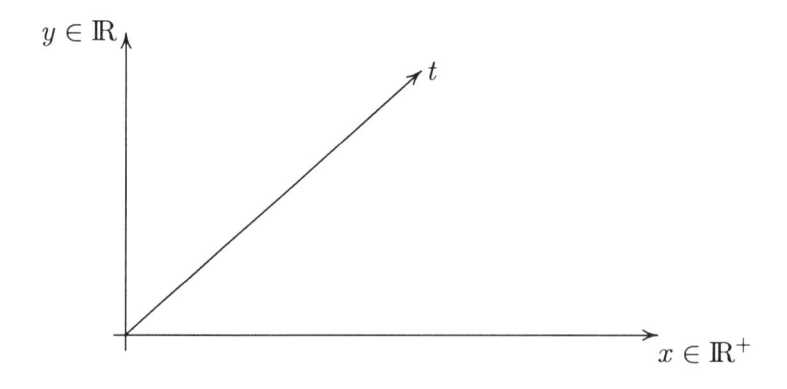

Figure 6.1.2

PROBLEM What are the necessary and sufficient conditions, such that the problem (6.1.6) has a solution?

6.1 Boundary conditions for linear systems

Let us assume that

$$A = \begin{pmatrix} A_0 & & \\ & A_I & \\ & & A_{II} \end{pmatrix},$$

where the eigenvalues of A_I are negative and those of A_{II} are positive. We assume that rank $A_I = l$, rank $A_{II} = l_2$ and rank $A_0 = l_0$, $A_0 \in \mathbb{R}^{l_0 \times l_0}$, where $l_0 + l + l_2 = n$. Then for $u := (u_1, ..., u_n)$ let

$$Z := \begin{pmatrix} u_1 \\ \vdots \\ u_{l_0} \end{pmatrix}, \quad v^I := \begin{pmatrix} u_{l_0+1} \\ \vdots \\ u_{l_0+l_1} \end{pmatrix}, \quad v^{II} := \begin{pmatrix} u_{l_0+l_1+1} \\ \vdots \\ u_n \end{pmatrix}.$$

Assume that the initial boundary value problem (6.1.6) can be written as

$$\begin{aligned} Lu &= F \quad \text{in } \Omega_T, \\ v^I - Sv^{II} &= g \quad \text{in } x = 0, \\ u(\cdot, 0) &= u_0 \quad \text{in } \Omega, \end{aligned} \tag{6.1.7}$$

where L is a differential operator of first order and $S \in \mathbb{R}^{l_1 \times l_2}$. Then the following result holds.

THEOREM 6.1.4 (Existence) *For all $T > 0$ let $F \in L^2(\Omega \times]0, T[)$, $u_0 \in L^2(\Omega)$ and $g \in L^2(\partial\Omega \times]0, T[)$ the initial boundary value problem (6.1.7) has a unique strong solution u. Moreover, for sufficiently large η we have*

$$\|u(t)\|_{0,\Omega,\eta} + \sqrt{\eta}\|u\|_{0,\Omega\times]0,t[,\eta} + \|v\|_{0,\partial\Omega\times]0,t[,\eta}$$
$$\leq C \left(\|u_0\|_{0,\Omega,\eta} + \frac{1}{\sqrt{\eta}} \|F\|_{0,\Omega\times]0,t[,\eta} + \|g\|_{0,\partial\Omega\times]0,t[,\eta} \right), \tag{6.1.8}$$

where v is the projection of u on

$$\operatorname{Ker}\left(\sum_j B_j n_j\right), \quad \text{where } n \text{ is the outer normal to } \partial\Omega,$$

and

$$\|u\|_{0,G,\eta} := \|e^{-\eta t/2} u\|_{L^2(G)}.$$

Proof *See [156, Theorem 1.12, page 621].*

REMARK 6.1.5 The statement in the theorem that u is a strong solution means in our case that $u \in L^2(\Omega \, times \,]0,T[)$ such that there exist $\{u_n\} \subset C_0^\infty(\bar{\Omega} \times [0,T])$ and $v_0 \in L^2(\partial\Omega \times]0,T[)$ with

$$\|u_n - u\|_{L^2(\Omega \times T[)} \to 0 ,$$
$$\|v_n - v_0\|_{L^2(\partial\Omega \times]0,T[)} \to 0 ,$$
$$\|Lu_n - F\|_{L^2(\Omega \times [0,T])} \to 0 ,$$
$$\|u_n(\cdot, 0) - u_0\|_{L^2(\Omega)} \to 0 ,$$
$$v_0^I - Sv_0^{II} = g .$$

DEFINITION 6.1.6 (Well-posedness) The system (6.1.6) is well posed if any solution of (6.1.6) satisfies an estimate like (6.1.8) (i.e. existence, uniqueness and continuous dependence on the initial values).

In order to get necessary conditions for well-posedness, let us consider a special case given by

$$C = 0, \quad A, B_j, E = \text{const}, \quad F = 0, \quad u_1 = 0 .$$

Then it remains to consider

$$\partial_t u = A \partial_x u + \sum_j B_j \partial_j u \quad \text{in } \mathbb{R}^+ \times \mathbb{R}^m \times \mathbb{R}^+ , \tag{6.1.9}$$

$$u(x, y, 0) = u_0(x, y) , \tag{6.1.10}$$

$$Eu(0, y, t) = 0 . \tag{6.1.11}$$

LEMMA 6.1.7 For given $\omega \in \mathbb{R}^m$ let $\varphi \in L^2(\mathbb{R})$ be an eigenfunction with eigenvalue s (i.e. solution) of

$$s\varphi = A\varphi' + i \sum_j \omega_j B_j \varphi ,$$

$$E\varphi(0) = 0 . \tag{6.1.12}$$

Then $u(x, y, t) := \varphi(x) e^{i\omega y + st}$ is a solution of (6.1.9) and (6.1.11).

6.1 Boundary conditions for linear systems

Proof It is easy to get

$$\partial_t u - A\partial_x u - \sum_j B_j \partial_j u = us - A\varphi' e^{i\omega y + st} - \sum_j B_j i\omega_j u$$

$$= e^{i\omega y + st}\left(s\varphi - A\varphi' - \sum_j B_j i\omega_j \varphi\right) = 0,$$

$$Eu = E(\varphi(x)e^{i\omega y + st})\Big|_{x=0} = e^{i\omega y + st} E\varphi(0) = 0.$$

$\square$

THEOREM 6.1.8 (Ill-posedness of (6.1.9) – (6.1.11)) *Assume that there is an $s \in \mathbb{C}$ with $\mathrm{Re}(s) > 0$ and $\omega \in \mathbb{R}^m$ such that*

$$u(x, y, t) := \varphi(x)e^{i\omega y + st}$$

satisfies (6.1.9), (6.1.11) and

$$\varphi \neq 0 \quad \text{and} \quad |\varphi(x)| \leq c e^{-\beta x} \quad \text{as } x \to +\infty \text{ for } \beta > 0.$$

Then (6.1.9)–(6.1.11) is ill posed in the sense of Definition 6.1.6, i.e. no stability estimate of the form (6.1.8) holds.

Proof The proof is based on a scaling argument. We let $u_\alpha(x, y, t) := \varphi(\alpha x)e^{i\alpha\omega y + \alpha s t}$, $\alpha \in \mathbb{R}$. Then u_α also satisfies (6.1.9), (6.1.11) and

$$u_\alpha(x, y, 0) = \varphi(\alpha x)e^{i\alpha\omega y} =: u_{1\alpha}(x, y).$$

With respect to x the function u_α has a finite L^2 norm because of the assumption on φ. In order to get finite L^2 norms with respect to y, we truncate u_α:

$$u_{\alpha,\lambda}(x, y, t) := \psi\left(\frac{y}{\lambda}\right) u_\alpha \quad \text{for } \lambda \in \mathbb{R}^+$$

with a cut-off function $\psi \in C_0^\infty(\mathbb{R}^m)$ such that

$$\psi(y) = \begin{cases} 1 & \text{if } |y| \leq 1, \\ 0 & \text{if } |y| \geq 2. \end{cases}$$

432   6 Initial boundary value problems for conservation laws

Then $u_{\alpha,\lambda}$ satisfies an inhomogeneous equation

$$\partial_t u_{\alpha,\lambda} - A\partial_x u_{\alpha,\lambda} - \sum B_j \partial_j u_{\alpha,\lambda}$$
$$= \psi\left(\frac{y}{\lambda}\right)\partial_t u_\alpha - A\psi\left(\frac{y}{\lambda}\right)\partial_x u_\alpha - \sum B_j \psi\left(\frac{y}{\lambda}\right)\partial_j u_\alpha$$
$$- \sum B_j \frac{1}{\lambda}(\partial_j \psi)\left(\frac{y}{\lambda}\right)u_\alpha = -\frac{1}{\lambda}\sum B_j (\partial_j \psi)\left(\frac{y}{\lambda}\right)u_\alpha \qquad (6.1.13)$$

and the following initial and boundary conditions:

$$u_{\alpha,\lambda}(x,y,0) = \psi\left(\frac{y}{\lambda}\right)u_\alpha(x,y,0) = \psi\left(\frac{y}{\lambda}\right)u_{1\alpha}(x,y),$$

$$Eu_{\alpha,\lambda}(0,y,0) = \psi\left(\frac{y}{\lambda}\right)Eu_\alpha(0,y,t) = 0.$$

Let $w(x,y,t) = u_{\alpha,\lambda} = \psi(y/\lambda)\varphi(\alpha x)e^{i\alpha\omega y + \alpha st}$, multiply (6.1.13) by $\bar{w}$ and integrate with respect to

$$\int := \int_{\mathbb{R}^+}\int_{\mathbb{R}^m}\int_0^t \ldots d\tau\,dy\,dx.$$

We obtain

$$\int \partial_t w \cdot \bar{w} - \int A\partial_x w \bar{w} - \sum_j B_j \int \partial_j w \bar{w}$$
$$= -\frac{1}{\lambda}\sum_j B_j \int \partial_j \psi u_\alpha \bar{w}$$

and, after integrating by parts,

$$\frac{1}{2}\int_{\mathbb{R}^m\times\mathbb{R}^+}|w(t)|^2 + \frac{1}{2}\int_0^t\int_{\mathbb{R}^m} Aw\cdot\bar{w}\Big|_{x=0}$$
$$= \frac{1}{2}\int_{\mathbb{R}^m\times\mathbb{R}^+}|w(0)|^2 - \frac{1}{\lambda}\sum_j B_j \int \partial_j \psi u_\alpha \bar{w}. \qquad (6.1.14)$$

6.1 Boundary conditions for linear systems

Now we obtain for the terms in (6.1.14)

$$T_1 := \frac{1}{2}\int_{\mathbb{R}^m}\int_{\mathbb{R}^+}|w(t)|^2 = \frac{1}{2}\int_{\mathbb{R}^m}\int_{\mathbb{R}^+}\left|\psi\left(\frac{y}{\lambda}\right)\varphi(\alpha x)e^{i\alpha wy+\alpha st}\right|^2$$

$$= \frac{1}{2}\int_{\mathbb{R}^m}\int_{\mathbb{R}^+}\left|\psi\left(\frac{y}{\lambda}\right)\varphi(\alpha x)\right|^2 dx\,dy\, e^{2\alpha\,\text{Re}(s)t}$$

$$= \frac{1}{2}\int_{\mathbb{R}^m}\int_{\mathbb{R}^+}|\psi(y)\varphi(x)|^2\,dx\,dy\,\frac{\lambda^m}{\alpha}e^{2\alpha\,\text{Re}(s)t},$$

$$T_2 := \frac{1}{2}\int_{\mathbb{R}^m}\int_{\mathbb{R}^+}|w(0)|^2 = \frac{1}{2}\int_{\mathbb{R}^m}\int_{\mathbb{R}^+}|\psi(y)\varphi(x)|^2\,dx\,dy\,\frac{\lambda^m}{\alpha},$$

$$T_3 := \frac{1}{2}\int_{\mathbb{R}^m}\int_0^t Aw\cdot\bar{w}\Big|_{x=0}$$

$$= \frac{1}{2}\int_{\mathbb{R}^m}\int_0^t\left|\psi\left(\frac{y}{\lambda}\right)e^{i\alpha wy+\alpha s\tau}\right|^2 \varphi(0)A\varphi(0)\,dy\,d\tau$$

$$= \frac{1}{2}\varphi(0)A\varphi(0)\lambda^m\int_{\mathbb{R}^m}|\psi(y)|^2\,dy\int_0^t e^{2\alpha\,\text{Re}(s)\tau}\,d\tau$$

$$= \frac{1}{2}\varphi(0)A\varphi(0)\lambda^m\|\psi\|^2_{L^2(\mathbb{R}^m)}\frac{e^{2\alpha\,\text{Re}(s)t}-1}{2\alpha\,\text{Re}(s)},$$

$$T_4 := -\frac{1}{\lambda}\sum_j\int B_j\partial_j\psi\left(\frac{y}{\lambda}\right)\varphi(\alpha x)e^{i\alpha wy+\alpha s\tau}\psi\left(\frac{y}{\lambda}\right)\varphi(\alpha x)e^{-i\alpha wy+\alpha s\tau}$$

$$= -\frac{1}{\lambda}\sum_j\underbrace{\int B_j\partial_j\psi\left(\frac{y}{\lambda}\right)\psi\left(\frac{y}{\lambda}\right)\frac{1}{\alpha}\|\varphi\|^2_{L^2(\mathbb{R}^+)}\int_0^t e^{2\alpha\,\text{Re}(s)\tau}\,d\tau}_{=0}$$

$$= 0.$$

This follows since supp ψ is bounded and B_j is symmetric. So we have

$$\frac{1}{2}e^{2\alpha t\,\text{Re}(s)}\frac{\lambda^m}{\alpha}\|\varphi\|^2_{L^2}\|\psi\|^2_{L^2} + \frac{1}{2}\varphi(0)A\varphi(0)\lambda^m\|\psi\|^2\frac{e^{2\alpha\,\text{Re}(s)t}}{2\alpha\,\text{Re}(s)}$$

434   6 Initial boundary value problems for conservation laws

$$= \frac{1}{2} \frac{\lambda^m}{\alpha} \|\psi\|_{L^2}^2 \|\varphi\|_{L^2}^2 \;. \tag{6.1.15}$$

This means that the left-hand side of (6.1.15) behaves as a function of λ and α like

$$\frac{\lambda^m}{\alpha} e^{2\alpha t \, \mathrm{Re}(s)} + \frac{\lambda^m}{\alpha}$$

and the right-hand side like

$$\frac{\lambda^m}{\alpha} \;.$$

It is obvious from this that for each fixed λ this leads for large α to a contradiction (since $\mathrm{Re}(s) > 0$). This means that we cannot expect an energy estimate of the form (6.1.8). □

The statement of Theorem 6.1.8 can also be expressed as

LEMMA 6.1.9 *The initial value problem (6.1.9)–(6.1.11) is ill posed if (6.1.12) has for some $w \in \mathbb{R}^m$ an eigenvalue $s \in \mathbb{C}$ such that $\mathrm{Re}(s) > 0$.*

Let us assume that A^{-1} exists. The eigenvalue problem (6.1.12) can also be written as

$$\varphi' = A^{-1}(s - i w \cdot B)\varphi =: M(w, s)\varphi \;.$$

LEMMA 6.1.10 *Let l denote the number of negative eigenvalues of A and let $\mathrm{Re}(s) > 0$. Then $M(w, s)$ has l eigenvalues with negative real part and $n - l$ eigenvalues with positive real part for any $w \in \mathbb{R}^m$.*

Proof First we show that M has no eigenvalue κ with $\mathrm{Re}(\kappa) = 0$. If $\kappa = i\sigma$ is an eigenvalue then

$$i\sigma w = M(w, s)w = A^{-1}(s - i w B)w$$

for some $w \neq 0$, and therefore

$$(i\sigma A + i w B)w = sw$$

6.1 Boundary conditions for linear systems

Now we know that the system is hyperbolic, i.e.

$$\sigma A + \omega B$$

has only real eigenvalues. Therefore $-is$ is real. This contradicts $\text{Re}(s) > 0$, and therefore $\text{Re}(\kappa) \neq 0$ if $Mw = \kappa w$.

Now let $s = \eta + i\xi$ with $\eta > 0$ and let κ be an eigenvalue of $M(\omega, s)$. Then $\kappa = \kappa(\eta, \xi, \omega)$ is a continuous function of $(\eta, \xi, \omega) \in \mathbb{R}^+ \times \mathbb{R} \times \mathbb{R}^m =: \Omega$ and $\text{Re}(\kappa)$ cannot change its sign on Ω, because then $\text{Re}(\kappa) = 0$ somewhere.

Let

$$N(\eta, \xi, \omega) = \sharp\{\kappa(\eta, \xi, \omega) \mid \kappa \text{ is eigenvalue of } M(\eta, \xi, \omega)$$
$$\text{and} \text{Re}(\kappa) < 0\}.$$

Then N is constant on Ω; in particular,

$$N(\eta, \xi, \omega) = N(1, 0, 0).$$

But $M(1, 0, 0) = A^{-1}$ and

$$\sharp\{\kappa \mid \kappa \text{ is an eigenvalue of } A \text{ with } \text{Re}(\kappa) < 0\}$$
$$= \sharp\{\kappa \mid \kappa \text{ is an eigenvalue of } A^{-1} \text{ with } \text{Re}(\kappa) < 0\}$$
$$= \sharp\{\kappa \mid \kappa \text{ is an eigenvalue of } M(1, 0, 0) \text{ with } \text{Re}(\kappa) < 0\}$$
$$= \sharp\{\kappa \mid \kappa \text{ is an eigenvalue of } M(\eta, \xi, \omega) \text{ with } \text{Re}\, \kappa < 0\}$$
$$= l.$$

This proves the lemma. □

Let $\kappa_1, \ldots, \kappa_l, \ldots, \kappa_n$ be the eigenvalues of M and let $w_1, \ldots, w_l$ be the corresponding eigenvectors such that

$$\text{Re}(\kappa_j) < 0 \quad \text{if } 1 \leq j \leq l.$$

Then the $\varphi_j(x) := e^{\kappa_j x} w_j$ are linearly independent solutions of the differential equation in (6.1.12), since

$$\varphi'_j(x) = \kappa_j \varphi_j(x) = M(\omega, s)\varphi_j(x) = A^{-1}(s - i\omega B)\varphi_j(x). \quad (6.1.16)$$

Now we can prove the following necessary condition for well-posedness.

6 Initial boundary value problems for conservation laws

THEOREM 6.1.11 (Necessary condition for well-posedness) *Let the initial boundary value problem be well posed. Then*

$$\det(E\varphi_1(0), ..., E\varphi_l(0)) \neq 0 \qquad (6.1.17)$$

for all $s \in \mathbb{C}$, $\mathrm{Re}(s) > 0$ and all $\omega \in \mathbb{R}^m$, where the φ_j are defined in (6.1.16).

Note $E \in \mathbb{R}^{l \times n}$, $\varphi \in \mathbb{R}^n$ and $E\varphi \in \mathbb{R}^l$.

Proof Let us assume that

$$\det(E\varphi_1(0), ..., E\varphi_l(0)) = 0.$$

Then there exists a $c = (c_1, ..., c_l) \neq 0$ such that

$$0 = \sum_{j=1}^{l} c_j E\varphi_j(0) \ .$$

Define $u_j(x, y, t) := \varphi_j(x)e^{i\omega y + st}$ and

$$u(x, y, t) := \sum_{j=1}^{l} c_j u_j(x, y, t) = e^{i\omega y + st}\varphi(x) \ ,$$

where $\varphi(x) = \sum_{j=1}^{l} c_j \varphi_j(x)$. Since $\mathrm{Re}(\kappa_j) < 0$ as $j = 1, ..., l$

$$|\varphi(x)| \leq c e^{-\beta x} \quad \text{as } x \to \infty.$$

Furthermore, u solves (6.1.9) and

$$0 = \sum_{j=1}^{l} c_j E\varphi_j(0) = E \sum_{j=1}^{l} c_j \varphi_j(0)$$

$$= E \sum_{j=1}^{l} c_j \varphi_j(0) e^{i\omega y + st}$$

$$= Eu(0, y, t).$$

In particular, $\varphi = \sum_{j=1}^{l} c_j \varphi_j \neq 0$. Then all the conditions of Theorem 6.1.8 are satisfied, and therefore the problem is ill posed. But this is a contradiction, and therefore

$$\det(E\varphi_1(0), ..., E\varphi_l(0)) \neq 0.$$ □

A condition that is a little bit stronger than (6.1.17) is also sufficient for well-posedness. This is the statement of the following remark.

REMARK 6.1.12 (Uniform Kreiss condition) *Let us assume that the $\varphi_1, \ldots, \varphi_l$ are orthonormal in $x = 0$. Then the condition*

$$\det(E\varphi_1(0), ..., E\varphi_l(0)) \geq \delta \qquad (6.1.18)$$

uniformly for all $w \in \mathbb{R}^m$ and $s \in \mathbb{C}$, $\mathrm{Re}(s) > 0$ is a sufficient condition for well posedness. The condition (6.1.18) is called the Uniform Kreiss condition.

Proof See [96, page 186], [156, Theorem 1] and [117].

Now we should like to consider initial boundary value problems for nonlinear scalar equations. It is not obvious a priori which boundary conditions are allowed for

$$\partial_t u + \mathrm{div}\, f(u) = 0 \quad \text{in } \Omega \times\,]0, T[=: \Omega_T\, ,$$
$$u(x, 0) = u_0(x)\, .$$

6.2 Boundary conditions for nonlinear scalar equations

For nonlinear problems the situation is much more complicated than in the linear case. Since the characteristics depend on the solution itself, it cannot be decided a priori which part of the boundary is an inflow one and which an outflow one. As in the definition of the entropy solution, we use the viscosity limit of a sequence of solutions of initial boundary value problems for the corresponding parabolic equation to get admissible boundary conditions for

438   6 Initial boundary value problems for conservation laws

the conservation law. The results that we shall present here are due to Bardos, LeRoux and Nedelec [16, 17].

In order to get an admissible condition on the boundary we consider the initial boundary value problem for

$$\begin{aligned}
\partial_t u_\varepsilon + \operatorname{div} f(u_\varepsilon) &= \varepsilon \Delta u_\varepsilon && \text{in } \Omega_T\,, \\
u_\varepsilon(x,0) &= u_0(x) && \text{in } \Omega\,, \\
u_\varepsilon(\cdot,t) &= 0 && \text{in } \partial\Omega \times\,]0,T[\,,
\end{aligned} \qquad (6.2.1)$$

and investigate what the boundary conditions are in the limit as $\varepsilon \to 0$.

DEFINITION 6.2.1 (Function space $\mathrm{BV}(\Omega)$) *The space of functions of bounded variation is defined as*

$$\mathrm{BV}(\Omega) := \{u \in L^1(\Omega) \mid \|u\|_{\mathrm{BV}(\Omega)} < \infty\}$$

where

$$\|u\|_{\mathrm{BV}(\Omega)} := \sup\left\{ \int_\Omega u \operatorname{div} \varphi \ \Big|\ \varphi \in C_0^1(\Omega, \mathbb{R}^n), |\varphi| < 1 \right\}.$$

THEOREM 6.2.2 (Compactness) *Let $\Omega \subset \mathbb{R}^n$ be open and bounded with smooth boundary $\partial\Omega$. Let $(f_k)_k$ be a sequence in $\mathrm{BV}(\Omega)$ such that*

$$\|f_k\|_{\mathrm{BV}(\Omega)} \leq \mathrm{const}$$

uniformly in k. Then there exists a subsequence $(f_k)_k$ and a function $f \in \mathrm{BV}(\Omega)$ such that

$$f_k \to f \quad \text{in } L^1(\Omega).$$

Proof See [64].

First we need two lemmas, which we shall present without proof.

6.2 Boundary conditions for nonlinear scalar equations

LEMMA 6.2.3 *There exists a trace map*

$$\gamma : \mathrm{BV}(\Omega \times {]0,T[}) \to L^\infty(\Omega \times \{0\}) \times L^\infty(\partial\Omega \times {]0,T]}) \,,$$

$$u \to \begin{pmatrix} u \mid_{t=0} \\ u \mid_\Gamma \end{pmatrix} ,$$

such that for all $h \in C^1(\mathbb{R})$ we have

$$\gamma[h(u)] = h[\gamma(u)].$$

Furthermore, we have

$$\int_\Omega |u(\cdot,t) - u(\cdot,0)| \to 0 \quad \text{as } t \to 0 \,.$$

Proof See [16] and [17, Lemma 1]. ∎

LEMMA 6.2.4 *For $v \in C^1(\bar{\Omega})$ we have*

$$\lim_{\varepsilon \to 0} \int_{\{x \in \Omega \mid |v(x)| < \varepsilon\}} \nabla v \, dx = 0 \,.$$

Proof See [207]. ∎

In the following theorem we get the convergence of u_ε.

THEOREM 6.2.5 *Let Ω be as in Theorem 6.2.2 and let the data in (6.2.1) be sufficiently smooth, such that $u_\varepsilon, \nabla u_\varepsilon \in C^2(\bar{\Omega}_T)$. Then the solutions u_ε of (6.2.1) are compact in $L^1(\Omega_T)$ and each limit u belongs to $\mathrm{BV}(\Omega_T)$. In particular, any u has traces in $\{t=0\}$ satisfying*

$$u(x,0) = u_0(x) \quad \text{for } x \in \Omega.$$

Proof The maximum principle implies that

$$\|u_\varepsilon\|_{L^\infty(\Omega_T)} \leq e^{CT} \|u_0\|_{L^\infty(\Omega)}. \tag{6.2.2}$$

Now we shall describe an L^1 estimate for $v = \partial_t u_\varepsilon$. Formally we have to multiply (6.1.2) with $|\partial_t u_\varepsilon|$ and to integrate with respect to Ω and $[0,t]$ in order to get this. In the following we shall give a rigorous proof of this. We differentiate (6.2.1) with respect to t and obtain for $v = \partial_t u_\varepsilon$

$$\partial_t v + \operatorname{div}[f'(u_\varepsilon)v] = \varepsilon \Delta v \quad \text{in } \Omega_T. \tag{6.2.3}$$

Let $g_\eta \in C^1(\mathbb{R})$ be an approximation of the sign function such that $g_\eta' \geq 0$ and $g_\eta(x) = -g_\eta(-x)$. Now we multiply (6.2.3) by $g_\eta(v)$ and integrate with respect to Ω:

$$\underbrace{\int_\Omega \partial_t v g_\eta(v)}_{I_1} + \underbrace{\int_\Omega \operatorname{div}[f'(u_\varepsilon)v]g_\eta(v)}_{I_2} = \varepsilon \underbrace{\int_\Omega g_\eta(v)\Delta v}_{I_3}.$$

We obtain

$$I_2 = -\int_\Omega f'(u_\varepsilon)v g_\eta'(v)\nabla v \longrightarrow 0 \quad \text{as } \eta \to 0$$

because of Lemma 6.2.4. Let $G_\eta(s) := \int_0^s g_\eta(\tau)d\tau$. Then we obtain

$$\int_0^t I_1 = \int_0^t \int_\Omega \frac{d}{dt} G_\eta(v) = \int_\Omega G_\eta(v(t)) - \int_\Omega G_\eta(v(0))$$

$$\longrightarrow \int_\Omega |v(t)| - \int_\Omega |v(0)|,$$

since $G_\eta(s) \to |s|$ as $\eta \to 0$. Furthermore,

$$I_3 = -\varepsilon \int_\Omega g_\eta'(v)\nabla v \cdot \nabla v \leq 0.$$

Therefore we obtain

$$\int_\Omega |\partial_t u_\varepsilon| = \int_\Omega |(v(t))| \leq \int_\Omega |v(0)|$$

$$= \int_\Omega \left|\varepsilon \Delta u_0(x) - \operatorname{div} f(u_0)\right| dx \leq \text{const} . \tag{6.2.4}$$

6.2 Boundary conditions for nonlinear scalar equations

Now we shall derive a similar estimate for $\|\nabla u_\varepsilon\|_{L^1}$. For this, we take the gradient of the differential equation in (6.2.1). For $w = \nabla u_\varepsilon$ we obtain

$$\partial_t w + \operatorname{div}[f'(u_\varepsilon)w] = \varepsilon \Delta w.$$

We take the scalar product of this equation with $(\partial_1 r(w), ..., \partial_n r(w))^t$, where

$$r(\xi) := \int_0^{|\xi|} g_\eta(\tau)\, d\tau$$

and integrate with respect to Ω:

$$\int_\Omega \partial_t w (\nabla r)(w) + \int_\Omega \operatorname{div}[f'(u_\varepsilon)w](\nabla r)(w)$$

$$= \varepsilon \int_\Omega \Delta w (\nabla r)(w) \, . \tag{6.2.5}$$

For the second term in (6.2.5) we get

$$\int_\Omega \operatorname{div}[f'(u_\varepsilon)w](\nabla r)(w)$$

$$= \int_\Omega \left\{ \operatorname{div} f'(u_\varepsilon) w \cdot (\nabla r)(w) + f'(u_\varepsilon)\nabla[r(w)] \right\}$$

$$= \int_\Omega \operatorname{div} f'(u_\varepsilon)[w \cdot (\nabla r)(w) - r(w)] + \int_{\partial\Omega} f'(u_\varepsilon) \cdot n r(w)$$

$$= \int_\Omega \operatorname{div} f'(u_\varepsilon)[|w|g_\eta(|w|) - r(w)] + \int_{\partial\Omega} f'(0) n r(w).$$

For the third term in (6.2.5) we obtain, also after integration by parts,

$$\int_\Omega \Delta w (\nabla r)(w) = \int_\Omega \sum_i \sum_k \partial_i^2 \partial_k u_\varepsilon (\partial_k r)(w)$$

$$= - \int_\Omega \sum_i \sum_k \sum_j \partial_i \partial_k u_\varepsilon\, \partial_j \partial_k r \partial_i u_\varepsilon + \int_{\partial\Omega} \sum_k n \cdot \nabla w_k \partial_k r$$

$$= - \int_\Omega \sum_{i,k,j} \partial_j \partial_k r\, \partial_i \partial_k u_\varepsilon\, \partial_j \partial_i u_\varepsilon + \int_{\partial\Omega} \partial_n r(w) \, .$$

Putting these results together, we obtain from (6.2.5)

$$\frac{d}{dt}\int_\Omega r(w) = -\varepsilon \int_\Omega \sum_{i,k,j} \partial_j\partial_k r\, \partial_i\partial_k u_\varepsilon\, \partial_j\partial_i u_\varepsilon$$

$$- \int_\Omega \operatorname{div} f'(u_\varepsilon)[|w|g_\eta(|w|) - r(w)]_1$$

$$- \int_{\partial\Omega} f'(0)nr(w) + \varepsilon \int_{\partial\Omega} \partial_n r(w) \,. \qquad (6.2.6)$$

Since r is convex, the first term on the right-hand side is nonpositive. The expression $[\ \ ]_1$ tends to zero as $\eta \to 0$, and therefore so does the second term on the right-hand side. It remains to estimate the integrals on the boundary $\partial\Omega$. From the differential equation (6.2.1) for u_ε, evaluated on the boundary $\partial\Omega$, we obtain

$$f'(0) \cdot \nabla u_\varepsilon = \varepsilon \Delta u_\varepsilon \,.$$

Now since $u_\varepsilon = 0$ on $\partial\Omega$, we know that there is an $\alpha : \partial\Omega \to \mathbb{R}$ such that

$$n(x) = \alpha(x)\nabla u_\varepsilon(x) \quad \text{for all } x \in \partial\Omega \,,$$

and this implies

$$[f'(0)n]\,(n\nabla u_\varepsilon) = f'(0)\nabla u_\varepsilon = \varepsilon\Delta u_\varepsilon,$$

$$f'(0)n = \varepsilon \frac{\Delta u_\varepsilon}{\partial_n u_\varepsilon}. \qquad (6.2.7)$$

Furthermore, we have $|w| = |\nabla u_\varepsilon| = |\partial_n u_\varepsilon|$ and

$$\partial_n r(w) = \partial_n \int_0^{|w|} g_\eta(\tau)\, d\tau$$

$$= \partial_n \int_0^{|\partial_n u_\varepsilon|} g_\eta(\tau)d\tau = g_\eta(|\partial_n u_\varepsilon|)\operatorname{sign}(\partial_n u_\varepsilon)\partial_n^2 u_\varepsilon$$

$$\longrightarrow \operatorname{sign}(\partial_n u_\varepsilon)\partial_n^2 u_\varepsilon \quad \text{as } \eta \to 0 \,. \qquad (6.2.8)$$

6.2 Boundary conditions for nonlinear scalar equations

From (6.2.7) we get

$$f'(0)n\ r(w) = \varepsilon \Delta u_\varepsilon \frac{r(\partial_n u_\varepsilon)}{\partial_n u_\varepsilon}$$

$$\to \varepsilon \Delta u_\varepsilon \frac{|\partial_n u_\varepsilon|}{\partial_n u_\varepsilon} = \varepsilon \Delta u_\varepsilon \operatorname{sign}(\partial_n u_\varepsilon) . \qquad (6.2.9)$$

Using (6.2.8) and (6.2.9) in (6.2.6) we obtain as $\eta \to 0$

$$\int_{\partial \Omega} [\varepsilon \partial_n r(w) - f'(0) n\ r(w)] \longrightarrow \varepsilon \int_{\partial \Omega} \operatorname{sign}(\partial_n u_\varepsilon)(\partial_n^2 u_\varepsilon - \Delta u_\varepsilon) ,$$

which can be estimated by

$$\varepsilon \int_{\partial \Omega} |\partial_n^2 u_\varepsilon - \Delta u_\varepsilon| . \qquad (6.2.10)$$

In normal and tangential coordinates $(\sigma, \tau) = (\sigma(x,y), \tau(x,y))$ with respect to $\partial \Omega$, Δu_ε can be written on $\partial \Omega$ in $\mathbb{R}^2$ as

$$\Delta u_\varepsilon = |\nabla \tau|^2 \partial_\tau^2 u_\varepsilon + |\nabla \sigma|^2 \partial_\sigma^2 u_\varepsilon + 2 \partial_\sigma \partial_\tau u_\varepsilon\ \nabla \tau \nabla \sigma$$
$$+ \partial_\tau u_\varepsilon \Delta \tau + \partial_\sigma u_\varepsilon \Delta \sigma$$
$$= \partial_\sigma^2 u_\sigma + \Delta \sigma \partial_\sigma u_\varepsilon ,$$

since $u_\varepsilon = 0$ on $\partial \Omega$, $\nabla \tau \cdot \nabla \sigma = 0$ and $n \nabla \sigma = \partial_n \sigma = 1$, $\nabla \sigma = n$ and $|\nabla \sigma| = 1$. This implies that (6.2.10) can be estimated by

$$c \varepsilon \int_{\partial \Omega} |\partial_n u_\varepsilon| . \qquad (6.2.11)$$

For the general case we have to straighten the boundary $\partial \Omega$ and obtain an estimate like (6.2.11) where c depends on the curvature of $\partial \Omega$ (see e.g. [76]). Using (6.2.10), we finally get

$$\lim_{\eta \to 0} \int_0^T \int_{\partial \Omega} [\varepsilon \partial_n r(w) - f'(0) n r(w)]$$

$$\leq c\varepsilon \int_0^T \int_{\partial\Omega} |\partial_n u_\varepsilon| \leq c_1\varepsilon \|\Delta u_\varepsilon\|_{L^1(\Omega_T)}$$

$$\leq c_2[\|\partial_t u_\varepsilon\|_{L^1(\Omega_T)} + \|\nabla u_\varepsilon\|_{L^1(\Omega_T)}] \qquad (6.2.12)$$

$$\leq c\left(T + \int_0^T \|\nabla u_\varepsilon\|_{L^1(\Omega)}\, d\tau\right).$$

The last two estimates are based on (6.2.1) and (6.2.4). Then we derive from (6.2.6) and (6.2.12) as $\eta \to 0$

$$\|\nabla u_\varepsilon(\cdot, t)\|_{L^1(\Omega)} \leq \|\nabla u_0\|_{L^1(\Omega)} + c\left(T + \int_0^T \|\nabla u_\varepsilon\|_{L^1(\Omega)}\, dt\right). \quad (6.2.13)$$

In order to get an estimate for $\|\nabla u_\varepsilon(\cdot, t)\|_{L^1(\Omega)}$ from (6.2.13), we need the following Gronwall inequality.

LEMMA 6.2.6 (Gronwall inequality) *Let $\beta > 0$, $\alpha \in \mathbb{R}$, $\varphi \in C^0([0,T])$ and*

$$\varphi(t) \leq \alpha + \beta \int_0^t \varphi(s)\, ds.$$

Then

$$\varphi(t) \leq \alpha e^{\beta t}.$$

Proof See [218] and [244].

We apply the Gronwall inequality to (6.2.13) with $\varphi(t) := \|\nabla u_\varepsilon(\cdot, t)\|_{L^1(\Omega)}$ and obtain, using (6.2.4) and (6.2.2),

$$\|u_\varepsilon\|_{L^\infty(\Omega_T)} + \|\nabla u_\varepsilon\|_{L^1(\Omega_T)} + \|\partial_t u_\varepsilon\|_{L^1(\Omega_T)} \leq c(T). \qquad (6.2.14)$$

This estimate and the compactness Theorem 6.2.2 imply that there is a subsequence u_ε and a $u \in BV(\Omega_T)$ such that

$$u_\varepsilon \to u \quad \text{in } L^1(\Omega_T). \qquad (6.2.15)$$

6.2 Boundary conditions for nonlinear scalar equations

Because of

$$\|u(\cdot,0) - u_0\|_{L^1(\Omega)} \leq \|u(\cdot,0) - u(\cdot,t)\|_{L^1(\Omega)}$$
$$+ \|u(\cdot,t) - u_\varepsilon(\cdot,t)\|_{L^1(\Omega)} + \|u_\varepsilon(\cdot,t) - u_0\|_{L^1(\Omega)}$$

and Lemma 6.2.3, $u(\cdot,0)$ has initial values u_0. This completes the proof of Theorem 6.2.5. □

6.2.7 (Motivation of the boundary condition) For

$$\partial_t u + a \partial_x u = 0 \quad \text{in } \mathbb{R}^+ \times \mathbb{R}^+$$

we can impose boundary conditions $u = g$ in $x = 0$ if $a > 0$, and this is not possible if $a < 0$. But in any case the condition

$$\min_{k \in I(\gamma u, g)} \{-a|\gamma u - k|\} = 0 \quad \text{(pointwise in } (x,t) \in \mathbb{R}^+ \times \mathbb{R}^+\text{)},$$

where γ is the trace operator as defined in Lemma 6.2.3, is satisfied. There we have used the notation

$$I(a,b) := \Big[\min\{a,b\}, \max\{a,b\}\Big].$$

If $a > 0$, the minimum is equal to zero if and only if $\gamma u = g$, i.e. u has the prescribed boundary values. If $a < 0$, the minimum is obtained for $k = \gamma u$ independent of g and is equal to 0.

The generalization to the nonlinear equation with homogeneous boundary condition $a = 0$ is

$$\min_{k \in I(\gamma u, a)} \{\operatorname{sign}(\gamma u - a)[f(\gamma u) - f(k)] \cdot n\} = 0. \tag{6.2.16}$$

Let us consider this condition in $\{x = 0\}$ for the special case where $n = \binom{-1}{0}$ and $f = \binom{f_1}{f_1}$. Then (6.2.16) can be written as

$$\min_{k \in I(\gamma u, 0)} \left\{ \operatorname{sign}(\gamma u) \frac{f_1(\gamma u) - f_1(k)}{\gamma u - k} (\gamma u - k)(-1) \right\} = 0. \tag{6.2.17}$$

Notice that for $k \in I(\gamma u, 0)$

$$\text{sign}(\gamma u)(\gamma u - k)(-1) \leq 0 .$$

Then if the characteristics are leaving $\mathbb{R}^+$ in $\{x = 0\}$, we have

$$\frac{f_1(\gamma u) - f_1(k)}{\gamma u - k} < 0 ,$$

and the condition (6.2.17) can be satisfied for $k = \gamma u$.

If the characteristic is entering $\mathbb{R}^+$ in $\{x = 0\}$, we have

$$\frac{f_1(\gamma u) - f_1(k)}{\gamma u - k} > 0 ,$$

and (6.2.17) is satisfied if and only if $\gamma u = 0$.

The generalization to nonlinear problems with inhomogeneous boundary condition a is given by

$$\min_{k \in I(\gamma u, a)} \{\text{sign}(\gamma u - a)[f(\gamma u) - f(k)] \cdot n\} = 0. \tag{6.2.18}$$

Now we can define a weak solution of the initial boundary value problem

$$\partial_t u + \text{div } f(u) = 0 \quad \text{in } \Omega_T ,$$
$$u(x, 0) = u_0(x) \quad \text{in } \Omega ,$$
$$\min_{k \in I(\gamma u, 0)} \{\text{sign}(\gamma u)[f(\gamma u) - f(k)] \cdot n\} = 0 .$$

DEFINITION 6.2.8 (Weak solution of the initial boundary value problem, see [16] and [17]) *A function $u \in BV(\Omega_T)$ is a weak solution of (6.2.19) if for all $k \in \mathbb{R}$ and all non–negative test functions $\varphi \in C_0^2(\bar{\Omega} \times]0, T[)$, it satisfies the inequality*

$$\int_{\Omega_T} \{|u - k|\partial_t \varphi + \text{sign}(u - k)[f(u) - f(k)]\nabla \varphi\}$$

$$+ \int_0^T \int_{\partial \Omega} \text{sign}(k)[f(\gamma u) - f(k)] \cdot n\varphi \geq 0 \tag{6.2.19}$$

6.2 Boundary conditions for nonlinear scalar equations

and the initial condition

$$u(x,0) = u_0(x) \quad \text{in } \Omega$$

almost everywhere in Ω.

REMARK 6.2.9 *If u is sufficiently smooth, we choose $k = \inf u$ and $k = \sup u$ to show that u is a classical solution in Ω_T. Now we can give the main result of this section.*

THEOREM 6.2.10 (Existence and uniqueness for the initial boundary value problem) *The initial boundary value problem (6.2.19) admits a unique solution in the sense of Definition 6.2.8 that is given by the vanishing viscosity method.*

Proof Let $u \in BV(\Omega)$ be the L^1 limit of a sequence of viscosity solutions u_ε as stated in Theorem 6.2.5. In this theorem we have shown that u satisfies the initial condition. Let $k \in \mathbb{R}$ and $\varphi \in C_0^2(\bar{\Omega} \times {]0, T[})$, $\varphi \geq 0$, and multiply (6.2.1) by $g_\eta(u_\varepsilon - k)\varphi$, with g_η as in the proof of Theorem 6.2.5. Then we obtain, using integration by parts,

$$\int_{\Omega_T} \int_k^{u_\varepsilon} g_\eta(v - k)\partial_t \varphi + \int_{\Omega_T} g_\eta(u_\varepsilon - k)[f(u_\varepsilon) - f(k)]\nabla\varphi$$

$$+ \int_{\Omega_T} [f(u_\varepsilon) - f(k)]\nabla u_\varepsilon g'_\eta(u_\varepsilon - k)\varphi$$

$$= \varepsilon \int_{\Omega_T} |\nabla u_\varepsilon|^2 g'_\eta(u_\varepsilon - k)\varphi + \varepsilon \int_{\Omega_T} g_\eta(u_\varepsilon - k)\nabla u_\varepsilon \nabla\varphi$$

$$+ \varepsilon \int_{\partial\Omega} \int_0^T g_\eta(k)\varphi \partial_n u_\varepsilon - \int_{\partial\Omega} \int_0^T g_\eta(k)[f(0) - f(k)]n\varphi.$$

As $\eta \to 0$, we obtain

$$\int_{\Omega_T} |u_\varepsilon - k|\partial_t \varphi + \int_{\Omega_T} \text{sign}(u_\varepsilon - k)[f(u_\varepsilon) - f(k)]\nabla\varphi$$

448   6 Initial boundary value problems for conservation laws

$$\geq \varepsilon \int_{\Omega_T} \text{sign}(u_\varepsilon - k) \nabla u_\varepsilon \nabla \varphi + \varepsilon \int_{\partial\Omega} \int_0^T \text{sign}(k) \varphi \partial_n u_\varepsilon$$

$$- \int_{\partial\Omega} \int_0^T \text{sign}(k)[f(0) - f(k)] n \varphi. \tag{6.2.20}$$

Now we have to control the limit as $\varepsilon \to 0$. The left-hand side in (6.2.20) converges to the corresponding terms in (6.2.19). The estimate (6.2.14) implies that we have for the first term on the right-hand side in (6.2.20)

$$\varepsilon \left| \int_{\Omega_T} \text{sign}(u_\varepsilon - k) \nabla u_\varepsilon \nabla \varphi \right| \leq \varepsilon C \|\nabla \varphi\|_{L^\infty(\Omega_T)}.$$

The second term on the right-hand side in (6.2.20) is more delicate. In order to estimate, it we introduce the function $\rho_\delta \in C^2(\bar{\Omega})$ satisfying

$$\rho_\delta = 1 \quad \text{on } \{x \in \Omega \mid \text{dist}(x, \partial\Omega) < \tfrac{1}{2}\delta\},$$
$$\rho_\delta = 0 \quad \text{on } \{x \in \Omega \mid \text{dist}(x, \partial\Omega) > \delta\},$$
$$0 \leq \rho_\delta \leq 1 \quad \text{on } \Omega, \quad \|\nabla \rho_\delta\|_{L^\infty(\Omega)} \leq \frac{c}{\delta}.$$

Obviously we have

$$\varepsilon \int_{\partial\Omega} \int_0^T \varphi \partial_n u_\varepsilon = \varepsilon \int_{\Omega_T} [\Delta u_\varepsilon \varphi \rho_\delta + \nabla u_\varepsilon \nabla (\varphi \rho_\delta)].$$

The equation for u_ε implies

$$\varepsilon \int_{\partial\Omega} \int_0^T \varphi \partial_n u_\varepsilon = \int_{\Omega_T} \{[\partial_t u_\varepsilon + \text{div}\, f(u_\varepsilon)] \varphi \rho_\delta + \varepsilon \nabla u_\varepsilon \nabla (\varphi \rho_\delta)\}$$

$$= - \int_{\Omega_T} [u_\varepsilon \partial_t \varphi + f(u_\varepsilon) \nabla \varphi] \rho_\delta - \int_{\Omega_T} f(u_\varepsilon) \varphi \nabla \rho_\delta$$

$$+ \int_{\partial\Omega} \int_0^T f(u_\varepsilon) \varphi \rho_\delta n + \varepsilon \int_{\Omega_T} \nabla u_\varepsilon \nabla (\varphi \rho_\delta).$$

6.2 Boundary conditions for nonlinear scalar equations

First we let ε converge to zero, and in the limit we have to replace u_ε by u. The first term

$$\int_{\Omega_T} [u\partial_t \varphi + f(u)\nabla\varphi]\rho_\delta$$

converges to zero as $\delta \to 0$. The remaining terms converge to

$$\int_{\partial\Omega} \int_0^T [f(0) - f(\gamma u)]\varphi n \,.$$

Here we have used the fact that

$$\int_{\Omega_T} f(u)\varphi\nabla\rho_\delta \to \int_{\partial\Omega}\int_0^T f(\gamma u)\varphi n \qquad (6.2.21)$$

as $\delta \to 0$. This follows from the following lemma.

LEMMA 6.2.11 *Let $u \in \mathrm{BV}(\Omega)$. Then there exists a Radon measure $\mu = \mu(u)$ such that we have for all $\Phi \in C^1(\bar\Omega)^n$*

$$\int_\Omega \Phi \, d\mu = -\int_\Omega u \,\mathrm{div}\, \Phi + \int_{\partial\Omega} \gamma u \Phi \,.$$

A similar result holds if we replace Ω by Ω_T.

Proof See [16] and [17].

Because of (6.2.13), we also know that $f(u)$ is in $\mathrm{BV}(\Omega_T)$, and therefore the left-hand side in (6.2.21) can be written as

$$-\int_{\Omega_T} f(u)\nabla\varphi\nabla\rho_\delta - \int_{\Omega_T} \varphi\rho_\delta \, d\nu + \int_{\Omega_T} f(\gamma u)\varphi n \,. \qquad (6.2.22)$$

Now, using Lebesgue's theorem, the first two terms in (6.2.22) converge to zero and we get (6.2.21). We finally obtain from (6.2.20)

$$\int_{\Omega_T} |u-k|\partial_t \varphi + \int_{\Omega_T} \operatorname{sign}(u-k)[f(u)-f(k)]\nabla\varphi$$

$$\geq \int_{\partial\Omega} \int_0^T [f(0)-f(\gamma u)]\operatorname{sign}(k) n\varphi$$

$$-\int_{\partial\Omega} \int_0^T [f(0)-f(k)]\operatorname{sign}(k) n\varphi$$

$$= \int_{\partial\Omega} \int_0^T [f(k)-f(\gamma u)]\operatorname{sign}(k) n\varphi . \tag{6.2.23}$$

This finishes the existence part of the theorem.

Now we shall show that (6.2.16) is satisfied. Let $\psi \in C_0^2(\bar{\Omega} \times]0,T[)$, $\psi \geq 0$. We choose $\varphi = \psi \rho_\delta$ in (6.2.19) and obtain from (6.2.23)

$$\int_{\Omega_T} \{|u-k|\partial_t\varphi + \operatorname{sign}(u-k)[f(u)-f(k)]\nabla\varphi\}$$

$$+ \int_0^T \int_{\partial\Omega} \operatorname{sign}(k)[f(\gamma u)-f(k)]n\psi \geq 0 .$$

As $\delta \to 0$, the first integral on the left-hand side converges to zero (Lebesgue's theorem). The second integral will be approximated by

$$\int_{\Omega_T} g_\eta(u-k)[f(u)-f(k)]\nabla\varphi . \tag{6.2.24}$$

Because of (6.2.13), we can apply Lemma 6.2.11 as before, and we obtain that (6.2.24) equals

$$-\int_{\Omega_T} \psi\rho_\delta \, d\nu + \int_{\partial\Omega} \int_0^T g_\eta(u-k)[f(u)-f(k)]\psi .$$

6.2 Boundary conditions for nonlinear scalar equations

Now as $\delta \to 0$ the first term disappears and the second does not depend on δ. Then as $\eta \to 0$ we obtain

$$\int_{\Omega_T} \text{sign}(u-k)[f(u) - f(k)]\nabla\varphi$$

$$= \int_{\partial\Omega} \int_0^T \text{sign}(\gamma u - k)[f(\gamma u) - f(k)]\psi \,.$$

This means that

$$\int_0^T \int_{\partial\Omega} [\text{sign}(\gamma u - k) + \text{sign}(k)][f(\gamma u) - f(k)]n\psi \geq 0 \,,$$

and therefore

$$[\text{sign}(\gamma u - k) + \text{sign}(k)][f(\gamma u) - f(k)]n \geq 0 \qquad (6.2.25)$$

almost everywhere on $\partial\Omega \times]0,T[$. Since for $k \in I(\gamma u, 0)$ we have $\text{sign}(\gamma u - k) = \text{sign}(k)$, (6.2.25) reduces to

$$\min_{k \in I(\gamma u, 0)} \{\text{sign}(\gamma u)[f(\gamma u) - f(k)]n\} \geq 0 \,. \qquad (6.2.26)$$

But this implies

$$\min_{k \in I(\gamma u, 0)} \{\cdots\} = 0 \,,$$

and therefore (6.2.16).

Now let us assume that we have two solutions u and v. Then we get from (6.2.19) for $\varphi \in C_0^1(\Omega_T)$

$$\int_{\Omega_T} |u-v|\partial_t\varphi + \text{sign}(u-v)[f(u) - f(v)]\nabla\varphi \geq 0. \qquad (6.2.27)$$

Let $\psi \in C_0^1(]0,T[)$ and let ρ_δ be defined as above, and use

$$\varphi(x,t) := \psi(t)[1 - \rho_\delta(x)] \quad \text{on } \Omega_T$$

452 6 Initial boundary value problems for conservation laws

as a test function in (6.2.27). We get as $\delta \to 0$

$$\int_{\Omega_T} |u-v|\psi' - \int_{\partial\Omega_T} \text{sign}(\gamma u - \gamma v)[f(\gamma u) - f(\gamma v)]\psi n \geq 0.$$

Define

$$k(x,t) := \begin{cases} \gamma u(x,t) & \text{if } \gamma u \in I(0, \gamma v), \\ 0 & \text{if } 0 \in I(\gamma u, \gamma v), \\ \gamma v(x,t) & \text{if } \gamma v \in I(\gamma u, 0). \end{cases}$$

Then it is easy to see that

$$\begin{aligned} R &:= \text{sign}(\gamma u - \gamma v)[f(\gamma u) - f(\gamma v)] \\ &= \text{sign}(\gamma u - k)[f(\gamma u) - f(k)] + \text{sign}(\gamma v - k)[f(\gamma v) - f(k)]. \end{aligned}$$

Since $k \in I(\gamma u, 0)$ and $k \in I(\gamma v, 0)$, the inequality (6.2.26) implies that $R \geq 0$, and therefore

$$\int_{\Omega_T} |u-v|\psi' \geq 0;$$

furthermore $(t_1 \geq t_0)$,

$$\int_\Omega |u(t_1) - v(t_1)| \leq \int_\Omega |u(t_0) - v(t_0)|.$$

Letting $t_0 \to 0$ we obtain $u = v$. □

REMARK 6.2.12 *In [178] a solution concept for scalar conservation laws with boundary conditions that are weaker than the one given in (6.2.19) is considered.*

6.2 Boundary conditions for nonlinear scalar equations

DEFINITION 6.2.13 (Finite volume scheme for initial boundary value problems, see [21]) *Let us consider the following initial boundary value problem. For given $f \in C^1(\mathbb{R})$, $a \in C^\infty(\bar{\Omega})$ and $u_0 \in BV(\Omega)$ find $u \in BV(\Omega_T)$ such that*

$$\partial_t u + \operatorname{div} f(u) = 0 \quad \text{in } \Omega \times \mathbb{R}^+ ,$$
$$u(x,0) = u_0(x) \quad \text{in } \Omega ,$$

and for all $c \in \mathbb{R}$ and $(x,t) \in \partial\Omega \times \mathbb{R}^+$

$$\{\operatorname{sign}[u(x,t) - c] - \operatorname{sign}[a(x,t) - c]\}$$
$$[f(u(x,t)) - f(c)]n \geq 0 . \tag{6.2.28}$$

We assume that Ω is a polygonal domain such that for any h

$$\bigcup_{j \in I} T_j^h = \Omega$$

for a suitable index set I. Furthermore, we use the same notation as in Chapter 3. Then the finite volume method for this initial boundary value problem can be defined as follows [21]:

$$u_j^0 := \frac{1}{|T_j|} \int_{T_j} u_0(x) \, dx , \quad j \in I , \tag{6.2.29}$$

$$u_j^{n+1} = u_j^n - \frac{\Delta t}{|T_j|} \sum_{l \in N_j} g_{jl}(u_j^n, v_{jl}^n) ,$$

where

$$v_{jl}^n = \begin{cases} u_{jl}^n & \text{if } S_{jl} \cap \partial\Omega = \emptyset , \\ a_{jl}^n & \text{if } S_{jl} \subset \partial\Omega \end{cases}$$

and

$$a_{jl}^n = \frac{1}{|S_{jl}|} \int_{S_{jl}} a(\cdot, t^n) \, d\sigma .$$

REMARK 6.2.14 *The boundary conditions in (6.2.28) and (6.2.18) are equivalent.*

For this scheme the following result has been proved in [21].

THEOREM 6.2.15 (Convergence of finite volume methods for initial boundary value problems) *Let g_{jl} satisfy the conditions (3.2.8)–(3.2.10) and let g_{jl} be monotone, $u_0 \in BV(\Omega)$, $a \in C^\infty(\bar{\Omega})$, $f \in C^1(\mathbb{R})$ and let $u_h(x,t) = u_j^n$ if $x \in T_j$ and $t^n \le t < t^{n+1}$ be the approximate solution as defined in (6.2.29). Then if $\Delta t/h$ remains sufficiently small, we have for $1 \le p < \infty$*

$$u_h \longrightarrow u \quad \text{in } L^p_{\text{loc}}(\Omega \times \mathbb{R}^+)$$

and u is the solution as defined in Theorem 6.2.10.

Another idea for the formulation of an initial boundary value problem was studied in [56] and [57]. In this approach systems are included. Dubois and LeFloch consider the following initial boundary value problem for $u \in \mathbb{R}^m$ and given $u_0 \in L^1(\mathbb{R}) \cap L^\infty(\mathbb{R})$ and $g \in \mathbb{R}$:

$$\partial_t u + \partial_x f(u) = 0 \quad \text{in } \mathbb{R} \times \mathbb{R}^+ , \tag{6.2.30}$$
$$u(x,0) = u_0(x) \quad \text{in } \mathbb{R} , \tag{6.2.31}$$
$$u(0,t) \in V(g) \quad \text{for all } t \in \mathbb{R}^+ , \tag{6.2.32}$$

where $V(g)$ is defined as follows. First we fix some notation. We denote the solution of the Riemann problem $R(u_l, u_r)$

$$\partial_t u + \partial_x f(u) = 0 \quad \text{in } \mathbb{R} \times \mathbb{R}^+ ,$$
$$u(x,0) = \begin{cases} u_l & \text{if } x < 0 , \\ u_r & \text{if } x \ge 0 \end{cases}$$

by $u(x,t) = w(x/t; u_l, u_r)$. Since u depends only on x/t, the limit

$$\lim_{x \to 0+} w(x/t; u_l, u_r) , \quad t > 0 ,$$

does not depend on t and will therefore be denoted by

$$w(0+; u_l, u_r) := \lim_{x \to 0+} w(x/t; u_l, u_r) .$$

6.2 Boundary conditions for nonlinear scalar equations

Let $\mathcal{U}$ be the set such that

$$\mathcal{U} = \{(v_1, v_2) \mid R(v_1, v_2) \text{ is solvable}\}$$

and

$$V(v_1) = \{w(0+; v_1, v_2) \mid (v_1, v_2) \in \mathcal{U}\} .$$

Then the condition (6.2.32) means that there exists a constant $\bar{g} \in \mathbb{R}^m$ such that the solution $w(x/t, g, \bar{g})$ of the Riemann problem $R(g, \bar{g})$ evaluated in $x = 0$ is equal to $u(0, t)$, i.e. there exists a $\bar{g} \in \mathbb{R}^m$ such that

$$u(0, t) = w(0+, g, \bar{g})$$

In [56, Theorem 2.1], it is shown that the initial boundary value problem (6.2.30)–(6.2.32) is well posed. In particular, the following theorem is proved.

THEOREM 6.2.16 (Well-posedness of (6.2.30)–(6.2.32), [56]) *Let u_0 and g be constant such that $(u_0, g) \in \mathcal{U}$. Then the problem (6.2.30)–(6.2.32) has a unique solution in the class of functions that consists of constant states separated by at most m elementary waves.*

REMARK 6.2.17 *Since the initial boundary value problem (6.2.30)–(6.2.32) can also be used for scalar conservation laws, we now have two formulations of boundary conditions: this one and (6.2.19). Dubois and LeFloch [56, Proposition 2.2] show that the two formulations are equivalent for scalar conservation laws with arbitrary (also nonconvex) C^1-class flux functions and for linear systems.*

REMARK 6.2.18 *The existence of a solution of*

$$\partial_t u + \mathrm{div}(au) = f \quad \text{in } \Omega \times \mathbb{R}^+ ,$$
$$u = 0 \quad \text{on } \partial_- \Omega \times \mathbb{R}^+ ,$$
$$u(x, 0) = u_0(x) \quad \text{in } \Omega ,$$

where $\partial_- \Omega = \{x \in \partial \Omega \mid \nu(x) \cdot a < 0\}$ and $\nu(x)$ is the outer normal in $x \in \partial \Omega$ to Ω, is investigated in [179, Corollary 4.12.10].

7 Convection-dominated problems

7.1 Singular perturbation problems for elliptic equations

The numerical methods for solving the conservation laws described in ealier chapters are also useful for computing approximate solutions for convection-dominated diffusion problems

$$-\varepsilon \Delta u + \operatorname{div} f(u) + c(x,u) = g \quad \text{in } \Omega,$$
$$u = 0 \quad \text{on } \partial\Omega. \tag{7.1.1}$$

Problems of this type are also called singular perturbation problems, since the type or the order of the differential equation changes if ε tends to zero. They are the simplest model problems for the incompressible Navier–Stokes equation with large Reynolds number. In particular, the one-dimensional form of (7.1.1) has been studied very extensively. Let us consider a one-dimensional example.

EXAMPLE 7.1.1 The exact solution of

$$-\varepsilon u'' - u' = 0 \quad \text{in }]0,1[,$$
$$u(0) = 1, \quad u(1) = 0 \tag{7.1.2}$$

is given by

$$u(x) := \frac{v(x)}{v(0)}, \tag{7.1.3}$$

where $v(x) := 1 - e^{(1-x)/\varepsilon}$. The exact solution is plotted in Figure 7.1.1 for $\varepsilon = 0.01$

7.1 Singular perturbation problems for elliptic equations

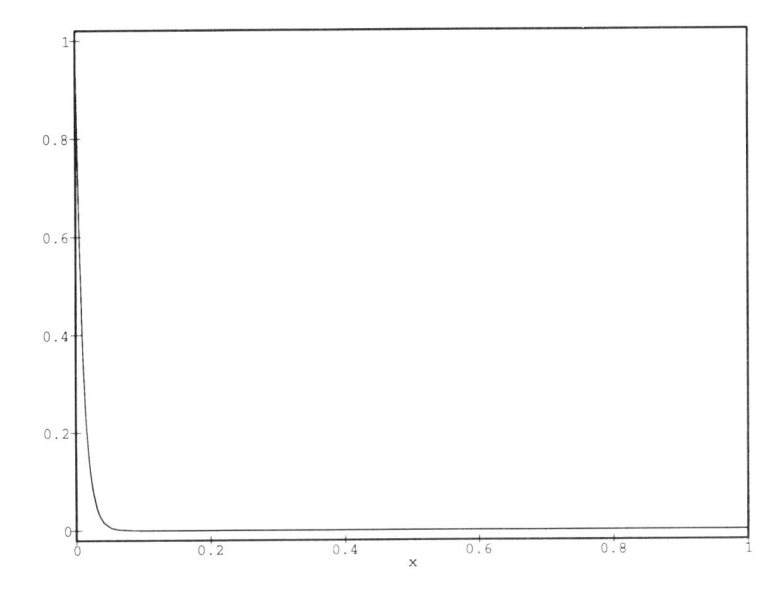

Figure 7.1.1

The limit problem for $\varepsilon \to 0$ is

$$-u' = 0 \quad \text{in }]0,1[,$$
$$u(0) = 1, \quad u(1) = 0 .$$

Obviously this boundary value problem has no solution. But there are functions $u_0(x) = 1$ and $u_1(x) = 0$ on $[0,1]$ satisfying

$$-u_0' = 0 \quad \text{in }]0,1[\quad \text{and} \quad u_0(0) = 1$$

and

$$-u_1' = 0 \quad \text{in }]0,1[\quad \text{and} \quad u_1(1) = 0 .$$

Since the characteristics go from the left to the right side, the solution of (7.1.2) will converge on $]0,1[$ to u_1 a.e. A heuristic argument for this can be obtained as follows. The solution of (7.1.2) can be considered as a stationary solution of

$$\partial_t w - \varepsilon \partial_x^2 w - \partial_x w = 0 \quad \text{in }]0,1[\times \mathbb{R}^+ ,$$
$$w(x,0) = \bar{w}(x) \quad \text{for } x \in [0,1] , \tag{7.1.4}$$

458   7 Convection-dominated problems

$$\left.\begin{array}{l} w(0,t) = 1 ,\\ w(1,t) = 0 \end{array}\right\} \quad \text{for } t \in \mathbb{R}^+ , \qquad (7.1.5)$$

where $\bar{w}$ is just the solution of (7.1.2). We know from Theorem 2.1.7 that the solution $w(=w_\varepsilon)$ of (7.1.4), (7.1.5) converges to w_0 a.e. in $]0,1[\times \mathbb{R}^+$, where w_0 is the solution of

$$\partial_t w_0 - \partial_x w_0 = 0 \quad \text{in }]0,1[\times \mathbb{R}^+ ,$$
$$w_0(x,0) = \bar{w}(x) \quad \text{in } [0,1] .$$

For this problem boundary values can only be prescribed in $x = 1$, since the characteristics go from the left to the right side. These arguments can be made rigorous with the results of Chapter 6. Therefore we know that we have for $u_\varepsilon := u$ (u solution of (7.1.2))

$$u_\varepsilon \longrightarrow u_1 \quad \text{a.e. on } [0,1]$$

and

$$u_\varepsilon(0) = 1 , \quad u_\varepsilon(1) = 0 ,$$

but $u_1 = 0$. Therefore there will be a steep gradient of u_ε in a neighbourhood of $x = 0$. This neighbourhood is called the boundary layer of the problem. For numerical algorithms this steep gradient causes similar problems to shocks, and therefore the schemes developed for conservation laws can be applied to problems like (7.1.1). In particular, we consider the linear version of (7.1.1) in the form

$$-\varepsilon \Delta u + b \cdot \nabla u + cu = f \quad \text{in } \Omega ,$$
$$u = 0 \quad \text{on } \partial \Omega . \qquad (7.1.6)$$

EXAMPLE 7.1.2 (Engquist–Osher scheme) Let us assume a triangulation of $\Omega = \Omega_h$ consisting of equal-sided triangles. Using the same notation as in §3.2 and the Engquist–Osher scheme for the discretization of the convection term, we obtain the following discretization of (7.1.6):

$$-\varepsilon \frac{1}{|T_j|} \sum_{l=1}^{3} \frac{|S_{jl}|}{|d_{jl}|}(u_{jl} - u_j) + \frac{1}{|T_j|} \sum_{l=1}^{3} g_{jl}(u_j, u_{jl}) + c_j u_j = f_j$$

7.1 Singular perturbation problems for elliptic equations

where

$$g_{jl}(u,v) = |S_{jl}| \left[(b \cdot n_{jl})^+ u + (b \cdot n_{jl})^- v \right]$$

and

$$f_j := \frac{1}{|T_j|} \int_{T_j} f, \quad c_j := \frac{1}{|T_j|} \int_{T_j} c.$$

For notation see Figure 7.1.2.

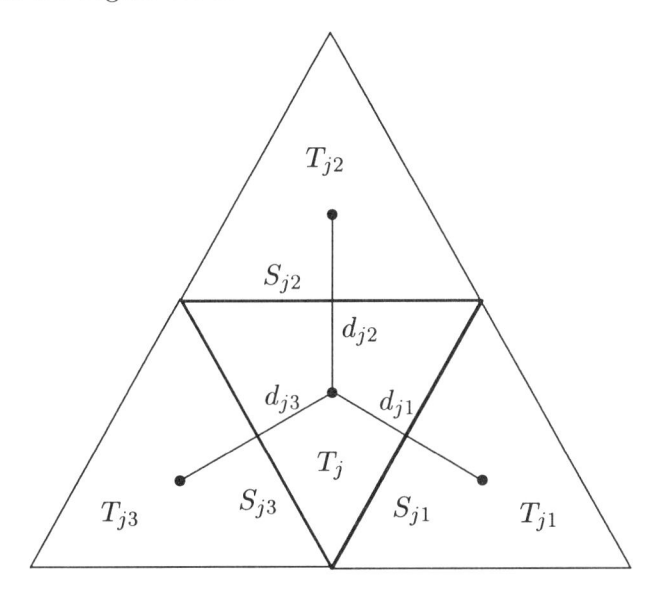

Figure 7.1.2

There are several other methods for solving (7.1.1) numerically (see [3, 4, 114, 144, 151, 165, 166, 167, 177, 196, 197, 198, 241] for 1-D and [113, 147, 168, 172, 195, 242, 253] for 2-D). In the following two examples we mention the streamline diffusion method (see [147]) and the Il'in scheme in the form as used by Angermann [7]–[10].

EXAMPLE 7.1.3 (Streamline diffusion method) Let

$$V := \{v \in H^1(\Omega) | v = 0 \text{ on } \partial\Omega\}$$

460   7 Convection-dominated problems

and let V_h be a finite-dimensional subspace of V such that $v|_{T_j}$ is polynomial for all $v \in V_h$. Then find $u_h \in V_h$ such that

$$\varepsilon(\nabla u_h, \nabla v_h) + (b\nabla u_h + cu_h, v_h)$$
$$+ \sum_{j \in I} \delta_j(-\varepsilon\Delta u_h + b \cdot \nabla u_h + cu_h, b \cdot \nabla v_h)_{T_j}$$
$$= (f, v_h) + \sum_{j \in I}(f, b \cdot \nabla v_h)_{T_j}$$

for all $v_h \in V_h$ and suitable δ_j. Here (u, v) denotes the usual scalar product in $L^2(\Omega)$ and $(u, v)_{T_j}$ in $L^2(T_j)$. This method is similar to that described in §2.6 with δ_j instead of h for time-dependent problems. The following theorem has been proved in [147].

THEOREM 7.1.4 (Convergence) *Let $\Omega \subset \mathbb{R}^2$ be sufficiently smooth and assume that for each $\varepsilon \in [\varepsilon_1, \varepsilon_2]$, $\varepsilon > 0$, there exists a unique solution $u_\varepsilon \in L^\infty \cap H^k$ of (7.1.6). Furthermore, we assume for the unique factors δ_i that there exist constants C_i and C independent of ε and h such that*

(i) $0 \leq \varepsilon\delta_i \leq C_i h_i^2$,

(ii) $\delta_i \leq Ch_i$,

where $h_i := \operatorname{diam} T_i$ and $h = \max h_i$. Then there exists for all $\varepsilon \in [\varepsilon_1, \varepsilon_2]$ a discrete solution $u_{h,\varepsilon} \in V$ such that

$$\|u_\varepsilon - u_{h,\varepsilon}\|_{\varepsilon,\delta} \leq Ch^k(\sqrt{\varepsilon} + \sqrt{h}) ,$$

where C depends on $[\varepsilon_1, \varepsilon_2]$ and

$$\|v\|_{\varepsilon,\delta}^2 = \varepsilon\|\nabla v\|_{L^2(\Omega)}^2 + \sum_i \delta_i \|b \cdot \nabla v\|_{L^2(T_i)} .$$

EXAMPLE 7.1.5 (Exponential scheme) This algorithm is defined on the dual cells, and we use the following notation:

x_j, x_l: neighbouring nodes of the grid;

$D_j := \{x \in \Omega | \operatorname{dist}(x, x_j) < \operatorname{dist}(x, x_l)$ for all nodes $x_l, l \neq j\}$ = dual cell around x_j;

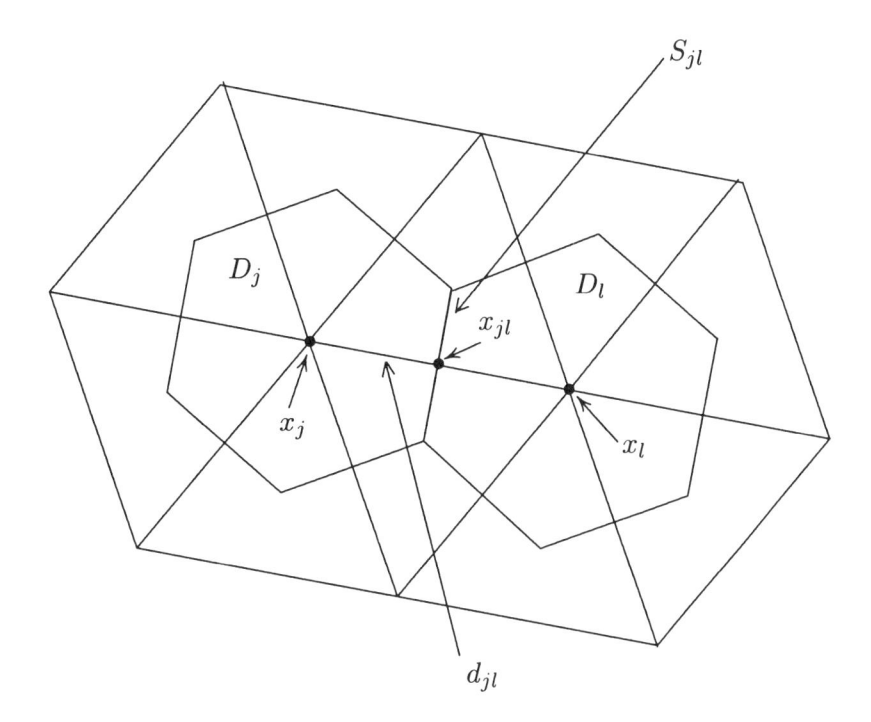

7.1 Singular perturbation problems for elliptic equations

Figure 7.1.3

S_{jl}: joint edge of D_j and D_l;

n_{jl}: outer normal of D_j on S_{jl};

$d_{jl} := |x_j - x_l|$.

Let

$$N_{jl} := \frac{1}{|S_{jl}|} \int_{S_{jl}} n_{jl} b, \quad E(y) := \exp\left(\frac{N_{jl}}{\varepsilon} y\right). \tag{7.1.7}$$

In order to motivate the scheme, we integrate (7.1.6) with respect to D_j and obtain

$$-\int_{\partial D_j} n(\varepsilon \nabla u - bu) - \int_{D_j} (\nabla \cdot bu - cu) = \int_{D_j} f,$$

462   7 Convection-dominated problems

or

$$-\sum_{l\in N_j}\int_{S_{jl}} n_{jl}(\varepsilon\nabla u - bu) - \int_{D_j}(\nabla\cdot bu - cu) = \int_{D_j} f , \qquad (7.1.8)$$

where N_j denotes the neighbouring nodes of x_j. Define

$$w(t) := u(x_j + tn_{jl}) \quad \text{for } 0 \le t \le t_0 := (x_l - x_j)n_{jl} .$$

Then we have

$$-\varepsilon w' + n_{jl}\cdot bw = -n_{jl}(\varepsilon\nabla u - bu) , \qquad (7.1.9)$$
$$w(0) = u(x_j) , \qquad (7.1.10)$$
$$w(t_0) = u(x_l) . \qquad (7.1.11)$$

The term $n_{jl}(\varepsilon\nabla u - bu)$ is just the same as in (7.1.8). Now we have to find a suitable approximation for this term. Let u_h denote the discrete solution for which we are going to derive necessary conditions. We replace the operator $-\varepsilon w' + n_{jl}bw$ in (7.1.9) by $-\varepsilon w' + N_{jl}w$, where N_{jl} is defined in (7.1.7). For a given constant $r_l \in \mathbb{R}$ consider the general solution of the ordinary differential equation

$$-\varepsilon w' + N_{jl}w = r_l ,$$

which is given by

$$w(t) = C_l E(t) + \frac{r_l}{N_{jl}} .$$

We assume that C_l and r_l are free parameters. Now we choose special values for C_l and r_l such that w satisfies the conditions

$$w(0) = u_{h_j}, \quad w(t_0) = u_{h_l} ,$$

which correspond to the conditions (7.1.10) and (7.1.11). We obtain

$$C_l = \frac{u_{h_j} - u_{h_l}}{E(0) - E(t_0)} , \quad r_l = \frac{u_{h_j}E(t_0) - u_{h_l}E(0)}{E(t_0) - E(0)} N_{jl} .$$

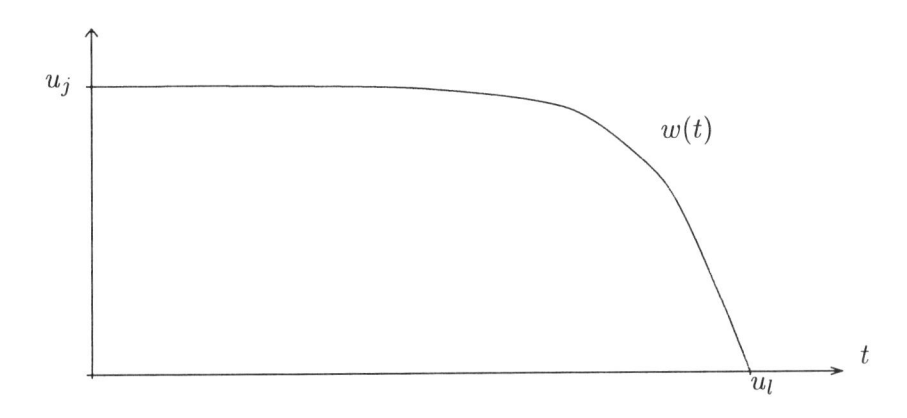

7.1 Singular perturbation problems for elliptic equations 463

Figure 7.1.4

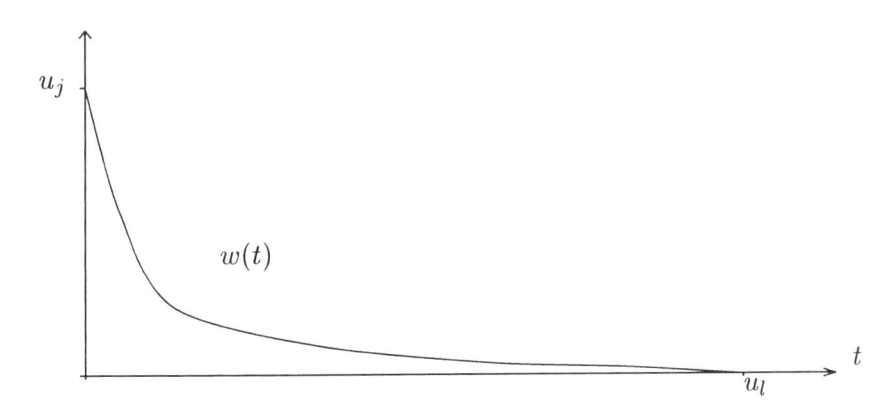

Figure 7.1.5

7 Convection-dominated problems

If $N_{jl} > 0$, $w(t)$ behaves as in Figure 7.1.4, and if $N_{jl} < 0$, it behaves as in Figure 7.1.5. Therefore for the first term in (7.1.8) we use the approximation

$$-\sum_{l \in N_j} \int_{S_{jl}} n_{jl}(\varepsilon \nabla u - bu) \sim \sum_{l \in N_j} r_l |S_{jl}|$$

$$= \sum_{l \in N_j} \frac{u_{h_j} E(t_0) - u_{h_l} E(0)}{E(t_0) - E(0)} N_{jl} |S_{jl}| \,. \tag{7.1.12}$$

For the term $\int_{D_j} \nabla \cdot bu$ we use the approximation

$$-\int_{D_j} \nabla \cdot bu \sim -u_{h_j} \int_{D_j} \nabla \cdot b$$

$$= -u_{h_j} \sum_{l \in N_j} |S_{jl}| n_{jl} b$$

$$\sim -u_{h_j} \sum_{l \in N_j} N_{jl} |S_{jl}| \,. \tag{7.1.13}$$

Putting (7.1.12) and (7.1.13) together, we obtain

$$-\sum_{l \in N_j} \int_{S_{jl}} n_{jl}(\varepsilon \nabla u - bu) - \int_{D_j} \nabla \cdot bu$$

$$\sim \sum_{l \in N_j} \left[\frac{u_{h_j} E(t_0) - u_{h_l} E(0)}{E(t_0) - E(0)} - u_{h_j} \right] N_{jl} |S_{jl}|$$

$$= \sum_{l \in N_j} \{\varepsilon - [1 - r(z)] z \varepsilon\} (u_{h_j} - u_{h_l}) \frac{|S_{jl}|}{d_{jl}} \,, \tag{7.1.14}$$

where

$$z := \frac{N_{jl} d_{jl}}{\varepsilon}, \quad B(z) := \frac{z}{e^z - 1}, \quad r(z) := 1 - \frac{1 - B(z)}{z} \,.$$

The derivation of the last equality in (7.1.14) needs some elementary calculations. Then the discrete version of the problem (7.1.6) is

$$\sum_{l \in N_j} \{\varepsilon - [1 - r(z)] z \varepsilon\} (u_{h_j} - u_{h_l}) \frac{|S_{jl}|}{d_{jl}} + c_j u_{h_j} |D_j| = f_j |D_j| \,,$$

7.1 Singular perturbation problems for elliptic equations

where

$$f_j := \frac{1}{|D_j|} \int_{D_j} f, \quad c_j := \frac{1}{|D_j|} \int_{D_j} c.$$

The corresponding variational form is as follows. Find $u_h \in V_h$ such that for all $v_h \in V_h$

$$a_h(u_h, v_h) = (f, v_h)_h,$$

where

$$a_h(u_h, v_h) := \sum_{j \in I} \sum_{l \in N_j} \left(\{\varepsilon - [1 - r(z)]z\varepsilon\}(u_{h_j} - u_{h_l}) \frac{|S_{jl}|}{d_{jl}} \right.$$

$$\left. + c_j u_{h_j} |D_j| \right) v_{hj} \qquad (7.1.15)$$

and

$$(f, v_h)_h := \sum_{j \in I} f_j v_{h_j} |D_j|.$$

The following result has been proved in [7].

THEOREM 7.1.6 *Let $\partial\Omega, f$ and c be sufficiently smooth and let $c - \frac{1}{2} \text{div } b > c_0 > 0$ such that (7.1.6) has a solution $u_\varepsilon \in H_0^1(\Omega) \cap H^2(\Omega)$. Furthermore, we assume that for $0 < h < h_0$ the problem (7.1.15) has a unique solution denoted by u_h. Then we have*

$$\sqrt{\varepsilon} \|\nabla(u_\varepsilon - u_h)\|_{L^2(\Omega)} + \|u_\varepsilon - u_h\|_{L^2(\Omega)} \le c \frac{h}{\sqrt{\varepsilon}} \|u_\varepsilon\|_{H^2(\Omega)}.$$

Now we should like to report on some numerical experiments comparing these three schemes, described in Examples 7.1.2, 7.1.3 and 7.1.5 (see [211]). The schemes have been applied to the following test problem:

$$-\varepsilon \Delta u + \begin{pmatrix} 1 \\ 1 \end{pmatrix} \cdot \nabla u + 2u = f \quad \text{in } \Omega :=]0, 1[\times]0, 1[,$$

$$u = 0 \quad \text{on } \partial\Omega.$$

466   7 Convection-dominated problems

The function f is given by the exact solution

$$u(x,y) = xy\left(1 - e^{(x-1)/\varepsilon}\right)\left(1 - e^{(y-1)/\varepsilon}\right).$$

The computations have been performed on a grid of the form described in Figure 7.1.6 for different parameters of ε and h. The L^2 error as well as the experimental order of convergence are given in Tables 7.1.1 – 7.1.3.

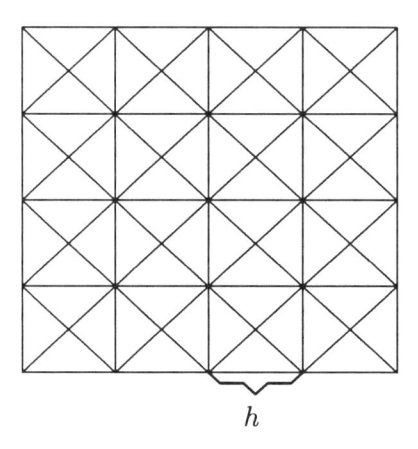

Figure 7.1.6 Criss-cross.

In Figures 7.1.7 – 7.1.9 the exact solution and different numerical solutions are plotted along a line parallel to the diagonal of the square, shifted by 0.03 in the x-direction. It turns out that for large ε, i.e. $\varepsilon \approx 1$, the streamline diffusion and the Il'in scheme are the best ones, with an EOC of 2. But for small ε the Engquist–Osher scheme has an EOC of nearly 1, while the EOCs of the others go down to $\frac{1}{2}$. Also the L^2 errors of the Engquist–Osher scheme are much better than for the others.

REMARK 7.1.7 *For further investigations of convection-dominated diffusion problems we refer to [151, 165, 166, 167, 168, 3, 100, 232, 12, 256, 195, 13, 148, 196, 257, 145, 181, 199].*

REMARK 7.1.8 *In [189] the singular perturbed boundary value problem*

$$\alpha\nabla u - \sigma_1\partial_x^2 u - \sigma_2\partial_y^2 u = 0$$

7.1 Singular perturbation problems for elliptic equations

h	$\varepsilon = 1.0\,E{-}0$	$\varepsilon = 1.0\,E{-}1$	$\varepsilon = 1.0\,E{-}2$	$\varepsilon = 1.0\,E{-}3$
$\frac{1}{4}$	4.15 E−3 (−)	2.13 E−2 (−)	2.61 E−2 (−)	2.12 E−2 (−)
$\frac{1}{8}$	2.12 E−3 (0.970)	1.50 E−2 (0.510)	1.47 E−2 (0.817)	1.19 E−2 (0.837)
$\frac{1}{16}$	1.06 E−3 (0.997)	9.03 E−3 (0.730)	8.68 E−3 (0.764)	7.44 E−3 (0.676)
$\frac{1}{32}$	5.30 E−4 (1.002)	4.90 E−3 (0.882)	8.74 E−3 (−)	6.57 E−3 (0.181)
$\frac{1}{64}$	2.65 E−4 (1.002)	2.53 E−3 (0.953)	6.67 E−3 (0.390)	4.89 E−3 (0.425)
$\frac{1}{128}$	1.32 E−4 (1.001)	1.29 E−3 (0.977)	4.23 E−3 (0.656)	2.70 E−3 (0.857)
$\frac{1}{256}$	6.60 E−5 (1.001)	6.47 E−4 (0.989)	2.36 E−3 (0.843)	2.95 E−3 (−)
$\frac{1}{512}$			1.24 E−3 (0.931)	2.41 E−3 (0.292)

h	$\varepsilon = 1.0\,E{-}4$	$\varepsilon = 1.0\,E{-}5$	$\varepsilon = 1.0\,E{-}6$	$\varepsilon = 1.0\,E{-}7$
$\frac{1}{4}$	2.13 E−2 (−)	2.14 E−2 (−)	2.14 E−2 (−)	2.14 E−2 (−)
$\frac{1}{8}$	1.16 E−2 (0.874)	1.17 E−2 (0.874)	1.17 E−2 (0.874)	1.17 E−2 (0.874)
$\frac{1}{16}$	6.08 E−3 (0.938)	6.07 E−3 (0.940)	6.08 E−3 (0.940)	6.08 E−3 (0.940)
$\frac{1}{32}$	3.16 E−3 (0.945)	3.10 E−3 (0.971)	3.10 E−3 (0.971)	3.10 E−3 (0.971)
$\frac{1}{64}$	1.82 E−3 (0.795)	1.57 E−3 (0.984)	1.57 E−3 (0.986)	1.57 E−3 (0.986)
$\frac{1}{128}$	1.54 E−3 (0.693)	7.97 E−4 (0.975)	7.87 E−4 (0.993)	7.87 E−4 (0.993)
$\frac{1}{256}$	1.82 E−3 (−)	4.38 E−4 (0.864)	3.95 E−4 (0.995)	3.94 E−4 (0.996)
$\frac{1}{512}$	1.76 E−3 (0.054)	3.36 E−4 (0.384)	1.99 E−4 (0.987)	1.97 E−4 (0.998)

Table 7.1.1 1st–orderEngquist–Osher.

h	$\varepsilon = 1.0$ E-0	$\varepsilon = 1.0$ E-1	$\varepsilon = 1.0$ E-2	$\varepsilon = 1.0$ E-3
$\frac{1}{4}$	1.80 E-3 (–)	9.14 E-2 (–)	2.10 E-1 (–)	2.11 E-1 (–)
$\frac{1}{8}$	4.55 E-4 (1.984)	3.74 E-2 (1.288)	1.52 E-1 (0.464)	1.62 E-1 (0.378)
$\frac{1}{16}$	1.14 E-4 (1.996)	1.14 E-2 (1.711)	9.82 E-2 (0.633)	1.20 E-1 (0.434)
$\frac{1}{32}$	2.85 E-5 (1.999)	3.03 E-3 (1.914)	5.42 E-2 (0.856)	8.68 E-2 (0.467)
$\frac{1}{64}$	7.14 E-6 (2.000)	7.70 E-4 (1.977)	2.34 E-2 (1.214)	6.03 E-2 (0.525)
$\frac{1}{128}$			7.58 E-3 (1.626)	
$\frac{1}{256}$			2.06 E-3 (1.878)	
$\frac{1}{512}$			5.27 E-4 (1.967)	

h	$\varepsilon = 1.0$ E-4	$\varepsilon = 1.0$ E-5	$\varepsilon = 1.0$ E-6	$\varepsilon = 1.0$ E-7
$\frac{1}{4}$	2.10 E-1 (–)	2.10 E-1 (–)	2.10 E-1 (–)	2.10 E-1 (–)
$\frac{1}{8}$	1.62 E-1 (0.379)	1.62 E-1 (0.379)	1.62 E-1 (0.379)	1.62 E-1 (0.379)
$\frac{1}{16}$	1.19 E-1 (0.437)	1.19 E-1 (0.438)	1.19 E-1 (0.438)	1.19 E-1 (0.438)
$\frac{1}{32}$	8.63 E-2 (0.469)	8.63 E-2 (0.469)	8.62 E-2 (0.469)	8.62 E-2 (0.469)
$\frac{1}{64}$	6.17 E-2 (0.483)	6.16 E-2 (0.485)	6.16 E-2 (0.485)	6.16 E-2 (0.485)

Table 7.1.2 SDM Lube.

7.1 Singular perturbation problems for elliptic equations

h	$\varepsilon = 1.0\,\mathrm{E}{-}0$	$\varepsilon = 1.0\,\mathrm{E}{-}1$	$\varepsilon = 1.0\,\mathrm{E}{-}2$	$\varepsilon = 1.0\,\mathrm{E}{-}3$
$\frac{1}{4}$	1.05 E−3 (–)	4.87 E−2 (–)	1.50 E−1 (–)	1.49 E−1 (–)
$\frac{1}{8}$	2.62 E−4 (1.998)	1.59 E−2 (1.612)	1.03 E−1 (0.546)	1.12 E−1 (0.412)
$\frac{1}{16}$	6.56 E−5 (2.000)	4.30 E−3 (1.889)	5.92 E−2 (0.797)	8.19 E−2 (0.451)
$\frac{1}{32}$	1.64 E−5 (2.000)	1.10 E−3 (1.972)	2.67 E−2 (1.151)	5.90 E−2 (0.473)
$\frac{1}{64}$	4.10 E−6 (2.000)	2.75 E−4 (1.993)	8.84 E−3 (1.593)	4.01 E−2 (0.558)
$\frac{1}{128}$			2.42 E−3 (1.866)	2.39 E−2 (0.748)
$\frac{1}{256}$			6.21 E−4 (1.964)	1.16 E−2 (1.039)
$\frac{1}{512}$			1.56 E−4 (1.991)	4.18 E−3 (1.474)

h	$\varepsilon = 1.0\,\mathrm{E}{-}4$	$\varepsilon = 1.0\,\mathrm{E}{-}5$	$\varepsilon = 1.0\,\mathrm{E}{-}6$	$\varepsilon = 1.0\,\mathrm{E}{-}7$
$\frac{1}{4}$	1.49 E−1 (–)	1.49 E−1 (–)	1.49 E−1 (–)	1.49 E−1 (–)
$\frac{1}{8}$	1.11 E−1 (0.415)	1.11 E−1 (0.416)	1.11 E−1 (0.416)	1.11 E−1 (0.416)
$\frac{1}{16}$	8.11 E−2 (0.458)	8.10 E−2 (0.459)	8.10 E−2 (0.459)	8.10 E−2 (0.459)
$\frac{1}{32}$	5.82 E−2 (0.478)	5.81 E−2 (0.479)	5.81 E−2 (0.480)	5.81 E−2 (0.480)
$\frac{1}{64}$	4.16 E−2 (0.487)	4.14 E−3 (0.490)	4.14 E−2 (0.490)	4.14 E−2 (0.490)
$\frac{1}{128}$	2.96 E−2 (0.489)	2.94 E−2 (0.494)	2.94 E−2 (0.495)	2.94 E−2 (0.495)
$\frac{1}{256}$	2.11 E−2 (0.488)	2.08 E−2 (0.489)	2.08 E−2 (0.497)	2.08 E−2 (0.497)
$\frac{1}{512}$	1.47 E−2 (0.527)	1.48 E−2 (0.496)	1.47 E−2 (0.498)	1.47 E−2 (0.499)

Table 7.1.3 Angermann.

470   7 Convection-dominated problems

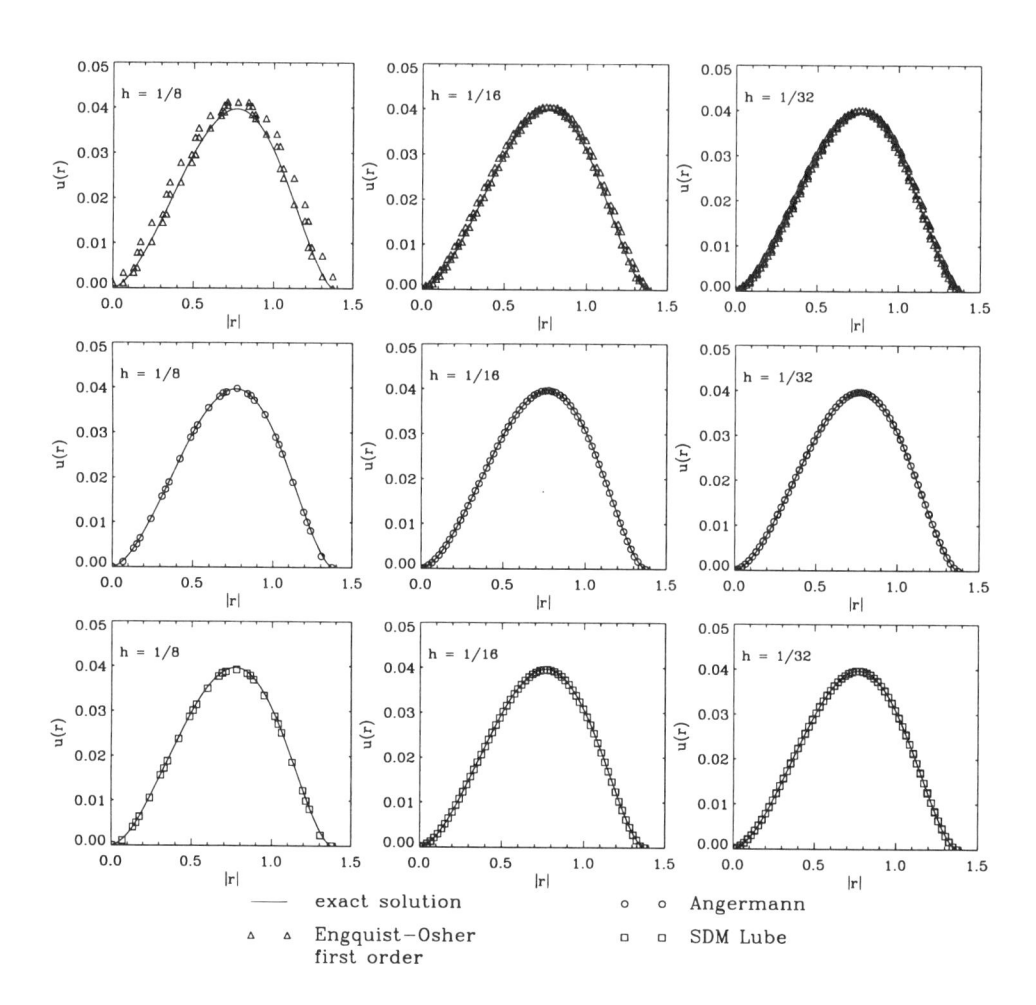

Figure 7.1.7 $\varepsilon = 1.0$

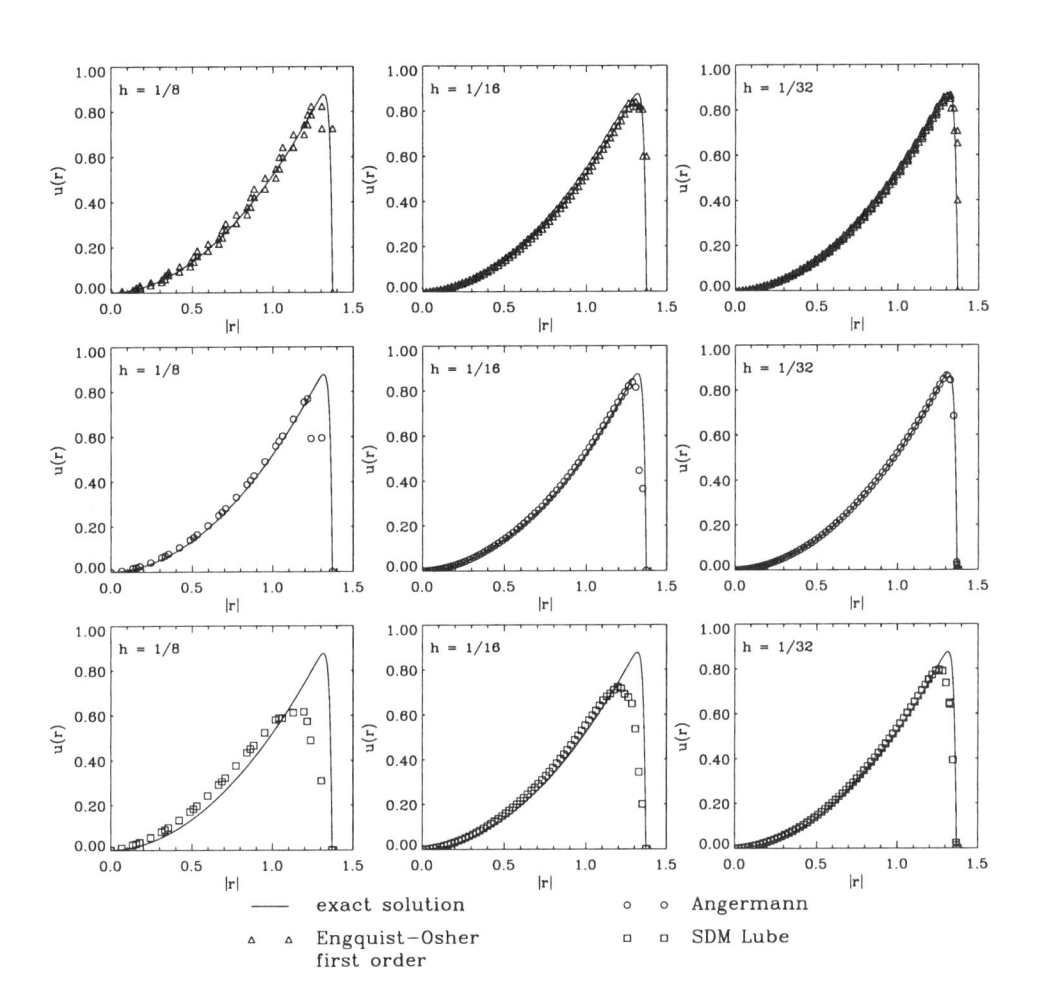

7.1 Singular perturbation problems for elliptic equations

— exact solution
△ △ Engquist–Osher first order
○ ○ Angermann
□ □ SDM Lube

Figure 7.1.8 $\varepsilon = 10^{-2}$

472   7 Convection-dominated problems

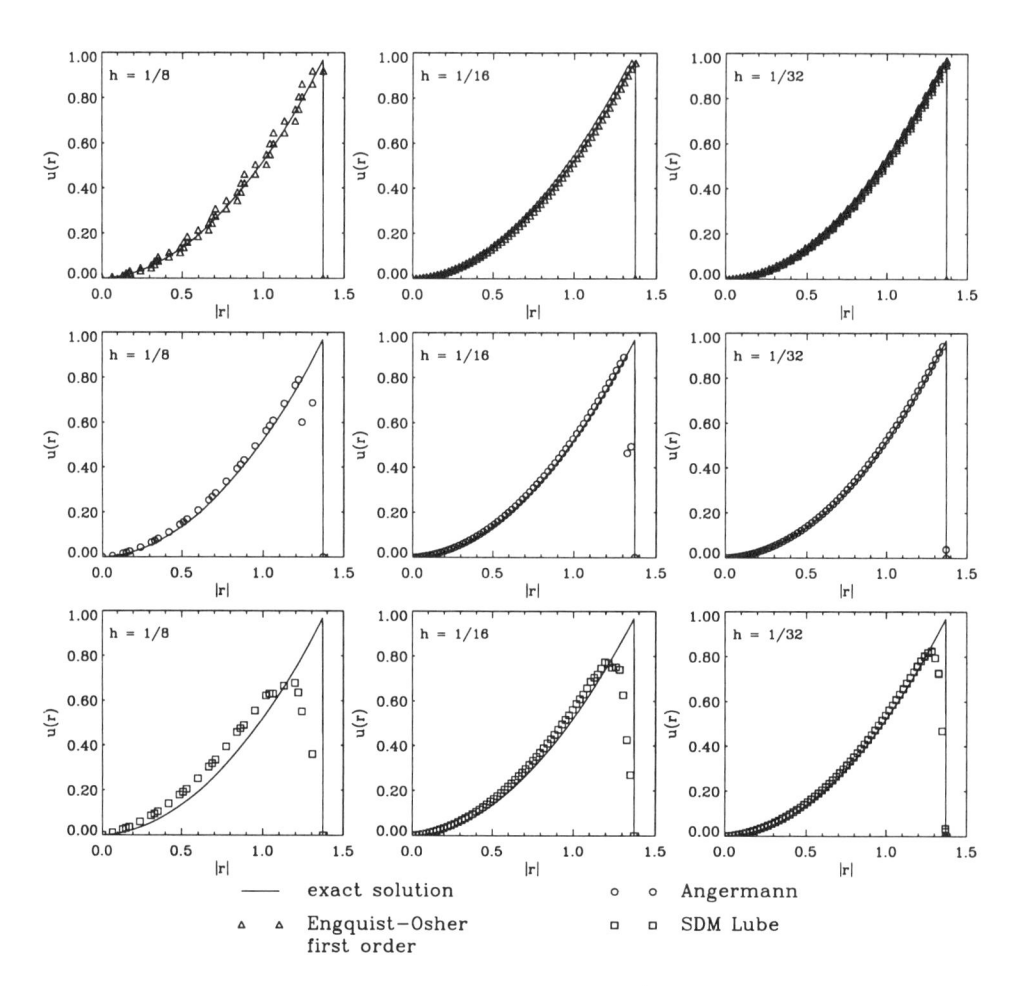

Figure 7.1.9 $\varepsilon = 10^{-4}$

for small σ_1 and σ_2 is considered. The discrete solution u_h is defined by a discontinuous Galerkin method, and the following error estimate is shown:

$$\|u_h - u\|_{L^2(\Omega)} \le ch^{n+\frac{1}{2}} \|u\|_{H^{n+1,2}(\Omega)} .$$

EXAMPLE 7.1.9 (Semiconductor equations) The system of semiconductor equations is a mathematical model for the time evolution of the electrostatic potential ψ, the electron concentration n and the hole concentration p in a given domain Ω. For a given doping function D, a net generation–recombination rate of electrons and holes R, a permittivity ε, and given constants μ_n and μ_p, we have to find functions ψ, n and p such that

$$-\varepsilon\Delta\psi = p - n + D \quad \text{in } \Omega \times]0, T[,$$
$$\partial_t n - \mu_n \Delta n + \text{div}(n\nabla\psi) = -R \quad \text{in } \Omega \times]0, T[,$$
$$\partial_p - \mu_p \Delta p - \text{div}(p\nabla\psi) = -R \quad \text{in } \Omega \times]0, T[.$$

Suitable boundary and initial conditions for ψ, n and p have to be added. If μ_n (or μ_p) is small the convection term $\nabla n \cdot \nabla \psi$ (or $\nabla p \cdot \nabla \psi$) may become dominating, and upwind discretizations have to be used. A common discretization for the system is the Scharfetter–Gummel discretization [209], an exponentially fitted upwind discretization. For further details we refer to [159], [200] and [71].

7.2 Discretization of the compressible Navier–Stokes equations

In Chapters 2 – 6 we have considered conservation laws that correspond to inviscid flows. From the numerical point of view the same problem concerning the discretization of the convection term arises for viscous flows with small viscosity or large Reynolds number. Therefore in this section we shall show how to discretize the compressible Navier–Stokes equations [66]. In particular, we shall discretize the viscous terms by a finite element and the convection part by a finite volume method.

7 Convection-dominated problems

Let $\Omega \subset \mathbb{R}^2$ be a bounded domain. Then the compressible flow in Ω is described by the following system of nonlinear partial differential and algebraic equations (see [66]):

$$\partial_t \begin{pmatrix} \rho \\ \rho u \\ \rho v \\ e \end{pmatrix} + \partial_x \begin{pmatrix} \rho u \\ \rho u^2 + p \\ \rho u v \\ u(e+p) \end{pmatrix} + \partial_y \begin{pmatrix} \rho v \\ \rho u v \\ \rho v^2 + p \\ v(e+p) \end{pmatrix} \quad (7.2.1)$$

$$+ \partial_x \begin{pmatrix} 0 \\ \tau_{11} \\ \tau_{12} \\ \tau_{11} u + \tau_{12} v + k \partial_x \Theta \end{pmatrix} + \partial_y \begin{pmatrix} 0 \\ \tau_{21} \\ \tau_{22} \\ \tau_{21} u + \tau_{22} v + k \partial_y \Theta \end{pmatrix} = 0,$$

in $\Omega \times]0, T[$,

$$p = (\gamma - 1) \left[e - \frac{\rho}{2}(u^2 + v^2) \right], \quad (7.2.2)$$

$$e = \rho \left[c_\nu \Theta + \frac{1}{2}(u^2 + v^2) \right]. \quad (7.2.3)$$

The definition of ρ, u, v, p and e is given in Chapter 1. Furthermore, τ_{ij} are the components of the viscous part of the stress tensor ($v_1 = u, v_2 = v$):

$$\tau_{ij} = \lambda \operatorname{div} \begin{pmatrix} u \\ v \end{pmatrix} \delta_{ij} + \mu (\partial_j v_i + \partial_i v_j), \quad (7.2.4)$$

Θ is the absolute temperature, c_ν is the specific heat at constant volume, k is the heat conductivity, and λ and μ are viscosity coefficients. We assume that

$$c_\nu, k, \mu > 0 \quad \text{and} \quad \lambda = -\tfrac{2}{3}\mu. \quad (7.2.5)$$

For the system (7.2.1)–(7.2.3) we consider the initial values

$$\begin{pmatrix} \rho \\ u\rho \\ v\rho \\ e \end{pmatrix}_{t=0} = \begin{pmatrix} \rho_0 \\ u_0 \rho_0 \\ v_0 \rho_0 \\ e_0 \end{pmatrix} \quad (7.2.6)$$

7.2 Discretization of the compressible Navier–Stokes equations

and the boundary conditions

$$\rho = \bar{\rho}, \quad \begin{pmatrix} u \\ v \end{pmatrix} = \begin{pmatrix} \bar{u} \\ \bar{v} \end{pmatrix}, \quad \Theta = \bar{\Theta}, \quad \text{on } \Gamma_i \tag{7.2.7}$$

$$\begin{pmatrix} u \\ v \end{pmatrix} = 0, \quad \partial_n \Theta = 0 \quad \text{on } \Gamma_w, \tag{7.2.8}$$

$$\sum_{i=1}^{2} \tau_{ij} n_i = 0, \quad j = 1, 2, \quad \partial_n \Theta = 0, \text{on } \Gamma_0 \tag{7.2.9}$$

where $\partial \Omega = \Gamma_i \cup \Gamma_w \cup \Gamma_0$. The parts Γ_i, Γ_0 and Γ_w of the boundary are related to the inflow, outflow boundary and impermeable walls.

At present no existence and uniqueness results have been obtained for the problem (7.2.1)–(7.2.9). For small data or small time interval $]0, T[$ there are some results (see [65]). Only recently has the existence of a global solution for large data been proved for the system (7.2.1), (7.2.4) with Dirichlet boundary conditions on $\partial \Omega$ and with the energy equation in (7.2.1) replaced by $p = p(\rho)$ (see [143]). The basic idea for the discretization of (7.2.1)–(7.2.9) is the splitting of the whole system (7.2.1) into

$$\frac{1}{2} \partial_t \begin{pmatrix} \rho \\ \rho u \\ \rho v \\ e \end{pmatrix} + \partial_x \begin{pmatrix} \rho u \\ \rho u^2 + p \\ \rho u v \\ u(e+p) \end{pmatrix} + \partial_y \begin{pmatrix} \rho v \\ \rho u v \\ \rho v^2 + p \\ v(e+p) \end{pmatrix} = 0, \tag{7.2.10}$$

$$\frac{1}{2}\partial_t \begin{pmatrix} \rho \\ \rho u \\ \rho v \\ e \end{pmatrix} + \partial_x \begin{pmatrix} 0 \\ \tau_{11} \\ \tau_{12} \\ \tau_{11} u + \tau_{12} v + k \partial_x \Theta \end{pmatrix}$$

$$+ \partial_y \begin{pmatrix} 0 \\ \tau_{21} \\ \tau_{22} \\ \tau_{21} u + \tau_{23} v + k \partial_y \Theta \end{pmatrix} = 0. \tag{7.2.11}$$

476 7 Convection-dominated problems

We assume that a triangulation $\mathcal{T}_h$ of Ω is given. For the discretization of (7.2.10) we use the finite volume methods on the dual cells as described in §5.1

$$W_j^{n+1} = W_j^n - \frac{\Delta t}{|T_j|} \sum_{l=1}^{k} g_{jl}(W_j^n, W_l^n), \qquad (7.2.12)$$

where g_{jl} is defined as in (5.1.13) and T_j is the dual cell around the node j (see Example 7.1.5 for the definition of the dual cells).

Now we shall describe how to discretize (7.2.11). By $P_i, i \in I$, we denote the vertices of all triangles $T \in \mathcal{T}_h$. For the discretization we shall use conforming piecewise-linear finite elements. Let

$$X_h := \{\varphi_h \in C^0(\bar{\Omega}_h) \big| \varphi_h|_T \text{ is linear for each } T \in \mathcal{T}_h\}^4,$$

$$V_h := \{\varphi_h = (\varphi_1, \varphi_2, \varphi_3, \varphi_4) \in X_h , \varphi_i = 0 \text{ on } \Gamma_i \cup \Gamma_w\},$$

$$W_h := \{w_h \in X_h | w_h \text{ satisfies the}$$
$$\text{boundary conditions (7.2.7)–(7.2.9) }\}.$$

The approximate solution on the $(n+1)$th time level will be denoted by w_h^{n+1}. Then we seek a function

$$w_h^{n+1} \in W_h \qquad (7.2.13)$$

such that

$$\int_{\Omega_h} w_h^{n+1} \varphi_h = \int_{\Omega_h} w_h^n \varphi_h - \Delta t \int_{\Omega_h} \sum_{i=1}^{2} R_i(w_h^n, \nabla w_h^n) \partial_i \varphi_h \qquad (7.2.14)$$

for all $\varphi_h \in V_h$. Here we have used the notation

$$R_i(w, \nabla w) = (0, \tau_{i1}, \tau_{i2}, \tau_{i1} u + \tau_{i2} v + k \partial_i \Theta).$$

7.2 Discretization of the compressible Navier–Stokes equations

The integrals in (7.2.14) are approximated by the following quadrature formula

$$\int_T F \sim \frac{1}{3}|T|\sum_{i=1}^{3} F(P_T^i)$$

for $F \in C^0(T)$ and $T \in \mathcal{T}_h$ with vertices P_T^1, P_T^2 and P_T^3. Then the left-hand side in (7.2.14) has to be replaced by

$$(w_h, \varphi_h)_h := \frac{1}{3} \sum_{T \in \mathcal{T}_h} |T| \sum_{i=1}^{3} w_h(P_T^i)\varphi_h(P_T^i)$$

for $w_h \in W_h, \varphi_h \in V_h$ and correspondingly the second integral in (7.2.14) by $a_h(w_h, \varphi_h)$. Then the scheme for the viscous part (7.2.11) can be written as

$$(w_h^{n+1}, \varphi_h)_h = (w_h^n, \varphi_h)_h - \Delta t a_h(w_h, \varphi_h),$$

where a_h is the corresponding bilinear form (for more details see [66]). Now we have to describe how to discretize the whole system (7.2.10), (7.2.11). We shall present three different forms: the inviscid–viscous operator splitting, the explicit discretization of the whole problem and the semi-implicit one. Let us start with the inviscid–viscous operator splitting.

We assume that the values w_j^n for all nodes P_j are already known. Then we have to define the value w_j^{n+1} of the new time level $n + 1$. In the first step we solve the convection part using (7.2.12):

$$w_j^{n+\frac{1}{2}} = w_j^n - \frac{\Delta t}{|T_j|} \sum_{l=1}^{k} g_{jl}(w_j^n, w_{jl}^n). \qquad (7.2.15)$$

Now let

$$w_h^{n+\frac{1}{2}} \in X_h \text{ such that } w_h^{n+\frac{1}{2}}(P_i) = w_i^{n+\frac{1}{2}}$$

for all nodes P_i.

In the third step we define $w_h^{n+1} \in W_h$ such that

$$(w_h^{n+1}, \varphi_h)_h = (w_h^{n+\frac{1}{2}}, \varphi_h)_h - \Delta t a_h(w_h^{n+\frac{1}{2}}, \varphi_h) \quad \text{for all } \varphi_h \in V_h.$$

Let $w_i^{n+1} := w_h^{n+1}(P_i)$, and repeat the whole procedure to get the values of the next time level.

Now we shall describe the explicit discretization. We define for $w_h, \varphi_h \in X_h$

$$b_h(w_h, \varphi_h) := \sum_j \varphi_h(P_j) \sum_{l=1}^k g_{jl}(w_h(P_i), w_h(P_j)) \, .$$

Then (7.2.15) can be written as

$$(w_h^{n+\frac{1}{2}}, \varphi_h)_h = (w_h^n, \varphi_h)_h - \Delta t b_h(w_h^k, \varphi_h)$$

for all $\varphi_h \in V_h$. Then a discretization of the complete system (7.2.1) is given in the following form. In the explicit case we find $w_h^{n+1} \in W_h$ such that

$$(w_h^{n+1}, \varphi_h)_h = (w_h^n, \varphi_h)_h - \Delta t \left[b_h(w_h^n, \varphi_h) + a_h(w_h^n, \varphi_h) \right]$$

for all $\varphi_h \in V_h$.

Similarly, the semi-implicit case is given as follows: find $w_h^{n+1} \in W_h$ such that

$$(w_h^{n+1}, \varphi_h)_h = (w_h^n, \varphi_h)_h - \Delta t \left[b_h(w_h^n, \varphi_h) + a_h(w_h^{n+1}, \varphi_h) \right] \, .$$

List of figures

The figures and the corresponding numerical results in this book are due to the following collaborators

1.0.5	T. Geßner	7
1.0.6	T. Geßner	7
2.2.5	T. Makeben	51
2.3.2	G. Lorse	71
2.3.3	T. Makeben	71
2.3.4	M. Küther	88
2.3.5	M. Küther	88
2.3.6	M. Küther	89
2.5.1	G. Lorse	102
2.5.2	T. Makeben	102
2.5.4	G. Lorse	107
2.5.7	G. Lorse	111
2.5.8	G. Lorse	111
3.7.1	G. Lorse	242
3.7.2	G. Lorse	242
3.7.3	G. Lorse	242
3.7.4	G. Lorse	243
3.7.5	G. Lorse	243
3.7.6	G. Lorse	244
3.7.7	G. Lorse	244
3.7.8	G. Lorse	244
3.7.11	G. Lorse	249
3.7.12	G. Lorse	250
3.7.13	G. Lorse	250
3.7.14	G. Lorse	251
3.7.15	G. Lorse	251

3.7.16	G. Lorse	252
3.7.17	G. Lorse	252
3.7.18	G. Lorse	253
3.7.19	G. Lorse	253
3.9.3	M. Schmieder	264
3.9.4	M. Schmieder	265
3.9.5	M. Schmieder	265
3.9.6	M. Schmieder	266
3.9.7	M. Schmieder	266
3.9.8	M. Schmieder	267
3.9.9	M. Schmieder	267
3.9.10	R. Beinert	269
3.9.11	R. Beinert	269
3.9.12	R. Beinert	271
3.9.13	R. Beinert	272
3.9.15	G. Lorse	274
3.9.16	G. Lorse	275
4.4.2	G. Lorse	356
4.4.3	G. Lorse	356
4.4.4	G. Lorse	357
4.4.5	G. Lorse	357
5.3.3	M. Wierse	397
5.3.4	M. Schmieder	399
5.3.5	M. Schmieder	400
5.3.6	M. Schmieder	401
5.3.7	M. Schmieder	401
5.3.8	M. Schmieder	402
5.3.9	M. Schmieder	402
5.3.10	M. Schmieder	403
5.3.11	M. Schmieder	403

List of figures

5.3.12	M. Schmieder	404
5.3.13	M. Schmieder	404
5.3.14	T. Geßner	405
5.3.15	J. Becker	406
5.3.16	J. Becker	406
5.3.17	J. Becker	407
5.3.18	J. Becker	408
5.3.19	T. Geßner	409
5.3.20	T. Geßner	411
5.3.21	T. Geßner	412
5.3.22	T. Geßner	413
5.3.23	T. Geßner	414
5.3.24	M. Wierse	416
5.3.25	M. Wierse	417
5.3.26	M. Wierse	418
5.3.27	M. Wierse	419
5.3.28	M. Wierse	419
5.3.29	M. Wierse	420
5.3.30	M. Wierse	420
5.3.31	M. Wierse	421
5.3.32	M. Wierse	421
7.1.6	A. Schneider	466
7.1.7	A. Schneider	470
7.1.8	A. Schneider	471
7.1.9	A. Schneider	472

References

[1] Abgrall, R.: *On essentially non–oscillatory schemes on unstructured meshes: Analysis and implementation.* ICASE Report No. 92-74 (1992), Langley Research Center, Hampton.

[2] Abgrall, R., Lafon, F.C.: *ENO schemes on unstructured meshes.* In: Computational Fluid Dynamics, von Kármán Institute for Fluid Dynamics, Lecture Series 1993-04 (1993).

[3] Abrahamsson, L., Osher, S.: *Monotone difference schemes for singular perturbation problems.* SIAM J. Numer. Anal. 9 (1982), 979–992.

[4] Adam, D., Felgenhauer, A., Roos, H.-G., Stynes, M.: *A nonconforming finite element method for a singularly perturbed boundary problem.* Computing 54 (1995), 1–25.

[5] Agmon, S., Douglis, A., Nirenberg, L.: *Estimates near the boundary for solutions of elliptic partial differential equations satisfying general boundary conditions I.* Commun. Pure Appl. Math. 12 (1959), 623–727.

[6] Agmon, S., Douglis, A., Nirenberg, L.: *Estimates near the boundary for solutions of elliptic partial differential equations satisfying general boundary conditions II.* Commun. Pure Appl. Math. 17 (1964), 35–92.

[7] Angermann, L.: *Numerical solution of second-order elliptic equations on plane domains.* RAIRO Modélisation Math. Anal. Numér. 25, (1991), 169–191.

[8] Angermann, L.: *A modified error estimator of Babuška–Rheinboldt's type for singularly perturbed elliptic problems.* ISAM '91, Numerical Methods in Singularly Perturbed Problems, 1–12.

[9] Angermann, L.: *Addendum to the paper: Numerical solution of second-order elliptic equations on plane domains.* RAIRO Modélisation Math. Anal. Numér. 27, (1993), 1–7.

[10] Angermann, L.: *Balanced a-posteriori error estimates for finite volume type discretizations of convection-dominated elliptic problems.* Institut für Angewandte Mathematik, Universität Erlangen–Nürnberg 157, (1994).

[11] Apel, T., Dobrowolski, M.: *Anisotropic interpolation with applications to the finite element method.* Computing 47 (1992), 277–293.

[12] Apel, T., Lube, G.: *Local inequalities for anisotropic finite elements and their application to convection-diffusion problems.* Preprint SPC 94-26 (1994), Technische Universität Chemnitz–Zwickau, Fakultät für Mathematik.

[13] Apel, T., Lube, G.: *Anisotropic mesh refinement in stabilized Galerkin methods.* Preprint SPC 95–1 (1995), Technische Universität Chemnitz–Zwickau, Fakultät für Mathematik.

[14] Apel, T., Nicaise, S.: *Elliptic problems in domains with edges: anisotropic regularity and anisotropic finite element meshes.* Preprint SPC 94-16 (1994), Technische Universität Chemnitz–Zwickau, Fakultät für Mathematik.

[15] Bänsch, E.: *Local mesh refinement in 2 and 3 dimensions.* IMPACT Comput. Sci. Eng. 3 (1991), 181–191.

[16] Bardos, C., LeRoux, A. Y., Nedelec, J. C.: *First order quasilinear equations with boundary conditions.* Commun. Partial Differ. Equations 4 (1979), 1017–1034.

[17] Bardos, C., Brézis, D. Brézis, H.: *Perturbations singulières et prolongements maximaux d'operateurs positifs.* Arch. Ration. Mech. Anal. 53 (1973), 69–100.

[18] Barth, T. J.; Jesperson, D. C.: *The design and application of upwind schemes on unstructured meshes.* AIAA-98-0366.

[19] Becker, J.: *Finite Volumen Verfahren in 2-D für Systeme von hyperbolischen Differentialgleichungen mit Flußfunktion von Osher und Solomon.* Diplomthesis, IAM Univ. Bonn, 1995.

[20] Beinert, R., Kröner, D.: *Finite volume methods with local mesh refinement in 2-D.* Notes Numer. Fluid Mech. 46 (1994), 38–53.

[21] Benharbit, S., Chalabi, A., Vila, J.P.: *Numerical viscosity and convergence of finite volume methods for conservation laws with boundary conditions.* Preprint 1993, Toulouse.

[22] Berger, M., Colella, P.: *Local adaptive mesh refinement for shock hydrodynamics.* J. Comput. Phys. 82 (1989), 64–84.

[23] Bey, J., Wittum, G.: *Downwind numbering: a robust multigrid method for convection diffusion problems on unstructured grids.* ICA Bericht 95/2, Univ. Stuttgart.

[24] Bikker, S., Greza, H., Koschel, W.: *Parallel computing and multigrid solutions on adaptive unstructured meshes* In: Proc. International Workshop on Numerical Methods for the Navier–Stokes Equations, Heidelberg, 1994.

[25] Billey, V.: *Résolution des équations d'Euler par des méthodes d'éléments finis.* Thèse, L'Université Pierre et Marie Curie, Paris VI, 1984.

[26] Chorin, A. J.: *Random choice solution of hyperbolic systems.* J. Comput. Phys. 22 (1976), 517–533.

[27] Chorin, A. J., Marsden, J.E.: *A mathematical introduction to fluid mechanics,* 3rd ed Springer, New York, 1992.

[28] Ciarlet, P.G.: *The finite element methods for elliptic problems.* North-Holland, Amsterdam ,1987.

[29] Cockburn, B., Coquel, F., LeFloch, Ph., Shu, C.-W.: *Convergence of finite volume methods.* IMA Preprint No. 771, 1991.

[30] Cockburn, B., Hou, S., Shu, C.-W.: *The Runge–Kutta local projection discontinuous Galerkin finite element method for conservation laws IV: the multidimensional case.* Math. Comput. 54 (1990), 545–581.

[31] Cockburn, B.: *Quasimonotone schemes for scalar conservation laws I.* SIAM J. Numer. Anal. 26 (1989), 1325–1341.

[32] Cockburn, B.: *Quasimonotone schemes for scalar conservation laws II.* SIAM J. Numer. Anal. 27 (1990), 247–258.

[33] Cockburn, B.: *Quasimonotone schemes for scalar conservation laws III.* SIAM J. Numer. Anal. 27 (1990), 259–276.

[34] Cockburn, B., Coquel, F., LeFloch, P.: *An error estimate for finite volume methods for multidimensional conservation laws.* Math. Comput. 63 (1994), 77–103.

[35] Cockburn, B., Gremaud, P.-A.: *Error estimates for finite element methods for scalar conservation laws.* SIAM J. Numer. Anal. 33 (1996), 522–554.

[36] Cockburn, B., Shu, C.-W.: *TVB Runge–Kutta local projection discontinuous Galerkin finite element method for conservation laws II: general framework.* Math. Comput. 52 (1989), 411–435.

[37] Cockburn, B., Lin, S.-Y., Shu, C.-W.: *TVB Runge–Kutta local projection discontinuous Galerkin finite element method for conservation laws III: one-dimensional systems.* J. Comput. Phys. 84 (1989), 90–113.

[38] Cockburn, B., Shu, C.-W.: *The P^1-RKDG method for two-dimensional Euler equations of gas-dynamics.* ICASE–Report No. 91-32, (1991).

[39] Colella, P., Woodward, P.R.: *The piecewise parabolic method for gas-dynamical simulation.* J. Comput. Phys. 54 (1984), 174–201.

[40] Conway, E., Smoller, J.: *Global solutions of the Cauchy problem for quasi-linear first-order equations in several space variables.* Commun. Pure Appl. Math. 19 (1966), 95–105.

[41] Coquel, F., LeFloch, P.: *Convergence of finite difference schemes for conservation laws in several space dimensions: the corrected antidiffusive flux approach.* Math. Comput. 57 (1991), 169–210.

[42] Coquel, F., LeFloch, P.: *Convergence of finite difference schemes for conservation laws in several space dimensions: a general theory.* SIAM J. Numer. Anal. 30 (1993), 675–700.

[43] Courant, R., Friedrichs, K.O.: *Supersonic flow and shock waves.* Springer, New York, 1976.

[44] Courant, R., Friedrichs, K.O., Lewy, H.: *Über die partiellen Differenzengleichung der mathematischen Physik.* Math. Ann. 100 (1928), 32–74.

[45] Crandall, M., Majda, A.: *Monotone difference approximations for scalar conservation laws.* Math. Comput. 34 (1980), 1–21.

[46] Crandall, M., Majda, A.: *The method of fractional steps for conservation laws.* Numer. Math. 34 (1980), 285–314.

[47] Dafermos, C.M.: *Polygonal approximations of solutions of the initial value problem for a conservation law.* J. Math. Anal. Appl. 38 (1972), 33–41.

[48] Deconinck, H., Struijs, R.: *Multidimensional upwind schemes for the Euler equations using fluctuation distribution on a grid consisting of triangles.* In: Proceedings of the Eighth GAMM-Conference on Numerical Methods in Fluid Mechanics, Notes on Numerical Fluid Mechanics, Vol. 29, Vieweg, Braunschweig, 1989, 533–543

[49] Deconinck, H., Struijs, R.: *A comparison of fluctuation splitting on triangular meshes for scalar equations and application to the Euler equations.* Preprint, 1989.

[50] Deconinck, H., Roe, P.L., Struijs, R.: *A conservative linearisation of the multidimensional Euler equations.* Preprint, 1991.

[51] Deconinck, H., Roe, P.L., Struijs, R., Bourgois, G.: *Compact advection schemes on unstructured grids.* In: Computational Fluid Dynamics, von Kármán Institute for Fluid Dynamics, Lecture Series 1993-04, (1993).

[52] De Vore, R.: *Hyperbolic conservation laws.* Preprint 1993, Aachen.

[53] DiPerna, R.J.: *Measure–valued solutions to conservation laws.* Arch. Ration. Mech. Anal. 88 (1985), 223–270.

[54] DiPerna, R.J.: *Convergence of approximate solutions to conservation laws.* Arch. Ration. Mech. Anal. 82 (1983), 27–70.

[55] Dörfler,W.:*A convergent adaptive algorithm for Poisson's equation.* SIAM J. Numer. Anal. 33 (1996), 1106–1124.

[56] Dubois, F., LeFloch, P.: *Boundary conditions for nonlinear hyperbolic systems of conservation laws.* J. Differential Equations 71 (1988), 93–122.

[57] Dubois, F., LeFloch, P.: *Condition a la limite pour un system de lois de conservation.* C.R. Acad. Sci. Paris, 304 (1987) Série I, n°3, 75–78.

[58] Durlofsky, L.J., Engquist, B., Osher, S.: *Triangle based adaptive stencils for the solution of hyperbolic conservation laws.* J. Comput. Phys. 98 (1992), 64–73.

[59] Einfeldt, B., Munz, C.D., Roe, P.L., Sjörgreen, B.: *On Godunov-type methods near low densities.* J. Comput. Phys. 92 (1991), 273–295.

[60] Elman, H., C., Chernesky, M., P.: *Ordering effects on relaxation methods applied to the discrete convection–diffusion equation.* In: Recent Advances in Iterative Methods, ed. Golub, G. et al., IMA Vol. 60. Springer, Berlin, 1994, pp.45–58.

[61] Engquist, B., Majda, A.: *Absorbing boundary conditions for the numerical simulation of waves.* Math. Comput. 31 (1977), 629–651.

[62] Engquist, B., Osher, S.: *One-sided difference approximations for nonlinear conservation laws.* Math. Comput. 36 (1981), 321–351.

[63] Eriksson, K., Johnson, C.: *Adaptive finite element methods for parabolic problems I, A linear model problem.* SIAM J. Numer. Anal. 28 (1991), 43–77.

[64] Evans, L. C., Gariepy, R. P.: *Measure theory and fine properties of functions.* CRC Press, Boca Raton 1992.

[65] Feistauer, M.: *Mathematical methods in fluid dynamics.* Pitman, Longman, Harlow 1993.

[66] Feistauer, M., Felcman, J., Lukácová–Medvidová, M.: *Combined finite element–finite volume solution of compressible flow.* Submitted to J. Comput. Appl. Math.

[67] Felcman, J. Dolejsi, V., Feistauer, M.: *Adaptive finite volume method for the numerical solution of the compressible Euler equations.* In: Computational Fluid Dynamics '94, Proc. 2nd European CFD Conference. Wiley, Chichester, 1994, pp. 894–901.

[68] Frink, N. T., Parikh, P., Pirzadeh, S.: *A fast upwind solver for the Euler equations on three-dimensional unstructured meshes.* AIAA-91-0102.

[69] Fukuda, H.; Yamamoto, K.: *An upwind method using unstructured triangular meshes.* In: Proc. 4th International Symposium on Computational Fluid Dynamics, 1991.

[70] Fezoui, L., Stoufflet, B.: *A class of implicit upwind schemes for Euler simulations with unstructured meshes.* J. Comput. Phys. 84 (1989), 174–206.

[71] Gajewski, H., Gärtner, K.: *On the discretization of van Roosboeck's equations with magnetic field.* Technical Report No. 94/14 ETH Zürich, Integrated Systems Laboratory.

[72] Geiben, M.: *Convergence of MUSCL-Type upwind finite volume schemes on unstructured triangular grids.* SFB256, Preprint 318 (1993), Bonn.

[73] Geiben, M., Kröner, D., Rokyta, M.: *A Lax–Wendroff type theorem for cell-centered, finite volume schemes in 2-D.* SFB256, Preprint 278 (1991), Bonn.

[74] Geßner, T.: *Zeitabhängige Adaption für Finite Volumen Verfahren höherer Ordnung am Beispiel der Euler-Gleichungen der Gasdynamik.* Diplomthesis, IAM Univ. Bonn, 1994.

[75] Giaquinta, M.: *Introduction to regularity theory for nonlinear elliptic systems.* Lectures in Mathematics, ETH Zürich, Birkhäuser, Basel 1993.

[76] Gilbarg, D., Trudinger, N.S.: *Elliptic partial differential equations of second order.* Springer Berlin, 1977.

[77] Giusti, E.: *Minimal surfaces and functions of bounded variations.* Birkhäuser, Boston 1984.

[78] Glimm, J.: *Solutions in the large for nonlinear hyperbolic systems of equations.* Commun. Pure Appl. Math. 18 (1965), 697–715.

[79] Glinsky, N., Fézoui, L., Ciccoli, M. C., Desidéri, J.-A.: *Non-equilibrium hypersonic flow computations by implicit second-order upwind finite-elements.* In: Proceedings of the Eighth GAMM-Conference on Numerical Methods in Fluid Mechanics, Notes on Numerical Fluid Mechanics, Vol. 29, Vieweg, Braunschweig, 1990.

[80] Godlewski, E., Raviart, P.-A.: *Hyperbolic systems of conservation laws.* Mathematiques et Applications, Ellipses-Edition, Marketing, 1991.

[81] Godunov, S.K.: *Finite difference method for numerical computations of discontinuous solutions of the equations of fluid dynamics.* Mat. Sbornik 47 (1959), 271–306.

[82] Göhner, U., Warnecke, G.: *A second order finite difference error indicator for adaptive transonic flow computations.* Numer. Math. 70 (1995), 129–161.

[83] Goodman, J.B., LeVeque, R.J.: *On the accuracy of stable schemes for 2D scalar conservation laws.* Math. Comput. 45 (1985), 15–21.

[84] Guderley, K.G.: *Starke kugelige und zylindrische Verdichtungsstöße in der Nähe des Kugelmittelpunktes bzw. der Zylinderachse.* Luftfahrtforschung 19 (1992), 302–312.

[85] Hackbusch, W.: *On first and second order box schemes.* Computing 41 (1989), 277–296.

[86] Harten, A.: *Multi-dimensional ENO schemes for general geometries.* ICASE Report 91-76 (1991), Langley Research Center, Hampton.

[87] Harten, A., Engquist, B., Osher, S., Chakravarthy, S.R.: *Uniformly high order accurate essentially non-oscillatory schemes, III.* J. Comput. Phys. 71 (1987), 231–303.

[88] Harten, A., Lax, P.D.: *A random choice finite difference scheme for hyperbolic conservation laws.* SIAM. J. Numer. Anal. 18 (1981), 289–315.

[89] Harten, A.: *On a class of high resolution total-variation-stable finite-difference schemes.* SIAM J. Numer. Anal. 21 (1984), 1–23.

[90] Harten, A.: *High resolution schemes for hyperbolic conservation laws.* J. Comput. Phys. 49 (1983), 357–393.

[91] Harten, A., Lax, P.D., van Leer, B.: *On upstream differencing and Godunov-type schemes for hyperbolic conservation laws.* SIAM Rev. 25 (1983), 35–61.

[92] Harten, A., Hyman, J.M., Lax, P.D.: *On finite-difference approximations and entropy conditions for shocks.* Commun. Pure Appl. Math. 29 (1976), 297–322.

[93] Heinrich, B.: *Coercive and inverse-isotone discretizations of diffusion-convection problems.* Akademie der Wissenschaften der DDR Karl-Weierstrass-Institut für Mathematik, Preprint P-Math-19/88.

[94] Heinrich, B.: *Finite difference methods on irregular networks.* International series of numerical mathematics Vol. 82, Birkhäuser, Basel, 1987.

[95] Hersch, R.: *Mixed problems in several variables.* J. Math. Mech. 12 (1963), 317–334.

[96] Higdon, R.L.: *Initial–boundary value problems for linear hyperbolic systems.* SIAM Rev. 28 (1986), 177–217.

[97] Hirsch, C.: *Numerical computation of internal and external flows, Vols I+II.* Wiley, Chichester, 1988.

[98] Holden, H.,Holden, L.: *On scalar conservation laws in one dimension.* In: Ideas and Methods in Mathematics and Physics, ed. S. Alberverio et al. Cambridge University Press, 1992, pp. 480–509.

[99] Holden, H.,Holden, L.,Høegh–Krohn, R.: *A numerical method for first order nonlinear scalar conservation laws in one dimension.* Comput. Math. Appl. 15 (1988), 595–602.

[100] Ikeda, T. *Maximum principle in finite element models for convection–diffusion phenomena.* Mathematics Studies Lecture Notes in Numerical and Applied Analysis, Vol.4, North-Holland, Amsterdam, 1983.

[101] Jaffre, J. Johnson, C., Szepessy, A.: *Convergence of the discontinous Galerkin finite element method for hyperbolic conservation laws.* INRIA Preprint, 1992.

[102] Fey, M., Jeltsch, R.: *A new multidimensional Euler-scheme.* Seminar für Angewandte Mathematik, ETH Zürich. Research Report No. 92-09 (1992).

[103] Fey, M., Jeltsch, R.: *A simple multidimensional Euler-scheme.* Seminar für Angewandte Mathematik, ETH Zürich. Research Report No. 92-10 (1992). Also in Proc. 1st European Computational Fluid Dynamics Conference, Brussels, 7–11 September 1992, ed. Hirsch C.. To appear in: Science Publishers.

[104] Jeltsch, R., Botta, N.: *A numerical method for unsteady flows.* ETH Zürich, Research Report 94-11, 1994.

[105] Johnson, C., Rannacher, R.: *On error control in CFD.* Preprint 1994-07, Dept. of Math., Chalmers University of Technology, Göteborg.

[106] Johnson, C.: *Adaptive finite element methods for diffusion and convection problems.* In: Comput. Meth. Appl. Mech. Engng., Proceedings from the Workshop on Reliability in Computational Mechanics, Austin, November 1989.

[107] Johnson, C., Pitkäranta, J.: *An analysis of the discontinuous Galerkin method for a scalar hyperbolic equation.* Math. Comput. 46 (1986), 1–26.

[108] Johnson, C., Szepessy, A.: *Adaptive finite element methods for conservation laws based on a posteriori error estimates.* Preprint 1992-31. Dept. of Math., Chalmers Univ. Techenology, Göteborg.

[109] Johnson, C.: *Numerical solution of partial differential equations by the finite element method.* Cambridge University Press, 1994.

[110] Johnson, C., Saranen, J.: *Streamline diffusion methods for the incompressible Euler and Navier–Stokes equations.* Math. Comput. 47 (1986), 1–18.

[111] Johnson, C., Szepessy, A.: *On the convergence of a finite element method for a nonlinear hyperbolic conservation law.* Math. Comput. 49 (1987), 427–444.

[112] Johnson, C., Szepessy, A., Hansbo, P.: *On the convergence of shock-capturing streamline diffusion finite element methods for hyperbolic conservation laws.* Math. Comput. 54 (1990), 107–129.

[113] Johnson, C., Schatz, A.H., Wahlbin, L.B.: *Crosswind smear and pointwise errors in streamline diffusion finite element methods.* Math. Comput. 49 (1987), 25–38.

[114] Kellog, R.B., Tsan, A.: *Analysis of some difference approximations for a singular perturbation problem without turning points.* Math. Comput. 32 (1978), 1025–1039.

[115] Keyfitz, B., Kranzer; H. C.: *A viscosity approximation to a system of conservation laws with no classical Riemann solution.* In: Nonlinear hyperbolic problems, ed C. Carasso et al., Lecutre Notes in Math. No. 1402, Springer Berlin, 1988, pp. 185–198.

[116] Kornhuber, R., Roitzsch, R.: *Adaptive Finite-Element Methoden für konvektionsdominierte Randwertprobleme bei partiellen Differentialgleichungen.* Konrad-Zuse-Zentrum, Preprint SC 88-9, Berlin 1988.

[117] Kreis, H.O.: *Initial boundary value problems for hyperbolic systems.* Commun. Pure Appl. Math. 23 (1970), 277–298.

[118] Kröner, D., Luckhaus, S.: *Flow of oil and water in a porous medium.* J. Differential Equations 55 (1984), 276–288.

[119] Kröner, D.: *Directionally adapted upwind schemes in 2-D for the Euler equations.* Notes on Numerical Fluid Mechanics, Finite Approximations in Fluid Mechanics II, DFG-Priority Research Program, Results 1986–1988, Vieweg.

[120] Kröner, D., Rokyta, M.: *Convergence of upwind finite volume schemes for scalar conservation laws in two dimensions.* SIAM J. Numer. Anal. 31 (1994), 324–343.

[121] Kröner, D., Noelle, S., Rokyta, M.: *Convergence of higher order upwind finite volume schemes on unstructured grids for scalar conservation laws in several space dimensions.* Numer. Math. 71 (1995), 527–560.

[122] Kröner, D.: *Absorbing boundary conditions for the linearized Euler equations in 2-D.* Math. Comput. 57 (1991), 153–167.

[123] Kruzkov, S.N.: *First order quasiliner equations in several independent variables.* Mat. Sbornik 81 (1970) 123 (Russian) and Math. USSR Sbornik 10 (1970), 217–243.

[124] Kuznetsov, N. N.: *Accuracy of some approximate methods for computing the weak solutions of a first-order quasi-linear equation.* USSR. Comput. Math. Math. Phys. 16/6 (1976), 105–119.

[125] Lax, P.D.: *Weak solutions of nonlinear hyperbolic equations and their numerical computation.* Commun. Pure Appl. Math. 7 (1954), 159–193.

[126] Lax, P.D.: *Hyperbolic systems of conservation laws and the mathematical theory of shock waves.* Conf. Board. Math. Sci. Regional Conf. Series in Appl. Math. 11, SIAM, Philadelphia, 1972.

[127] Lax, P.D.: *Shock waves and entropy.* In: Proc. Symposium at the University of Wisconsin, ed. E. H. Zarantello, 1971, pp. 603–634.

[128] Lax, P., Wendroff, B.: *Systems of conservation laws.* Commun. Pure Appl. Math. 13 (1960), 217–237.

[129] Leer, B. van: *Towards the ultimate conservative difference scheme. I.* In: Lecture Notes in Physics, No. 18, Springer, Berlin, 1973.

[130] Leer, B. van: *Towards the ultimate conservative difference scheme. II. Monotonicity and conservation combined in a second-order-scheme.* J. Comput. Phys. 14 (1974), 361–370.

[131] Leer, B. van: *Towards the ultimate conservative difference scheme. III. Upstream-centered finite-difference schemes for ideal compressible flow.* J. Comput. Phys. 23 (1977), 263–275.

[132] Leer, B. van: *Towards the ultimate conservative difference scheme. IV. A new approach to numerical convection.* J. Comput. Phys. 23 (1977), 276–299.

[133] Leer, B. van: *Towards the ultimate conservative difference scheme. V. A second-order sequel to Godunov's method.* J. Comput. Phys. 32 (1979), 101–136.

[134] Leer, B. van: *Flux vector splitting for the Euler equations.* In: Proc. 8th International Conference on Numerical Methods in Fluid Dynamics. Springer, Berlin 1992.

[135] Leonard, B.P.: *A stable and accurate convective modelling procedure based on quadratic upstream interpolation.* Comput. Meth. Appl. Mech. Engng. 19 (1979), 59–98.

[136] Le Roux, A.Y.: *Convergence of an accurate scheme for first order quasi linear equations.* RAIRO Anal. Numér. 15 (1981), 151–170.

[137] LeVeque, R.J.: *High resolution finite volume methods on arbitrary grids via wave propagation.* J. Comput. Phys. 78 (1988), 36–63.

[138] LeVeque, R.J.: *Large time step shock-capturing techniques for scalar conservation laws.* SIAM J. Numer. Anal. 22 (1985), 1051–1073.

[139] LeVeque, R.J.: *Convergence of a large time step generalization of Godunov's method for conservation laws.* Commun. Pure Appl. Math. 37 (1984), 463–477.

[140] LeVeque, R.J.: *Cartesian grid methods for flow in irregular regions.* Preprint.

[141] LeVeque, R.J.: *Numerical methods for conservation laws.* Lectures in Mathematics, ETH Zürich, Birkhäuser, Basel, 1990.

[142] Liu T.P.: *Admissible solutions of hyperbolic conservation laws.* Mem. Am. Math. Soc. 240, Vol. 30 (1981).

[143] Lions, P.L.: *Global existence of solutions for isentropic compressible Navier–Stokes equations.* Math. Prob. Mech. 316 (1993), 1335–1340.

[144] Lorenz, J.: *Numerical solutions of a singular perturbation problem without turning points.* In: Lecture Notes in Mathematics, No. 1017, Springer, Berlin, 1983, pp. 433–439.

[145] Lorenz, J.: *Nonlinear boundary value problems with turning points and properties of difference schemes.* In: Lecture Notes in Mathematics, No. 942, Springer, Berlin, 1980, pp. 150–169.

[146] Lorse, G.: *Essentially Non-Oscillatory Verfahren. Eine anwendungs-orientierte Einführung.* Diplomthesis, Bonn, 1995.

[147] Lube, G.: *Streamline diffusion finite element method for quasilinear elliptic problems.* Numer. Math. 61 (1992), 335–357.

[148] Lube, G., Weiss, D.: *Stabilized finite element methods for singularly perturbed parabolic problems.* NAM-Bericht Nr. 71 (1994), Universität Göttingen, Inst. für Numerische und Angewandte Mathematik.

[149] Lucier, B.J.: *Error bounds for the method of Glimm, Godunov and LeVeque.* SIAM J. Numer. Anal. 22 (1985), 1074–1081.

[150] Hong Luo, Baum, J.D., Löhner, R.: *An efficient spatial adaption algorithm on 3-D unstructured meshes for the Euler equations.* In: Proc. 5th International Symposium on CFD, 1993, Sendai, Japan.

[151] Mackenzie, J.A., Morton, K.W.: *Finite volume solutions of convection–diffusion test problems.* Math. Comput. 60 (1992), 189–220.

[152] Mackenzie, J.A., Süli, E., Warnecke, G.: *A posteriori analysis for Petrov-Galerkin approximation of Friedrichs Systems.* Oxford Univ. Comp. Lab. Report 95/01.

[153] Majda, A.: *Compressible fluid flow and systems of conservation laws in several space variables.* Springer, Berlin (1994).

[154] Majda, A., Osher, S.: *Numerical viscosity and the entropy condition.* Commun. Pure Appl. Math. 32 (1979), 797–838.

[155] Majda, A., Osher, S.: *A systematic approach for correcting nonlinear instabilities.* Numer. Math. 30 (1978), 429–452.

[156] Majda, A., Osher, S.: *Initial–boundary value problems for hyperbolic equations with uniformly characteristic boundary.* Commun. Pure Appl. Math. 28 (1975), 607–675.

[157] Mavriplis, D.J.: *Accurate multigrid solutions of the Euler equation on unstructured and adaptive grids.* AIAA J. 28 (1990), 2.

[158] Mavriplis, D.J.: *Unstructured mesh generation and adaptivity.* ICASE Report No. 96-26 (1995).

[159] Molenaar, J.: *Multigrid methods for semiconductor device simulations.* Centrum voor Wiskunde en Informatica CWI TRACT, Amsterdam (1993)

[160] Morton, K.W., Paisley, M.F.: *A finite volume scheme with shock fitting for the steady Euler equations.* J. Comput. Phys. 80 (1989), 168–203.

[161] Munz, C.-D.: *On the numerical dissipation of high resolution schemes for nonlinear hyperbolic conservation laws.* Manuscript, Kernforschungszentrum Karlsruhe, 1987.

[162] Murat, F.: *A review of compensated compactness.* In: Contributions to the calculus of variations, ed. L.Cesari, Longman, Harlow, 1987, pp. 148–183.

[163] Natanson, I.P.: *Theorie der Funktion einer rellen Veränderlichen.* Akademie-Verlag XII, Berlin, 1975.

[164] Nessyahu, H., Tassa, T., Tadmor, E., *The convergence rate of Godunov type schemes.* SIAM J. Numer. Anal. 31 (1994), 1–16.

[165] Niijima, K.: *An error analysis for a difference scheme of exponential type applied to a nonlinear singular perturbation problem without turning points.* J. Comput. Appl. Math. 15 (1986), 93–101.

[166] Niijima, K.: *A uniformly convergent differential scheme for a semilinear singular perturbation problem.* Numer. Math. 43 (1984), 175–198.

[167] Niijima, K.: *On a difference scheme of exponential type for a nonlinear singular perturbation problem.* Numer. Math. 46 (1988), 521–539.

[168] Niijima, K.: *Pointwise error estimates for a streamline diffusion finite element scheme.* Numer. Math. 56 (1990), 707–719.

[169] Noelle, S.: *Convergence of higher order finite volume schemes on irregular grids.* Adv. Comput. Math. 3, No. III (1995), 197–218.

[170] Noelle, S.: *A note on entropy inequalities and error estimates for higher order accurate finite volume schemes on irregular families of grids.* SFB256, Preprint 400 (1995), Bonn.

[171] Oleinik, O.A.: *Discontinuous solutions of non-linear differential equations.* Am. Math. Soc. Transl. Ser. 2, 26 (1963), 95–172.

[172] O'Riordon, E., Stynes, M.: *A globally uniformly convergent finite element method for a singularly perturbed elliptic problem in two dimensions.* Math. Comput. 57 (1991), 47–62.

[173] Osher, S.: *Riemann solvers, the entropy condition, and difference approximations.* SIAM J. Numer. Anal. 21 (1984), 217–235.

[174] Osher, S.: *Convergence of generalized MUSCL schemes.* SIAM J. Numer. Anal. 22 (1985), 947–961.

[175] Osher, S., Solomon, F.: *Upwind difference schemes for hyperbolic systems of conservation laws.* Math. Comput. 38 (1982), 339–374.

[176] Osher, S., Tadmor, E.: *On the convergence of difference approximations to scalar conservation laws.* Math. Comput. 50 (1988), 19–51.

[177] Osher, S.: *Nonlinear singular perturbation problems and one sided difference schemes.* SIAM J. Numer. Anal. 18 (1981), 129–144.

[178] Otto, F.: *First order equations with boundary conditions.* SFB 256, Preprint 234 (1992), Bonn.

[179] Pazy, A.: *Semigroups of linear operators and applications to partial differential equations.* Springer, Berlin, 1983.

[180] Perthame, B.: *Convergence of N-schemes for linear advection equations.* Trends of Applications of Mathematics in Mechanics, A collection of selected papers presented at the 9th Symposium, Lisbon, Portugal, July 1994, Pitman, Longmann, New York (1995), 323–333.

[181] Pironneau, O.: *On the transport–diffusion algorithms and its applications to the Navier–Stokes equations.* Numer. Math. 38 (1982), 309–332.

[182] Polthier, K., Rumpf, M.: *A concept for timedependent processes.* SFB256, Report No. 13 (1994), Bonn.

[183] Powell, K.G., Leer, B. van: *A genuinely multi-dimensional upwind cell-vertex scheme for the Euler equations.* AIAA 89-0095, 9–12 January, 1989, Reno, Nevada.

[184] Prigogine, I., Herman, R.: *Kinetic theory of vehicular traffic.* American Elsevier, New York, 1971.

[185] Pulliam, T.H., Chaussee, D.S.: *A diagonal form of an implicit approximate-factorization algorithm.* J. Comput. Phys. 39 (1981), 347–363.

[186] Rauch, J.: *BV estimates fail for most quasilinear hyperbolic systems in dimension greater than one.* Commun. Math. Phys. 106 (1986), 481–484.

[187] Roe, P.,L.: *Approximate Riemann solvers, parameter vectors, and difference schemes.* J. Comput. Phys. 43 (1981), 357–372.

[188] Rohde, C.: *Analysis der Stromlinien-Diffusionsmethode mit der Theorie der maßwertigen Lösungen.* Diplomthesis (1992), Hannover.

[189] Richter, G.R.: *The discontinuous Galerkin method with diffusion* Math. Comput. 58 (1992), 631–643.

[190] Richtmyer, R.D., Morton, K.W.: *Difference methods for initial-value problems.* Wiley, New York, 1967.

[191] Riedel, U.: *Numerische Simulation reaktiver Hyperschallströmungen mit detaillierten Reaktionsmechanismen.* PhD–Thesis, Heidelberg 1992.

[192] Riemann, B.: *Über die Fortpflanzung ebener Luftwellen von endlicher Schwingungsweite.* In: Bernhard Riemann's gesammelte mathematische Werke und Wissenschaftlicher Nachlass, Teubner, Leibzig, 1892, pp. 157–175.

[193] Roe, P. L.: *Linear advection schemes on triangular meshes.* CoA Report No. 8720, 1987.

[194] Rogerson, A., Meiburg, E.: *A numerical study of the convergence properties of ENO schemes.* SIAM J. Sci. Comput. 5 (1990), 151–167.

[195] Roos, H.-G., Adam, D., Felgenhauer, A.: *A novel nonconforming uniformly convergent finite element method in two-dimensions.* Preprint 1994, Universität Dresden, Inst. of Numerical Mathematics.

[196] Roos, H.-G.: *Ten ways to generate the Il'in and related schemes.* J. Comput. Appl. Math. 53 (1994), 43–59.

[197] Roos, H.-G.: *A second order monotone upwind scheme.* Computing 36 (1986), 57–67.

[198] Roos, H.-G.: *Higher order uniformly convergent methods for singular perturbation problems.* Comput. Meth. Appl. Mech. Engng. 116 (1994), 273–280.

[199] Roos, H.-G., Vulanovic̀, R.: *A higher order uniform convergence results for a turning point problem.* Z. Anal. Anwend. 12 (1993), 723–728.

[200] Roosbroeck, W.V.van: *Theory of flow of electrons and holes in germanium and other semiconductors.* Bell Syst. Tech. J. 19 (1950), 560–607.

[201] Rossow, C.: *Berechnung von Strömungsfeldern durch Lösung der Eulergleichungen mit einer erweiterten Finite-Volumen Diskretisierungsmethode.* DLR Bericht, Braunschweig, 1988.

[202] Rossow, C.: *Comparison of cell centered and cell vertex finite volume schemes.* In: Proceedings of the 7th GAMM-Conference on Numerical Methods in Fluid Mechanics. Notes on Numerical Fluid Mechanics 20 (1988), pp. 327–335.

[203] Rumpf, M.: *A variational approach to optimal meshes.* SFB256 Preprint 331, Bonn (1994).

[204] Rumpf, M., Schmidt, A., et al.: *GRAPE, GRaphics Application and Programming Environment.* SFB256, No. 8, Bonn (1989).

[205] Rumpf, M., Geiben, M., Geßner, T.: *Moving and Tracing in Timedependent Vector Fields on Adaptive Meshes,* SFB256 Report, Bonn (1994).

[206] Rumpf, M., Schmidt, A., Siebert, K.: *Functions describing arbitrary meshes – a flexible interface between numerical data and visualization routines.* Computer Graphics Forum 15 (1996), 129–141; SFB 256 Report, Bonn (1995).

[207] Saks, S.: *Theory of the integral.* Warsaw, 1937.

[208] Sanders, R.: *On convergence of monotone finite difference schemes with variable spatial differencing.* Math. Comput. 40 (1983), 91–106.

[209] Scharfetter, D.L., Gummel, H.K.: *Large-signal analysis of a silicon read diode oscillator.* IEEE Trans. Electron Devices, ED-16 (1969), 64–77.

[210] Schmieder, M.: *Adaptive numerische Verfahren mit rechteckigen Elementen zur Lösung hyperbolischer Differentialgleichungen.* Diplomthesis, IAM Univ. Bonn 1996.

[211] Schneider, A.: *Upwind-Diskretisierung für singulär gestörte Probleme mit Krylowmethoden.* Diplomthesis, Freiburg 1995.

[212] Schroll, H.J.: *Konvergenz finiter Differenzenverfahren für nichtlineare hyperbolisch-parabolische Systeme.* Dissertation ETH Zürich, Nr.10273.

[213] Schulz-Rinne, C.W.: *Classification of the Riemann problem for two-dimensional gas dynamics.* SIAM J. Math. Anal. 24 (1993), 76–88.

[214] Schulz-Rinne, C.W., Collins, J.P., Glaz, H.M.: *Numerical solutions of the Riemann problem for two-dimensional gas dynamics.* SIAM J. Sci. Comput. 14 (1993), 1394–1414.

[215] Shu, C.-W.: *TVB uniformly high-order schemes for conservation laws.* Math. Comput. 49 (1987), 105–121.

[216] Shu, C.W.: *Numerical experiments on the accuracy of ENO and modified ENO schemes.* SIAM J. Sci. Comput. 5 (1990), 127–149.

[217] Siebert, K.: *An a posteriori error estimator for anisotropic refinement.* SFB 256, Preprint 313, Bonn (1993).

[218] Smoller, J.: *Shock waves and reaction-diffusion equations.* Springer, New York, 1983.

[219] Struijs, R., Deconinck, H., de Palma, P.:*Progress on multidimensional upwind schemes for unstructured grids.* AIAA Paper 91-1550, 1991.

[220] Struijs, R., Deconinck, H.: *Multidimensional upwind schemes for the Euler equations using fluctuation distribution.* VKI, 1989/90.

[221] Sod, G.A.: *Numerical methods in fluid dynamics.* Cambridge University Press, 1984.

[222] Sonar, T.: *Multivariante Rekonstruktionsverfahren zu numerischen Berechnung hyperbolischer Erhaltungsgleichungen.* Forschungsbericht 95-02 (1995), Institut für Strömungsmechanik Göttingen, Deutsche Forschungsanstalt für Luft und Raumfahrt.

[223] Sonar, T.: *Entropy production in second order three-point schemes.* Numer. Math. 62 (1992), 371–390.

[224] Sonar, T., Süli, E.: *A dual graph-norm refinement indicator for finite volume approximations of the Euler equations.* Oxford Univ. Comput. Lab. Report 94/9.

[225] Sonar, T.: *Strong and weak norm error indicators based on the finite element residual for compressible flow computations.* DLR-Preprint 92-07, Göttingen.

[226] Steger, J.L., Warming, R.F.: *Flux vector splitting of the inviscid gasdynamic equations with application to finite-difference methods.* J. Comput. Phys. 40 (1981), 263–293.

[227] Strang, G.: *On the construction and comparison of difference schemes.* SIAM J. Numer. Anal. 5 (1968), 506–517.

[228] Stoker, J.J.: *Water waves. The mathematical theory with application.* Wiley, New York, 1992.

[229] Süli, E.: *Finite volume methods on distorted meshes: stability, accuracy, adaptivity.* Oxford Univ. Comput. Lab. Report 89/6.

[230] Sweby, P.K.: *High resolution schemes using flux limiters for hyperbolic conservation laws.* SIAM J. Numer. Anal. 21 (1984), 995–1011.

[231] Szepessy, A.: *Convergence of a shock–capturing streamline diffusion finite element method for a scalar conservation law in two space dimensions.* Math. Comput. 53 (1989), 527–545.

[232] Tabata, M.: *Uniform convergence of the upwind finite element approximation for semilinear parabolic problems.* J. Math. Kyoto Univ. 18 (1978), 327–351.

[233] Tadmor, E.: *Numerical viscosity and the entropy condition for conservative difference scheme.* ICASE Report 172141, NASA, Langley, 1983.

[234] Tadmor, E.: *Local error estimates for discontinuous solutions of nonlinear hyperbolic equations.* SIAM J. Numer. Anal. 28 (1991), 891–906.

[235] Tartar, L.: *The compensated compactness method applied to systems of conservation laws.* In: Systems of nonlinear PDE, ed. J.M. Ball, NATO ASI Series, Oxford, pp. 263–285 (1983).

[236] Tartar, L.: *Compensated compactness and applications to partial differential equations.* In: Nonlinear Analysis and Mechanics: Herriot–Watt Symposium, Vol. 4, ed. R.J. Knops, Pitman, London, 1979, pp. 136–212.

[237] Vijayasundaram, G.: *Transonic flow simulations using an upstream centered scheme of Godunov in finite elements.* J. Comput. Phys. 63 (1986), 416–433.

[238] Vila J. P.: *Convergence and error estimates in finite volume schemes for general multi-dimensional scalar conservation laws I. Explicit monotone schemes.* RAIRO Anal. Numér. 28 (1994), 267–295.

[239] Vilsmeier, R., Hänel, D.: *Generation and adaption of 2D unstructured meshes.* In: Proc. 3rd International Conference on Numerical Grid Generation in CFD and Related fields, Barcelona, 1991.

[240] Volpert, A.I.: *The spaces BV and quasilinear equations.* Math. USSR-Sbornik 2 (1967),225–267.

[241] Vulanovič, R.: *A uniform numerical method for quasilinear singular perturbation problems without turning points.* Computing 41 (1989), 97–106.

[242] Vulanovič, R.: *Non equidistant finite difference methods for elliptic singular perturbation problems.* In: Computational Methods for Boundary and Internal Layers in Several Dimensions, ed. J.J.H. Miller, Boole Press, Dublin, 1991, pp. 203–223.

[243] Walkington, N.J.: *Convergence of nonconforming finite element approximations to the first order linear hyperbolic equations.* Math. Comput. 58 (1992), 671–691.

[244] Walter, W.: *Gewöhnliche Differentialgleichungen.* Springer, Berlin, 1972.

[245] Warnecke, G.: *Analytische Methoden in der Theorie der Erhaltungsgleichungen.* Habilitationsschrift, Stuttgart (1991).

[246] Wang, J., Warnecke, G.: *On entropy consistency of large time step schemes I. The Godunov and Glimm schemes.* SIAM J. Numer. Anal. 30 (1993), 1229–1251.

[247] Wang, J., Warnecke, G.: *On entropy consistency of large time step schemes II. Approximate Riemann solvers.* SIAM J. Numer. Anal. 30 (1993), 1252–1267.

[248] Warming, R.F., Beam, R.M.: *Upwind second order difference schemes and applications in aerodynamics.* AIAA J. 14 (1976), 1241–1249.

[249] Wierse, M.: *Solving the compressible Euler equations in timedependent geometries.* In: Proc. 5th International Conference on Hyperbolic Problems, Stony Brook, 1994.

[250] Wierse, M.: *Higher order upwind schemes on unstructured grids for the compressible Euler equations in timedependent geometries in 3-D.* PhD Thesis, University of Freiburg 1994.

[251] Wierse, M.: *An automotive application on unstructured adaptive grids.* In: Proc. ICFD Conference on Numerical Methods for Fluid Dynamics, Oxford, 1995.

[252] Wierse, A., Rumpf M.: *GRAPE, Eine interaktive Umgebung für Visualisierung und Numerik.* Informatik, Forschung und Entwicklung, Springer 7 (1992), 145–151.

[253] Wu Qi–guang, Sun Xias-di: *Numerical solutions of a singularly perturbed elliptic hyperbolic partial differential equation on a nonuniform discretization mesh.* Appl. Math. Mech. (English Edition) 13 (1992), 1081–1088.

[254] Yanenko, N.N., Vorozcov, E. V.: *Methods for localization of singularities in the numerical solutions of gas dynamic problems.* Springer, Berlin, 1990.

[255] Zajaczkowski, W., Kantiem, K.: *The existence and uniqueness of solutions of equations for ideal compressible polytropic fluids.* J. Appl. Anal., to appear.

[256] Zhou, G., Rannacher, R.: *Pointwise superconvergence of the streamline diffusion finite element method.* Interdisziplinäres Zentrum für Wissenschaftliches Rechnen, Universität Heidelberg, Preprint 94-72, (1994).

[257] Zhou, G.: *Local L^2-error analysis of the streamline diffusion FE-Method for nonstationary hyperbolic systems.* Preprint 94-07 (SFB 359), Interdiziplinäres Zentrum für wissenschaftliches Rechnen, Universität Heidelberg, 1994.

Index

a posteriori estimates, 259
a priori error estimate, 199
a priori estimates, 259
absorbing boundary conditions, 427
admissible linear reconstructions, 216
admissible stencils, 239
alignment, 268
alignment algorithm, 270
angle condition, 271
approximate Riemann solver, 333
Arzelá–Ascoli, 152

backward differences, 46
Besov spaces, 84
bisection, 262
blue refinement, 270
boundary conditions, 437
boundary conditions for linear systems, 422
bounded variation, 146
Buckley–Leverett equation, 5
Burger
    initial value problem of, 117
Burgers
    equation, 1
BV norms, 147
BV spaces, 143

cell entropy inequality, 225
cell-centred, 157
central differences, 31
centre of gravity of T_j, 156

CFL condition, 51
    Courant, Friedrichs, Lewy condition, 51
CFL number, 51
Chakravarthy and Osher limiter, 109
change of type, 304
characteristic, 16
characteristic variables, 335
characteristics, 1, 287
Chorin method, 319
classical solution, 16, 17
classical solutions, 15
coarsening, 263
compactness, 438
compactness in L^1, 147
compatibility condition, 183, 184, 217
compression shocks, 311
conservation, 176
conservation form, 146
conservation of energy, 8
conservation of mass, 7
conservation of momentum, 8
conservation property, 44, 160
conservative variables, 9
consistency, 93, 160
consistency with the entropy pair, 165
consistent, 43
consistent with the entropy condition, 58

contact discontinuity, 305–307, 315, 316
convection-dominated diffusion equation, 117
convergence for higher-order finite volume schemes, 217
convergence of monotone schemes, 70
convergence of monotone schemes in 2-D, 146
convergence of the streamline diffusion method, 121
convergence to the entropy solution, 58
Courant, Friedrichs, Lewy condition, 68
Crandall
    Theorem of Crandall and Tartar, 147

damping, 47
density, 277
diagonal form of the Euler equations, 286
dimensional splitting for linear equations, 143
dimensional splitting scheme, 144, 154
    convergence for, 144
Dirac measure, 166
discrete entropy condition, 57, 58
discrete Fourier transform, 94
discretization matrices, 384
distributional sense, 18
div–curl Lemma, 131
dual cells, 157, 245

E-scheme, 84
electron concentration, 473

elementary wave solutions, 315
elliptic, 304
Engquist–Osher, 466
Engquist–Osher scheme, 47, 182, 191, 355, 458
ENO, 238
ENO schemes, 117
entropy, 280
entropy condition, 25
    Lax entropy condition, 25
entropy condition for a k-shock, 298
entropy dissipation, 184
entropy fluxes, 184
entropy inequality, 226, 293
entropy pair, 165, 293
entropy pair for the p-System, 294
entropy solution, 27, 142
EOC
    experimental order of convergence, 221
equation of state, 277, 279
equations
    elliptic equations, 258
    parabolic equations, 258
error estimate, 78
error estimates, 191
error estimates for the linear case, 191
error estimators, 258
    for conservation laws, 260
    for convection-dominated problems, 259
essentially non-oscillatory, 238
Euler equations, 276, 291
Euler equations in 1-D, 278
Euler equations of gas dynamics, 373

existence for the conservation law, 22
existence for the Riemann problem, 349
experimental order of convergence (EOC), 221
exponentially fitted upwind discretization, 473
extrapolation techniques, 261

finite element method, 117
finite volume, 155, 473
finite volume methods, 158
finite volume methods for initial boundary value problems, 454
finite volume methods for systems, 372
finite volume scheme, 159
finite volume schemes in 2-D, 155
flow around a double ellipsoid, 407
flow around two cylinders, 408
fluctuation splitting, 270
fluctuation splitting schemes, 254
flux vector splitting, 374
flux vector splitting scheme, 353
forward differences, 46
forward-facing step, 405
forward-facing step in 2-D, 410
fractional step method, 143
functions of bounded variation, 438

Galerkin finite element method standard Galerkin finite element method, 118
Gauss's theorem, 8
genuinely nonlinear, 298
Glimm scheme, 325–327

Glimm's existence proof, 325
Godunov scheme, 66, 70, 325, 339
grid density $h(x)$, 258
Gronwall inequality, 444

Helly's selection principle, 39
higher-order, 98
higher-order discretization in time, 241
higher-order finite volume scheme, 215
higher-order schemes, 393
hole concentration, 473
hyperbolic, 278
hyperbolic systems, 278

Il'in scheme, 466
ill-posed, 437
ill-posedness, 431
implicit, 44
implicit schemes, 86
inhomogeneous boundary condition, 446
inhomogeneous equations, 154
initial boundary value problem, 446, 447
initial boundary value problems, 453
initial value problems for systems, 372
integral form of the conservation law, 42
intermediate values, 216
internal energy, 277
interpolation polynomial, 240
inviscid–viscous operator splitting, 477
isentropic gas, 11, 290

jump condition, 19

jump conditions, 282

k-Riemann invariant, 289
k-characteristic, 287, 292
k-shock, 298
Kruzkov entropy condition, 27, 142
Kuznetsov theory, 84

Lagrangian coordinates, 9
large-time-step method, 84
Lax
    Lax–Wendroff theorem, 334
Lax entropy condition, 25, 39, 295, 298
Lax theorem, 93
Lax–Friedrichs scheme, 46, 182, 191
Lax–Wendroff scheme, 100
    Stability in the linear case, 100
Lax–Wendroff theorem, 54, 121, 375
least square methods, 120
limiters, 109
    Chakravarthy and Osher, 109
    minmod, 109
    superbee of Roe, 109
    van Albada, 109
    van Leer, 109
linear advection equation, 120
linear hyperbolic system, 335
linear reconstruction, 213
linear system, 336
linear systems, 277, 335
linear transport equation, 85, 97
linearization of scalar equations, 340
linearly degenerate, 304

local
    truncation error, 44, 45
local existence, 16
local Lipschitz condition, 160
locally monotone, 146

Mach number, 7
Mach-three wind tunnel, 6
measure valued solution, 164, 165
mesh alignment, 270
mesh indicator, 261, 268
minmod, 109
monotone schemes, 65, 66
monotonicity, 174
monotonicity in 2-D, 146
moving piston, 417
Murat
    Theorem of Murat, 122
MUSCL type schemes, 116

Navier–Stokes equations, 473
no-flux boundary condition in $\mathbb{R}^2$, 425
nonconformal node, 262
nonexistence for general data, 307
nonpositive density for the Roe scheme, 350
numerical damping, 50
numerical entropy, 217
numerical entropy flux, 58
numerical flux, 44
numerical schemes
    E-scheme, 84
    TVD, 64
    dimensional splitting, 154
    Engquist–Osher, 47, 182, 191, 355
    ENO schemes, 117
    finite volume, 159

Index   507

flux vector splitting, 353
higher-order finite volume, 215
implicit, 86
in conservation form, 44, 53
Lax–Friedrichs, 182, 191
Lax–Wendroff, 100
monotone, 66
MUSCL type, 116
Osher–Solomon scheme, 355
QUICK, 116
Roe, 341
unstable, 91
numerical viscosity, 46, 79, 83

Oleinik entropy condition, 39
one-sided differences, 31
operator splitting, 477
order of convergence, 93
Osher–Solomon scheme, 355, 361, 365

p-system, 12, 294, 298, 304
pair of entropy functions, 23
piecewise-smooth solutions, 19
pointwise-consistent, 161
pressure, 277
primitive variables, 9, 279
probability measure, 164
problem P, 245
propagation speed, 19

QUICK scheme, 116

random choice method, 325
random numbers, 327
Rankine–Hugoniot, 33
Rankine–Hugoniot condition, 19, 21, 281
1-rarefaction, 315
rarefaction wave, 34, 301, 316

1-rarefaction wave, 313
3-rarefaction wave, 316
rarefaction waves, 292
reactive flows, 13
reconstruction, 213, 239
reconstructions, 239
representation formula of Lax, 35
Riemann invariant, 287, 309
Riemann invariants, 289
Riemann problem, 6, 307, 308, 319
Riemann problem for a linear system, 336
Riemann problem for the Euler equations of gas dynamics, 348
Riemann solver of Harten, Lax and van Leer, 352
Riemann solver of Roe, 341
Roe mean values, 347
Roe scheme, 341, 347, 349
rotating cone, 263
Runge–Kutta method, 216

Scharfetter–Gummel discretization, 473
schemes
    explicit, 85
    implicit, 85
selection principle of Helly, 60
self-adapting algorithms, 258
semiconductor equations, 473
shallow water waves, 12
shock, 25
    1-shock, 315
    3-shock, 310, 311
shock reflection, 401
shock tube problem, 6, 355
shock velocity, 25

shock-capturing finite element method, 140
singular perturbation, 456
solutions in the distributional sense, 281
specific internal energy, 277
stability, 92
stability condition, 94
stability estimate, 63
standard Galerkin, 117, 118
Steger and Warming, 375
streamline diffusion, 120, 466
streamline diffusion method, 117, 459
streamline diffusion shock capturing, 191, 259
strictly hyperbolic systems, 373
subsonic inflow, 425
sufficient conditions for convergence, 60
superbee of Roe, 109
supersonic flow around a cylinder, 399
supersonic inflow, 425
systems in 1-D, 333

Tartar
    compactness theorem of Tartar, 123
    Theorem of Crandall and Tartar, 147
Tartar's theorem, 165
total energy, 277
total variation, 59, 332
total variation diminishing, 40, 64
traffic flow, 4
travelling waves, 277
truncation error, 93
TVB schemes, 117

TVD, 40
    sufficient condition for TVD, 80
TVD scheme, 64
TVD operator, 73
TVD Runge–Kutta method, 216
two passing supersonic trains, 402

uniform Kreiss condition, 437
uniqueness of the entropy solution, 27
unstable schemes, 91
unstructured grid, 156
upwind scheme for linear systems, 335

van Albada limiter, 109
van Leer, 375
van Leer limiter, 109
variable extrapolation, 112
variational approach, 268
velocity, 277
Vijayasundaram, 355, 375
viscosity form, 335
viscosity limit, 22, 142
viscosity method, 22
viscous terms, 473
von Neumann condition, 94

Warming and Beam, 106, 109
wave equation, 12
weak law of large numbers, 331
weak solution, 18
well-posedness, 430, 436, 455
Wendroff
    Lax–Wendroff theorem, 334
wind tunnel, 6

Young measure, 166